Metal-Based Nanoparticles' Biological Activity and Pharmaceutical Applications

Metal-Based Nanoparticles' Biological Activity and Pharmaceutical Applications

Guest Editors

Elitsa L. Pavlova
Debbie C. Crans

MDPI

Basel • Beijing • Wuhan • Barcelona • Belgrade • Novi Sad • Cluj • Manchester

Guest Editors
Elitsa L. Pavlova
Faculty of Physics
Sofia University "St. Kliment Ohridski"
Sofia
Bulgaria

Debbie C. Crans
Department of Chemistry
Colorado State University
Fort Collins, CO
USA

Editorial Office
MDPI AG
Grosspeteranlage 5
4052 Basel, Switzerland

This is a reprint of the Special Issue, published open access by the journal *Pharmaceuticals* (ISSN 1424-8247), freely accessible at: https://www.mdpi.com/journal/pharmaceuticals/special_issues/4KE79171UI.

For citation purposes, cite each article independently as indicated on the article page online and as indicated below:

Lastname, A.A.; Lastname, B.B. Article Title. *Journal Name* **Year**, *Volume Number*, Page Range.

ISBN 978-3-7258-6572-7 (Hbk)
ISBN 978-3-7258-6573-4 (PDF)
https://doi.org/10.3390/books978-3-7258-6573-4

Cover image courtesy of Debbie C. Crans

Contents

About the Editors

Elitsa L. Pavlova

Elitsa L. Pavlova is an Assoc. Professor in Biophysics at the Faculty of Physics, Sofia University "St. Kliment Ohridski", Bulgaria. She received an M.Sc. degree in Biology, specializing in Cellular Biology and Ontogenesis, an M.Sc. degree in Biology and Chemistry, and a Ph.D. in "Human and Animal Physiology" from Sofia University "St. Kliment Ohridski" in 2001, 2002, and 2007, respectively. Her research is focused on the biomarkers of oxidative stress. Her expertise lies in the fields of nutritional supplements, antioxidants, and nanomaterials, as well as the evaluation of pharmaceuticals, incl. antivirals, their combinations, and their side effects. Her specialization includes training in Italy, Canada, Switzerland, the USA, and Bulgaria. In addition to her research, she serves as a reviewer for highly ranked and impactful international journals and international projects. Her work has been cited numerous times worldwide.

Debbie C. Crans

Debbie C. Crans is a University Distinguished Professor in the Chemistry Department and the Cell and Molecular Biology Program at Colorado State University (CSU) in Fort Collins, Colorado, USA. She was born in Copenhagen, Denmark, and received her initial degree in Biochemistry and Organic Chemistry from the University of Copenhagen, working with James P. Snyder. She received her Ph.D. from Harvard University, working with George M. Whitesides, and completed postdoctoral studies with Orville L. Chapman and Paul D. Boyer at the University of California, Los Angeles. She started as an Assistant Professor in Organic chemistry with expertise in biological, organic, and inorganic chemistry in January 1987 at CSU. She has been working to understand the mode of action of drugs as well as approaches to improve them using the synthesis of new compounds, spectroscopy to characterize the systems, structural approaches, biological assays to test their activities, and other methods as needed, such as nanoparticles for drug delivery systems. Currently she is particularly interested in metals in medicine, enzymes in signal transduction with effects in cancer, diabetes, tuberculosis, and substance use disorder, as well as in the electron transport pathway in bacteria.

Preface

The application of nanoparticles in human health is increasing as their biological and pharmaceutical activities are enhanced and their properties are being developed, so that the applications can target specific therapeutic aims. There is therefore a need for better development of these systems. One metal-based nanoparticle is already used in the clinic—Ferumoxytol, an iron oxide used for iron-deficiency anemia and explored for additional therapeutic applications. Many more metal-based nanoparticles are at various levels of clinical studies, including two gold nanoparticles, Aurimune (CYT-6091) and a spherical oligonucleotide conjugate (NU-0129), treating tumors, as well as a silver nanoparticle for local application and wound care. Additional metal-based nanoparticles include NBTXR3/Hensify®, a hafnium-oxide nanoparticle drug that obtained CE marking in Europe as "Hensify" and is in phase 2/3 for the treatment of soft-tissue sarcoma and multiple other tumors in the USA.

In this reprint we describe fundamental studies on applications of a range of different metal-based nanoparticles prepared from different metals, including gold, silver, copper, zinc, vanadium, and other rare earth metals. The manuscripts are prepared by active researchers working around the globe, exploring the topic of the biological activity and pharmaceutical applications of metal-based nanoparticles.

The definition of a metal nanoparticle is a nanoparticle that consists of a metal core. Such is Aurimune (CYT-6091), which is in fact a pegylated colloidal gold nanoparticle bearing recombinant human tumor necrosis factor-α (rhTNF). Aurimune has been designed to deliver rhTNF directly to tumor sites, to target the tumor vasculature and treat types of solid tumors such as thyroid and pancreatic cancer. However, metal nanoparticles are a small subgroup of metal-based nanoparticles, which can contain nanoparticles made of metal oxides, metal salts, metal complexes, metal–organic frameworks, and metal-doped polymers. Examples of drugs and vaccines in the clinical practice are nanoparticle liposomes, including the doxorubicin hydrochloride liposome (Doxil®/Caelyx®) used for the treatment of ovarian cancer and Kaposi's sarcoma. Another example includes paclitaxel bound to albumin nanoparticles (Abraxane®) used in breast, lung, and pancreatic cancers. An additional example is a combination of daunorubicin and cytarabine, co-encapsulated in liposomes, for certain acute myeloid leukemia subtypes. There is no doubt that these products stimulate researchers and illustrate the unlimited potential of metal-based nanoparticles.

In the presented papers we focus on the biological properties and the potential of various metal-based nanoparticles for therapeutic applications as well as their limitations. The reprint explores options in composition, shape, size, material, chemical properties, and preparation, as well as the resulting biological properties of new and known nanoparticles. It provides many examples of nanoparticles and their metabolism, including biological activities of the systems as well as their safety and toxicity. The studies also include investigating the redox properties of the metal-based nanoparticles and their potential involvement in processes and metabolic reactions controlled by reactive oxygen species.

Elitsa L. Pavlova and Debbie C. Crans
Guest Editors

pharmaceuticals

MDPI

Editorial

Metal Nanoparticles' Biological Activity and Pharmaceutical Applications

Elitsa L. Pavlova [1,*] and Debbie C. Crans [2]

1 Optics and Spectroscopy Department, Faculty of Physics, Sofia University "St. Kliment Ohridski", 5 James Bourchier Blvd., 1164 Sofia, Bulgaria
2 Department of Chemistry and Cell and Molecule Biology Program, Colorado State University, Fort Collins, CO 80532, USA; debbie.crans@colostate.edu
* Correspondence: elli_pavlova@abv.bg

Because of their variable physicochemical characteristics, metal nanoparticles are suitable and highly favorable for introduction to, and use in, living systems. The latest research on the design, synthesis, properties, biological evaluation, and pharmaceutical application of various metal nanoparticles is presented in the form of three reviews and nine original research articles in this Special Issue. The contributions highlight the importance of designed nanoparticle synthesis, including "green" and bio-inspired methods, as well as multiple approaches to enhancing their efficacy and safety through physicochemical and biological means [1–3]. The applications of metal nanoparticles in biology and medicine are described in articles ranging from toxicity studies to studies on antimicrobial activity against multiple Gram-positive and Gram-negative bacteria [4–7], diabetes [8], and several types of cancers [9,10], although many challenges with regard to design, toxicity, specificity, and cellular uptake remain and are addressed herein [9,11–20]. Nanoparticles prepared from silver, gold, manganese oxides, copper oxides, zinc oxides, zeolites, rare earth metals, and small polyoxovanadates, as well as nanoparticles laced with other metals or biomaterials, are described, along with their particular application and mechanisms of action. The properties of the nanoparticles depend strongly on their shape, size, and material, as well as the load they carry. The ability of metal nanoparticles to interact with cells, as well as their structures and cellular metabolites, is of particular importance, as highlighted in their applications as drug carriers and in the diagnosis and treatment of various diseases. Such representable examples are described in this Special Issue.

"Green" and bio-inspired syntheses, recently developed, address concerns about eco-safety and toxicity while enhancing the bioactivity of the newly composed metal nanoparticles [2,3]. Dolashka et al. [Contribution 1] utilized the mucus of the garden snail, *Cornu aspersum*, as a natural reducing and stabilizing agent to produce copper oxide nanoparticles. The newly synthesized nanoparticles were spherical, ~150 nm, and demonstrated superior antimicrobial activity against multiple Gram-positive (*Staphylococcus*) and Gram-negative bacteria (*Escherichia coli* and *Salmonella*) when compared to the mucus alone. Importantly, proteins present in the snail mucus contributed to nanoparticle stabilization and enhanced their bactericidal efficacy.

Alsaleh et al. [Contribution 2] synthesized zinc oxide-doped manganese oxide nanoparticles with improved catalytic properties and demonstrated their relatively low toxicity in murine macrophages. The doping altered nanoparticle uptake and increased the generation of reactive oxygen species and the activation of the cell antioxidant system. The obtained findings suggest that ZnO (1%) doping impacts cellular bio-interaction and the consequent toxicological outcomes.

Asif and colleagues [Contribution 3] investigated enhanced doping methods by synthesizing aluminum-doped zinc oxide nanocomposites. Notably, the $Zn_{0.75}Al_{0.25}O$ formulation demonstrated the most pronounced antimicrobial efficacy, particularly targeting *Bacillus cereus*. In addition, these nanocomposites exhibited selective cytotoxicity against liver, breast, and ovarian cancer cell lines. Importantly, their minimal toxicity toward normal liver cells suggests potential for safer anticancer applications.

The antimicrobial properties of silver-based nanomaterials have been the subject of recent investigations [4–7]. In a study by Pavlova et al. [Contribution 4], the researchers investigated silver-exchanged zeolites with varied silver content and Si/Al ratios. Their findings demonstrated a clear proportional relationship between the release of active silver species and the materials' antibacterial and bactericidal efficacy. In general, as the zeolites released more silver, their effectiveness against bacterial pathogens increased. However, the leaching of silver ions also raised important environmental concerns. The bactericidal concentrations of silver ions also caused significant mortality in *Daphnia magna*—a commonly used bioindicator in ecotoxicology. This result underscores the challenge of balancing antimicrobial efficacy with environmental safety, highlighting the potential ecological risks associated with spreading silver-based materials capable of producing effective antimicrobial concentrations.

Stoyanova and colleagues [Contribution 5] developed silver nanoparticles encapsulating mebeverine, an antispasmodic agent commonly used for irritable bowel syndrome, along with a structural analogue. Their in vitro studies indicated that cytotoxicity was concentration-dependent, but importantly, no genotoxic effects were observed at the concentrations investigated. These findings suggest that there is a high potential for the nanoparticle-based delivery system to be both safe and effective for potential therapeutic applications.

Rehman and colleagues [Contribution 6] utilized *Azadirachta indica* seed extract to biosynthesize silver nanoparticles exhibiting notable antidiabetic properties. In vitro analyses revealed that these nanoparticles significantly enhanced glucose uptake and adsorption and more effectively inhibited α-amylase activity compared to crude extracts. Extending their research in vivo, the team observed that diabetic mice treated with these nanoparticles experienced marked hypoglycemic effects and tissue regeneration in both the pancreas and liver. In general, these findings highlight the considerable therapeutic potential of "green"-synthesized silver nanoparticles derived from *Azadirachta indica*.

Originally, Rangel-López et al. [Contribution 7] introduced an innovative approach to synthesize gold nanoparticles, employing triple-negative breast cancer cell lysate as a reducing agent with immunomodulatory properties. This method resulted in nanoparticles that enhanced antigen presentation and stimulated proinflammatory cytokine secretion, while also increasing populations of CD8+ and CD22+ lymphocytes in murine models. These immunological effects translated into improved survival rates and inhibited tumor implantation, highlighting the nanoparticles' significant potential in advancing cancer immunotherapy strategies, in agreement with the suggestions made previously [9,10].

In another investigation, Gawas et al. [Contribution 8] successfully synthesized gold nanoparticles utilizing *Cordyceps militaris*, a medicinal mushroom, by a green, quality-by-design method. The obtained nanoparticles exhibited significant stability and biocompatibility in addition to pronounced antioxidant, antidiabetic, and antibacterial properties. FTIR spectroscopy analysis revealed that bioactive constituents derived from the mushroom played a pivotal role in the stabilization of the nanoparticles, thereby emphasizing the environmentally sustainable characteristics and multifunctional potential of these materials for prospective biomedical applications.

Panchal et al. [Contribution 9] conducted a comprehensive review focusing on the biomedical applications of plant-mediated silver, gold, and zinc oxide nanoparticles. Plant-based synthesis of metal-based nanoparticles utilizes natural extracts as reducing and stabilizing agents and, in this manner, minimize harmful leaching and other toxic by-products. They focused on their antioxidant, antimicrobial, and anticancer activities. The review emphasized the advantages of the "green" synthetic approaches using plant-based materials, which, in turn, result in enhanced therapeutic activity and improved safety compared to reported conventional synthetic approaches. Furthermore, the authors discussed the potential integration of metal nanoparticles with graphitic carbon nitride to create hybrid nanocomposites. Such materials could open new applications in drug delivery, diagnostics, and environmental remediation.

Burlec et al. [Contribution 10] comprehensively reviewed the synthesis, characterization, and therapeutic applications of mainly silver and gold nanoparticles. Their paper outlined the roles of nanoparticles in drug delivery, antimicrobial activity, cancer therapy, and diagnostic procedures, including data from over thirty clinical trials. Despite these promising findings, the authors direct the readers' attention to ongoing challenges related to biocompatibility and regulatory approval, which is delaying the approval for the use of these approaches and explains why so few of these metal nanoparticle systems are used clinically. As a result, this paper highlights the fact that challenges are not only scientific but also include legal limitations before metal nanoparticles technologies achieve broader clinical applications.

Barbosa and colleagues [Contribution 11] carried out an in-depth toxicity evaluation of a small, nanosized, mixed-valence polyoxovanadate cluster, employing both in vitro models (peripheral blood mononuclear cells and *Artemia salina*) and *in vivo* studies (mice). This cluster is unique because it is redox-active and hence differs from other polyoxometalate clusters. The study showed that it demonstrated pronounced cytotoxic effects in vitro, but the acute single-dose administration in mice was safe, suggesting acceptable short-term tolerability. Yet, with repeated administration over 28 days, marked metabolic disturbances emerged, particularly impairing hepatic and renal function. These adverse outcomes were linked to oxidative stress and lipid peroxidation, as has been reported for other polyoxovanadates compounds [21,22]. These findings underscore the necessity for caution regarding the biomedical application of redox active vanadates.

Titova and colleagues [Contribution 12] reviewed rare earth metal nanoparticles, such as those derived from gadolinium, europium, ytterbium, holmium, lutetium, dysprosium, erbium, terbium, thulium, scandium, yttrium, lanthanum, europium, neodymium, promethium, samarium, praseodymium, or cerium, which are increasingly being recognized for their potential in biomedicine. These systems provide many applications in imaging, cancer therapies, and regenerative medicine. However, the review also describes practical problems such as inconsistent toxicity profiles and the need for further modification techniques, such as surface functionalization, to make these materials safer for practical medical use.

In conclusion, optimizing the therapeutic impact of the metal nanoparticles requires a complex understanding and careful control of how the nanoparticles interact with biological systems, while minimizing potentially harmful side-effects and environmental risks. Despite the exciting progress presented in this Special Issue, several key challenges must be overcome before metal nanoparticles can be widely used in living systems [11–13]. Comprehensive toxicological profiling will provide information that will be very valuable.

The complexity of nanoparticle interactions with biological systems demands more detailed *in vivo* studies addressing biodistribution, long-term toxicity, immunogenicity, and clearance mechanisms. Differentiating between beneficial cytotoxicity against pathogens or cancer cells and harmful side-effects affecting healthy tissues is critical [9,14], and increased specificity is paramount. Increased standardization and reproducibility are needed, possibly through understanding the nature of the inhomogeneity of the samples and the correspondingly different responses to such samples. The variability in synthesis methods, nanoparticle size, shape, and surface chemistry complicate the comparison of the biological effects across studies. Establishing standardized protocols for synthesis, characterization, and biological evaluation will be essential in order to translate the laboratory findings into clinical practice [15,16].

Targeting and functionalization strategies remain key aspects that could provide improved results in the future. The functionalization of nanoparticles with targeting ligands, biomolecules, or drugs holds great promise for enhancing their specificity and reducing their systemic toxicity. Future research focusing on the optimization of this process in order to improve targeted delivery and controlled release could be important [17–20].

Oxidative stress involves not only the generation of reactive oxygen species but also their processing. Understanding and modulating these could be a key mechanism for antimicrobial and anticancer effects. Excessive amounts of free radicals formed because of oxidative stress may induce adverse side-effects. Fine-tuning the redox activity through doping and surface modification represents a potential approach for safer nanomedicines [23,24].

Environmental considerations may affect the various applications of nanoparticles and materials, necessitating regulatory frameworks to address the unique properties of these materials, including their potential environmental effects, bioaccumulation, and disposal. The development of eco-friendly "green" synthesis methods and the evaluation of environmental toxicity will enhance the development of sustainable nanotechnology applications [25].

Finally, integration with personalized medicine is an area of future importance. The diverse biological effects of nanoparticles, depending on size, composition, and functionalization, suggest that these materials have great potential for personalized therapies that satisfy individual patient needs. Future studies should explore their integration in genomics and proteomics [13,20,26] applications. Hence, there is no doubt that the application of these materials will increase in the future.

Finally, we would like to thank the authors, reviewers, and other contributors who helped make this Special Issue a reality. We look forward to learning more about future research directions in the second edition of this Special Issue, which we expect will further unravel the potential of metal nanoparticles in advancing pharmaceutical science.

Conflicts of Interest: The authors declare no conflict of interest.

List of Contributions:

1. Dolashka, P.; Marinova, K.; Petrov, P.; Petrova, V.; Ranguelov, B.; Kaynarov, D.; Stoycheva, I.; Pisareva, E.; Tomova, A.; Kosateva, A.; et al. Development of CuO Nanoparticles from the Mucus of Garden Snail Cornu aspersum as New Antimicrobial Agents. *Pharmaceuticals* **2024**, *17*, 506. https://doi.org/10.3390/ph17040506.
2. Alsaleh, N.B.; Aljarbou, A.M.; Assal, M.E.; Assiri, M.A.; Almutairi, M.M.; As Sobeai, H.M.; Alshamrani, A.A.; Almudimeegh, S.; Hatshan, M.R.; Adil, S.F. Synthesis, Characterization, and Toxicity Assessment of Zinc Oxide-Doped Manganese Oxide Nanoparticles in a Macrophage Model. *Pharmaceuticals* **2024**, *17*, 168. https://doi.org/10.3390/ph17020168.

3. Asif, M.; Tahir, M.; Jamil, F.; Sardar, H.; Rehman, J.; Dahlous, K.A. Synthesis, Characterization, and Evaluation of the Antimicrobial and Anticancer Activities of Zinc Oxide and Aluminum-Doped Zinc Oxide Nanocomposites. *Pharmaceuticals* **2024**, *17*, 1216. https://doi.org/10.3390/ph17091216.
4. Pavlova, E.L.; Nenova, E.P.; Yocheva, L.D.; Ivanova, I.A.; Georgiev, P.A. Antimicrobial and Oxidative Activities of Different Levels of Silver-Exchanged Zeolites X and ZSM-5 and Their Ecotoxicity. *Pharmaceuticals* **2024**, *17*, 1586. https://doi.org/10.3390/ph17121586.
5. Stoyanova, M.; Milusheva, M.; Georgieva, M.; Ivanov, P.; Miloshev, G.; Krasteva, N.; Mohammadi Ziarani, G.; Stojnova, K.; Tsoneva, S.; Todorova, M.; et al. Synthesis, Cytotoxic and Genotoxic Evaluation of Drug-Loaded Silver Nanoparticles with Mebeverine and Its Analog. *Pharmaceuticals* **2025**, *18*, 397. https://doi.org/10.3390/ph18030397.
6. Rehman, G.; Umar, M.; Shah, N.; Hamayun, M.; Ali, A.; Khan, W.; Khan, A.; Ahmad, S.; Alrefaei, A.F.; Almutairi, M.H.; et al. Green Synthesis and Characterization of Silver Nanoparticles Using Azadirachta indica Seeds Extract: In Vitro and In Vivo Evaluation of Anti-Diabetic Activity. *Pharmaceuticals* **2023**, *16*, 1677. https://doi.org/10.3390/ph16121677.
7. Ármides, M.; Ginette, D. Gold Nanoparticles Synthesized with Triple-Negative Breast Cancer Cell Lysate Enhance Antitumoral Immunity: A Novel Synthesis Method. *Pharmaceuticals* **2025**, *18*, 330. https://doi.org/10.3390/ph18030330.
8. Gawas, G.; Ayyanar, M.; Gurav, N.; Hase, D.; Murade, V.; Nadaf, S.; Khan, M.S.; Chikhale, R.; Kalaskar, M.; Gurav, S. Process Optimization for the Bioinspired Synthesis of Gold Nanoparticles Using Cordyceps militaris, Its Characterization, and Assessment of Enhanced Therapeutic Efficacy. *Pharmaceuticals* **2023**, *16*, 1311. https://doi.org/10.3390/ph16091311.
9. Panchal, P.; Rauwel, P.; Nehra, S.P.; Singh, P.; Karla, M.; Hermosa, G.; Rauwel, E. A Review on Biomedical Applications of Plant Extract-Mediated Metallic Ag, Au, and ZnO Nanoparticles and Future Prospects for Their Combination with Graphitic Carbon Nitride. *Pharmaceuticals* **2025**, *18*, 820. https://doi.org/10.3390/ph18060820.
10. Burlec, A.F.; Corciova, A.; Boev, M.; Mircea, C.; Cioanca, O.; Danila, G.; Danila, M.; Bucur, A.F.; Hancianu, M. Current Overview of Metal Nanoparticles' Synthesis, Characterization, and Biomedical Applications, with a Focus on Silver and Gold Nanoparticles. *Pharmaceuticals* **2023**, *16*, 1410. https://doi.org/10.3390/ph16101410.
11. Barbosa, M.D.; Lima, L.M.; Alves, W.A.; Lima, E.K.; Silva, L.A.; Silva, T.D.; Postal, K.; Ramadan, M.; Kostenkova, K.; Gomes, D.A.; et al. In Vitro, Oral Acute, and Repeated 28-Day Oral Dose Toxicity of a Mixed-Valence Polyoxovanadate Cluster. *Pharmaceuticals* **2023**, *16*, 1232. https://doi.org/10.3390/ph16091232.
12. Titova, S.A.; Kruglova, M.P.; Stupin, V.A.; Manturova, N.E.; Silina, E.V. Potential Applications of Rare Earth Metal Nanoparticles in Biomedicine. *Pharmaceuticals* **2025**, *18*, 154. https://doi.org/10.3390/ph18020154.

References

1. Altammar, K.A. A review on nanoparticles: Characteristics, synthesis, applications, and challenges. *Front. Microbiol.* **2023**, *14*, 1155622. [CrossRef]
2. Dhaka, A.; Chand Mali, S.; Sharma, S.; Trivedi, R. A review on biological synthesis of silver nanoparticles and their potential applications. *Results Chem.* **2023**, *6*, 101108. [CrossRef]
3. Pechyen, C.; Tangnorawich, B.; Toommee, S.; Marks, R.; Parcharoen, Y. Green synthesis of metal nanoparticles, characterization, and biosensing applications. *Sens. Int.* **2023**, *5*, 100287. [CrossRef]
4. Yaqoob, A.A.; Ahmad, H.; Parveen, T.; Ahmad, A.; Oves, M.; Ismail, I.M.; Qari, H.A.; Umar, K.; Mohamad Ibrahim, M.N. Recent Advances in Metal Decorated Nanomaterials and Their Various Biological Applications: A Review. *Front. Chem.* **2020**, *8*, 528583. [CrossRef]
5. Gomes, D.; Esteruelas, G.; Bonilla, L.; Laura, A.; Galindo, R.; Cano, A.; Espina, M.; Ettcheto, M.; Camins, A.; Silva, A.M.; et al. Metal-Based Nanoparticles as Antimicrobial Agents: An Overview. *Nanomaterials* **2020**, *10*, 292. [CrossRef]
6. Luo, L.; Huang, W.; Zhang, J.; Yu, Y.; Sun, T. Metal-Based Nanoparticles as Antimicrobial Agents: A Review. *ACS Appl. Nano Mater.* **2024**, *7*, 2529–2545. [CrossRef]

7. Zhang, S.; Lin, L.; Huang, X.; Lu, G.; Zheng, L.; Feng, Y. Antimicrobial Properties of Metal Nanoparticles and Their Oxide Materials and Their Applications in Oral Biology. *J. Nanomater.* **2022**, *2022*, 2063265. [CrossRef]
8. Feng, B.; Dong, Y.; Shang, B.; Zhang, B.; Crans, D.C.; Yang, X. Inhibitor Design for PTP1B and TCPTP: Exchangeable Vanadium Coordination Complexes on Graphene Quantum Dots. *Adv. Funct. Mater.* **2022**, *32*, 2108645. [CrossRef]
9. Al-Samydai, A.; Abu Hajleh, M.N.; Al-Sahlawi, F.; Nsairat, H.; Khatib, A.A.; Alqaraleh, M.; Ibrahim, A.K. Advancements of metallic nanoparticles: A promising frontier in cancer treatment. *Sci. Prog.* **2024**, *107*, 00368504241274967. [CrossRef]
10. Villalobos Gutiérrez, P.T.; Muñoz Carrillo, J.L.; Sandoval Salazar, C.; Viveros Paredes, J.M.; Gutiérrez Coronado, O. Functionalized Metal Nanoparticles in Cancer Therapy. *Pharmaceutics* **2023**, *15*, 1932. [CrossRef]
11. Morgan, R.N.; Aboshanab, K.M. Green Biologically Synthesized Metal Nanoparticles: Biological Applications, Optimizations and Future Prospects. *Future Sci. OA* **2024**, *10*, FSO935. [CrossRef]
12. Jamkhande, P.G.; Ghule, N.W.; Bamer, A.H.; Kalaskar, M.G. Metal nanoparticles synthesis: An overview on methods of preparation, advantages and disadvantages, and applications. *J. Drug Deliv. Sci. Technol.* **2019**, *53*, 101174. [CrossRef]
13. Hheidari, A.; Mohammadi, J.; Ghodousi, M.; Mahmoodi, M.; Ebrahimi, S.; Pishbin, E.; Rahdar, A. Metal-based nanoparticle in cancer treatment: Lessons learned and challenges. *Front. Bioeng. Biotechnol.* **2024**, *12*, 1436297. [CrossRef]
14. Ernsting, M.J.; Murakami, M.; Roy, A.; Li, S. Factors controlling the pharmacokinetics, biodistribution and intratumoral penetration of nanoparticles. *J. Control. Release* **2013**, *172*, 782–794. [CrossRef]
15. Hachhach, M.; Bayou, S.; El Kasmi, A.; Saidi, M.Z.; Akram, H.; Hanafi, M.; Achak, O.; El Moujahid, C.; Chafik, T. Towards Sustainable Scaling-Up of Nanomaterials Fabrication: Current Situation, Challenges, and Future Perspectives. *Eng* **2025**, *6*, 149. [CrossRef]
16. Kroll, A.; Pillukat, M.H.; Hahn, D.; Schnekenburger, J. Interference of engineered nanoparticles with in vitro toxicity assays. *Arch. Toxicol.* **2012**, *86*, 1123–1136. [CrossRef]
17. Chandrakala, V.; Aruna, V.; Angajala, G. Review on metal nanoparticles as nanocarriers: Current challenges and perspectives in drug delivery systems. *Emergent Mater.* **2022**, *5*, 1593–1615. [CrossRef]
18. Karami, M.H.; Abdouss, M.; Maleki, B. The state of the art metal nanoparticles in drug delivery systems: A comprehensive review. *Nanomed. J.* **2024**, *11*, 222–249. [CrossRef]
19. Shariatzadeh, S.; Moghimi, N.; Khalafi, F.; Shafiee, S.; Mehrabi, M.; Ilkhani, S.; Tosan, F.; Nakhaei, P.; Alizadeh, A.; Varma, R.S.; et al. Metallic Nanoparticles for the Modulation of Tumor Microenvironment; A New Horizon. *Front. Bioeng. Biotechnol.* **2022**, *10*, 847433. [CrossRef]
20. Truong, T.T.; Mondal, S.; Doan, V.H.M.; Tak, S.; Choi, J.; Oh, H.; Nguyen, T.D.; Misra, M.; Lee, B.; Oh, J. Precision-engineered metal and metal-oxide nanoparticles for biomedical imaging and healthcare applications. *Adv. Colloid Interface Sci.* **2024**, *332*, 103263. [CrossRef]
21. Dinda, R.; Garribba, E.; Sanna, D.; Crans, D.C.; Costa Pessoa, J. Hydrolysis Ligand Exchange Redox Properties of Vanadium Compounds: Implications of Solution Transfor-mation on Biological Therapeutic Environmental Applications. *Chem. Rev.* **2025**, *125*, 1468–1603. [CrossRef]
22. Aureliano, M.; De Sousa-Coelho, A.L.; Dolan, C.C.; Roess, D.A.; Crans, D.C. Biological Consequences of Vanadium Effects on Formation of Reactive Oxygen Species Lipid Peroxidation. *Int. J. Mol. Sci.* **2023**, *24*, 5382. [CrossRef]
23. Manke, A.; Wang, L.; Rojanasakul, Y. Mechanisms of Nanoparticle-Induced Oxidative Stress and Toxicity. *BioMed Res. Int.* **2013**, *2013*, 942916. [CrossRef]
24. Batir-Marin, D.; Boev, M.; Cioanca, O.; Lungu, I.-I.; Marin, G.-A.; Burlec, A.F.; Mitran, A.-M.; Mircea, C.; Hancianu, M. Exploring Oxidative Stress Mechanisms of Nanoparticles Using Zebrafish (*Danio rerio*): Toxicological and Pharmaceutical Insights. *Antioxidants* **2025**, *14*, 489. [CrossRef]
25. Rodríguez-Gómez, F.D.; Monferrer, D.; Penon, O.; Rivera-Gil, P. Regulatory Pathways and Guidelines for Nanotechnology-enabled Health Products: A Comparative Review of EU and US Frameworks. *Front. Med.* **2025**, *12*, 1544393. [CrossRef]
26. Shahalaei, M.; Azad, A.K.; Sulaiman, W.M.A.W.; Derakhshani, A.; Mofakham, E.B.; Mallandrich, M.; Kumarasamy, V.; Subramaniyan, V. A Review of Metallic Nanoparticles: Present Issues and Prospects Focused on the Preparation Methods, Characterization Techniques, and Their Theranostic Applications. *Front. Chem.* **2024**, *12*, 1398979. [CrossRef]

pharmaceuticals

Article

Development of CuO Nanoparticles from the Mucus of Garden Snail *Cornu aspersum* as New Antimicrobial Agents

Pavlina Dolashka [1], Karina Marinova [1], Petar Petrov [1], Ventsislava Petrova [2], Bogdan Ranguelov [3], Stella Atanasova-Vladimirova [3], Dimitar Kaynarov [1], Ivanka Stoycheva [1], Emiliya Pisareva [2], Anna Tomova [2], Angelina Kosateva [1], Lyudmila Velkova [1] and Aleksandar Dolashki [1,*]

1 Institute of Organic Chemistry with Center for Phytochemistry, Bulgarian Academy of Sciences, 1000 Sofia, Bulgaria; pda54@abv.bg (P.D.); karina.marinova@orgchm.bas.bg (K.M.); petar.petrov@orgchm.bas.bg (P.P.); mitkokaynarov@abv.bg (D.K.); stoycheva@orgchm.bas.bg (I.S.); angelina.kosateva@orgchm.bas.bg (A.K.); lyudmila.velkova@orgchm.bas.bg (L.V.)

2 Faculty of Biology (SU-BF), Sofia University "St. Kliment Ohridski", 1504 Sofia, Bulgaria; vpetrova@biofac.uni-sofia.bg (V.P.); episareva@uni-sofia.bg (E.P.); aatomova@biofac.uni-sofia.bg (A.T.)

3 Institute of Physical Chemistry "Rostislav Kaishev", Bulgarian Academy of Sciences, 1000 Sofia, Bulgaria; rangelov@ipc.bas.bg (B.R.); statanasova@ipc.bas.bg (S.A.-V.)

* Correspondence: aleksandar.dolashki@orgchm.bas.bg or adolashki@yahoo.com; Tel.: +359-887193423

Abstract: Several biologically active compounds involved in the green synthesis of silver and gold nanoparticles have been isolated from snail mucus and characterized. This paper presents a successful method for the application of snail mucus from *Cornu aspersum* as a bioreducing agent of copper sulfate and as a biostabilizer of the copper oxide nanoparticles (CuONPs-Muc) obtained. The synthesis at room temperature and neutral pH yielded nanoparticles with a spherical shape and an average diameter of 150 nm. The structure and properties of CuONPs-Muc were characterized using various methods and techniques, such as ultraviolet–visible spectroscopy (UV–vis), high-performance liquid chromatography (HPLC), one-dimensional polyacrylamide gel electrophoresis (1D-PAGE), up-conversion infrared spectroscopy Fourier transform (FTIR), scanning electron microscopy combined with energy dispersive spectroscopy (SEM/EDS), Raman spectroscopy and imaging, thermogravimetric analysis (TG-DSC), etc. Mucus proteins with molecular weights of 30.691 kDa and 26.549 kDa were identified, which are involved in the biogenic production of CuONPs-Muc. The macromolecular shell of proteins formed around the copper ions contributes to a higher efficiency of the synthesized CuONPs-Muc in inhibiting the bacterial growth of several Gram-positive (Bacillus subtilis NBIMCC2353, *Bacillus spizizenii* ATCC 6633, *Staphylococcus aureus* ATCC 6538, *Listeria innocua* NBIMCC8755) and Gram-negative (*Escherichia coli* ATCC8739, *Salmonella enteitidis* NBIMCC8691, *Salmonella typhimurium* ATCC 14028, *Stenotrophomonas maltophilia* ATCC 17666) bacteria compared to baseline mucus. The bioorganic synthesis of snail mucus presented here provides CuONPs-Muc with a highly pronounced antimicrobial effect. These results will expand knowledge in the field of natural nanomaterials and their role in emerging dosage forms.

Keywords: snail *Cornu aspersum*; copper nanoparticles; antimicrobial activity

1. Introduction

The synthesis of metal nanoparticles is attracting increasing interest [1,2]. Applied physicochemical methods have several disadvantages related to the release of toxic and harmful chemicals [3] as well as the use of complex equipment and a large amount of energy [4]. The solution to this problem is the application of a new method—green synthesis of metal nanoparticles, which is an ecological method that requires less energy and ecological solvents (water, ethanol, etc.) and does not generate toxic waste. Therefore, it has emerged as an attractive research area in recent years [5,6].

Various types of metal oxide nanoparticles (silver, gold, zinc, strontium, copper, etc.) with unique physical and chemical characteristics are used as adsorbents [7].

Silver (AgNPs) and gold (AuNPs) nanoparticles synthesized using plant extracts are of great interest to researchers [8,9]. Recently, Castillo-Henríquez et al. (2020) reported the green synthesis of gold and silver nanoparticles from plant extracts and their capacity as antimicrobial agents in the field of agriculture to combat bacterial and fungal pathogens. Moreover, this work makes a brief review of nanoparticles' contribution to water treatment and the development of "environmentally-friendly" nanofertilizers, nanopesticides, and nanoherbicides [8]. Lubis et al. [10] presented the use of *Persicaria odorata* leaf extract to reduce silver ions to AgNPs, while AuNPs were synthesized using an aqueous extract of *Limoniaacidissima* leaves, which have been shown to have an enhanced antimicrobial effects and to help in wound healing [11].

Strontium-based nanoparticles [12], as well as zinc oxide nanoparticles (ZnNPs), which can be used for wound healing as an adjuvant drug without toxic side effects, also have a strong regenerative effect on bone damaged by chemotherapeutic drugs [13]. Copper nanoparticles (CuNPs) also support the wound healing process and find biomedical applications because of their anticancer, antioxidant, anti-inflammatory, and neuroprotective effects [14,15]. Green biosynthesis of *Phragmanthera austroarabica* extract-derived CuONPs with highly effective antibacterial and antifungal effects against *Staphylococcus aureus, Escherichia coli, Penicillium chrysogenum,* and *Fusarium oxysporum* has been reported [16]. Biogenic methodologies used for the synthesis of nanoparticles are mostly limited to microorganisms and plant extracts, with very few studies conducted on body secretions of living animals. Data have been published on much more suitable sources than plants [11] for the synthesis of AgNPs and AuNPs from secretions of living organisms that are not killed but returned to nature, thus preserving biodiversity [17,18].

Also, the mucus from different snail species studied shows a wide variety of active substances, such as those collected from the feet of two land snails, *Lissachatina (Achatina) fulica* and *Hemiplecta distincta,* which have different composition, properties, and antimicrobial activity [19]. The published results of phenolic compounds in the beetle's guard gland extract showed that they could act as a stabilizing agent for the synthesis of AgNPs and also exhibited anticancer activity and cytotoxicity against a Dalton's lymphoma (DLA) cell line [20]. Also, ecofriendly synthesized biogenic AgNPs-SM from the mucus of *Achatina fulica* snail by Mane et al. showed strong antibacterial, antifungal, as well as anticancer activity against Hela cells [21]. Of particular interest is the potential of *Achatina fulica* mucilage in the green synthesis of bio-nanocomposites of copper oxide (SM-CuONC) and cobalt oxide (SM-Co_3O_4NC), with highly pronounced anticancer activity against human breast cancer (MDA-MB-231), colon cancer (HCT-15), cervical cancer (HeLa), etc. Both bionanocomposites showed better antioxidant activity and 100% larvicidal activity against mosquitoes [22].

Snails are a large and diverse group of organisms belonging to the Gastropod, a class of the phylum Mollusca. One of the most common land snail species is the brown garden snail *Cornu aspersum*, which is a rich source of biologically active natural substances such as peptides, proteins, glycoproteins, and hyaluronic and glycolic acid. They have long been documented to be of importance to health benefits, which is why they are incorporated into drugs to treat various human diseases [23]. For example, the chemical composition and mineral profile of *Helix aspersa* Müller mucus are related to its antimicrobial properties and its importance in wound healing [24–27]. Several results have been published on the toxicity of *H. aspersa* mucus extracts against eukaryotic cells. Trapella et al. reported a lack of toxicity and a proven cytostatic effect at all concentrations tested in vitro of mucus extracts (HelixComplex) on human dermal fibroblasts (MRC-5) [28]. This was also confirmed by a study by Gentili et al. on the non-toxic effect of HelixComplex on human keratinocytes [29].

Mucus has been found to protect cells from apoptosis and to significantly induce cell proliferation and migration through direct and indirect mechanisms. Research by Deng et al. showed strong adhesion to wet tissue of mucus extracts from two land snails, hemostatic effect, good biocompatibility, hemocompatibility, and in vivo pro-healing activity for skin wounds [30]. The mucus of the land snail *H. aspersa* was found to play a crucial role as a stabilizing agent in the synthesis of bioactive AgNPs [31]. Purified mucus from the snail *Helix aspersa* allows for the formation of biogenic AgNPs under very mild pH and temperature conditions: at room temperature and at neutral pH. The AgNPs produced inhibited bacterial growth and were used as a reference sample [31]. The synthesis of copper oxide nanoparticles (CuONPs) using *C. aspersum* mucus has not been documented in the literature. Moreover, CuONPs are among the group of oligodynamic noble metals [32] that have received the most recognition for their applications in molecular diagnostics, antimicrobial and antiaging therapies, and medicine [14,32].

Based on our previous studies, the present study is aimed at the environmentally friendly synthesis of CuONPs and reducing agents from *C. aspersum* snail mucus, and also at determining the bioactivity of mucus proteins incorporated into copper oxide nanoparticles with mucus components CuONPs-Muc.

2. Results

2.1. Isolation and Characterization of Mucus Extract from Garden Snail C. aspersum

2.1.1. Isolation of Mucus Extract from Garden Snail *C. aspersum*

This study analyzed an extract of mucus collected from the snails *C. aspersum*, grown in Bulgarian eco-farms, using patented technology [33]. After 24 h dialysis and concentration with a 20 kDa membrane, the snail mucus solution was purified of additives and enriched with proteins with molecular weight (MW) higher than 20 kDa. Comparative analysis of the UV spectra of the starting mucus (Figure 1) and the resulting fraction after dialysis and membrane concentration confirmed the purity of the fraction with predominant proteins with MW > 20 kDa.

The total concentration of protein in the sample was determined based on the obtained spectrum by measuring the absorbance at 280 nm (Figure 1), but this method is not the most accurate because it is based on the presence of the amino acids tryptophan (Trp), tyrosine (Tyr), and phenylalanine (Phe) in the protein composition. After applying the Bradford method, which is more sensitive than bicinchoninic acid (BCA) protein assay, a total protein concentration of 1.56 mg/mL in the snail mucus obtained was determined.

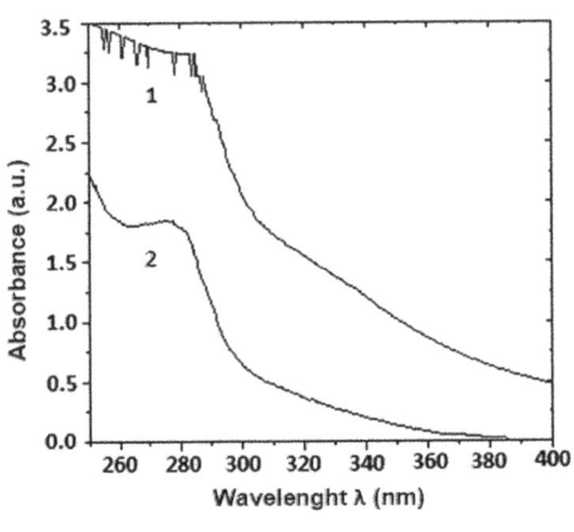

Figure 1. UV–vis spectra of *C. aspersum* mucus with MW > 20 kDa: (**1**) Crude extract; (**2**) Dialyzed and concentrated sample with a 20 kDa membrane.

2.1.2. Characterization of Mucus Extracts from Garden Snail *C. aspersum*

The mucus of *C. aspersum* snails is mainly a mixture of proteins, glycoproteins, acids, etc. Proteins in the fraction with MW > 20 kDa were characterized using 1D-polyacrylamide gel electrophoresis (1D-PAGE) and mass spectrometric analysis (Figure 2A,B).

The electrophoretic profile of the *C. aspersum* mucus fraction with MW > 20 kDa determined using 12% SDS-PAGE clearly shows eight major protein bands on the gel with MWs around 22, 26, 30, 35, 48, 58, 80, 170, and 250 kDa (Figure 2A). The exact MWs of 22 431, 26 539, 30 520, 35 528, 48 602, and 78 933 Da of the proteins in this fraction were determined from the MS spectrum using matrix-assisted laser desorption/ionization time-of-flight mass spectrometry (MALDI-Tof-MS) (Figure 2B).

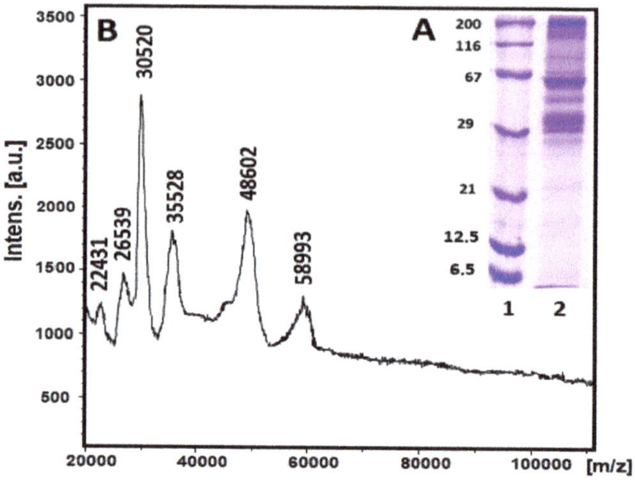

Figure 2. (**A**) 12% SDS–PAGE staining with Coomassie Brilliant Blue G-250: position 1, standard protein marker range 6.5–200 kDa (SERVA Electrophoresis GmbH, Heidelberg, Germany); position 2, extract of *C. aspersum* mucus with MW > 20 kDa; (**B**) MALDI-TOF-MS spectrum of *C. aspersum* mucus, recorded between 20 000 and 100 000 *m/z*.

2.2. *Green Synthesis and Characterization of CuONPs-Muc Using Snail Mucus Extract*

2.2.1. Green Synthesis of CuONPs-Muc Using Snail Mucus Extract with MW > 20 kDa

A dialysis-purified and concentrated fraction of *C. aspersum* mucus, mainly rich in proteins with MW > 20 kDa, was used in this study to synthesize CuONPs-Muc and evaluate their antimicrobial activity. The synthesis of CuONPs-Muc depends on various factors, such as the concentration of the starting mucus sample and of the copper ions. The CuONPs-Muc were obtained after mixing 100 mL of snail mucus fraction with MW > 20 kDa with 100 mL of 0.1 M $CuSO_4$ added gradually under continuous stirring at room temperature.

In the synthesis of nanoparticles, the reaction time is also an important parameter, and the synthesis of CuONPs-Muc was followed for 5 days. The visible color change of the $CuSO_4$ solution from pale blue to dark green indicated that the formation of CuONPs-Muc started 2 days after mixing, becoming more intense with time until the 3rd day.

After centrifugation at 4000 rpm and 12 000 rpm for 10 min and removal of excess products, washing three times with water and acetone, and drying at 70 °C for 12 h, the CuONPs obtained were subjected to analytical analyses using UV–vis, high-performance liquid chromatography (HPLC), one-dimensional polyacrylamide gel electrophoresis (1D-PAGE), Fourier transform infrared spectroscopy (FTIR), scanning electron microscopy (SEM), Raman spectroscopy and imaging, and thermogravimetric analysis (TG-DSC).

2.2.2. Characterization of CuONPs-Muc Using UV–Vis and Fluorescence Spectroscopy

The effect of reaction time on the formation of CuONPs-Muc was followed by recording UV–vis spectra at different times of incubation of the reaction mixture of mucus fraction with MW > 20 kDa and copper sulfate.

The recorded UV–vis spectra after 60 min, 3 h, 6 h, 24 h, 2 days, and 3 days confirmed the formation of CuONPs-Muc after 3 days of incubation by the most pronounced maximum at 456 nm (Figure 3). The formation of CuONPs-Muc was investigated using the exceedingly sensitive method of fluorescence spectroscopy. A comparison of the emission spectra of the mucus fraction with MW > 20 kDa used in the green synthesis and the fraction after obtaining CuONPs-Muc, shown in Figure 4A, after excitation at λ 280 nm reflects a change in intensity (h = 64 mm to h = 40 mm, respectively) and the position of the maximum fluorescence intensity (λ_{max}) (from 351 nm to 316 nm, respectively).

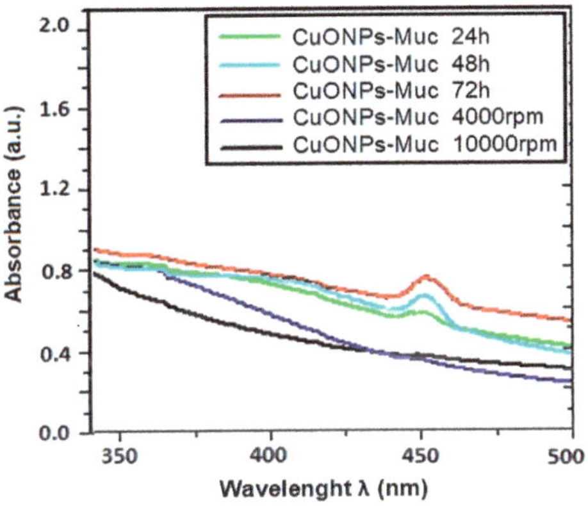

Figure 3. UV–vis spectra of the CuONPs-Muc synthesized at different reaction times. Wavelength absorption curves (Cu_2SO_4: extract 1:1, T = 20 °C) after 1 h and centrifugation at 4000 rpm; at 12 000 rpm; after 2 h, 48 h, and 72 h.

A change in the intensity (from h = 87 mm to h = 32 mm, respectively) and position of λ_{max} (from 355 nm to 341 nm, respectively) was also observed after excitation at λ 295 nm for both samples (Figure 4B). These changes resulted from a change in the location of some exposed indole cores of tryptophans from the surface of the proteins to the interior of the CuONPs-Muc formed.

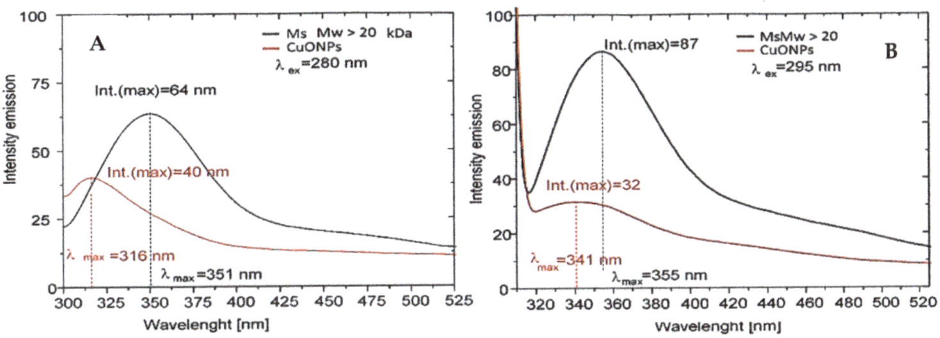

Figure 4. The comparative analysis of the emission spectra of the mucus fraction with MW > 20 kDa used in the green synthesis and the fraction after obtaining CuONPs-Muc (A_{280} = 0.1 in 0.1 M Tris buffer, pH 8). Emission spectra were recorded with a 1 cm quartz cuvette after excitation (**A**) at 280 nm and (**B**) at 295 nm.

2.2.3. Characterization of the Obtained CuONPs-Muc by Scanning Electron Microscopy Combined with Energy-Dispersive Spectroscopy (SEM/EDS)

SEM/EDS (JEOL JSM6390 and Oxford Instruments, Abingdon, UK) are used to obtain information on the morphology and shape of the CuONPs, as well as their elemental composition and size distribution.

The shape of the CuONPs-Muc is almost round, spherical; some of them are agglomerated in larger grains; nevertheless, individual particles could be distinguished from one another (Figure 5A). Thus, we were able to evaluate the overall size distribution, which clearly shows that the main sizes of the CuONPs-Muc are in the interval between 75 and 200 nm (Figure 5B).

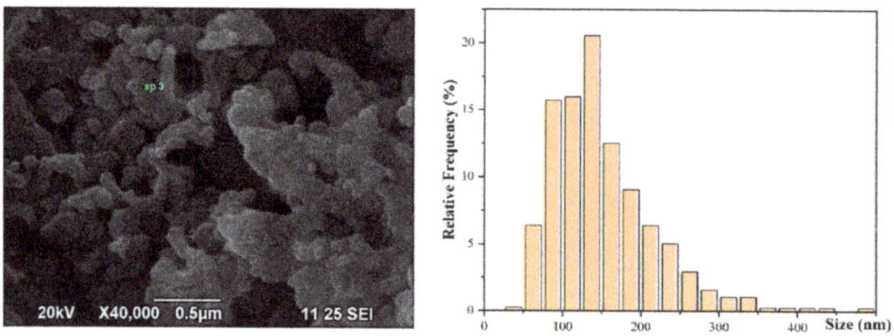

(A) (B)

Figure 5. (**A**) SEM images of CuONPs-Muc; (**B**) CuNP size distribution determined from SEM images.

EDS analysis showed the purity of the synthesized CuONPs-Muc, as copper (Cu) was obtained in a large amount, as shown in Figure 6.

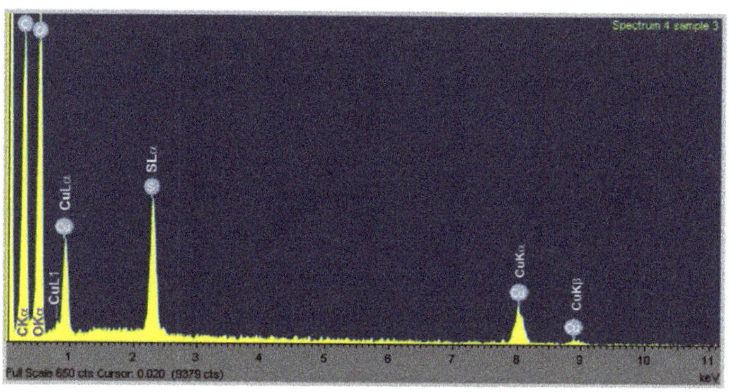

Figure 6. EDS spectra of Cu/Cu_2O nanoparticles synthesized using snail mucus with MW > 20 kDa.

2.2.4. Characterization of CuONPs-Muc Using Raman Spectroscopy and Imaging

As shown in Figure 7, the Raman spectra for the two samples, pure CuO and CuONPs-Muc formed from mucus with MW > 20 kDa with $CuSO_4$, were recorded at a 532 nm laser wavelength. The characteristic nanoscale peaks of CuO at 107, 284, and 550 cm^{-1} (Figure 7A) are observed, as well as peaks for the valence vibrations of the S-O bond from the $(SO_4)^{-2}$ ion, at 1015, 1044, and 1098 cm^{-1} (Figure 7B) [34].

2.2.5. Characterization of CuONPs-Muc Using IR Spectroscopy

To identify the proteins in the mucus fraction with MW > 20 kD involved in CuONP-Muc synthesis, FTIR spectra of the starting slime and of the resulting copper system synthesized with

$CuSO_4$ were recorded (Figure 8). Comparative analysis of the infrared spectra of the starting sample before and after dialysis showed several broad bands between 1453 and 1500 cm^{-1} that were not detected in the dialyzed sample and indicated the removal of additives in the starting mucus after dialysis.

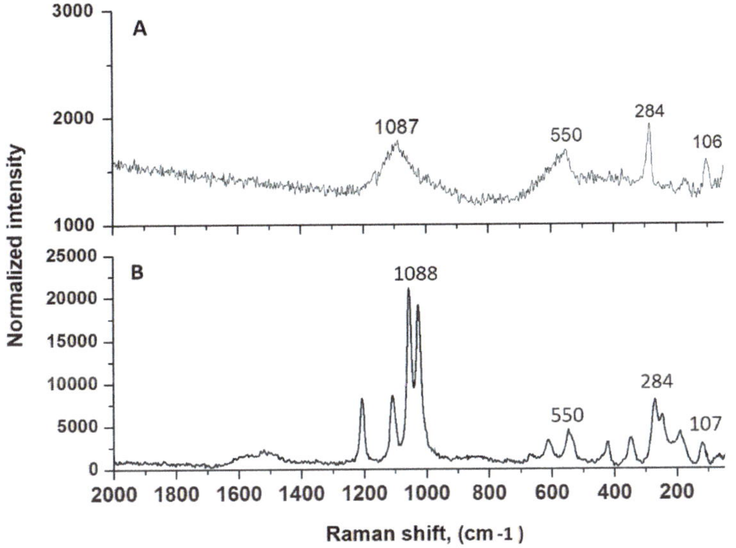

Figure 7. Raman spectra recorded on Raman Microscope Senterra II (Bruker) at a 532 nm laser wavelength for (**A**) pure CuO and (**B**) CuONPs-Muc from mucus with MW > 20 kDa with $CuSO_4$.

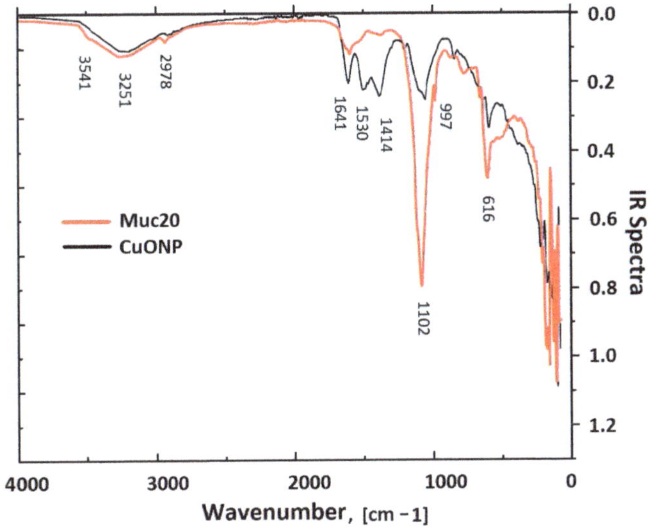

Figure 8. Infrared spectra of dialyzed snail mucus (red line) and dried CuONPs-Muc collected from the synthesis of *C. aspersum* mucus.

The FTIR spectra presented in Figure 8 show different vibrational bands between 500 and 4000 cm^{-1}, which reflect the changes in the functional groups of the initial sample and after CuONP-Muc formation.

The IR spectrum of the synthesized CuONPs-Muc proves the presence of an organic matrix surrounding the copper core with a broad band with a maximum at 3251 cm^{-1}, which reflects the stretching of oxygen–hydrogen (-OH) vibrations (Table 1). The additional band at 2978 cm^{-1} confirms the presence of traces of carbon, reflecting C–H stretching vibrations [35].

Two other peaks for the extract and nanoparticles, observed at 1641 and 1530 cm^{-1}, are related to C=O bond stretching vibration, which is typical of amides (amide I and amide II). Both bands were much more intense after dialysis, reflecting the increase in protein content of the sample. The absorption at 1641 cm^{-1} associated with amide I reflects the stretching vibrations of the C=O bond of the amide, while the absorption at 1530 cm^{-1} associated with amide II is mainly due to the bending vibrations of the NH bond.

Table 1. FT-IR characteristic peaks of the mucus extract of *C. aspersum* and CuONPs-Muc.

Wavenumber [cm^{-1}]	Interaction
616	Cu–O
1102	C–O bending
1414	C–N stretching
1530	NH
1641	O–H bending
2978	C–H stretching
3251	O–H stretching

Other important bands are observed at 661 and 616 cm^{-1} due to the stretching of Cu(I)–O particles, with the vibrational band of CuO at 616 cm^{-1} confirming the presence of Cu nanoparticles. The band at 1414 cm^{-1} in the spectra of both samples reflects the C–N stretching vibrations of the amines and carboxylate moieties associated with the acid derivatives of sugars and amino acid chains. The presence of glycosylated proteins is represented by the broad band centered at 1102 cm^{-1} associated with CO bonds [36].

2.3. *Characterization of CuONP-Muc Stability Using TG/DSC-MS Analysis*

The stability of the synthesized CuONPs-Muc was analyzed using thermogravimetric TG-DSC methods, and the interactions between the proteins from the mucus and copper ions were determined, as well as their thermal stability. The TG curves reflect changes in the weight of the sample depending on the temperature rise, while the processes occurring over time as a result of heating the material are traced by the DTG curves.

The TG and DSC results are presented in Figure 9 and Tables 2 and 3, and clearly show a great difference in the thermal degradation processes of pure mucus and CuONPs-Muc. Analyzing the TG-DSC results for both samples, two stages can be distinguished in the heating processes in an argon atmosphere in the range of 25–300 °C. The first stage covers 25 to 120 °C, with the endothermic effect at about 80 °C and the mass loss corresponding to evaporation of the physically adsorbed water (T_{max}, IDS = 73.3 °C, M_{loss}, IDS = 16.94%).

The endothermic peak for CuONPs-Muc in the same temperature range is also related to the desorption of physically adsorbed water, but it occurs at a lower temperature and with a greater mass loss compared to pure mucus (Tmax, IDS = 62.2 °C, M_{loss}, IDS = 23.47%). In both samples, after the endothermic peak due to the release of water, there follows an exothermic peak, which is much more pronounced in the pure mucus and is located at higher temperatures (Tpeak mucus = 135.8 °C, Tpeak composite = 91.5 °C).

The second stage in the temperature range of 120 to 300 °C is characterized by significant mass loss for pure mucus (M_{loss} = 75.75%) and CuONPs-Muc (M_{loss} = 69.44%) and Tpeak mucus = 183.9 °C, Tpeak composite = 158.9 °C, respectively, which continues up to 180 °C. These changes are related to the initial destruction of the material, with complete destruction of the nanoparticles formed occurring above 180 °C. The processes of degradation of protein molecules, which are endothermic, in the composite are associated with the removal of low molecular weight compounds from the surface of copper-containing particles (van der Waals forces). This is the reason for the less pronounced endothermic peak and the more pronounced mass loss (M_{loss} = 69.44%) of the composite sample up to 130 °C. After the separation of the low molecular weight compounds, complete destruction of the material occurs.

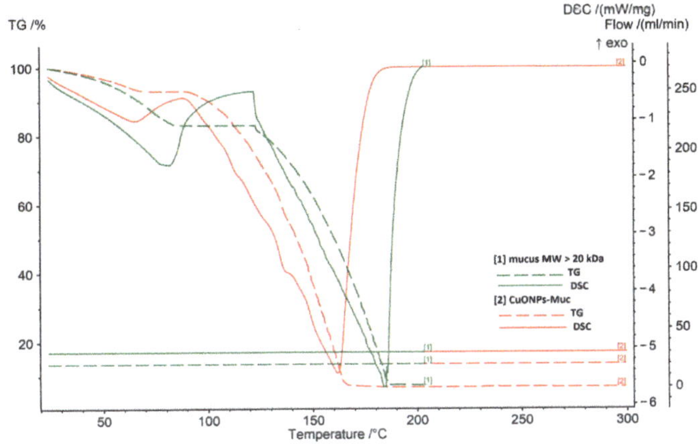

Figure 9. TG-DSC analysis of pure mucus sample with MW > 20 kDa and CuONP-Muc composites.

The output mucus contains low molecular weight and high molecular weight proteins. In high molecular weight proteins, the interactions between the side chains in the interior of the molecule are more pronounced, which is related to higher conformational stability compared to low molecular weight proteins. Therefore, the larger enthalpy values (ΔH) determined from the peak area of the DCS curves for the pure mucus sample (ΔH = −1216 J/g) reflect stronger intramolecular interactions in the protein (Table 3).

Table 2. TG –DTG data for pure mucus sample and CuONP composites.

Sample	TG				DTG		
	$M_{loss,IDS}$ [%] 30–120 [°C]	M_{loss1} [%] 120–200 [°C]	$M_{loss2.5}$ [%] 200–300 [°C]	$M_{loss,TOTAL}$ [%]	$T_{max,IDS}$ [°C]	T_{max1} [°C]	$T_{max2.5}$ [°C]
Mucus	16.94% (9.08 mg)	75.75% (40.64 mg)	0.03% (0.016 mg)	92.72%	73.3	123.0	183.0
CuONPs	23.47% (11.49 mg)	69.44% (34.05 mg)	0.04% (0.019 mg)	92.95%	62.2	134.9	160.9

$M_{loss,IDS}$, mass loss in initial decomposition step; M_{loss}, mass loss in main decomposition step; $M_{loss,TOTAL}$, total mass loss; $T_{max,IDS}$, maximum temperature in initial decomposition step; $T_{max1,2}$, temperature at the maximum degradation step.

Table 3. DSC data for pure mucus sample and CuONP composites.

Sample	DSC							
	T_{onset} [°C]	T_{peak} [°C]	T_{end} [°C]	ΔH J/g	T_{onset} [°C]	T_{peak} [°C]	T_{end} [°C]	ΔH J/g
Mucus	58.2	80.8	91.1	−289.2	127.9	183.9	189.0	−1216
CuONPs	35.3	65.2	85.4	−183.8	125.3	158.9	181.7	−1058

T_{onset}, temperature of the decomposition initiation; T_{peak}, peak maximum decomposition temperature; T_{end}, final decomposition temperature; ΔH, the heat generated during the decomposition reaction obtained by the integration of the thermal peaks (area).

2.4. Identification of Proteins Involved in Green Synthesis of CuONPs

To determine the involved proteins from the mucus in the formation of CuONPs-Muc, two comparative analyses were carried out using reversed-phase high-performance liquid chromatography (RP-HPLC) and one-dimensional polyacrylamide gel electrophoresis (1D-PAGE), comparing the changes in the proteins of the mucus fraction with MW > 20 kDa and of the supernatant after CuONP-Muc preparation and precipitation.

2.4.1. 1D-PAGE and RP-HPLC Analyses

The first analysis was carried out using 12% SDS-PAGE, and Figure 10B presents an electrophoretic probe of the starting fraction used for green synthesis (Figure 10A, line 2) and the supernatant after separation of the synthesized CuONPs-Muc. The original *C. aspersum* mucus sample (Figure 10A, lane 2) shows proteins with MW ~210 kDa, ~100 kDa, and ~60 kDa, which are well represented, as well as with MW between 55–65 kDa and 25–41 kDa.

A change in the two electrophoretic paths shows a decrease in the intensity of the band at 30 kDa and 26 kDa, which are very intense in the starting fraction, as well as a greatly increased intensity of the bands with MW 235 kDa and 160 kDa, presented in Figure 10A, line 3 (synthesized CuONPs). No or very weak change was observed for the proteins with MW ~40 kDa and ~60 kDa. Additional information on the involvement of proteins as a reducing agent is the comparative analysis of the changes in the chromatograms of the eluted proteins on a C18 column (8.00 × 250 mm) of RP-HPLC in the dialyzed mucus with MW > 20 kDa and the fraction after obtaining NPs. The comparative analysis presented in Figure 10B shows a significant change in peaks 3, 5, 7, and 9, which do not elute after NP synthesis. We can assume that these proteins are involved in the formation of CuONPs-Muc. Also, two new peaks (1 and 2) appear, which are probably related to the formation of CuONPs-Muc.

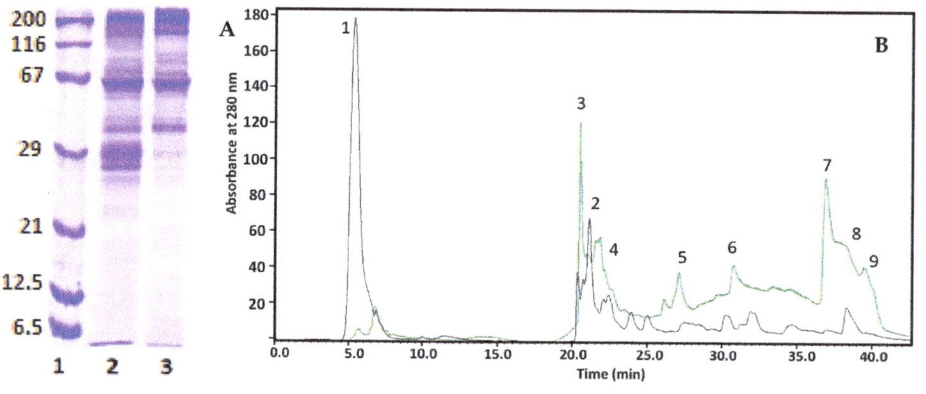

Figure 10. (**A**) 12% SDS-PAGE of line (**1**) standard protein marker in the range 6.5 kDa–200 kDa; line (**2**) mucus fraction with MW > 20 kDa; (**3**) the supernatant after precipitation of CuONPs-Muc. (**B**) HPLC chromatograms of mucus with MW > 20 kDa (black line) and the supernatant after centrifugation of CuONPs-Muc (green line), separated on a C18 column (8.00 × 250 mm).

2.4.2. Image Analysis of 10% SDS-PAGE Using ImageQuant™ TL v8.2.0 Software

The identification of proteins was carried out using 10% SDS-PAGE (Figure 11), and the exact MWs of the isolated fractions eluted as peaks 7 and 9 (Figure 10, lanes 2 and 3) were determined.

The electrophoretic profile showed the highest expression of proteins with MW mainly above 100 kDa, as well as decreased expression of proteins with MW 16.715 kDa, 22.474, 51.410 kDa, 65.132 kDa, 78.939 kDa, and 88.913 kDa.

The electrophoretic profile of fractions 7 and 9 eluted using RP-HPLC on a C18 column expressed two major bands with MW 30.59 and 26.55 kDa (positions 2 and 3 on 10% SDS-PAGE, Figure 11). In addition to these proteins (common to both fractions), the electrophoretic analysis of fraction 9 also included some poorly expressed proteins.

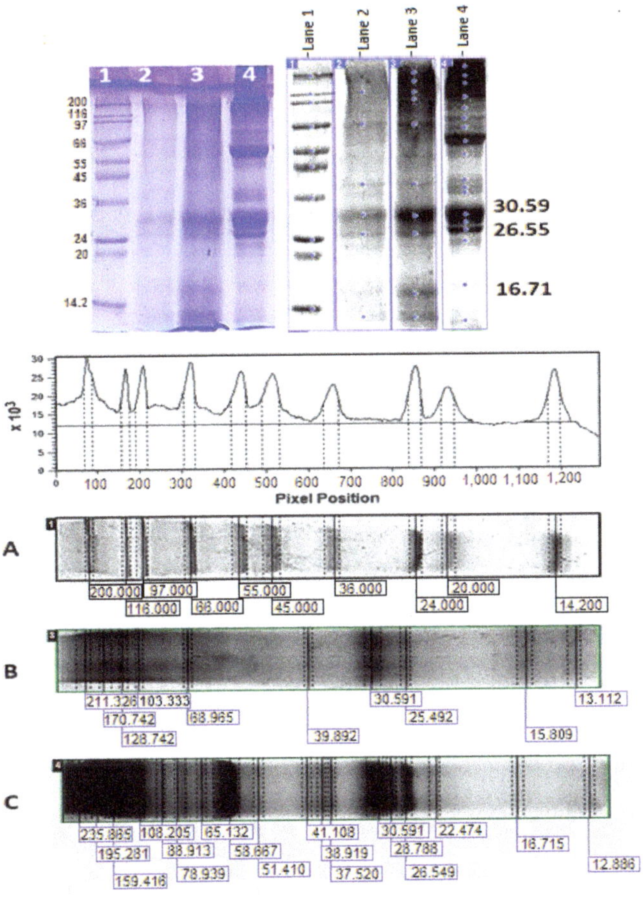

Figure 11. ImageQuant™ TL analysis of 10% SDS-PAGE, visualized with Coomassie G-250 of: (**A**) line 1—standard protein marker in the range 6.5–200 kDa (SigmaMarker™, Sigma-Aldrich, Saint Louis, MO, USA); (**B**) lane 2, peak 7 of the HPLC chromatogram of the supernatant of the *C. aspersa* mucus fraction with MW > 20 kDa after precipitation of CuONPs (Figure 10); (**C**) lane 3—peak 9.

The electrophoretic profile of fractions 7 and 9, isolated on a C18 column, expressed two main bands with MW 30.59 and 26.55 kDa. This is a confirmation of the result presented in Figure 10, where the difference between the supernatant of the *C. aspersa* mucus fraction with MW > 20 kDa is presented before and after CuONP-Muc precipitation.

2.5. Antibacterial Activity of CuONPs

Studying the antibacterial activities of pure mucus and synthesized CuONPs-Muc against the eight Gram-positive and Gram-negative pathogenic bacterial strains revealed that mucus itself has no or very weak inhibitory effect against the tested microorganisms (Figure 12).

Yet, its use as a reducing agent for synthesizing CuONPs-Muc significantly increased the microbial growth inhibition properties. CuONPs-Muc obtained by *C. aspersum* mucus fraction with Mm > 20 kD had noticeable antimicrobial effects on most of the tested Gram-positive and Gram-negative bacterial strains (Figure 13). However, the observed effect of CuONPs-Muc was significantly lower for Gram-negative bacteria. *E. coli* and *S. typhimurium* showed resistance to the CuONPs-Muc obtained, and no zones of growth inhibition were detected. The other *Salmonella* species, *S. enteritidis*, together with *Stenotrophomonas maltophilia*, showed lower susceptibility to the tested nanomaterials (Figure 13A).

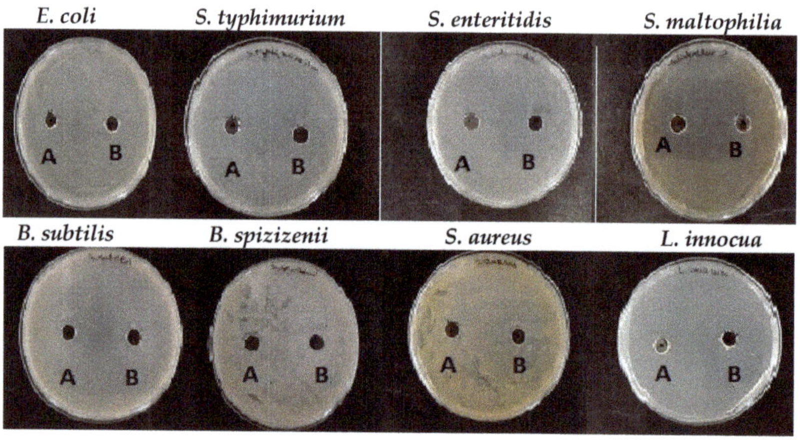

Figure 12. Antibacterial activity of fractions of (**A**) pure mucus from *C. aspersum* with MW < 20 kD and (**B**) fraction with MW > 20 kDa.

Much better antibacterial activity of CuONPs-Muc was found against Gram-positive test strains, and the measured growth inhibition zones varied between 19 ± 0.9 and 38 ± 1.6 mm (Figure 13B). The highest antibacterial potential of the CuONPs-Muc obtained was detected against both *Bacillus* species at 35 ± 1.2 mm and 38 ± 1.6 mm, respectively. Among all the Gram-positive bacteria tested, *Staphylococcus aureus* showed lower resistance to CuONPs-Muc; its growth was relatively slightly inhibited and the measured zones were like those of the Gram-negative bacteria (10 ± 0.4–19 ± 0.3 mm).

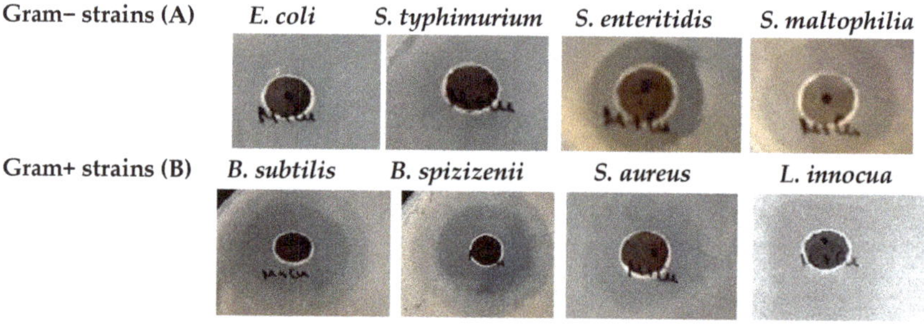

Figure 13. Bacterial growth inhibitory effect of CuONPs-Muc obtained with reducing agent mucus from *C. aspersum* with MW > 20 kDa; antimicrobial activity against (**A**) Gram-negative bacteria; (**B**) Gram-positive bacteria.

Next, the antibacterial effect of CuONPs-Muc was assessed compared to commonly used commercial antibiotics. The antibiograms performed revealed that the most effective against Gram-positive bacteria were vancomycin (VA) and erythromycin (E), with a growth inhibition zone varying from 12 ± 0.6 to 27 ± 1.0 mm (Figure 14A). At the same time, the CuONPs synthesized from the mucus fraction with MW > 20 kDa showed zones 30% bigger than those of the vancomycin (Figure 14A).

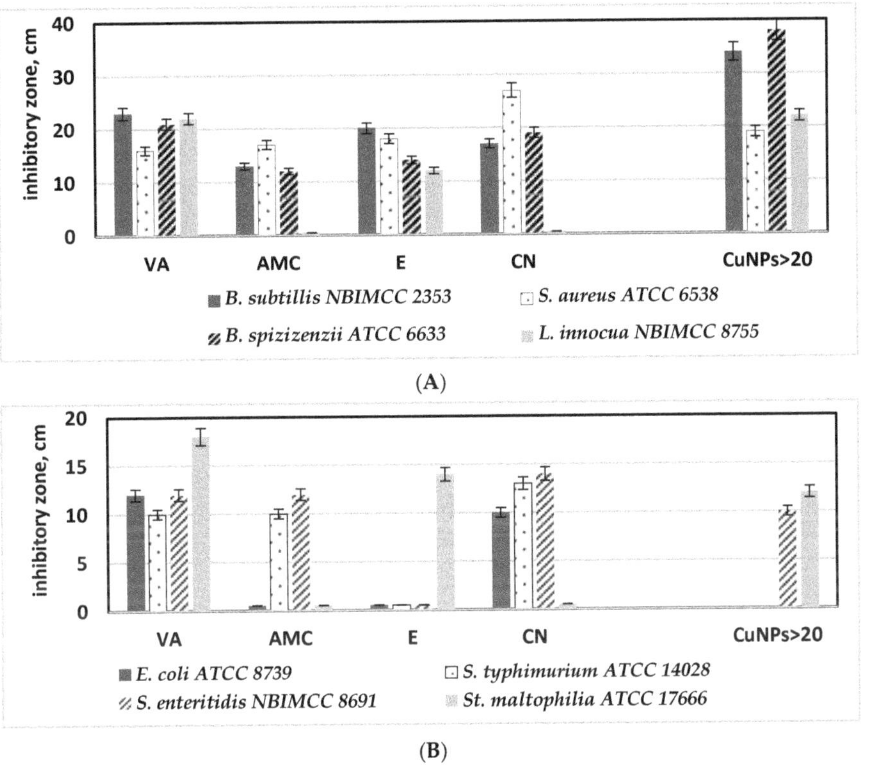

Figure 14. Comparative analysis between the efficacy of commonly used antibiotics and synthetized CuONPs-Muc against tested Gram-positive (**A**) and Gram-negative (**B**) pathogenic bacterial strains: vancomycin (VA), amoxiclav (AMC), erythromycin (E), and cephalexin (CN), CuONPs obtained with a fraction of mucus from *C. aspersum* with MW > 20 kDa (CuONPs > 20).

Similar results were observed in Gram-negative bacteria. The efficiency of the CuONPs-Muc against *Salmonella enteritidis* and *Stenotrophomonas maltophilia* is similar to those found for vancomycin, the most effective antibiotic against Gram-negative bacterial strains tested (Figure 14B).

3. Discussion

Nanoparticles occupy an increasingly important place in medical therapies as they increase the efficacy of drugs through their proper targeting and distribution in tissues [37–39]. Environmentally friendly technologies were used to generate nanoparticles of silver, zinc oxide, and other materials. A suitable source for green synthesis of NPs is the mucus from garden snails of the species "Helix", which is rich in bioactive substances with potential applications in the prevention and/or treatment of some human diseases.

Various mucus extraction techniques and methods have been used and patented, such as acidic or neutral stimulating solutions, which have been shown to affect the final composition of snail mucus [19,40,41]. To protect the substances present in the mucus, we used extraction without stimulants using our patented method [33]. The UV–vis spectra presented in Figure 1 of snail mucus before and after dialysis confirm the purity of the sample from impurities.

The mucus of land snails has been shown to contain between 75 and 95% water and a mixture of proteoglycans, glycosaminoglycans, glycoprotein enzymes, hyaluronic acid, proteins, antimicrobial peptides, amino acids, and metal ions [42]. However, based on information reported by Xie et al., for the ineffective participation of proteins with MW below 7 kDa in the synthesis of AgNPs [42], the mucus used for CuONPs was enriched in proteins with MW > 20 kDa after concentration on a 20 kDa membrane.

The distribution of proteins in the mucus fraction from snail *C. aspersum* with MW > 20 kDa demonstrated using electrophoretic profiling (Figure 2A) as well as MALDI analysis showed proteins with MWs of 22 431, 26 539, 30 520, 35 528, 48 602, and 78 933 kDa (Figure 2B). The results obtained are consistent with published information on *Actinia equina* mucus, which contains about 24.2% proteins with different MWs of 3.5, 14.3, 20.1, 29.0, 43.0, 66.0, and 97.4 kDa, as well as for mucus proteins from *Mastacembelus armatus* at 34 kDa, 45 kDa, and 144 kDa [43]. It was found that proteins with MW above 100 kDa in the mucus of *H. aspersa* and *H. pomatia* corresponded to glycoproteins [44].

Another published information related to the advantage of using the whole plant extract was used in CuONP synthesis, where all compounds in the extract participate in the process as reducing agents and also as capping agents in complex systems. If only individual isolated compounds are used in the process of NP synthesis, nanoparticles with a smaller size, shape, and uniformity are formed than when using the whole extract. Another important factor is the neutral pH of the medium, which is consistent with the behavior of some biomolecules in mucus, which may be inactive under acidic conditions and will not participate in nanoparticle formation. Based on all these factors, the applied efficient, simple, biogenic, and sustainable synthetic route conducted at room temperature showed CuONP-Muc formation after 5 days of incubation of a mixture of MW > 20 kDa fraction of garden snail mucus with 0.1 M solution of $CuSO_4$.

Evidence for the formation of copper nanoparticles and presentation of their physicochemical properties are the data obtained from the various methods and techniques applied, such as UV–vis, HPLC, 1D-PAGE, FTIR, SEM, Raman spectroscopy and imaging, and DTG. The green synthesis of CuONPs of the fraction with MW > 20 kDa was proven using the most pronounced maximum at 456 nm (Figure 3) of UV–vis spectra after 3 days of incubation, as well as the much more sensitive method of fluorescence spectroscopy.

It is known that amino acids, such as proline, tryptophan, or tyrosine, are often used as ligands in nanoparticle preparation processes [45]. Due to their ability to interact with hydrophobic regions in protein secondary structures, they disrupt the natural tendency of selected proteins to form oligomers by stacking β-sheet structures [46]. Aggregation and fibrillation of proteins controlled by nanomaterials can contribute to a change in the behavior of proteins, which is expressed in a change in fluorescence intensity or a shift in the fluorescence emission from some NPs as a result of interaction with proteins [47].

The observed change in the fluorescence analysis of the mucus fraction with MW > 20 kDa before and after the formation of CuONPs-Muc is due to the formation of a protein corona, which leads to a change in the conformation and environment of the tryptophan residues in the proteins.

The fluorescent amino acid Trp, Tyr, and Phe residues of the protein are indicators of the changes in the physical and chemical properties of the environment in which the fluorophore is located in the realized conformation. They reflect the changes that take place both in the environment and inside the protein molecule itself. These changes are expressed in the fluorescence spectra with a shift in the maximum and emission intensity [48]. Protein absorption at 280 nm is known to be mainly associated with Tyr and Trp residues, while at wavelengths greater than 295 nm, they mainly reflect the absorption of Trp residues [48].

Tryptophan residues located on the surface of proteins in the mucus fraction with MW > 20 kDa are responsible for the fluorescence emission at 351 nm (λ_{ex} = 295 nm). After the formation of biogenic nanoparticles, they find themselves in a more hydrophobic environment, buried in the protein corona of the CuONPs-Muc formed. As a result, a blue shift of the fluorescence emission at 341 nm (λ_{ex} = 295 nm) and a significant decrease in its intensity were observed. The present fluorescence study shows the reorganization of the secondary structure and spatial conformation (or three-dimensional conformation) of proteins in the mucus fraction with MW > 20 kDa, supporting the formation of CuONPs-Muc as reducing and stabilizing agents.

Comparative analysis of the emission spectra of the initial sample with MW > 20 kDa and after CuONP-Muc synthesis was performed under excitation at 280 nm and 295 nm. The emission spectra presented in Figure 2, recorded after excitation at 280 nm, reflect a change in intensity (from 64 mm to 40 mm, respectively) and the position of the fluorescence emission maximum, λ max (from 351 nm to 316 nm, respectively). This change can be explained by the location of the Tyr and Thr residues in the molecule, as λ_{max} = 351 nm reflects the fluorophores exposed on the surface of the proteins and access to them. The fluorescence intensity decreases in the sample after the formation of CuONPs-Muc, which is the result of the coagulation of the proteins in nanoparticles, as a result of which a part of the fluorophores are buried in the CuONPs-Muc thus formed. This is also confirmed by the shift of λ_{max} = 351 nm to 316 nm, which corresponds to the emission of "buried" fluorophores.

The stated assumption is also supported by the results presented in Figure 4B: emission spectra of both samples after excitation at 295 nm. According to the two-state hypothesis, for Trps located on the surface of the molecule, a change in intensity is observed (from 87 mm to 32 mm, respectively), which can be explained by the role of Trps and, more precisely, surface-localized Trp residues. Also, a shift in the wavelength of the fluorescence emission in the long-wave range from λ_{max} 355 nm to λ_{max} 341 nm was recorded, which is a sign of a realized structural conformation of the proteins. As a result of CuONP-Muc formation, the exposed indole rings of Trps from the surface of proteins are "buried" in the CuONPs-Muc formed, and only a fraction of the indole rings remain accessible, which are exposed on the surface of CuONPs.

SEM analysis (Figure 5) and Raman spectroscopy with peaks at 107, 284, and 550 cm^{-1} (Figure 7) confirm the formation of CuONPs-Muc. The images reveal that the CuONPs are agglomerated and form larger grains (Figure 5) with a size in the range of 100–150 nm. The high concentration of Cu, O, C, and S presented by the EDS analysis indicates the purity of the synthesized CuONPs-Muc, as well as the presence of unreacted copper ions. This result was also confirmed by the Raman spectra recorded at a wavelength of 532 nm, as peaks for valence vibrations of the S-O bond of $(SO_4)^{-2}$ ion were also observed at 1015, 1044, and 1098 cm^{-1}.

The formation of NPs is mainly the result of two types of interactions: between Cu ions and the side groups of proteins, as well as between the outer surfaces of the mucus proteins themselves. The formation of NPs is based on intermolecular forces such as van der Waals, electrostatic, covalent, hydrogen bonds, and π–π arrangement [45]. The crucial importance of proteins and enzymes for the creation and stability of nanoparticles has been demonstrated by FT-IR analysis of fungal biomass filtrate [49]. Snail mucus contains a wide variety of active compounds, such as allantoin, collagen, elastin, natural peptides, proteins, and enzymes (superoxide dismutase—SOD and glutathione S-transferase—GST); polyphenols; glycolic, hyaluronic, and lactic acid; vitamins A, C, and E; as well as metal ions—copper (Cu), iron (Fe), and zinc (Zn) [50]. The involvement of some of these proteins and enzymes in the green synthesis of CuONPs from the mucus of the garden snail *C. aspersa* was demonstrated by the changes in the functional groups on the surface of the CuONPs-Muc. The strong absorption peaks at 3251 cm^{-1}, 2924 cm^{-1}, 1641 cm^{-1}, 661 cm^{-1}, and 620 cm^{-1} show similar behavior with the different functional groups in the synthesized nanoparticles, such as OH groups of phenols and alcohols, the NH groups of amino acids in proteins, carboxyl (COO) residues, etc.

The observed IR spectra of amide I are associated with the C-O bond vibrations of α-helical (1630–1655 cm^{-1}), β-sheet (1660–1690 cm^{-1}), and disordered (1660–1670 cm^{-1}) structures that are used to study the secondary structure of proteins. The deformation N-H and valence C-N bonds of the peptide group of amide II are in the region of 1520–1580 cm^{-1}. The change in the absorption bands at 1641 cm^{-1} and 1530 cm^{-1}—well separated in the infrared spectrum of dialyzed mucus but, after the formation of CuONPs, merging and decreasing in intensity—confirms the involvement of glycoproteins in the synthesis of CuONPs. Another important evidence for the involvement of glycoproteins in the synthesized NPs is expressed by a decrease in the intensity of the peak at 1414 cm^{-1}, which reflects the carboxylate moieties associated with the acid derivatives of sugars and amino acid chains, as well as an increase in the intensity of the peak at 1102 cm^{-1}, related to CO bonds.

Supplementing the information on the involvement of proteins in agglomerates is TG/DSC-MS analysis, which reflects a difference in the stability of the forms. A lower stability of CuONPs-Muc in biological solutions (i.e., phosphate buffered saline) [51,52] and a greater thermal stability of the composite have been reported, which are associated with the adsorption of more high molecular weight compounds from the protein [53].

The information presented in Figure 9 and Tables 1 and 2, on the interactions between the protein and CuONPs-Muc, as well as its thermal stability, from TG-DSC analyses, shows a significant difference in the behavior of the two samples. An explanation for the faster decomposition of the composite compared to the pure mucus may be due to the mainly low molecular weight compounds adsorbed from the mucus that form the CuONPs-Muc. Moreover, the larger enthalpy values (ΔH) determined from the peak area of the DSC curves for the pure mucus sample ($\Delta H = -1216$ J/g) reflect more favorable intramolecular interactions in the protein.

This difference can be explained by the proteins in the mucus having different molecular weights. Protein size mainly affects the conformational change, folding and unfolding of proteins, and also the surface coverage of NPs. Proteins can be divided into three subcategories: small "hard" proteins with a rigid structure, large "soft" proteins with a flexible structure, and medium proteins. Small proteins do not undergo conformational changes upon contact with the substrate surface, whereas larger proteins undergo greater conformational changes, caused by their contact with the NP surface [54]. Therefore, the lower thermodynamic affinity of "stiff" proteins than that of larger proteins can be explained by the weaker conformational changes in the structure, reflected in a smaller number of binding points per protein molecule [55].

More precise information about the proteins that are involved in the formation of the NPs is presented using RP-HPLC analysis and 1D-PAGE, comparing the changes in the proteins of the mucus with MW > 20 kDa and of the supernatant, after obtaining and precipitation of CuONPs-Muc. The changes observed in peaks 3, 5, 7, and 9 were confirmed electrophoretically (Figure 10B) and reflect the proteins involved in the formation of CuONPs-Muc. These peaks were associated with proteins with MW of about 30 and 26 kDa on lane 2 at 12% SDS-PAGE, which were not observed on lane 3, suggesting the involvement of proteins as a reducing agent in CuONPs (Figure 10). Also, new peaks 1 and 2 appeared, which were probably related to the formation of CuONPs-Muc.

Additional information is provided by ImageQuant^{TM} TL analysis of the proteins with MW 30.691 kDa and 26.549 kDa, which were incorporated into CuONPs-Muc and eluted as peaks 7 and 9. Based on published data on the proteins contained in *C. aspersum* mucus, we can assume that the proteins with MW 26 549 kDa and 30 591 kDa most likely exhibit antioxidant activity and belong to the families of superoxide dismutase, glutathione peroxidase (GPx), and glutathione transferases (GST). Brieva et al. found that the mucus of *Cryptomphalus aspersa* (also known as *H. aspersa*) contains antioxidant superoxide dismutase and glutathione transferase activity [56].

Several studies in recent years have also confirmed the antioxidant properties of a *C. aspersum* mucus fraction with MW 10–30 kDa and a fraction with MW > 20 kDa [57–59].

Proteins from mollusks and gastropods are represented in the UniProt Knowledgebase (UniProtKB) as superoxide dismutase subunit [Cu-Zn]-like protein from *Pomacea canaliculata* with MW 15.6 kDa (XP_025098994) and 15.774 kDa from *Lymnaea stagnalis* (Q7YXL9) and glutathione S-transferases with MW 27.978 kDa (from *P. canaliculata*, A0A2T7PWN7); with MW 27.774 kDa (from *Haliotis rubra*, XM_025244070.1). The results obtained for a protein with an MW of 26.549 kDa agreed with an enzyme of the glutathione S-transferases family with an MW between 20.5 and 27.5 kDa.

The protein with MW 30.59 kDa is in good agreement with lectins and is also related to the antimicrobial properties of mucus [60]. The protein detected at 37.52 kDa is a major protein identified in *C. aspersum* mucus, with MW 37.4 kDa (QEG59312.1) [60]. Based on the presented results, we hypothesize that mucus proteins with MW 26.55 kDa and 30.59 kDa play a major role in forming CuONPs-Muc.

Many reports of antibacterial, antifungal, larvicidal, anti-termite [61–65], and anticancer [66,67] activities of CuONPs derived from plants, animals, and microorganisms. Furthermore, several studies conducted in recent years have demonstrated the advantages of employing snail mucus to synthesize Ag and AuNPs, which have been shown to have broad-spectrum antibacterial activity, anticancer potential, and anti-inflammatory properties [21,68]. Based on the presented investigations, it can be assumed that CuONPs obtained from the *C. aspersum* mucus with MW > 20 kDa also show in vitro antibacterial activity. However, higher susceptibility was detected for Gram-positive species, which data corresponds to other authors' results [69,70]. This is probably due to the specificity in cell wall chemical composition and structure of Gram-positive bacterial cells—stronger negative surface charge, allowing positive NPs to attract, and the presence of teichoic acid, whose phosphate chains interact with NPs, thus preventing their aggregation. The presence of a high number of pores on Gram+ cell walls additionally allows penetration of NPs into the cells and causes cell death [60,71].

The detected efficacy of the synthesized CuONPs-Muc is lower against Gram-negative bacterial strains. However, these findings are similar to the studies of other authors who revealed that the minimum inhibitory concentrations (MICs) of ultra-small CuO nanoparticles against *E. coli* is 40% higher than those for *S. aureus* [72]. It is assumed that the key factors responsible for bacterial growth inhibition by CuONPs are the electrostatic interactions and cell wall adherence. As a consequence, ROS are produced due to the Cu^{2+} dissociation. These ions have the ability to infiltrate cells and break their membranes, which can lead to bacterial cell leakage and a disturbance in the interior content of the cell [73]. Considering that the envelope of Gram-negative bacterial cells consists of both inner and outer membrane, it could be assumed that higher concentrations of CuONPs are required to hinder cellular growth.

Surprisingly, the efficacy of the tested CuONPs-Muc was similar to or even higher than that of commonly used antimicrobial drugs. Antibiotics such as vancomycin and erythromycin are often used to treat infections brought on by Gram-positive bacteria that are resistant to many medications, particularly methicillin-resistant *Staphylococcus aureus* [74,75]. Vancomycin has been utilized extensively in hospital settings during the past few decades as a result of the constantly increasing number of multidrug-resistant strains each year [72]. To address these serious problems in the postantibiotic era, new strategies for the treatment of microbial infections are urgently needed. One such new type of antibacterial agent is nanomaterials.

The antibacterial effect of metal nanoparticles is based on a set of their specific properties, such as the ability to destroy or prevent the formation of microbial biofilms, generation of reactive oxygen species (ROS) that can damage DNA, RNA, and proteins, inhibition of replication by binding to DNA, etc., thus killing microorganisms in ways against which the emergence of resistance is limited [69,73]. Therefore, the extensive inhibitory antibacterial potential exhibited by biosynthesized CuONPs using snail mucus makes them promising preparations for future medical applications, such as alternatives to antiseptics and antibiotics.

Our research will continue the topic of obtaining biogenic metal nanoparticles (from copper, zinc, and silver ions) in the mucus matrix of the garden snail *C. aspersum* under different conditions, including in the presence of ascorbic acid, as well as their characterization and biomedical applications. In order to clarify the mechanism of antimicrobial activity of the new biogenic nanoparticles obtained, the cellular redox status, the level of transcription, and the levels of apoptotic cell markers in the tested pathogenic microorganisms will be monitored before and after inoculation with novel NPs-Muc. Combining proteomics, mass spectrometry, and bioinformatics, the changes in the proteome maps of bacterial and fungal cells before and after treatment with the most active NPs-Muc will be tracked and the key proteins responsible for the relevant biological activities will be identified. In this way, our studies will lead to obtaining important new information that will serve to develop a new generation of antimicrobial therapeutics as an alternative to conventional antibiotic treatment and drug multiresistance.

4. Materials and Methods

4.1. Sample Preparation

The crude snail mucus was obtained from certified environmentally clean snail farms on the territory of Bulgaria using a patented method of extraction and filtration of mucus from the garden snails *C. aspersa*, so the snails survived without disturbing their biological functions [33]. After several steps of filtration, the snail mucus solution was freed of chloride ions and additives after dialysis for 24 h. The crude mucus extract was separated into two fractions after ultrafiltration using a membrane system (ÄKTA pure™) with a column at 20 kDa. A peptide fraction with MW above 20 kDa was used for the synthesis of CuO-NPs-Muc from snail mucus. The protein concentration was determined using the Bradford method [76].

4.2. 1D Polyacrylamide Gel Electrophoresis (1D-PAGE)

Protein fractions were analyzed by 12% and 10% sodium dodecyl sulfate-polyacrylamide gel electrophoresis (SDS-PAGE), according to the Laemmli method with modifications [77] and visualized by staining with Coomassie Brilliant Blue G-250. Molecular markers of standard proteins with a molecular weight from 6.5 kDa to 200 kDa (SERVA Prestained SDS PAGE Protein Marker 6.5–200 kDa, liquid mix, SERVA Electrophoresis GmbH, Heidelberg, Germany, and SigmaMarker™, Sigma-Aldrich, Saint Louis, MO, USA) were used. Because the commercial Laemmli buffer (Leammli Sample Buffer (2×) for SDS PAGE, SERVA Electrophoresis GmbH, Heidelberg, Germany) does not contain any reduction reagent, 10 mM DTT was added as a reducing sample buffer (concentrations refer to 1× sample buffer). Each sample is mixed with an equal volume of the sample buffer thus prepared, after which the samples are denatured by boiling at 100 °C for 5 min. Equal volumes containing 20 μg/lane of the samples dissolved in Laemmli sample buffer and protein standard mixture were run at a voltage of 145 V using the Bio-Rad feeder (Bio-Rad, Hercules, CA, USA).

4.3. MALDI-MS Analysis of Proteins

The molecular masses of proteins in the mucus fraction with MW > 20 kDa from snail *C. aspersum* were measured using an Autoflex™III, High-Performance MALDI-TOF & TOF/TOF System (Bruker Daltonics, Billerica, MA, USA), which uses a 200 Hz frequency-tripled Nd–YAG laser operating at a wavelength of 355 nm. The analysis was carried out using α-cyano-4-hydroxycinnamic acid as a matrix. In total, 2.0 μL of the sample was mixed with 2.0 μL of matrix solution (7 mg/mL of α-cyano-4-hydroxycinnamic acid (CHCA) in 50% CN containing 0.1% TFA), and only 1.0 μL of the mixture was spotted on a 192-well stainless steel target plate. They were dried at room temperature and subjected to mass analysis. A total of 3500 shots were acquired in the MS mode, and collision energy of 4200 was applied. The mass spectrometer was externally calibrated with a mixture of angiotensin I, Glu-fibrinopeptide B, ACTH (1–17), and ACTH.

4.4. The Synthesis of CuONPs-Muc

One gram of lyophilized mucus fraction with MW > 20 kDa was dissolved in 40 mL of distilled water (a concentration of 25 mg/mL). A 0.1 M solution of $CuSO_4$ was prepared. All solutions were moderately stirred with a magnetic stirrer until complete dissolution (about 30 min). Then, 100 mL of 0,1 M $CuSO_4$ solution was added dropwise to 100 mL of snail mucus fraction with MW > 20 kDa under constant stirring at room temperature. The whole mixture, with an initial pH of 5.3, was left in a dark environment with moderate agitation (250 rpm) for 5 days at room temperature (around 20 °C). After the 5-day incubation, samples were centrifuged at 4000 rpm for 10 min. Afterwards, supernatants were separated and centrifuged at 12 000 rpm for 10 min to remove the excess products formed.

4.5. Characterization of CuONPs-Muc Using Ultraviolet–Visible Spectroscopy Analysis

UV–vis absorption measurements were carried out using a UV–vis spectrometer (Shimadzu™ UVmini-1240, Shimadzu Corporation, Kyoto, Japan), using symmetric quartz cuvettes (Quartz Glass High Performance 200 nm–2500 nm, Hellma® absorption cuvettes) with 10 mm optical length at room temperature (25 °C) in the range of 250–500 nm. All absorption spectra were corrected by subtracting the absorption spectrum of the buffer solution in the same wavelength range. During the 5-day incubation, the UV–vis spectrum of the samples was measured at 24 h, 48 h, and 72 h to follow the dynamics of nanoparticle formation. Additionally, UV–vis spectra were also measured after the centrifugations at 4000 rpm and 12 000 rpm (VWR® Micro Star 30R Microcentrifuge, VWR International, Tokyo, Japan).

4.6. Characterization of CuONPs-Muc Using Fluorescence Spectroscopy

Fluorescence studies of samples containing a mucus fraction with MW > 20 kDa before and after the CuONPs-Muc were obtained by green synthesis were performed using a Shimadzu RF-6000 spectrofluorometer (Shimadzu Corporation, Kyoto, Japan) using a xenon lamp as the excitation source and a fluorescent cuvette with a path length of 10 × 10 mm (Hellma 101-QS, Floressan Quartz Küvet), at band ex. = 3 nm and band em. = 20 nm. Fluorescence emission spectra were measured after excitation at λ_{ex} 280 nm and at λ_{ex} 295 nm. For each excitation wavelength, the emission was recorded between 220 and 700 nm, with a step increment of 0.5 nm. The changes observed in the emission spectra of a sample containing a mucus fraction with MW > 20 kDa and a sample containing the CuNPs obtained by green synthesis were compared.

4.7. Characterization of CuONPs-Muc Using Raman Spectroscopy and Imaging

Raman spectra were recorded using Raman Microscope Senterra II (Bruker). The intensity of Raman bands is determined by the change in polarizability during normal vibration. Using this technique, it is possible to determine crystal forms, chemical compositions, intermolecular interactions, the degree of ordering, or the spatial distribution of stresses in the tested materials [78]. The samples were placed onto glass (approximately 10 mg) and analyzed using the vertical 20× objective in an 180° backscattering arrangement. The Raman spectrometer parameters used to analyze the CuONPs-Muc include a 532 nm laser wavelength and an exposure time of 100 s; the resolution was 4 cm^{-1} for all samples, and laser power was 6.5 mW.

4.8. Characterization of CuONPs-Muc Using FTIR and TG-DSC

Pallets centrifuged multiple times were subjected to FT-IR (Midac MC 2000 Spectrometer), where the FT-IR spectrum was examined in a range of 6000–300 cm^{-1} wavelength interval for the CuONPs-Muc and mucus extract [79].

Thermal analysis (TG) and differential scanning calorimetry (DSC) were performed with a STA 449 Jupiter F3 Netzsch apparatus (Selb, Germany). Two samples were investigated using TG-DSC analysis: a pure mucus sample and composite CuONPs-Muc. The analyses were performed under the following conditions: heating rate 10°K/min in an argon atmosphere with a flow rate of 30 mL/min, temperature range from 25 to 600 °C, and sample weight ~50 mg. A comparison was made between the results obtained from the study of the thermal behavior of the pure mucus sample and CuONPs-Muc in a programmed mode of operation. The ongoing phase transitions in both samples as a result of heating to 600 °C were also compared. Additionally, from the DSC analysis of the mucus and the CuONPs-Muc, the amount of heat ΔH released (absorbed) during a given process was calculated for each of the transitions: Tonset, the lowest temperature at which the reaction starts; Tpeak, the maximum transition temperature; Tend, the final temperature of the process. Using TG analysis, the maximum mass loss (Mtotal) was determined for both samples as a result of heating.

4.9. Image Analysis of 10% SDS-PAGE by ImageQuant™ TL v8.2.0 Software

After capturing the SDS-PAGE obtained on Image Scanner III (GE Healthcare, Chicago, IL, USA), the image was analyzed using the '1D gel analysis' utility of the Image Quant TL v8.2 software (GE Healthcare Bio-Sciences AB, Uppsala Sweden), which is highly automated software for image analysis. All bands were identified manually using a pen tool, including those in the standard protein marker. To compensate for the intensity of the image background, the background was modified with the "image rectangle" setting. Analysis of the molecular weight of each band was performed using the data for the protein marker in the range 6.5–200 kDa (SigmaMarker™, Sigma-Aldrich, Saint Louis, MO, USA). Automatically, horizontal bands were drawn to the individual bands of the MW marker and calculated using the cubic curve spline. Based on the precalculated number of bands in the marker, the number of bands tested was determined [80].

4.10. RP-HPLC Analysis of CuONPs-Muc

Reverse-phase high-performance liquid chromatography (RP-HPLC) on ZURA®; HPLC system (KNAUER, Berlin, Germany) was used to detect and identify which proteins of the mucus was capping/reducing the CuO ions. The chromatographic separation was achieved on a Eurospher II 100 C18 (250 × 8 mm) column (KNAUER, Berlin, Germany) by gradient of water–acetonitrile (with 0.1% TFA) for 55 min, which was adjusted as follows: 0–10 min, 100% buffer A (MiliQ-water with 0.1% TFA); 11–15 min, 85% buffer A; 16–40 min, 40% buffer A; 41–45 min, 0% buffer A; 46–47 min, 0% buffer A; 48–49 min, 100% buffer A; 50–55 min, 100% buffer A. The injection volume was 1 mL sample, and the flow rate was 1.5 mL/min at 25 °C. The detection wavelength was 280 nm.

The samples were centrifuged briefly and filtered through 0.2 µm syringe filters (Minisart NML, hydrophilic, cellulose acetate) before injection. The concentration of the used mucus fraction with MW > 20 kDa was 0.02 g/mL lyophilized sample in Milli-Q water.

4.11. Antibacterial Activity of CuONPs

Gram-positive (*B. subtilis* NBIMCC2353, *B. spizizenii* ATCC 6633, *S. aureus* ATCC 6538, *L. innocua* NBIMCC8755) and Gram-negative strains (*E. coli ATCC8739*, *S. enteritidis* NBIMCC8691, *S. typhimurium* ATCC 14028, *S. maltophilia* ATCC 17666) were selected as representative bacterial models to test the antibacterial properties of the samples. The antibacterial activity of mucus and NP samples was studied by the agar well diffusion method, and after 24 h of cultivation, the zones of cell growth inhibition were measured (d, mm). The results were compared to vancomycin antibacterial activity determined using the agar disk diffusion method (30 µg/disk).

The microbial growth inhibitory effect of CuONPs-Muc, obtained from mucus fraction with MW > 20 kDa, was studied using the agar well diffusion method [81]. As a negative control, a fraction of *C. aspersum* mucus was used. For positive control, the antibacterial activity of different commercially available antibiotics was studied: vancomycin (VA), amoxiclav (AMC), erythromycin (E), and cephalexin (CN)—all with concentrations of 30 µg/disk.

Nutrient agar plates were inoculated with 0.1 mL cell suspension (0.5 McFarland) by spreading it over the entire agar surface. Holes (diameter, 8 mm) were punched in agar with a cork borer, and the wells formed were filled with 0.1 mL of each mucus fraction or CuONPs-Muc sample. Then, the agar plates were stored for 4 h at fridge temperature (for antimicrobial agents' diffusion) and, after that, were transferred at 37 °C for 24 h. The antimicrobial activities of NPs were evaluated by measuring the zone of growth inhibition (diameter, mm), and the presented values were averaged from three different experiments. Microbial susceptibility to different antibiotics was tested according to the CLSI M39 guidelines [82].

5. Conclusions

The present research presents for the first time an environmentally friendly method without toxic waste for synthesizing biogenic CuONPs-Muc from a mucus fraction with MW > 20 kDa of the garden snail *C. aspersum*. The biosynthesized CuONPs-Muc were proven and characterized by applying various modern methods (UV–vis and fluorescence spectroscopy, RP-HPLC analysis, electrophoretic analysis, mass spectrometry, SEM and EDS, FTIR and TG-DSC analyses). The CuONPs-Muc showed higher inhibition of the bacterial growth of various Gram-positive (*B. subtilis* NBIMCC2353, *B. spizizenii* ATCC 6633, *S. aureus* ATCC 6538, *L. innocua* NBIMCC8755) and Gram-negative (*E coli* ATCC8739, *S. enteritidis* NBIMCC8691, *S. typhimurium* ATCC 14028, *S. maltophilia* ATCC 17666) bacteria compared to the initial mucus fraction with MW > 20 kDa.

The results of the in vitro studies show that a combination of new biomolecules with antimicrobial properties from the mucus of the snail *C. aspersum* and an ecological approach—biological synthesis of metal nanoparticles—can lead to new effective antimicrobial therapeutics against high-risk pathogens.

Research in this area will continue to obtain and characterize biogenic metal nanoparticles synthesized from the mucus of the snail *C. aspersum* under different conditions, including in the presence of ascorbic acid, and their biomedical applications. To elucidate the mechanism of the antimicrobial activity of the new biogenic nanoparticles obtained, the cellular redox status, the level of transcription, and the levels of apoptotic cell markers in the tested pathogenic microorganisms will be followed before and after inoculation with new NPs-Muc.

Author Contributions: Conceptualization, P.D. and A.D.; investigations K.M., P.P., V.P., S.A.-V., D.K., L.V., I.S., E.P., A.T. and A.K.; methodology, P.D., A.D., V.P. and B.R.; writing—original draft preparation, A.D., K.M., P.P., V.P., S.A.-V., D.K., I.S., E.P., A.T., A.K., L.V. and P.D.; writing—review and editing, P.D., A.D., B.R., V.P. and L.V. All authors have read and agreed to the published version of the manuscript.

Funding: This research was carried out with the support of the project KP-06 PN61-8/2022 funded by the Bulgarian National Science Fund.

Institutional Review Board Statement: Not applicable.

Informed Consent Statement: Not applicable.

Data Availability Statement: Data is contained within the article.

Acknowledgments: Project BG05M2OP001-1.002-0019: 'Clean Technologies for Sustainable Environment—Waters, Waste, Energy for a Circular Economy', financed by the Operational programme "Science and Education for Smart Growth" 2014–2020, co-financed by the European union through the European structural and investment funds.

Conflicts of Interest: The authors declare no conflicts of interest. The funders had no role in the design of the study; in the collection, analysis, or interpretation of data; in the writing of the manuscript; or in the decision to publish the results.

References

1. Kumar, B.; Jalodia, K.; Kumar, P.; Gautam, H. Recent advances in nanoparticle-mediated drug delivery. *J. Drug Deliv. Sci. Technol.* **2017**, *41*, 260–268. [CrossRef]
2. Britto-Hurtado, R.; Cortez-Valadez, M. Chapter 4—Green synthesis approaches for metallic and carbon nanostructures. In *Green Functionalized Nanomaterials for Environmental Applications*; A volume in Micro and Nano Technologies; Shanker, U., Hussain, C.M., Rani, M., Eds.; Elsevier: Amsterdam, The Netherlands, 2022; pp. 83–127.
3. Hoag, G.E.; Collins, J.B.; Holcomb, J.L.; Hoag, J.R.; Nadagouda, M.N.; Varma, R.S. Degradation of bromothymol blue by 'greener' nano-scale zero-valent iron synthesized using tea polyphenols. *J. Mater. Chem.* **2009**, *19*, 8671–8677. [CrossRef]
4. Horwat, D.; Zakharov, D.; Endrino, J.; Soldera, F.; Anders, A.; Migot, S.; Karoum, R.; Vernoux, P.; Pierson, J. Chemistry, phase formation, and catalytic activity of thin palladium-containing oxide films synthesized by plasma-assisted physical vapor deposition. *Surf. Coat. Technol.* **2011**, *205*, S171–S177. [CrossRef]

5. Alsammarraie, K.; Wang, W.; Zhou, P.; Mustapha, A.; Lin, M. Green synthesis of silver nanoparticles using turmeric extracts and investigation of their antibacterial activities. *Colloids Surf B Biointerfaces* **2018**, *171*, 398–405. [CrossRef]
6. Ying, S.; Guan, Z.; Ofoegbu, P.C.; Clubb, P.; Rico, C.; He, F.; Hong, J. Green Synthesis of Nanoparticles: Current Developments and Limitations. *Environ. Technol. Innov.* **2022**, *26*, 102336. [CrossRef]
7. Bedlovičová, Z. Chapter 21—Green synthesis of silver nanoparticles using actinomycetes. In *Green Synthesis of Silver Nanomaterials;* A volume in Nanobiotechnology for Plant Protection; Abd-Elsalam, K.A., Ed.; Elsevier: Amsterdam, The Netherlands, 2022; pp. 547–569.
8. Castillo-Henríquez, L.; Alfaro-Aguilar, K.; Ugalde-Álvarez, J.; Vega-Fernández, L.; Montes de Oca-Vásquez, G.; Vega-Baudrit, J.R. Green Synthesis of Gold and Silver Nanoparticles from Plant Extracts and Their Possible Applications as Antimicrobial Agents in the Agricultural Area. *Nanomaterials* **2020**, *10*, 1763. [CrossRef] [PubMed]
9. Rónavári, A.; Igaz, N.; Adamecz, D.I.; Szerencsés, B.; Molnar, C.; Kónya, Z.; Pfeiffer, I.; Kiricsi, M. Green Silver and Gold Nanoparticles: Biological Synthesis Approaches and Potentials for Biomedical Applications. *Molecules* **2021**, *26*, 844. [CrossRef]
10. Lubis, F.A.; Malek, N.A.; Sani, N.S.; Jemon, K. Biogenic Synthesis of Silver Nanoparticles Using Persicaria odorata Leaf Extract: Antibacterial, Cytocompatibility, and In Vitro Wound Healing Evaluation. *Particuology* **2022**, *70*, 10–19. [CrossRef]
11. Batool, Z.; Muhammad, G.; Iqbal, M.M.; Aslam, S.M.; Raza, A.M.; Sajjad, N.; Abdullah, M.; Akhtar, N.; Syed, A.; Elgorban, A.M.; et al. Hydrogel assisted synthesis of gold nanoparticles with enhanced microbicidal and in vivo wound healing potential. *Sci. Rep.* **2022**, *12*, 6575. [PubMed]
12. Chitra, S.; Mathew, N.K.; Jayalakshmi, S.; Balakumar, S.; Rajeshkumar, S.; Ramya, R. Strategies of Bioceramics, Bioactive Glasses in Endodontics: Future Perspectives of Restorative Dentistry. *Biomed. Res. Int.* **2022**, *2022*, 2530156. [CrossRef] [PubMed]
13. Zhang, H.; Ma, W.; Ma, H.; Qin, C.; Chen, J.; Wu, C. Spindle-Like Zinc Silicate Nanoparticles Accelerating Innervated and Vascularized Skin Burn Wound Healing. *Adv. Healthc. Mater.* **2022**, *11*, e2102359. [CrossRef] [PubMed]
14. Anandhi, P.; Rajeshkumar, S. Copper nanoparticles in wound healing: A review. *J. Surv. Fish. Sci.* **2023**, *10*, 49–63.
15. Alhalili, Z. Green synthesis of copper oxide nanoparticles CuO NPs from Eucalyptus Globoulus leaf extract Adsorption and design of experiments. *Arab. J. Chem.* **2022**, *15*, 103739. [CrossRef]
16. Alahdal, F.A.M.; Qashqoosh, M.T.A.; Manea, Y.K.; Mohammed, R.K.A.; Naqvi, S. Green synthesis and characterization of copper nanoparticles using *Phragmanthera austroarabica* extract and their biological/environmental applications. *Sustain. Mater. Technol.* **2023**, *35*, e00540. [CrossRef]
17. Rizzi, V.; Gubitosa, J.; Fini, P.; Nuzzo, S.; Agostiano, A.; Cosma, P. Snail Slime-Based Gold Nanoparticles: An Interesting Potential Ingredient in Cosmetics as an Antioxidant, Sunscreen, and Tyrosinase Inhibitor. *Photochem. Photobiol. B* **2021**, *224*, 112309. [CrossRef] [PubMed]
18. Gubitosa, J.; Rizzi, V.; Fini, P.; Laurenzana, A.; Fibbi, G.; Veiga-Villauriz, C.; Fanelli, F.; Fracassi, F.; Onzo, A.; Bianco, G.; et al. Biomolecules from Snail Mucus (*Helix Aspersa*) Conjugated Gold Nanoparticles, Exhibiting Potential Wound Healing and Anti-Inflammatory Activity. *Soft Matter* **2020**, *16*, 10876–10888. [CrossRef] [PubMed]
19. Noothuan, N.; Apitanyasai, K.; Panha, S.; Tassanakajon, A. Snail Mucus from the Mantle and Foot of Two Land Snails, *Lissachatina fulica* and *Hemiplecta distincta*, Exhibits Different Protein Profile and Biological Activity. *BMC Res. Notes* **2021**, *14*, 138. [CrossRef] [PubMed]
20. Ajaykumar, P.; Sabira, O.; Sebastian, M. A novel approach for the biosynthesis of silver nanoparticles using the defensive gland extracts of the beetle, *Luprops tristis* Fabricius. *Sci. Rep.* **2023**, *13*, 10186. [CrossRef] [PubMed]
21. Mane, P.C.; Sayyed, S.A.R.; Kadam, D.D.; D Shinde, M.; Fatehmulla, A.; Aldhafiri, A.M.; Alghamdi, E.A.; Amalnerkar, D.P.; Chaudhari, R.D. Terrestrial snail-mucus mediated green synthesis of silver nanoparticles and *in vitro* investigations on their antimicrobial and anticancer activities. *Sci. Rep.* **2021**, *11*, 13068–13083. [CrossRef] [PubMed]
22. Mane, C.; Kadam, D.; Khadse, N. Green Adeptness in Synthesis of Non-Toxic Copper and Cobalt Oxide Nanocomposites with Multifaceted Bioactivities. *Cancer Nano* **2023**, *14*, 79. [CrossRef]
23. Wargala, E.; Zalewska, A.; Sławska, M.; Kot, I. Snail Mucus as an Innovative Ingredient Used in the Cosmetology and Medical Industry. *Aesth. Cosmetol. Med.* **2023**, *12*, 45–49. [CrossRef]
24. Aouji, M.; Rkhaila, A.; Bouhaddioui, B.; Khalid, G.; Lrhorfi, L.A.; Bengueddour, R. Antioxidant activity, biochemical composition and physicochemical properties of *Helix aspersa* Muller snail slime. *Molecules* **2023**, *28*, 6323. [CrossRef] [PubMed]
25. Aouji, M.; Rkhaila, A.; Bouhaddioui, B.; Ziraric, M.; Harifid, H.; Taboza, Y.; Lrhorfia, A.L.; Bengueddour, R. Chemical composition, mineral profile, anti-bacterial and wound healing properties of snail slime of *Helix aspersa* Müller. *BioMedicine* **2023**, *13*, 4. [CrossRef] [PubMed]
26. Dolashki, A.; Velkova, L.; Daskalova, E.; Zheleva, N.; Topalova, Y.; Atanasov, V.; Voelter, W.; Dolashka, P. Antimicrobial Activities of Different Fractions from Mucus of the Garden Snail *Cornu aspersum*. *Biomedicines* **2020**, *8*, 315. [CrossRef] [PubMed]
27. Belouhova, M.; Daskalova, E.; Yotinov, I.; Topalova, Y.; Velkova, L.; Dolashki, A.; Dolashka, P. Microbial diversity of garden snail mucus. *Microbiologyopen* **2020**, *11*, e1263. [CrossRef] [PubMed]
28. Trapella, C.; Rizzo, R.; Gallo, S.; Alogna, A.; Bortolotti, D.; Casciano, F.; Zauli, G.; Secchiero, P.; Voltan, R. HelixComplex snail mucus exhibits pro-survival, proliferative and pro-migration effects on mammalian fibroblasts. *Sci. Rep.* **2018**, *8*, 17665. [CrossRef] [PubMed]

29. Gentili, V.; Bortolotti, D.; Benedusi, M.; Alogna, A.; Fantinati, A.; Guiotto, A.; Turrin, G.; Cervellati, C.; Trapella, C.; Rizzo, R.; et al. HelixComplex snail mucus as a potential technology against O3 induced skin damage. *PLoS ONE* **2020**, *15*, e0229613. [CrossRef] [PubMed]
30. Deng, T.; Gao, D.; Song, X.; Zhou, Z.; Zhou, L.; Tao, M.; Jiang, Z.; Yang, L.; Luo, L.; Zhou, A.; et al. A natural biological adhesive from snail mucus for wound repair. *Nat. Commun.* **2023**, *14*, 396. [CrossRef] [PubMed]
31. Di Filippo, M.F.; Di Matteo, V.; Dolci, L.S.; Albertini, B.; Ballarin, B.; Cassani, M.C.; Passerini, N.; Gentilomi, G.A.; Bonvicini, F.; Panzavolta, S. Effectiveness of Snail Slime in the Green Synthesis of Silver Nanoparticles. *Nanomaterials* **2022**, *12*, 3447. [CrossRef] [PubMed]
32. Woźniak-Budych, M.J.; Staszak, K.; Staszak, M. Copper and Copper-Based Nanoparticles in Medicine-Perspectives and Challenges. *Molecules* **2023**, *28*, 6687. [CrossRef] [PubMed]
33. Dolashka, P.; Atanasov, D. Device for Collecting Extracts from Garden Snail. BG Useful model 2097. 2015. Available online: https://portal.bpo.bg/bpo_online/-/bpo/utility-model-detail (accessed on 6 April 2024).
34. Berger, J. Infrared and Raman spectra of $CuSO_4$, $5H_2O$; $CuSO_4$, $5D_2O$; and $CuSeO_4$, $5H_2O$. *J. Raman Spectrosc.* **1976**, *5*, 103–114. [CrossRef]
35. Márquez, A.; Berger, T.; Feinle, A.; Hüsing, N.; Himly, M.; Duschl, A.; Diwald, O. Bovine Serum Albumin Adsorption on TiO_2 Colloids: The Effect of Particle Agglomeration and Surface Composition. *Langmuir* **2017**, *33*, 2251–2258. [CrossRef] [PubMed]
36. Umar, A.; Alshahrani, A.; Algarni, H.; Kumar, R. CuO Nanosheets as Potential Scaffolds for Gas Sensing Applications. *Sens. Actuators B Chem.* **2017**, *250*, 24–31. [CrossRef]
37. Saied, E.; Hussein, A.; Al-Askar, A.; Elhussieny, N.; Hashem, A. Therapeutic Effect of Biosynthesized Silver Nanoparticles on Hypothyroidism Induced in Albino Rats. Electron. *Biotechnol* **2023**, *65*, 14–23.
38. Tahir, B.; Sohaib, M.; Sagir, M.; Rafique, M. Role of Nanotechnology in Photocatalysis. *Encycl. Smart Mater.* **2022**, *2*, 578–589.
39. Ellijimi, C.; Hammouda, B.; Othman, H. *Helix aspersa maxima* mucus exhibits antimelanogenic and antitumoral effects against melanoma cells. *Biomed. Pharmacother.* **2018**, *101*, 871–880. [CrossRef] [PubMed]
40. Ekin, İ.; Şeşen, R. Molluscs: Their Usage as Nutrition, Medicine, Aphrodisiac, Cosmetic, Jewelry, Cowry, Pearl, Accessory and So on from the History to Today. *Middle East J. Sci.* **2018**, *4*, 45–51. [CrossRef]
41. Giovanni, C.; Filippo, F. Antimicrobial properties of terrestrial snail and slug mucus. *J. Complement. Integr. Med.* **2018**, *15*, 20170168.
42. Xie, J.; Lee, J.Y.; Wang, D.I.C.; Ting, Y.P. Silver Nanoplates: From Biological to Biomimetic Synthesis. *ACS Nano* **2007**, *1*, 429–439. [CrossRef]
43. Uthayakumar, V.; Ramasubramanian, V.; Senthilkumar, D.; Priyadarisini, V.B.; Harikrishanan, R. Biochemical characterization, antimicrobial and hemolytic studies on skin mucus of freshwater spiny eel *Mastacembelus armatus*. *Asian Pac. J. Trop. Biomed.* **2012**, *2*, s863–s869. [CrossRef]
44. Gabriel, U.I.; Mirela, S.; Ionel, J. Quantification of mucoproteins (glycoproteins) from snails mucus, *Helix aspersa* and *Helix pomatia*. *J. Agroaliment. Process. Technol.* **2011**, *17*, 410–413.
45. Drozd, M.; Duszczyk, A.; Ivanova, P.; Pietrzak, M. Interactions of proteins with metal-based nanoparticles from a point of view of analytical chemistry—Challenges and opportunities. *Adv. Colloid Interface Sci.* **2022**, *304*, 102656. [CrossRef] [PubMed]
46. Wang, S.; Zheng, J.; Ma, L.; Petersen, R.B.; Xu, L.; Huang, K. Inhibiting protein aggregation with nanomaterials: The underlying mechanisms and impact factors. *Biochim. Biophys. Acta-Gen. Subj.* **2022**, *1866*, 130061. [CrossRef] [PubMed]
47. John, R.; Mathew, J.; Mathew, A.; Aravindakumar, C.T.; Aravind, U.K. Probing the Role of Cu(II) Ions on Protein Aggregation Using Two Model Proteins. *ACS Omega* **2021**, *6*, 35559–35571. [CrossRef]
48. Royer, C.A. Probing protein folding and conformational transitions with fluorescence. *Chem. Rev.* **2006**, *106*, 1769–1784. [CrossRef] [PubMed]
49. Saravanan, A.; Kumar, S.; Karishma, S.; Vo, D.N.; Jeevanantham, S.; Yaashikaa, P.R.; George, C.S. A Review on Biosynthesis of Metal Nanoparticles and Its Environmental Applications. *Chemosphere* **2021**, *264 Pt 2*, 128580. [CrossRef]
50. Rashad, M.; Sampò, S.; Cataldi, A.; Zara, S. Biological activities of gastropods secretions: Snail and slug slime. *Nat. Prod. Bioprospect.* **2023**, *13*, 42. [CrossRef] [PubMed]
51. Ortelli, S.; Costa, A.L.; Blosi, M.; Brunelli, A.; Badetti, E.; Bonetto, A.; Hristozov, D.; Marcomini, A. Colloidal characterization of CuO nanoparticles in biological and environmental media. *Environ. Sci. Nano* **2017**, *4*, 1264–1272. [CrossRef]
52. Baldisserri, C.; Costa, A.L. Electrochemical detection of copper ions leached from CuO nanoparticles in saline buffers and biological media using a gold wire working electrode. *Nanopart. Res.* **2016**, *18*, 96. [CrossRef]
53. Chrzanowska, A.; Derylo-Marczewska, A. Mesoporous silica/protein biocomposites: Surface, topography, thermal properties. *Int. J. Biol. Macromol.* **2019**, *139*, 531–542. [CrossRef] [PubMed]
54. Andrade, J.D.; Hlady, V.; Wei, A.P. Adsorption of complex proteins at interfaces. *Pure Appl. Chem.* **1992**, *64*, 1777–1781. [CrossRef]
55. Dee, K.C.; Puleo, D.A.; Bizios, R. Chapter 3—Protein-surface interactions. In *An Introduction to Tissue-Biomaterial Interactions*; John Wiley & Sons, Inc.: Hoboken, NJ, USA, 2002; pp. 37–52.
56. Brieva, A.; Philips, N.; Tejedor, R.; Guerrero, A.; Pivel, J.P.; Alonso-Lebrero, J.L.; Gonzalez, S. Molecular basis for the regenerative properties of a secretion of the mollusk *Cryptomphalus aspersa*. *Ski. Pharmacol. Physiol.* **2008**, *21*, 15–22. [CrossRef] [PubMed]

57. Kostadinova, N.; Voynikov, Y.; Dolashki, A.; Krumova, E.; Abrashev, R.; Kowalewski, D.; Stevanovic, S.; Velkova, L.; Velikova, R.; Dolashka, P. Antioxidative screening of fractions from the mucus of garden snail *Cornu aspersum*. *Bul. Chem. Com.* **2018**, *50*, 176–183.
58. Tancheva, L.; Lazarova, M.; Velkova, L.; Dolashki, A.; Uzunova, D.; Minchev, B.; Petkova-Kirova, P.; Hassanova, Y.; Gavrilova, P.; Tasheva, K.; et al. Beneficial effects of snail *Helix aspersa* extract in an experimental model of Alzheimer's type dementia. *J. Alzheimers Dis.* **2022**, *88*, 155–175. [PubMed]
59. Petrov, L.; Kachaunov, M.; Alexandrova, A.; Tsvetanova, E.; Georgieva, A.; Dolashki, A.; Velkova, L.; Dolashka, P. Snail Mucus Protective Effect on Ethanol-Induced Gastric Ulcers in Mice. *Life* **2022**, *12*, 1106. [CrossRef] [PubMed]
60. Pitt, S.; Hawthorne, J.A.; Garcia-Maya, M.; Alexandrovich, A.; Symonds, R.C.; Gunn, A. Identification and characterisation of anti—*Pseudomonas aeruginosa* proteins in mucus of the brown garden snail, *Cornu aspersum*. *Br. J. Biomed. Sci.* **2019**, *76*, 129–136. [CrossRef] [PubMed]
61. Tahir, A.; Quispe, C.; Herrera-Bravo, J.; Iqbal, H.; ul Haq, Z.; Anum, F.; Javed, Z.; Sehar, A. Green Synthesis, Characterization and Antibacterial, Antifungal, Larvicidal and Anti-Termite Activities of Copper Nanoparticles Derived from *Grewia asiatica* L. *Bull. Natl. Res. Cent.* **2022**, *46*, 188. [CrossRef]
62. Das, K.R.; Pachapur, V.L.; Lonappan, L.; Naghdi, M.; Pulicharla, R.; Maiti, S.; Cledon, M.; Dalila, L.M.A.; Srama, S.J.; Brar, S.K. Biological synthesis of metallic nanoparticles: Plants, animals and microbial aspects. *Nanotechnol. Environ. Eng.* **2017**, *2*, 18. [CrossRef]
63. Antonio-Pérez, A.; Durán-Armenta, L.F.; Pérez-Loredo, M.G.; Torres-Huerta, A.L. Biosynthesis of Copper Nanoparticles with Medicinal Plants Extracts: From Extraction Methods to Applications. *Micromachines* **2023**, *14*, 1882. [CrossRef] [PubMed]
64. Akintelu, S.A.; Folorunso, A.S.; Folorunso, F.A.; Oyebamiji, A.K. Green synthesis of copper oxide nanoparticles for biomedical application and environmental remediation. *Heliyon* **2020**, *6*, e04508. [CrossRef] [PubMed]
65. Luque-Jacobo, C.M.; Cespedes-Loayza, A.L.; Echegaray-Ugarte, T.S.; Cruz-Loayza, J.L.; Cruz, I.; de Carvalho, J.C.; Goyzueta-Mamani, L.D. Biogenic Synthesis of Copper Nanoparticles: A Systematic Review of Their Features and Main Applications. *Molecules* **2023**, *28*, 4838. [CrossRef] [PubMed]
66. Gu, J.; Aidy, A.; Goorani, S. Anti-Human Lung Adenocarcinoma, Cytotoxicity, and Antioxidant Potentials of Copper Nanoparticles Green-Synthesized by *Calendula Officinalis*. *J. Exp. Nanosci.* **2022**, *17*, 285–296. [CrossRef]
67. Siddiqui, M.A.; Alhadlaq, H.A.; Ahmad, J.; Al-Khedhairy, A.A.; Musarrat, J.; Ahamed, M. Copper Oxide Nanoparticles Induced Mitochondria Mediated Apoptosis in Human Hepatocarcinoma Cells. *PLoS ONE* **2013**, *8*, e69534. [CrossRef]
68. Valeria, C.; Salvatore, P.; Luca, V.; Maria, G.; Ludovica, M.; Cristina, S.; Lucia, M.; Angela, C.; Valeria, S. Innovative snail-mucus-extract (SME)-coated nanoparticles exhibit anti-inflammatory and anti-proliferative effects for potential skin cancer prevention and treatment. *RSC Adv.* **2024**, *14*, 7655–7663. [CrossRef] [PubMed]
69. Rosli, N.A.; Teow, Y.H.; Mahmoudi, E. Current approaches for the exploration of antimicrobial activities of nanoparticles. *Sci. Technol. Adv. Mater.* **2021**, *22*, 885–907. [CrossRef] [PubMed]
70. Wang, L.; Hu, C.; Shao, L. The antimicrobial activity of nanoparticles: Present situation and prospects for the future. *Int. J. Nanomed.* **2017**, *12*, 1227–1249. [CrossRef] [PubMed]
71. Sarwar, A.; Katas, H.; Samsudin, S.N.; Zin, N.M. Regioselective sequential modification of chitosan via azide-alkyne click reaction: Synthesis, characterization, and antimicrobial activity of chitosan derivatives and nanoparticles. *PLoS ONE* **2015**, *10*, e0123084. [CrossRef] [PubMed]
72. Mancuso, G.; Midiri, A.; Gerace, E.; Biondo, C. Bacterial Antibiotic Resistance: The Most Critical Pathogens. *Pathogens* **2021**, *10*, 1310–1323. [CrossRef]
73. Singh, P.; Garg, A.; Pandit, S.; Mokkapati, V.R.S.S.; Mijakovic, I. Antimicrobial effects of biogenic nanoparticles. *Nanomaterials* **2018**, *8*, 1009. [CrossRef] [PubMed]
74. Selim, S. Mechanisms of gram-positive vancomycin resistance (Review). *Biomed. Rep.* **2022**, *16*, 7–12. [CrossRef] [PubMed]
75. Trifi, A.; Tlili, B.; Kallel Sellami, M.; Feki, M.; Mehdi, A.; Seghir, E.; Messaoud, L.; Abdellatif, S.; Ben Lakhal, S. Immunologic effect and clinical impact of erythromycin in septic patients: A randomized clinical trial. *J. Crit. Care* **2024**, *81*, 154533. [CrossRef] [PubMed]
76. Bradford, M.M. A rapid and sensitive method for the quantitation of microgram quantities of protein utilizing the principle of protein-dye binding. *Anal. Biochem.* **1976**, *72*, 248–254. [CrossRef] [PubMed]
77. Laemmli, U.K. Cleavage of structural proteins during the assembly of the head of bacteriophage T4. *Nature* **1970**, *227*, 680–685. [CrossRef] [PubMed]
78. Xu, J.F.; Ji, W.; Shen, Z.X.; Li, W.S.; Tang, S.H.; Ye, X.R.; Xin, X.Q. Raman spectra of CuO nanocrystals. *J. Raman Spectrosc.* **1999**, *30*, 413–415. [CrossRef]
79. Umer, A.; Naveed, S.; Rafique, M.S.; Imran, M. A green method for the synthesis of Copper Nanoparticles using L-ascorbic acid. *Matéria (Rio J.)* **2014**, *19*, 197–203. [CrossRef]
80. Lincz, L.F.; Scorgie, F.E.; Garg, M.B.; Gilbert, J.; Sakoff, J.A. A simplified method to calculate telomere length from Southern blot images of terminal restriction fragment lengths. *Biotechniques* **2020**, *68*, 28–34. [CrossRef] [PubMed]

81. Balouiri, M.; Sadiki, M.; Ibnsouda, S. Methods for in vitro evaluating antimicrobial activity: A review. *J. Pharm. Anal.* **2016**, *6*, 71–79. [CrossRef] [PubMed]
82. Clinical and Laboratory Standards Institute. *Analysis and Presentation of Cumulative Antimicrobial Susceptibility Test Data*, 5th ed.; Approved guideline; Clinical and Laboratory Standards Institute: Wayne, PA, USA, 2022.

pharmaceuticals

MDPI

Article

Synthesis, Characterization, and Toxicity Assessment of Zinc Oxide-Doped Manganese Oxide Nanoparticles in a Macrophage Model

Nasser B. Alsaleh [1,*,†], Anas M. Aljarbou [1,†], Mohamed E. Assal [2], Mohammed A. Assiri [1], Mohammed M. Almutairi [1], Homood M. As Sobeai [1], Ali A. Alshamrani [1], Sultan Almudimeegh [1], Mohammad R. Hatshan [2] and Syed F. Adil [2,*]

1 Department of Pharmacology and Toxicology, College of Pharmacy, King Saud University, P.O. Box 2457, Riyadh 11451, Saudi Arabia; 442106265@student.ksu.edu.sa (A.M.A.); moassiri@ksu.edu.sa (M.A.A.); malmotyre@ksu.edu.sa (M.M.A.); hassobeai@ksu.edu.sa (H.M.A.S.); aaalshamrani@ksu.edu.sa (A.A.A.); salmudimeegh@ksu.edu.sa (S.A.)

2 Department of Chemistry, College of Science, King Saud University, P.O. Box 2455, Riyadh 11451, Saudi Arabia; masl@ksu.edu.sa (M.E.A.); mhatshan@ksu.edu.sa (M.R.H.)

* Correspondence: nbalsaleh@ksu.edu.sa (N.B.A.); sfadil@ksu.edu.sa (S.F.A.)

† These authors contributed equally to this work.

Abstract: The doping of engineered nanomaterials (ENMs) is a key tool for manipulating the properties of ENMs (e.g., electromagnetic, optical, etc.) for different therapeutic applications. However, adverse health outcomes and the cellular biointeraction of doped ENMs, compared to undoped counterparts, are not fully understood. Previously, we have shown that doping manganese oxide nanoparticles with ZnO (ZnO-MnO_2 NPs) improved their catalytic properties. In this study, we assessed the toxicity of ZnO-MnO_2 NPs in Raw 264.7 cells. NPs were prepared via an eco-friendly, co-precipitation method and characterized by several techniques, including transmission and scanning electron microscopy, X-ray diffraction, and Fourier transform infrared. The physicochemical properties of ZnO-MnO_2 NPs, including size, morphology, and crystalline structure, were almost identical to MnO_2 NPs. However, ZnO-MnO_2 NPs showed slightly larger particle aggregates and negative charge in cell culture media. Exposure to ZnO-MnO_2 NPs resulted in lower toxicity based on the cell viability and functional assay (phagocytosis) data. Exposure to both NPs resulted in the activation of the cell inflammatory response and the generation of reactive oxygen species (ROS). Despite this, exposure to ZnO-MnO_2 NPs was associated with a lower toxicity profile, and it resulted in a higher ROS burst and the activation of the cell antioxidant system, hence indicating that MnO_2 NP-induced toxicity is potentially mediated via other ROS-independent pathways. Furthermore, the cellular internalization of ZnO-MnO_2 NPs was lower compared to MnO_2 NPs, and this could explain the lower extent of toxicity of ZnO-MnO_2 NPs and suggests Zn-driven ROS generation. Together, the findings of this report suggest that ZnO (1%) doping impacts cellular biointeraction and the consequent toxicological outcomes of MnO_2 NPs in Raw 264.7 cells.

Keywords: nanomaterials; doping; toxicity; immune cell

1. Introduction

Engineered nanomaterials (ENMs), precisely manmade materials within the nanoscale (1–100 nm), are being developed for various applications across different industrial sectors [1]. Given the vast potential for the tunability of inorganic ENM physicochemical properties (e.g., optical, electromagnetic, etc.), there has been an increase in the development of inorganic ENMs over the past years, including biomedical applications, such as nanomedicine and tissue engineering [2,3]. As a result, human exposure to ENMs, including intentional exposure such as in the case of nanomedicine, will be unavoidable.

However, to date, there remains uncertainty regarding the safety profile and potential toxicological manifestations associated with exposure to ENMs [4,5].

The area of material science with a focus on synthesizing nanocomposites of different inorganic materials or metal-organic frameworks has gained attention over the past few years [6,7]. The doping of inorganic ENMs, which is the intentional introduction of a trace of inorganic material into the crystal lattice of inorganic ENMs, is a key tool for manipulating ENM physicochemical properties (e.g., electromagnetic, photocatalytic, optical, etc.) towards desired applications [8]. Indeed, previous efforts have demonstrated that the doping of inorganic ENMs may increase cytotoxicity against cancer cells [9], improve antimicrobial properties [10], enhance magnetic properties [11], and increase the capacity for drug loading and delivery [12]. Despite doping's potential in improving the biomedical applications of ENMs, there remains uncertainty and inconsistency in the literature with regard to the impact of doping on ENM toxicological outcomes. Currently, there is inconsistency with regard to the toxicological outcomes of doped materials as doping has been shown to influence ENM toxicity both positively and negatively [13,14]. In addition, the therapeutic applications of ENM doping are often pursued without enough assessment of the potential toxicity [14]. Therefore, there is a need for more research to improve our understanding of the nature of cellular biointeractions and the toxicity of doped ENMs in comparison with their undoped counterparts [15].

Manganese is an essential micronutrient for basic cell biochemical reactions involved in growth, differentiation, and homeostasis [16]. In the past few years, manganese oxide nanoparticles (MnO_x NPs) have attracted attention for applications across multiple disciplines including catalysis, solar cells, and water treatment, as well as in biomedical applications including bioimaging and biosensors, drug delivery, and cancer theranostics [17–20]. Due to the high tunability of their physicochemical properties (e.g., size, shape, surface functionalization, etc.), low cost and high abundancy, the sustainability of synthesis via green methods, biocompatibility, and low toxicity profile, MnO_x NPs represent an excellent material for a wide range of biomedical applications [17]. It is also recognized that a key attractive property of using MnO_x NPs in cancer therapy stems from acting as a transition metal and hence participating in Fenton-like reactions to generate free radicals [21–23]. Previous reports demonstrated in vivo toxicity following exposure to MnO_x NPs with increased toxicity outcomes in comparison to bulk materials [24–26]. An important toxicological aspect to consider is the biodegradability of ENMs [27]. Indeed, a major drawback of numerous inorganic ENMs with key biomedical applications (e.g., gold, silver, silica, etc.) is their biopersistence [28]. More studies are now focused on synthesizing biodegradable materials, as well as understanding the underlying mechanisms of biodegradability, including Mn-based ENMs [28–31].

Currently, there is a growing interest in exploiting and fine-tuning the redox properties of MnO_x NPs via doping with other metals [32–34]. Nevertheless, the safety profile and toxicological outcomes of such platforms are yet to be studied. We have previously shown that doping MnO_x NPs with ZnO improved their catalytic properties, where doping with ZnO (1%) resulted in the highest catalysis [35]. Here, we investigated whether doping with ZnO (1%) could affect the toxicity outcomes of MnO_2 NPs in Raw 264.7 cells, an established macrophage-like model. To our knowledge, this is the first report assessing the toxicity of ZnO-doped MnO_2 NPs and their toxicological consequence.

2. Results and Discussion

2.1. Nanoparticle Characterization

The synthesis of MnO_2 NPs and ZnO-MnO_2 NPs was carried out through an eco-friendly, facile and straightforward co-precipitation procedure, with calcination at 400 °C in a muffle furnace to obtain the desired NPs. The crystallinity, composition, and morphology of fabricated NPs were assessed using XRD, TGA, FT-IR, EDX, SEM, and HR-TEM. The surface morphology of the fabricated NPs was studied using TEM and SEM. The representative images showed that both NPs have similar particle shapes (Figure 1). The HR-TEM images of ZnO-MnO_2 NPs displayed well-defined nano-sized polycrystalline particles with evident interplanar lattice fringes of 0.26 nm that are related to the plane (2 0 1) of MnO_2 NPs, whereas the morphology of the MnO_2 NPs was not well defined according to the TEM images (Figure 1).

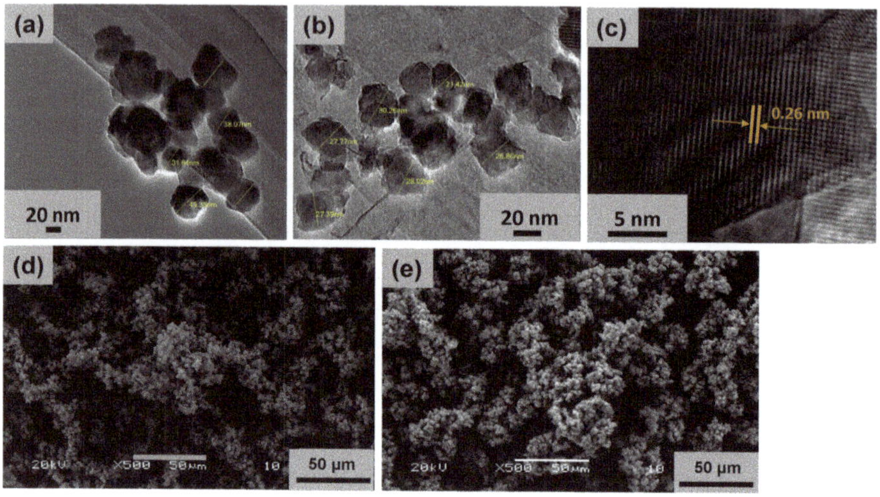

Figure 1. Representative transmission and scanning electron microscopy (SEM and TEM) images of the nanoparticles. Representative TEM images of (**a**) MnO_2 NPs and (**b**) ZnO-MnO_2 NPs were obtained after dispersion of the NPs via ultrasonication before microscopy analysis. (**c**) TEM image of ZnO-MnO_2 NPs at higher magnification showing interplanar lattice fringes of 0.26 nm (arrows). Representative SEM images of (**d**) MnO_2 NPs and (**e**) ZnO-MnO_2 NPs.

The crystal structure of the NPs was assessed using XRD analysis. The XRD patterns for the synthesized MnO_2 NPs and ZnO-MnO_2 NPs are presented in Figure 2A. The obtained diffraction peaks distinctly show that the fabricated materials are crystalline in nature. The MnO_2 NPs annealed at 400 °C exhibited a pattern that is highly in accordance with the standard XRD results of pyrolusite manganese dioxide (JCPDS-No. 24-0735). As anticipated, ZnO-MnO_2 NPs showed a similar XRD pattern to that of MnO_2 NPs, in addition to the diffraction signal labelled with star (*), which is possibly ascribed to the existence of ZnO NPs. The thermal properties of the synthesized NPs were examined using thermal gravimetric analysis (TGA) (Figure 2B). The TGA graph of MnO_2 NPs demonstrated thermal stability with overall weight loss around 18% and up to 800 °C, while for the ZnO-MnO_2 NPs, it was observed that total weight loss was only about 8% at 800 °C, indicating that ZnO-MnO_2 NPs is more thermally stable than pure MnO_2 NPs. The FT-IR analysis was employed to identify the surface functionalities of the as-obtained materials, and the results of the synthesized MnO_2 NPs and ZnO-MnO_2 NPs were demonstrated in Figure 2C. The FT-IR peak located at nearly 3425 cm^{-1} and 1640 cm^{-1} is associated with the characteristic vibrations of hydroxyl groups [36]. In addition, the intense peaks at 602 cm^{-1} are associated with the characteristic of Mn–O vibrations [37,38]. The EDX plot of the synthesized ZnO-MnO_2 NPs and MnO_2 NPs is shown in Figure 2D. The left-side EDX figure demonstrates the mass percentage composition of Mn and O elements in MnO_2 NPs. The right side illustrates the EDX plot of the synthesized ZnO-MnO_2 NPs, which indicated that the values obtained for the Mn and O composition elements are similar to the values obtained for MnO_2 NPs, along with the appearance of a small peak representing the amount of Zn, hence reflecting the successful inclusion of ZnO NPs in the system [35]. The values in tabular format are shown in Table S1 in the Supplementary Materials.

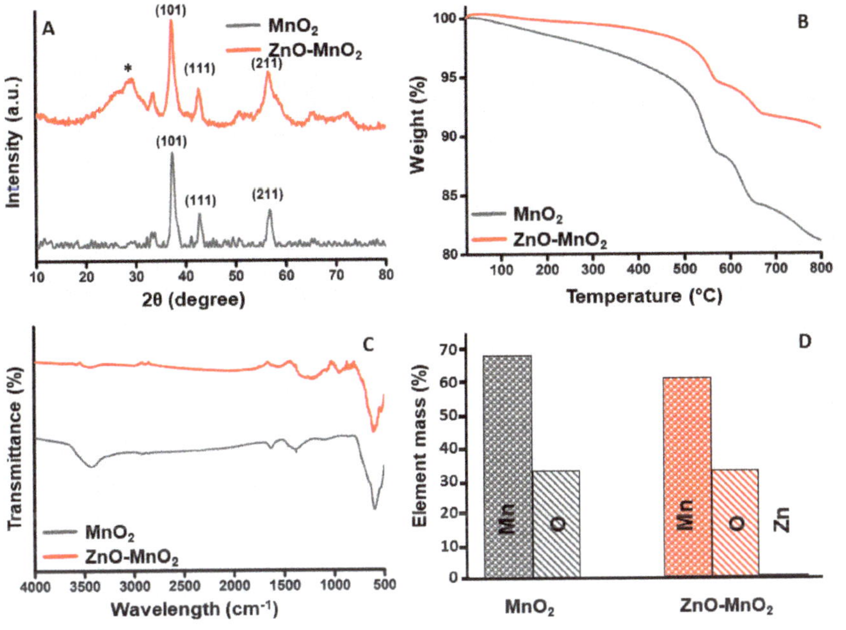

Figure 2. Characterization of nanoparticles. (**A**) Comparative X-ray diffractogram for the MnO_2 NPs (grey line) or ZnO-MnO_2 NPs (red line) scanned between 2θ 10 and 80 (* indicates diffraction signal); (**B**) results from the thermogravimetric analysis of MnO_2 NPs (grey line) or ZnO-MnO_2 NPs (red line) in the range from 25 °C to 800 °C; (**C**) Fourier transform infrared spectrum of MnO_2 NPs (grey line) or ZnO-MnO_2 NPs (red line) in the range from 400 cm^{-1} to 4000 cm^{-1}; (**D**) graphical illustration of elemental analysis of the MnO_2 NPs (grey) or ZnO-MnO_2 NPs (red).

2.2. Nanoparticle Preparation for In Vitro Studies and Characterization in Vehicle and Cell Culture Media (DMEM)

For in vitro experiments, NP stocks were prepared in ultrapure water (UPW; vehicle) and then diluted in serum-free cell culture media. Both NPs demonstrated larger hydrodynamic size (d_h), an indication of particle aggregation and/or agglomeration (Table 1). This is typically observed in nanotoxicological studies, particularly for uncoated inorganic NPs. Nevertheless, the polydispersity indices (PDIs) of both NP types in both UPW and DMEM were equal to or less than 0.3 (Table 1). The suspended MnO_2 NPs had a hydrodynamic size of 197 ± 35 ($d_{h\text{-}UPW}$) and 483 ± 15 in cell culture media ($d_{h\text{-}DMEM}$) while ZnO-MnO_2 NPs had a hydrodynamic size of 245 ± 7 in UPW ($d_{h\text{-}UPW}$) and 546 ± 30 in cell culture media ($d_{h\text{-}DMEM}$) (Table 1). The zeta potential (ζ) of the MnO_2 NPs was −16.2 ± 3.7 in UPW (ζ_{UPW}) and −4.8 ± 1 in cell culture media (ζ_{DMEM}), while for the ZnO-MnO_2 NPs, it was −20 ± 5.2 in UPW (ζ_{UPW}) and −5.5 ± 1.8 in cell culture media (ζ_{DMEM}) (Table 1).

Table 1. DLS measurements.

Nanoparticle—Media	Hydrodynamic Size (d_h) (nm)	Zeta Potential (ζ) (mV)	PDI
MnO_2 NPs—UPW	197 ± 35	−16.2 ± 3.7	0.20
MnO_2 NPs—DMEM	483 ± 15	−4.8 ± 1	0.25
ZnO-MnO_2 NPs—UPW	245 ± 7	−20 ± 5.2	0.25
ZnO-MnO_2 NPs—DMEM	546 ± 30	−5.5 ± 1.8	0.29

UPW: ultrapure water. PDI: polydispersity index.

2.3. Cell Viability following Nanomaterial Treatment

The toxicity of MnO_x NPs has been previously reported by several studies [24–26]. For instance, an early study showed that 28-day oral repeated exposure to MnO_2 NPs in Wistar rats was associated with larger tissue distribution and increased toxicity (e.g., DNA damage, biochemical and histopathological alterations, etc.) in comparison with bulk MnO_2 [24]. Another study has shown that exposure of Wistar rats to MnO_2 NPs via intratracheal instillation for 6 weeks resulted in size-dependent toxicity, with the smaller NPs resulting in more pronounced toxicity [26]. Together, these reports suggest that exposure to MnO_x NPs may induce toxicity, particularly following chronic exposure [24–26]. However, the toxicity and biointeraction of MnO_x NPs at cellular and molecular levels remain largely unknown. It is worth mentioning that previous works have exploited the property of MnO_x NPs to induce Fenton-like reactions and generate ROS (e.g., platforms used in photodynamic therapy) [21–23]. Such studies suggest that the toxicity of MnO_x NPs is largely driven via ROS generation, leading to oxidative stress and cellular damage. However, such a hypothesis is yet to be validated, and the safety profile for such platforms on normal cells remains to be studied.

Here, the data indicated concentration-dependent toxicity after treatment with both MnO_2 NPs and ZnO-MnO_2 NPs (Figure 3). However, the reduction in cell viability was significantly lower following exposure to ZnO-MnO_2 NPs in comparison with undoped MnO_2 NPs at several concentration points (12.5, 25, and 50 μg/mL). It is worth noting that ZnO-MnO_2 NPs appeared to have a tendency to form larger agglomerates and/or aggregates and have larger negative charge upon suspension in cell culture media (DMEM). Such formation of NP agglomerates and/or aggregates typically occurs with metal and metal oxide ENMs, particularly those with no or minimal surface coating [39–41]. Despite the fact that changes in hydrodynamic size and surface charge after doping were within a relatively close range, the impact on toxicity cannot be ruled out. However, other factors (e.g., Mn ion release) might have been influenced by doping and contributed to the observed difference in toxicity. Indeed, a previous report has demonstrated that ZnO-doped TiO_2 NPs (1, 5, and 10%) did not influence the hydrodynamic size or surface charge, yet they had a significant impact on toxicity. However, the toxicity was potentially mediated via the release of dopant ions [9]. Based on viability data, a dose of 12.5 μg/mL was used throughout subsequent experiments.

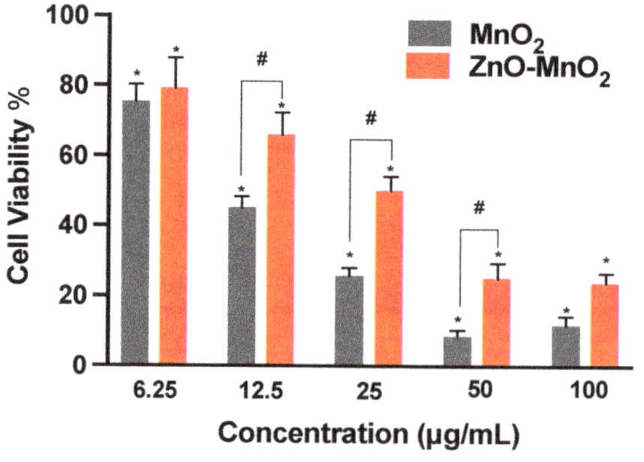

Figure 3. Viability of Raw 264.7 cells following exposure to nanoparticles. Cellular viability was estimated based on the formation of formazan salt after 24 h exposure to MnO_2 NPs (grey bars) or ZnO-MnO_2 NPs (red bars) within a range of 6.25–100 μg/mL in serum-free media. Independent experiments were carried out at least 3 times ($n \geq 3$). A p value of less than 0.05 ($p < 0.05$) indicates statistical significance between control and treatment groups (*) or between treatment groups (#).

2.4. Cell Functional Competence following Nanomaterial Treatment

Measuring cell functional competence following exposure to ENMs is an important parameter in the safety assessment of ENMs [42,43]. Macrophage is a key cell type for the elimination of invading pathogens and foreign bodies, including ENMs [44]. They are also a key component in engaging adaptive immunity [44]. Ample evidence has demonstrated the detrimental consequences of ENMs on macrophage health and function [45,46]. Earlier studies have demonstrated that treatment with even subtoxic levels of ENMs could result in adverse consequences with respect to macrophage function [43,46–48]. As part of assessing NP safety in Raw 264.7 cells, we measured cell phagocytic capacities based on the uptake of fluorescently labeled beads following exposure to the NPs for 2 h, followed by the washing of NPs and incubation with fluorescently labeled beads for 4 h. It is worth noting that the exposure time was shortened to prevent overwhelming the cell's functional capacity while the dose was kept at the same concentration points as in the viability experiments. The data demonstrated that exposure to both types of NPs was associated with a dose-dependent trend of toxicity relative to the cell capacity for phagocytosis (Figure 4). However, exposure to MnO_2 NPs resulted in a larger suppression of the cell's phagocytic capacity compared to ZnO-MnO_2 NPs. These results are in line with the viability data indicating that ZnO doping ameliorated the toxicity of MnO_2 NPs in Raw 264.7 cells.

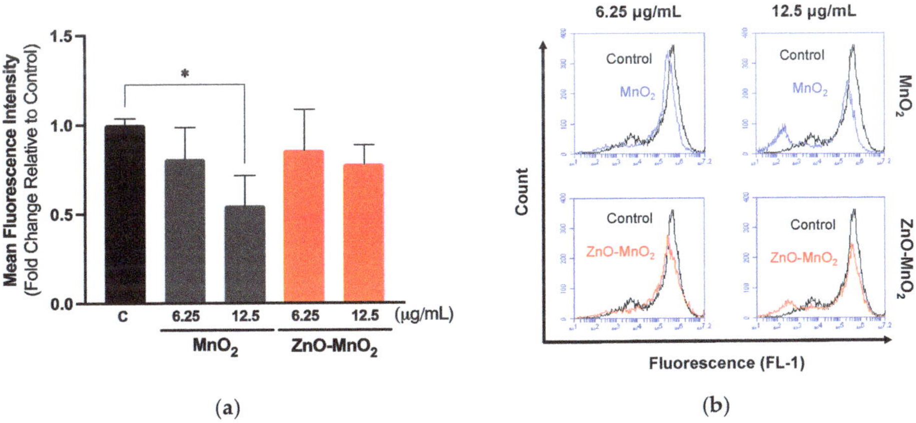

Figure 4. Functional capacity of Raw 264.7 cells following exposure to nanoparticles. (**a**) Cell phagocytosis capacity was estimated based on the internalization of fluorescently labeled latex beads (4 h) following exposure to 6.25 or 12.5 μg/mL of MnO_2 NPs (grey bars) or ZnO-MnO_2 NPs (red bars) for 2 h. (**b**) Representative histograms of the data from the flow cytometer. Independent experiments were carried out at least 3 times ($n \geq 3$). A p value of less than 0.05 ($p < 0.05$) indicates statistical significance between control and treatment groups (*).

2.5. Inflammatory Response and Reactive Oxygen Species Generation following Nanomaterial Treatment

The activation of cell inflammatory pathways is a critical cellular response to toxic insults [49]. It is important for overcoming stress, and it is part of the repair mechanism following exposure to environmental insults, including ENMs [49]. It also represents *Tier-2* of the previously proposed "hierarchical oxidative stress" model following exposure to ENMs [50]. The data of this study suggest that exposure to both NPs activated the inflammatory response, including the expression of *TNFα* and *IL-1β* genes (Figure 5a). Exposure to ZnO-MnO_2 NPs had a lower activation of *TNFα* compared to MnO_2 NPs. However, ZnO-MnO_2 NPs appeared to have a stronger activation of *IL-1β*. These results may suggest that doping with ZnO might have altered MnO_2 NP-mediated cellular interaction and biological outcomes. However, more work is needed to confirm such a hypothesis. For instance, the screening of a panel of inflammatory markers (e.g., pro- vs. anti-inflammatory) based on cytokine release will help delineate the impact of NPs on inflammatory pathways (i.e., doped vs. undoped).

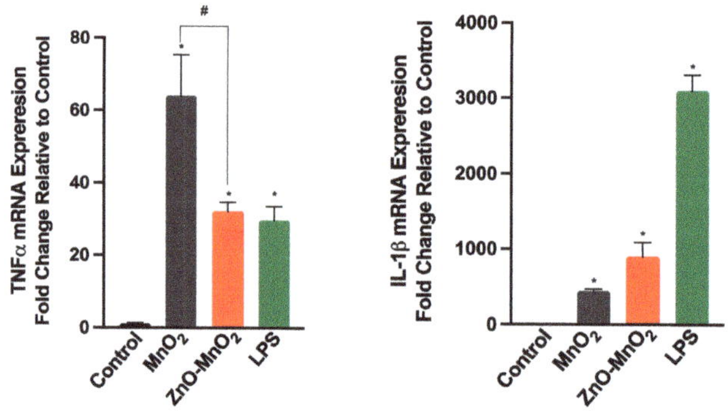

(a)

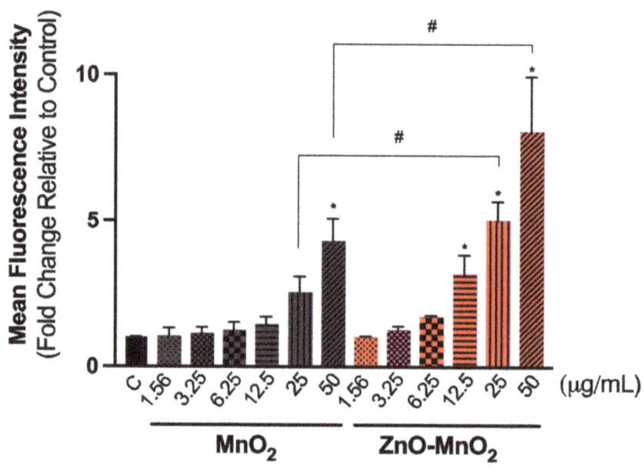

(b)

Figure 5. Inflammatory response and reactive oxygen generation following exposure to the nanoparticles. (**a**) *TNF-α* (left panel) and *IL-1β* (right panel) gene expression following exposure to NPs for 6 h. (**b**) The generation of intercellular ROS was assessed based on the fluorescence signal of dichlorofluorescin diacetate (H2DCFDA) following exposure to a range of concentrations (1.56–50 μg/mL) of MnO_2 NPs (grey bars) or ZnO-MnO_2 NPs (red bars) for 60 min. Treatment with lipopolysaccharide (LPS) (1 μg/mL) was included for comparison purposes. Independent experiments were carried out at least 3 times ($n \geq 3$). A p value of less than 0.05 ($p < 0.05$) indicates statistical significance between control and treatment groups (*) or between treatment groups (#).

A major paradigm of ENM toxicity is driven via the ENM-induced generation of reactive oxygen species (ROS), eventually leading to oxidative damage [50]. MnO_x NPs have been shown to generate ROS by acting as a transition metal, resulting in Fenton-like reactions [21–23]. The doping and intentional introduction of chemical defects into metal oxide ENMs have been successfully utilized to manipulate their properties towards a wide range of biomedical applications [8,15]. The unintentional formation of chemical defects could also result during the large-scale production of ENMs [13]. From a toxicological perspective, earlier research has established that the doping of metal oxide ENMs can lead to changes in (i.e., narrowing or expansion) their band gap and associated electron transfer to cellular redox species (i.e., the overlap between intracellular redox potential and ENM energy bands) and, hence, the generation of ROS and subsequent oxidative stress-mediated toxicity [51,52]. Indeed, a previous report showed that chemical defects present in ZnO NPs contributed to their cellular toxicity via NP–protein charge interaction or NP-induced ROS generation [53]. Nevertheless, doping could also ameliorate the toxicity of NPs via other mechanisms, such as increasing the material stability and reducing dissolution and the release of ions, which have been shown to contribute largely to the toxicity of some NPs, such as ZnO and CuO [54,55]. Our results indicated that exposure to NPs for 60 min led to ROS burst (Figure 5b). However, ROS levels were significantly higher in cells treated with ZnO-MnO_2 NPs. This was to our surprise since MnO_2 NPs resulted in increased toxicity; hence, we anticipated that it would induce higher levels of ROS compared to doped NPs. Although we did not measure the NP band gap in this study, the possibility of a change in the band gap upon doping cannot be ruled out. However, even in the case of a change in the NP band gap and associated ROS generation, the toxicity of undoped MnO_2 NPs was significantly higher than ZnO-doped MnO_2 NPs. This suggests that exposure to undoped MnO_2 NPs induced toxicity through other ROS-independent pathways. Indeed, the previous literature has shown that ENMs could induce toxicity relative to mammalian and eukaryotic cells via mechanisms that are independent of ROS generation. These mechanisms include the ROS-independent induction of cell cycle arrest, apoptosis, mitochondrial damage (respiratory chain complex IV), autophagy, and inflammatory cytokines [56–59]. Studies have also demonstrated a role for ENM concentration in mediating ROS-dependent versus -independent toxicity [58,59].

2.6. Antioxidant Response following Nanomaterial Treatment

We measured the protein expression of nuclear factor-erythroid 2-related factor 2 (Nrf-2), a central regulator of the cell antioxidant system, following exposure to NPs to assess the activation of the cell antioxidant system. Consistent with ROS data, exposure to ZnO-MnO_2 NPs appeared to induce the higher expression of Nrf-2 proteins compared to MnO_2 NPs (Figure 6). Together, the data indicate that doping with ZnO led to the generation of more ROS and the consequent activation of the cell antioxidant system, as evidenced by the higher expression of Nrf-2.

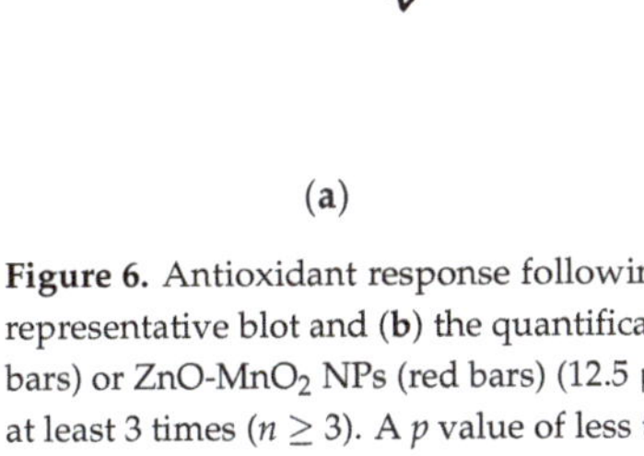

(a)

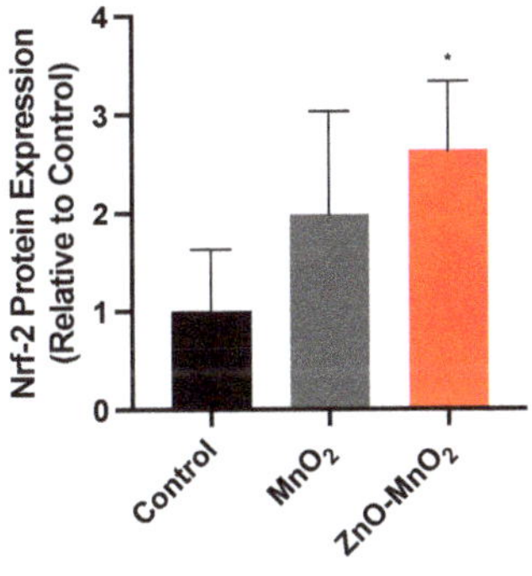

(b)

Figure 6. Antioxidant response following exposure to nanoparticles. (**a**) Nrf-2 protein expression representative blot and (**b**) the quantification of Western blots after treatment with MnO_2 NPs (grey bars) or ZnO-MnO_2 NPs (red bars) (12.5 μg/mL) for 24 h. Independent experiments were carried out at least 3 times ($n \geq 3$). A p value of less than 0.05 ($p < 0.05$) indicates statistical significance between control and treatment groups (*).

2.7. Cellular Internalization following Nanomaterial Treatment

The cellular internalization of ENMs is one of the key parameters in the assessment of ENM safety. We measured NP internalization via ICP-MS, a gold standard technique for measuring metal content. The data indicated that treatment with ZnO-MnO_2 NPs resulted in a significant reduction in NP internalization compared to MnO_2 NPs based on the Mn content (i.e., a difference of approximately 50% ppb between the doped and undoped NPs at 12.5 µg/mL) (Figure 7). Importantly, our data also showed increased cellular Zn content following exposure to the doped NPs compared to their undoped control (Figure S1). This could suggest the release of Zn ions that had subsequently generated ROS and influenced the cell molecular response (e.g., ROS generation and inflammatory pathways) as previously reported [9]. However, unlike the reported data where Zn doping increased toxicity, our findings showed less toxicity following exposure to ZnO-doped MnO_2 NPs. A possible explanation for the observed data is that doping in our case might have influenced the cellular internalization of NPs, interaction with and damage of intracellular organelles, and/or release of Mn ions. The cellular internalization of ENMs is a key parameter for understanding their biointeraction. ENM toxicity is often driven via cellular internalization, regardless of whether it is a result of active or passive cellular uptake (e.g., cell membrane damage) [60–62]. Subsequently, toxicity could then arise from lysosomal-mediated ion release, interactions with cellular organelles and the generation of ROS, and consequent macromolecular damage and disruption of organelle function [60,61]. Currently, it is hard to explain such differences in cellular internalization since the hydrodynamic size and surface charge of both NPs were within a close range; however, the impact of the relatively larger hydrodynamic size and smaller negative charge of the doped NPs on their cellular internalization cannot be ruled out. Further studies, such as studying multiple doping contents (e.g., 1% vs. 5% vs. 7%, etc.) and their impact on hydrodynamic size and surface charge, are needed to understand the impact of doping on NP physicochemical properties, cellular internalization, and toxicity.

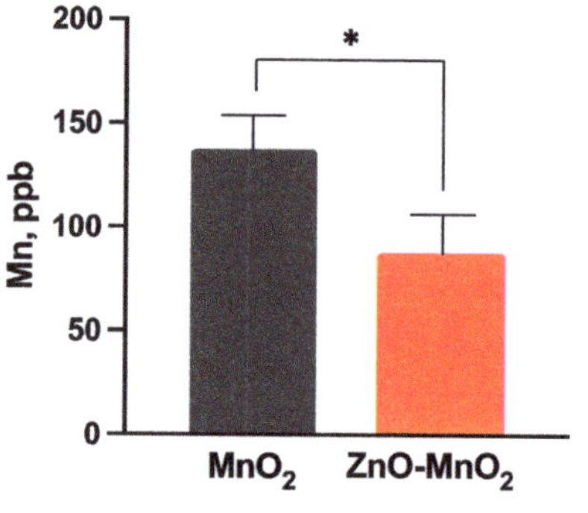

Figure 7. Nanoparticle internalization following exposure to nanoparticles. The cell internalization of the nanoparticle was measured based on the Mn content following treatment with MnO_2 NPs (grey bars) or ZnO-MnO_2 NPs (red bars) (12.5 µg/mL) for 24 h. Independent experiments were carried out at least 3 times ($n \geq 3$). A p value of less than 0.05 ($p < 0.05$) indicates statistical significance between control and treatment groups (*).

3. Materials and Methods

3.1. Nanoparticle Synthesis and Characterization

ZnO-MnO_2 NPs were synthesized using an eco-friendly and facile co-precipitation procedure. Stoichiometric amounts of manganese(II)nitrate-tetrahydrate ($Mn(NO_3)_2{\cdot}4H_2O$) and zinc nitrate hexahydrate ($Zn(NO_3)_2{\cdot}6H_2O$) were dissolved in distilled water. The mixture of solutions of $Mn(NO_3)_2{\cdot}4H_2O$ and $Zn(NO_3)_2{\cdot}6H_2O$ (100 mL) was stirred and transferred to a three-neck round-bottomed flask. The resulting solution was heated to 90 °C and vigorously stirred using a magnetic bar, and a solution of $NaHCO_3$ with a concentration of 0.50 M was added dropwise until the pH value of the solution reached around 9. The solution was continuously stirred at the same temperature of 90 °C for 3 h, and stirring occurred overnight at room temperature. The resultant solution was filtered using centrifugation, and the material obtained was dried at a temperature of 65 °C overnight. The obtained powder was annealed at 400 °C. The resulting powder was then annealed at 400 °C in a muffle furnace. Pure MnO_2 NPs were synthesized in the same procedure, except that we only used manganese(II)nitrate-tetrahydrate ($Mn(NO_3)_2 \cdot 4H_2O$). The fabricated NPs were analyzed using various characterization techniques, and all instrument details can be found in our previously reported publication [63].

3.2. Nanoparticle Preparation for Biological Studies

For biological experiments, the NP stock was prepared by suspending NPs in ultrapure water (UPW; vehicle) at 1 mg/mL. The suspended NPs were briefly sonicated using a probe sonicator (7 cycles of 10 sec on/10 sec off). DLS analysis is shown in Table 1.

3.3. Cell Culture and Viability Studies

The Raw 264.7 cell line (TIB-71) was purchased from the American Type Culture Collection (ATCC, Manassas, VA, USA). Cells were cultivated at 37 °C and 5% CO_2 using standard aseptic techniques. Cells were grown in Dulbecco's Modified Eagle's Medium (DMEM) and supplemented with 10% fetal bovine serum (FBS), 100 U penicillin/mL, and 100 μg streptomycin/mL (Gibco, Waltham, MA, USA).

The viability of cells was estimated via the MTT assay. Cells were grown into 96-well cell culture plates and allowed to develop to 70% confluency. Cells were treated with NPs at a range of concentrations (6.25, 12.5, 25, 50, or 100 μg/mL) for 24 h. The MTT reagent was added to each well, and the plate was shaken for 5 min before incubating at 37 °C for 45 min until color formation. After discarding the supernatants from each well, isopropyl alcohol was added to dissolve the formazan crystals, and the plate was shaken for 10 min. Absorbance in each well was then measured in a Synergy HT system at 570 nm (BioTek, Winooski, VT, USA).

3.4. Cell Phagocytosis

Cell phagocytic activity was determined as previously prescribed, based on measuring the engulfed latex beads with a fluorophore-labeled rabbit IgG, as directed by the manufacturer (Cayman, Ann Arbor, MI, USA) [43,64]. In summary, cells were plated in a 24-well plate and cultivated until 80% confluency; the media were then discarded, and the cells were exposed to 6.25 and 12.5 μg/mL of NPs for 2 h. Following this, the cells were washed and treated with new latex bead-containing culture media that delivered particles at a 200:1 cell ratio. The cells were incubated at 37 °C for 4 h before being gently rinsed with a cell-based assay buffer, collected via centrifugation, and resuspended in a cell-based assay buffer. Cells were gated to remove debris or dead cells, and total fluorescence was recorded at 10,000 events per sample using the BD Accuri™ C6 flow cytometer.

3.5. Gene Expression

Real-time PCR was used to quantify gene expression via the amplification of mRNA transcripts 7500 Fast RT-PCR (Applied Biosystems, Foster City, CA, USA). The cells were grown into 12-well culture plates to 70% confluency before being exposed to the NPs for 6 h at 12.5 μg/mL. Total RNA was extracted using TRIzol™ LS Reagent (ThermoFisher Scientific, Pleasanton, CA, USA) based on the manufacturer's protocol. Jenway Genova Nano (Cole-Parmer, Vernon Hills, IL, USA) was used to measure RNA concentration and quality. Using RT Master Mix HY-K0510 (MedChemExpress, South Brunswick Township, NJ, USA) and a Veriti thermocycler (Applied Biosystems, Woburn, MA, USA), mRNA was reverse transcribed into cDNA. Real-time PCR was carried out using an SYBR Green qPCR Master Mix (Low ROX) HY-K0522 (MedChemExpress, NJ, USA), reverse and forward mouse primers (Table 2) for tumor necrosis factor-alpha (*TNF-α*), Interleukin-1 beta (*IL-1β*), and *GAPDH*, which were obtained from integrated DNA technology (USA). The $\Delta\Delta C_T$ technique was used to calculate gene expression in comparison to a housekeeping gene (*GAPDH*).

Table 2. List of primers.

Gene	Primer Sequence
TNF-α	Forward 5′-ACTGAACTTCGGGGTGATTG-3′ Reverse 5′- GCTTGGTGGTTTGCTACGAC-3′
IL-1β	Forward 5′-CTATGGCAACTGTCCCTGAA-3′ Reverse 5′-GGCTTGGAAGCAATCCTTAATC-3′
GAPDH	Forward 5′-GGAAAGCTGTGGCGTGAT-3′ Reverse 5′-AAGGTGGAAGAATGGGAGTT-3′

3.6. ROS Generation

The quantification of reactive oxygen species (ROS) was measured via dichlorofluorescin diacetate (H2DCFDA) (ThermoFisher Scientific in Waltham, MA, USA). Briefly, Cells were grown into 96-well cell culture plates and allowed to develop to 70% confluency. Cells were treated with NPs within a range of concentrations (1.56–50 µg/mL). The cells were washed 2× with PBS, followed by the addition of H2DCFDA (5 µM, in PBS). This mixture was incubated with the cells for 60 min at a temperature of 37 °C, while ensuring protection from direct light exposure. Fluorescence was quantified via a spectrophotometer with an excitation–emission wavelength of 495–527 nm (Winooski, VT, USA).

3.7. Western Blot Analysis

Protein expression was measured by Western blotting. Cells were cultured in 6-well plates to a confluency of 70–80%. The cells were then treated with NPs (12.5 µg/mL) for 24 h. The cells were rinsed 3× with ice-cold PBS and lysed using a RIPA lysis solution (MedChemExpress, NJ, USA) supplemented with a protease inhibitor cocktail (Sigma-Aldrich, Burlington, MA, USA). The collected samples were sonicated and centrifuged at 10,000 rpm for 10 min at 4 °C. The total protein was quantified using the BCA assay (ThermoFisher Scientific, MA, USA). A concentration of 15 µg/mL of protein was loaded onto precast SDS-polyacrylamide gels after boiling with Laemmli Buffer (Bio-Rad, Hercules, CA, USA) for 5 min. The proteins were then transferred to PVDF membranes using electrophoresis. The membranes were blocked with nonfat milk for 1 h, followed by incubation with an Nrf2 primary antibody (ProteinTech's, Rosemont, IL, USA) overnight, washing, and incubation with a secondary antibody for 1 h at room temperature. Membranes were developed with Pierce Chemiluminescent Substrate. Image J software (https://imagej.net/ij/download.html, accessed on 1 January 2024) was used to quantify the intensity of bands relative to the housekeeping protein (Beta-Actin).

3.8. Nanoparticle Internalization

The cellular internalization of Mn was quantified via inductively coupled plasma mass spectrometry (ICP-MS). Briefly, cells were grown on 24-well cell culture plates and allowed to develop to 70% confluency. Cells were then treated with NPs (12.5 µg/mL) for 24 h. Later, samples were pelleted via centrifugation (5000 rpm, 5 min at 4 °C) and washed 3× with ice-cold phosphate-buffered saline (PBS, pH 7.4). The pellet was then dissolved in 10 mL 1% HNO_3, and NP internalization was measured using an Elan 9000 ICP-MS instrument (Perkin Elmer, Waltham, CT, USA). The process of instrument calibration included the use of a solution containing manganese (Mn) and zinc (Zn) at a concentration of 1 part per billion (ppb). The same standard solution was used for the purpose of optimizing nebulizer gas flow, mass calibration, and resolution. A multi-element internal solution with a concentration of 20 parts per billion (ppb) was used.

3.9. Statistical Analysis

All experiments were conducted at least in triplicate ($n \geq 3$). Data are shown as mean ± standard error of mean (SEM). A p value of less than 0.05 ($p < 0.05$) indicates statistical significance. Significant difference between two treatment groups was determined via Student's *t*-test. Significant differences between more than two treatment groups were determined via one-way analysis of variance (ANOVA) with Tukey's post hoc testing.

4. Conclusions

In conclusion, the main finding of this report indicates that doping with ZnO (1%) has the potential to ameliorate the toxicity of MnO_2 NPs in Raw 264.7 cells. Although doped NPs induced higher ROS generation and antioxidant responses, they reduced cellular toxicity compared to undoped NPs, as manifested by the reduction in cell viability and cell functional capacity. This suggests that undoped MnO_2 NPs induced toxicity via ROS-independent mechanisms (e.g., ion release, cellular internalization and direct interaction with and the disruption of organelle function, ROS-independent activation of cellular apoptosis and necrosis, etc.). Future work is needed to delineate the exact molecular mechanisms of toxicity between doped and undoped NPs.

Supplementary Materials: The following supporting information can be downloaded at: https://www.mdpi.com/article/10.3390/ph17020168/s1, Figure S1: Zn content following exposure to nanomaterials. Zn content following treatment with MnO_2 NPs (grey bars) or ZnO-MnO_2 NPs (red bars) (12.5 μg/mL) for 24 h. Independent experiments were carried out at least 3 times ($n \geq 3$). A p value of less than 0.05 ($p < 0.05$) indicates statistical significance between control and treatment groups (*). Table S1: Elemental composition of the NPs.

Author Contributions: Conceptualization, N.B.A., M.A.A., M.M.A., H.M.A.S., A.A.A. and S.A.; formal analysis, A.M.A.; funding acquisition, N.B.A.; investigation, A.M.A. and M.E.A.; methodology, N.B.A., M.A.A., M.M.A., H.M.A.S., A.A.A. and S.A.; resources, N.B.A.; supervision, N.B.A., M.R.H. and S.F.A.; validation, A.M.A.; visualization, A.M.A., M.E.A. and S.F.A.; writing—original draft, N.B.A. and S.F.A.; writing—review and editing, N.B.A. and S.F.A. All authors have read and agreed to the published version of the manuscript.

Funding: The authors extend their appreciation to the Deputyship for Research and Innovation, "Ministry of Education" in Saudi Arabia for funding this research work through project number: IFKSUDR_H244.

Institutional Review Board Statement: Not applicable.

Informed Consent Statement: Not applicable.

Data Availability Statement: Data are contained within the article.

Conflicts of Interest: The authors declare no conflicts of interest.

References

1. Vance, M.E.; Kuiken, T.; Vejerano, E.P.; McGinnis, S.P.; Hochella, M.F., Jr.; Rejeski, D.; Hull, M.S. Nanotechnology in the real world: Redeveloping the nanomaterial consumer products inventory. *Beilstein J. Nanotechnol.* **2015**, *6*, 1769–1780. [CrossRef]
2. Jamkhande, P.G.; Ghule, N.W.; Bamer, A.H.; Kalaskar, M.G. Metal nanoparticles synthesis: An overview on methods of preparation, advantages and disadvantages, and applications. *J. Drug Deliv. Sci. Technol.* **2019**, *53*, 101174. [CrossRef]
3. Singh, A.V.; Laux, P.; Luch, A.; Sudrik, C.; Wiehr, S.; Wild, A.M.; Santomauro, G.; Bill, J.; Sitti, M. Review of emerging concepts in nanotoxicology: Opportunities and challenges for safer nanomaterial design. *Toxicol. Mech. Methods* **2019**, *29*, 378–387. [CrossRef]
4. Liu, S.; Xia, T. Continued Efforts on Nanomaterial-Environmental Health and Safety Is Critical to Maintain Sustainable Growth of Nanoindustry. *Small* **2020**, *16*, e2000603. [CrossRef] [PubMed]
5. Oksel Karakus, C.; Bilgi, E.; Winkler, D.A. Biomedical nanomaterials: Applications, toxicological concerns, and regulatory needs. *Nanotoxicology* **2021**, *15*, 331–351. [CrossRef] [PubMed]
6. Sharma, G.; Kumar, A.; Sharma, S.; Naushad, M.; Dwivedi, R.P.; ALOthman, Z.A.; Mola, G.T. Novel development of nanoparticles to bimetallic nanoparticles and their composites: A review. *J. King Saud Univ.-Sci.* **2019**, *31*, 257–269. [CrossRef]
7. Yang, J.; Yang, Y.W. Metal-Organic Frameworks for Biomedical Applications. *Small* **2020**, *16*, e1906846. [CrossRef]
8. Peng, X.; Ai, F.; Yan, L.; Ha, E.; Hu, X.; He, S.; Hu, J. Synthesis strategies and biomedical applications for doped inorganic semiconductor nanocrystals. *Cell Rep. Phys. Sci.* **2021**, *2*, 100436. [CrossRef]
9. Ahamed, M.; Khan, M.A.; Akhtar, M.J.; Alhadlaq, H.A.; Alshamsan, A. Role of Zn doping in oxidative stress mediated cytotoxicity of TiO_2 nanoparticles in human breast cancer MCF-7 cells. *Sci. Rep.* **2016**, *6*, 30196. [CrossRef]
10. Naskar, A.; Lee, S.; Kim, K.S. Antibacterial potential of Ni-doped zinc oxide nanostructure: Comparatively more effective against Gram-negative bacteria including multi-drug resistant strains. *RSC Adv.* **2020**, *10*, 1232–1242. [CrossRef]

11. Saha, P.; Rakshit, R.; Mandal, K. Enhanced magnetic properties of Zn doped Fe_3O_4 nano hollow spheres for better bio-medical applications. *J. Magn. Magn. Mater.* **2019**, *475*, 130–136. [CrossRef]
12. Cui, H.; Zhu, G.; Qiu, L.; Ye, X. Facile synthesis of Mg-doped calcium silicate porous nanoparticles for targeted drug delivery and ovarian cancer treatment. *Ceram. Int.* **2021**, *47*, 24942–24948. [CrossRef]
13. Saleh, N.B.; Milliron, D.J.; Aich, N.; Katz, L.E.; Liljestrand, H.M.; Kirisits, M.J. Importance of doping, dopant distribution, and defects on electronic band structure alteration of metal oxide nanoparticles: Implications for reactive oxygen species. *Sci. Total Environ.* **2016**, *568*, 926–932. [CrossRef] [PubMed]
14. Gulati, K.; Scimeca, J.C.; Ivanovski, S.; Verron, E. Double-edged sword: Therapeutic efficacy versus toxicity evaluations of doped titanium implants. *Drug Discov. Today* **2021**, *26*, 2734–2742. [CrossRef] [PubMed]
15. Shenoy, R.U.K.; Rama, A.; Govindan, I.; Naha, A. The Purview of Doped Nanoparticles: Insights into Their Biomedical Applications. In *OpenNano*; Elsevier: Amsterdam, The Netherlands, 2022; p. 100070.
16. Avila, D.S.; Puntel, R.L.; Aschner, M. Manganese in health and disease. *Interrelat. Between Essent. Met. Ions Hum. Dis.* **2013**, *13*, 199–227. [CrossRef]
17. Hoseinpour, V.; Ghaemi, N. Green synthesis of manganese nanoparticles: Applications and future perspective—A review. *J. Photochem. Photobiol. B Biol.* **2018**, *189*, 234–243. [CrossRef] [PubMed]
18. Cai, X.; Zhu, Q.; Zeng, Y.; Zeng, Q.; Chen, X.; Zhan, Y. Manganese Oxide Nanoparticles As MRI Contrast Agents In Tumor Multimodal Imaging And Therapy. *Int. J. Nanomed.* **2019**, *14*, 8321–8344. [CrossRef] [PubMed]
19. Ding, B.; Zheng, P.; Ma, P.; Lin, J. Manganese Oxide Nanomaterials: Synthesis, Properties, and Theranostic Applications. *Adv. Mater.* **2020**, *32*, e1905823. [CrossRef]
20. Qian, X.; Han, X.; Yu, L.; Xu, T.; Chen, Y. Manganese-based functional nanoplatforms: Nanosynthetic construction, physiochemical property, and theranostic applicability. *Adv. Funct. Mater.* **2020**, *30*, 1907066. [CrossRef]
21. Lin, L.S.; Song, J.; Song, L.; Ke, K.; Liu, Y.; Zhou, Z.; Shen, Z.; Li, J.; Yang, Z.; Tang, W. Simultaneous Fenton-like ion delivery and glutathione depletion by MnO_2-based nanoagent to enhance chemodynamic therapy. *Angew. Chem.* **2018**, *130*, 4996–5000. [CrossRef]
22. Xu, J.; Han, W.; Yang, P.; Jia, T.; Dong, S.; Bi, H.; Gulzar, A.; Yang, D.; Gai, S.; He, F. Tumor microenvironment-responsive mesoporous MnO_2-coated upconversion nanoplatform for self-enhanced tumor theranostics. *Adv. Funct. Mater.* **2018**, *28*, 1803804. [CrossRef]
23. Zhang, H.; Zeng, W.; Pan, C.; Feng, L.; Ou, M.; Zeng, X.; Liang, X.; Wu, M.; Ji, X.; Mei, L. SnTe@ MnO_2-SP nanosheet–based intelligent nanoplatform for second near-infrared light–mediated cancer theranostics. *Adv. Funct. Mater.* **2019**, *29*, 1903791. [CrossRef]
24. Singh, S.P.; Kumari, M.; Kumari, S.I.; Rahman, M.F.; Mahboob, M.; Grover, P. Toxicity assessment of manganese oxide micro and nanoparticles in Wistar rats after 28 days of repeated oral exposure. *J. Appl. Toxicol.* **2013**, *33*, 1165–1179. [CrossRef]
25. Katsnelson, B.A.; Minigaliyeva, I.A.; Panov, V.G.; Privalova, L.I.; Varaksin, A.N.; Gurvich, V.B.; Sutunkova, M.P.; Shur, V.Y.; Shishkina, E.V.; Valamina, I.E.; et al. Some patterns of metallic nanoparticles' combined subchronic toxicity as exemplified by a combination of nickel and manganese oxide nanoparticles. *Food Chem. Toxicol.* **2015**, *86*, 351–364. [CrossRef]
26. Mate, Z.; Horvath, E.; Kozma, G.; Simon, T.; Konya, Z.; Paulik, E.; Papp, A.; Szabo, A. Size-Dependent Toxicity Differences of Intratracheally Instilled Manganese Oxide Nanoparticles: Conclusions of a Subacute Animal Experiment. *Biol. Trace Elem. Res.* **2016**, *171*, 156–166. [CrossRef] [PubMed]
27. Su, S.; Kang, P.M. Systemic Review of Biodegradable Nanomaterials in Nanomedicine. *Nanomaterials* **2020**, *10*, 656. [CrossRef] [PubMed]
28. Bourquin, J.; Milosevic, A.; Hauser, D.; Lehner, R.; Blank, F.; Petri-Fink, A.; Rothen-Rutishauser, B. Biodistribution, Clearance, and Long-Term Fate of Clinically Relevant Nanomaterials. *Adv. Mater.* **2018**, *30*, e1704307. [CrossRef] [PubMed]
29. Tian, K.; Yu, Y.; Qiu, Q.; Sun, X.; Meng, F.; Bi, Y.; Gu, J.; Wang, Y.; Zhang, F.; Huo, H. Mechanisms of BPA Degradation and Toxicity Resistance in Rhodococcus equi. *Microorganisms* **2022**, *11*, 67. [CrossRef] [PubMed]
30. Jain, P.; Jangid, A.K.; Pooja, D.; Kulhari, H. Design of manganese-based nanomaterials for pharmaceutical and biomedical applications. *J. Mater. Chem. B* **2024**, *12*, 577–608. [CrossRef] [PubMed]
31. Olivito, F.; Jagdale, P.; Oza, G. Synthesis and Biodegradation Test of a New Polyether Polyurethane Foam Produced from PEG 400, (L)-Lysine Ethyl Ester Diisocyanate (L-LDI) and Bis-hydroxymethyl Furan (BHMF). *Toxics* **2023**, *11*, 698. [CrossRef] [PubMed]
32. Gao, X.; Di Bernardo, I.; Kreider, P.; Tran-Phu, T.; Cai, X.; Wang, N.; Zhu, Y.; Venkataraman, M.B.; Lipton-Duffin, J.; Bayon, A. Lattice expansion in optimally doped manganese oxide: An effective structural parameter for enhanced thermochemical water splitting. *ACS Catal.* **2019**, *9*, 9880–9890. [CrossRef]
33. Huang, K.Z.; Zhang, H. Direct electron-transfer-based peroxymonosulfate activation by iron-doped manganese oxide (δ-MnO_2) and the development of galvanic oxidation processes (GOPs). *Environ. Sci. Technol.* **2019**, *53*, 12610–12620. [CrossRef] [PubMed]
34. Balaghi, S.E.; Triana, C.A.; Patzke, G.R. Molybdenum-doped manganese oxide as a highly efficient and economical water oxidation catalyst. *ACS Catalysis* **2020**, *10*, 2074–2087. [CrossRef]
35. Adil, S.F.; Assal, M.E.; Kuniyil, M.; Khan, M.; Shaik, M.R.; Alwarthan, A.; Labis, J.P.; Siddiqui, M.R.H. Synthesis and comparative catalytic study of zinc oxide (ZnO x) nanoparticles promoted $MnCO_3$, MnO_2 and Mn_2O_3 for selective oxidation of benzylic alcohols using molecular oxygen. *Mater. Express* **2017**, *7*, 79–92. [CrossRef]

36. Gac, W. The influence of silver on the structural, redox and catalytic properties of the cryptomelane-type manganese oxides in the low-temperature CO oxidation reaction. *Appl. Catal. B Environ.* **2007**, *75*, 107–117. [CrossRef]
37. Dubal, D.; Dhawale, D.; Salunkhe, R.; Pawar, S.; Lokhande, C. A novel chemical synthesis and characterization of Mn_3O_4 thin films for supercapacitor application. *Appl. Surf. Sci.* **2010**, *256*, 4411–4416. [CrossRef]
38. Maslen, E.; Streltsov, V.; Streltsova, N.; Ishizawa, N. Electron density and optical anisotropy in rhombohedral carbonates. III. Synchrotron X-ray studies of $CaCO_3$, $MgCO_3$ and $MnCO_3$. *Acta Crystallogr. Sect. B Struct. Sci.* **1995**, *51*, 929–939. [CrossRef]
39. Xia, T.; Kovochich, M.; Liong, M.; Madler, L.; Gilbert, B.; Shi, H.; Yeh, J.I.; Zink, J.I.; Nel, A.E. Comparison of the mechanism of toxicity of zinc oxide and cerium oxide nanoparticles based on dissolution and oxidative stress properties. *ACS Nano* **2008**, *2*, 2121–2134. [CrossRef]
40. Eom, H.J.; Choi, J. Oxidative stress of CeO_2 nanoparticles via p38-Nrf-2 signaling pathway in human bronchial epithelial cell, Beas-2B. *Toxicol. Lett.* **2009**, *187*, 77–83. [CrossRef]
41. Zhang, T.; Gaffrey, M.J.; Thomas, D.G.; Weber, T.J.; Hess, B.M.; Weitz, K.K.; Piehowski, P.D.; Petyuk, V.A.; Moore, R.J.; Qian, W.J.; et al. A proteome-wide assessment of the oxidative stress paradigm for metal and metal-oxide nanomaterials in human macrophages. *NanoImpact* **2020**, *17*, 100194. [CrossRef]
42. DeLoid, G.; Casella, B.; Pirela, S.; Filoramo, R.; Pyrgiotakis, G.; Demokritou, P.; Kobzik, L. Effects of engineered nanomaterial exposure on macrophage innate immune function. *NanoImpact* **2016**, *2*, 70–81. [CrossRef]
43. Alsaleh, N.B.; Assiri, M.A.; Aljarbou, A.M.; Almutairi, M.M.; As Sobeai, H.M.; Alshamrani, A.A.; Almudimeegh, S. Adverse Responses following Exposure to Subtoxic Concentrations of Zinc Oxide and Nickle Oxide Nanoparticles in the Raw 264.7 Cells. *Toxics* **2023**, *11*, 674. [CrossRef]
44. Shapouri-Moghaddam, A.; Mohammadian, S.; Vazini, H.; Taghadosi, M.; Esmaeili, S.A.; Mardani, F.; Seifi, B.; Mohammadi, A.; Afshari, J.T.; Sahebkar, A. Macrophage plasticity, polarization, and function in health and disease. *J. Cell Physiol.* **2018**, *233*, 6425–6440. [CrossRef]
45. Figueiredo Borgognoni, C.; Kim, J.H.; Zucolotto, V.; Fuchs, H.; Riehemann, K. Human macrophage responses to metal-oxide nanoparticles: A review. *Artif. Cells Nanomed. Biotechnol.* **2018**, *46*, 694–703. [CrossRef]
46. Duan, J.; Zhang, T.; Gaffrey, M.J.; Weitz, K.K.; Moore, R.J.; Li, X.; Xian, M.; Thrall, B.D.; Qian, W.J. Stochiometric quantification of the thiol redox proteome of macrophages reveals subcellular compartmentalization and susceptibility to oxidative perturbations. *Redox Biol.* **2020**, *36*, 101649. [CrossRef]
47. Chen, Q.; Wang, N.; Zhu, M.; Lu, J.; Zhong, H.; Xue, X.; Guo, S.; Li, M.; Wei, X.; Tao, Y.; et al. TiO_2 nanoparticles cause mitochondrial dysfunction, activate inflammatory responses, and attenuate phagocytosis in macrophages: A proteomic and metabolomic insight. *Redox Biol.* **2018**, *15*, 266–276. [CrossRef] [PubMed]
48. Alsaleh, N.B.; Minarchick, V.C.; Mendoza, R.P.; Sharma, B.; Podila, R.; Brown, J.M. Silver nanoparticle immunomodulatory potential in absence of direct cytotoxicity in RAW 264.7 macrophages and MPRO 2.1 neutrophils. *J. Immunotoxicol.* **2019**, *16*, 63–73. [CrossRef] [PubMed]
49. Peters, A.; Nawrot, T.S.; Baccarelli, A.A. Hallmarks of environmental insults. *Cell* **2021**, *184*, 1455–1468. [CrossRef] [PubMed]
50. Nel, A.; Xia, T.; Madler, L.; Li, N. Toxic potential of materials at the nanolevel. *Science* **2006**, *311*, 622–627. [CrossRef] [PubMed]
51. George, S.; Pokhrel, S.; Ji, Z.; Henderson, B.L.; Xia, T.; Li, L.; Zink, J.I.; Nel, A.E.; Mädler, L. Role of Fe doping in tuning the band gap of TiO_2 for the photo-oxidation-induced cytotoxicity paradigm. *J. Am. Chem. Soc.* **2011**, *133*, 11270–11278. [CrossRef]
52. Zhang, H.; Ji, Z.; Xia, T.; Meng, H.; Low-Kam, C.; Liu, R.; Pokhrel, S.; Lin, S.; Wang, X.; Liao, Y.P.; et al. Use of metal oxide nanoparticle band gap to develop a predictive paradigm for oxidative stress and acute pulmonary inflammation. *ACS Nano* **2012**, *6*, 4349–4368. [CrossRef] [PubMed]
53. Persaud, I.; Raghavendra, A.J.; Paruthi, A.; Alsaleh, N.B.; Minarchick, V.C.; Roede, J.R.; Podila, R.; Brown, J.M. Defect-induced electronic states amplify the cellular toxicity of ZnO nanoparticles. *Nanotoxicology* **2020**, *14*, 145–161. [CrossRef] [PubMed]
54. Xia, T.; Zhao, Y.; Sager, T.; George, S.; Pokhrel, S.; Li, N.; Schoenfeld, D.; Meng, H.; Lin, S.; Wang, X.; et al. Decreased dissolution of ZnO by iron doping yields nanoparticles with reduced toxicity in the rodent lung and zebrafish embryos. *ACS Nano* **2011**, *5*, 1223–1235. [CrossRef] [PubMed]
55. Naatz, H.; Lin, S.; Li, R.; Jiang, W.; Ji, Z.; Chang, C.H.; Koser, J.; Thoming, J.; Xia, T.; Nel, A.E.; et al. Safe-by-Design CuO Nanoparticles via Fe-Doping, Cu-O Bond Length Variation, and Biological Assessment in Cells and Zebrafish Embryos. *ACS Nano* **2017**, *11*, 501–515. [CrossRef] [PubMed]
56. Peng, Q.; Huo, D.; Li, H.; Zhang, B.; Li, Y.; Liang, A.; Wang, H.; Yu, Q.; Li, M. ROS-independent toxicity of Fe_3O_4 nanoparticles to yeast cells: Involvement of mitochondrial dysfunction. *Chem. Biol. Interact.* **2018**, *287*, 20–26. [CrossRef]
57. Yusuf, A.; Brophy, A.; Gorey, B.; Casey, A. Liposomal encapsulation of silver nanoparticles enhances cytotoxicity and causes induction of reactive oxygen species-independent apoptosis. *J. Appl. Toxicol.* **2018**, *38*, 616–627. [CrossRef] [PubMed]
58. Mohammadinejad, R.; Moosavi, M.A.; Tavakol, S.; Vardar, D.O.; Hosseini, A.; Rahmati, M.; Dini, L.; Hussain, S.; Mandegary, A.; Klionsky, D.J. Necrotic, apoptotic and autophagic cell fates triggered by nanoparticles. *Autophagy* **2019**, *15*, 4–33. [CrossRef] [PubMed]
59. Refsnes, M.; Skuland, T.; Lilleaas, E.; Ovrevik, J.; Lag, M. Concentration-dependent cytokine responses of silica nanoparticles and role of ROS in human lung epithelial cells. *Basic Clin. Pharmacol. Toxicol.* **2019**, *125*, 304–314. [CrossRef]
60. Sabella, S.; Carney, R.P.; Brunetti, V.; Malvindi, M.A.; Al-Juffali, N.; Vecchio, G.; Janes, S.M.; Bakr, O.M.; Cingolani, R.; Stellacci, F.; et al. A general mechanism for intracellular toxicity of metal-containing nanoparticles. *Nanoscale* **2014**, *6*, 7052–7061. [CrossRef]

61. Alsaleh, N.B.; Brown, J.M. Immune responses to engineered nanomaterials: Current understanding and challenges. *Curr. Opin. Toxicol.* **2018**, *10*, 8–14. [CrossRef]
62. Cortez-Jugo, C.; Czuba-Wojnilowicz, E.; Tan, A.; Caruso, F. A Focus on "Bio" in Bio-Nanoscience: The Impact of Biological Factors on Nanomaterial Interactions. *Adv. Healthc. Mater.* **2021**, *10*, e2100574. [CrossRef]
63. Assal, M.E.; Kuniyil, M.; Khan, M.; Shaik, M.R.; Al-Warthan, A.; Siddiqui, M.R.H.; Labis, J.P.; Adil, S.F. Comparative catalytic evaluation of nano-ZrO_x promoted manganese catalysts: Kinetic study and the effect of dopant on the aerobic oxidation of secondary alcohols. *Adv. Mater. Sci. Eng.* **2017**, *2017*, 3958319. [CrossRef]
64. Almomen, A.; Alsaleh, N.B.; El-Toni, A.M.; El-Mahrouky, M.A.; Alhowyan, A.A.; Alkholief, M.; Alshamsan, A.; Khurana, N.; Ghandehari, H. In Vitro Safety Assessment of In-House Synthesized Titanium Dioxide Nanoparticles: Impact of Washing and Temperature Conditions. *Int. J. Mol. Sci.* **2023**, *24*, 9966. [CrossRef]

pharmaceuticals

MDPI

Article

Synthesis, Characterization, and Evaluation of the Antimicrobial and Anticancer Activities of Zinc Oxide and Aluminum-Doped Zinc Oxide Nanocomposites

Muhammad Asif [1], Muhammad Fakhar-e-Alam [1,*], Muhammad Tahir [2], Farah Jamil [3], Hassan Sardar [1], Javed Rehman [4,5] and Kholood A. Dahlous [6,*]

1 Department of Physics, Government College University Faisalabad, Faisalabad 38000, Pakistan
2 Department of Chemistry, Quaid-e-Azam University, Islamabad 45320, Pakistan
3 Department of Applied Chemistry, Government College University Faisalabad, Faisalabad 38000, Pakistan
4 State Key Laboratory of Metastable Materials Science and Technology, School of Materials Science and Engineering, Yanshan University, Qinhuangdao 066004, China
5 MEU Research Unit, Middle East University, Amman 11831, Jordan
6 Department of Chemistry, College of Science, King Saud University, Riyadh 11451, Saudi Arabia
* Correspondence: fakharphy@gmail.com (M.F.-e.-A.); kdahloos@ksu.edu.sa (K.A.D.)

Abstract: In this research, we developed undoped and aluminum-doped zinc oxide for antimicrobial and anticancer activities. This study focuses on the synthesis, characterization, and biological activities of zinc oxide nanoparticles (ZnO NPs) and aluminum-doped zinc oxide nanocomposites ($Zn_{1-x}Al_xO$ NCs) at varying concentrations (x = 0, 0.25, 0.5, and 0.75 wt%) using the coprecipitation method. Various characterization techniques such as XRD, UV-Vis, FTIR, EDX, and SEM were performed to analyze the crystal structure, optical properties, functional group identification, elemental composition, and surface morphology. The antimicrobial activity test showed that $Zn_{0.75}Al_{0.25}O$ NCs exhibited the strongest inhibition zone against *Bacillus cereus* compared to *Staphylococcus aureus* > *Pasteurella multocida* > *Escherichia coli*. Moreover, the cytotoxicity and cell viability of liver cancer (HepG-2), breast cancer (MCF-7), ovarian cancer (SKOV3), and normal liver cell lines) were evaluated using the MTT assay, demonstrating that $Zn_{0.75}Al_{0.25}O$ NCs not only enhance cell destruction but also show low cytotoxicity and high biocompatibility at low concentrations. These results suggest that $Zn_{0.75}Al_{0.25}O$ NCs could be a promising candidate for in vivo anticancer applications and should be further investigated.

Keywords: ZnO NPs; Al-doped ZnO NCs; antimicrobial, anticancer activities

1. Introduction

Nanotechnology offers numerous advantages in every field of life, including information technology, medical technology, energy, global security, environmental science, and food safety. However, its impact is particularly effective in the field of nanomedicine, where it holds great potential for cancer diagnosis and treatment. Cancer, a leading cause of mortality and healthcare expenditure, affects millions globally, manifesting in various forms such as breast, liver, lung, and brain cancer [1–3].

According to a 2020 WHO report, cancer is the second foremost reason for death globally, resulting in over 10 million deaths annually. Liver cancer (HepG-2) is one of them, and unfortunately, it is on the rise [4,5]. Notably, 85% of cancer cases occur in developing nations. Pakistan has the highest death rate of liver cancer among Asian nations [6,7]. The major risk factors for HepG-2 cancer include infection with human papillomavirus (HPV) and hepatitis (HCV, HBV), as well as exposure to fatty acids, contaminated food, nicotine, and alcohol, which contribute to 25% of liver cancer deaths. However, these factors may vary from country to country [8,9]. Moreover, bacterial infections and their resistance have severely obstructed public health, causing widespread morbidity and mortality [10]. The treatment of bacterial infections often involves traditional medicines, which can have harmful side effects due to overuse and misuse, ultimately impacting patients' health and well-being [11–13].

Metal oxide nanoparticles (MO NPs) and their nanocomposites (especially ZnO NPs, Al_2O_3 NPs, and their nanocomposites) have demonstrated high efficiency against antimicrobial resistance (AMR), multidrug resistance (MDR), as well as in cancer diagnostics and treatment, and inhibiting both Gram-positive and Gram-negative bacteria [14,15].

A wide range of MO NPs, including ZnO NPs, manganese dioxide, magnesium oxide, aluminum oxide, silver oxide, and selenium oxide nanoparticles, as well as their nanocomposites, have gained significant attention in the biomedical field for applications such as nanomedicine, drug delivery systems (DDS), biosensing, photothermal therapy (PTT), hyperthermia therapy, photodynamic therapy (PDT), diagnostics, and cancer treatment. This interest is driven due to their exceptional physiochemical and biological properties [16–21]. For example, Rana et al. [22] described that ZnO NPs and Al-doped ZnO NCs exhibit broad-spectrum antimicrobial activity and promising efficacy against Gram-negative and Gram-positive bacteria. They also exhibit enhanced cytotoxicity toward various cancerous cell lines, including liver, breast, lung, and colon cancer cells. It can also boost cell death by inhibiting cancer growth cells. Yu et al. [23] reported that Al-doped ZnO NCs show great antibacterial activity against various bacterial strains, including *E. coli* and *P. aeruginosa*, especially for *S. aureus*. Moreover, they can also decrease antimicrobial resistance (AMR). Eddy et al. [24] described that Al-doped ZnO NCs exhibited broad-spectrum antimicrobial activity and promising efficacy against Gram-negative and Gram-positive bacteria. These nanocomposites can enhance cytotoxicity, reduce toxicity, and have minimal side effects on various cancerous cell lines, including liver, breast, lung, and colon cancer cells. Also, it was found to boost cell death while inhibiting cancer cell growth [12].

There are several reputable techniques for developing MO NPs, such as auto-combustion, coprecipitation, sol–gel, or hydrothermal methods, etc. [25,26]. These methods are ideal due to their simplicity, ease of use, eco-friendliness, and ability to fabricate nanoparticles with the desired morphology. Furthermore, the size, shape, and properties of nanoparticles can be easily modified by adjusting reaction parameters, including temperature and time [27–29].

In this research, undoped and aluminum-doped zinc oxide were successfully fabricated and characterized using various characterization techniques. These nanocomposites not only inhibit bacterial growth and enhance cell destruction but also show low cytotoxicity and high biocompatibility at low concentrations. Additionally, to address the problems of AMR and MDR barriers using these nanomaterials. These findings suggest that $Zn_{0.75}Al_{0.25}O$ NCs could be promising candidates for in vivo anticancer applications.

2. Results and Discussions

2.1. XRD Analysis

XRD analysis was employed to investigate the structural properties of ZnO NPs and Al-doped ZnO nanocomposites fabricated at varying concentrations (0.25%, 0.5%, and 0.75%). The XRD pattern revealed no additional phases, and the intensity of the samples increased by increasing aluminum concentration, as shown in Figure 1. The XRD patterns of ZnO NPs and Al-doped ZnO NCs showed three distinct peaks at 31.5°, 34.08°, and 36.01°, corresponding to the Miller indices (100), (002), and (101), respectively. These peaks matched well with the JCPDS card no. 36-1451, verifying the hexagonal wurtzite phase structure of ZnO NPs and Al-doped ZnO NCs with complete formation and no additional peaks. Moreover, low-intensity peaks were observed at 47.33°, 56.39°, 62.69°, 67.75°, and 68.89°, corresponding to the Miller indices (102), (110), (103), (112), and (201), respectively [16,19]. Table 1 shows that the crystallite size of Al-doped ZnO NPs does not change significantly with increasing Al concentration. The non-significant changes were observed in crystallite size with increasing Al concentration, which can be significant in various applications, such as electronic and optoelectronic devices, energy storage devices, and photocatalytic activities, particularly in biomedical applications. Furthermore, due to their small size, nanoparticles can be effectively used for targeted drug delivery [26,27]. The average crystallite sizes of ZnO NPs and Al-doped ZnO NCs were calculated using Scherrer's Equation (1).

$$C.S = \frac{k\lambda}{\beta cos\theta} \tag{1}$$

where,
$C.S$ = crystallite size;
k = shape factor (usually its value is 0.94);
λ = 0.154 nm, X-ray wavelength;
β = FWHM (full-width half maximum) in radian;
$Cos\theta$ = Bragg angle in degrees.

Table 1. ZnO NPs and Al-doped ZnO NCs at varying concentrations (0.25%, 0.5%, and 0.75%).

(h, k, l) Miller Indices	2θ (Angle of Diffraction)	C.S (nm)
	ZnO NPs	
100	31.5°	39.37
101	36.01°	45.25
110	56.39°	47.1
112	67.75°	44.54
	$Zn_{0.75}Al_{0.25}O$ NCs	
100	31.5°	36.67
101	36.01°	41.65
110	56.39°	43.71
112	67.75°	42.64
	$Zn_{0.5}Al_{0.5}O$ NCs	
100	31.5°	35.17
101	36.01°	39.29
110	56.39°	42.31
112	67.75°	40.54
	$Zn_{0.25}Al_{0.75}O$ NCs	
100	31.5°	32.87
101	36.01°	33.26
110	56.39°	35.58
112	67.75°	31.29

2.2. Fourier Transformation Infrared Spectroscopy

Figure 2 displays the FTIR spectra of ZnO NPs and Al-doped ZnO NCs at varying concentrations (0.25%, 0.5%, and 0.75%) over the 400–4000 cm^{-1} range. The FTIR spectra disclose the Zn-O stretching mode, thereby representing a band between 441 and 665 cm^{-1}, confirming the formation of ZnO NPs. Moreover, the broad band at 3349 cm^{-1} corresponds to the O-H stretching mode of the hydroxyl group, while peaks at 890 cm^{-1} and 3349 cm^{-1} indicate -OH bending and stretching vibrational modes. These peaks originate from the H_2O during Zn-O formation and moisture contents in the atmosphere. However, the intensity and magnitude of the –OH emission at 3349 cm^{-1} vary due to the use of different reducing agents. The peaks at 1422 cm^{-1} and 1642 cm^{-1} correspond to C=O and C–H vibrational modes, respectively, which are due to the presence of aldehyde or ketone groups partially substituted by zinc source nitrate ions ($Zn(NO_3)_2$). The absorption at 1376 cm^{-1}

suggests the presence of an aromatic ring at a moderate level. The observed peaks at 1642, 1460, 1422, 1376, and 865 cm^{-1} are associated with the asymmetrical and symmetrical zinc carboxylate [29–31].

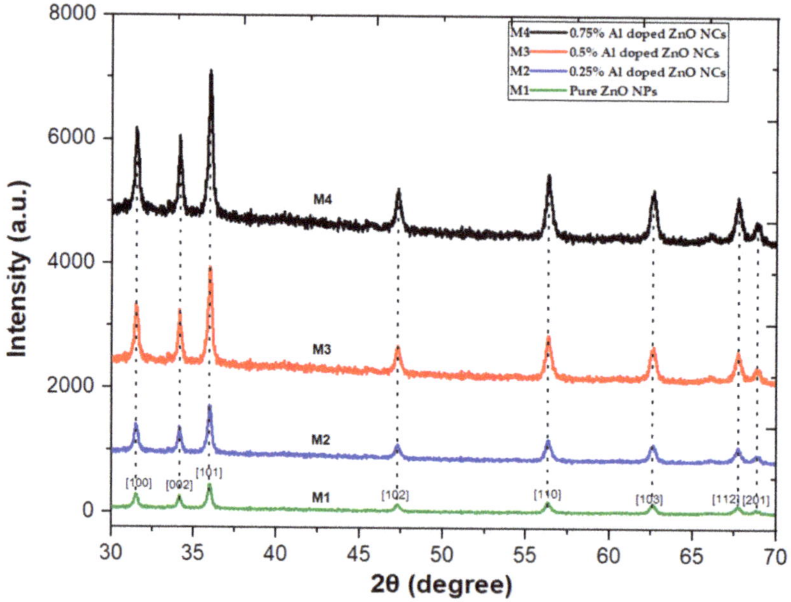

Figure 1. XRD spectrum of ZnO NPs and Al-doped ZnO NCs at varying concentrations 0.25%, 0.5%, and 0.75%.

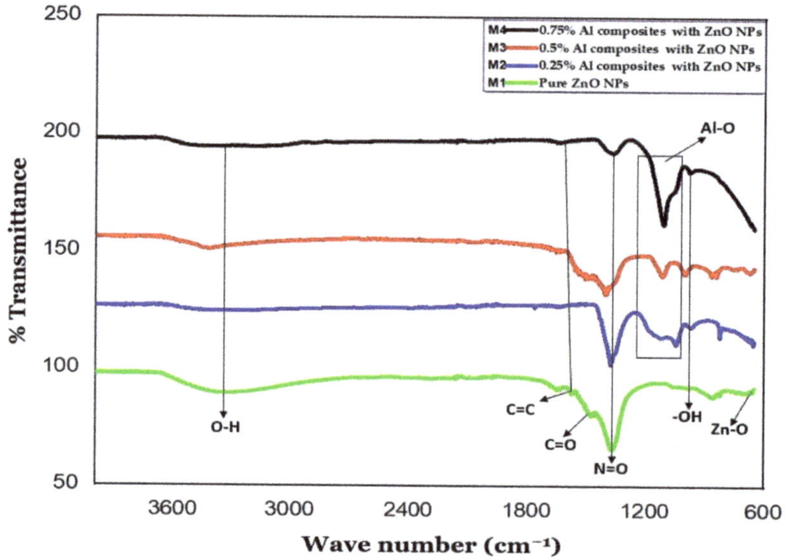

Figure 2. FTIR spectrum of ZnO NPs and Al-doped ZnO NCs at varying concentrations 0.25%, 0.5%, and 0.75%.

2.3. UV-Visible Spectroscopy

The UV-visible spectra in Figure 3 illustrate the absorption bands of ZnO NPs and Al-doped ZnO NCs at varying concentrations (0.25%, 0.5%, and 0.75%). The presence of aluminum causes a longer wavelength (red shift). Because of their high exciton binding energy (60 mV), ZnO NPs and their Al-doped ZnO NCs exhibit substantial exciton absorption in their spectra, even when synthesized at ambient temperature. Specifically, the absorption peak of ZnO NPs and Al-doped ZnO NCs occurs at 363 nm and ≥363 [19,32,33].

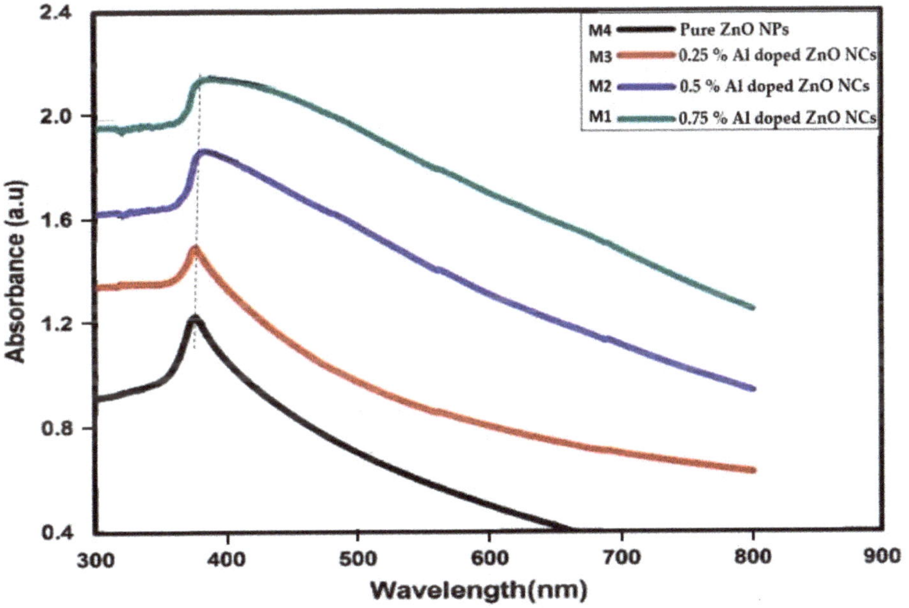

Figure 3. UV-visible spectrums of ZnO NPs and Al-doped ZnO NCs at varying concentrations 0.25%, 0.5%, and 0.75%.

2.4. Scanning Electron Microscopy and Zeta Potential Analysis

SEM analysis was conducted to investigate the surface morphology of the ZnO NPs and Al-doped ZnO NCs at varying concentrations (0.25%, 0.5%, and 0.75%), as shown in Figure 4a–d. With increasing Al^{3+} concentration, prominent changes in particle shape and size were observed. The SEM images demonstrate irregular shapes with small and large spherical grains, also observed in the Al-doped ZnO NCs. The spherical grains were circulated with red circles. In particular, the grain size increased with the increase in aluminum content. All the SEM images were taken at a resolution of 3 µm. Figure 4a–d displays SEM images of ZnO NPs, $Zn_{0.75}Al_{0.25}O$, $Zn_{0.5}Al_{0.5}O$, and $Zn_{0.25}Al_{0.75}O$ NCs, revealing average crystallite sizes of 43.5, 40.5, 39, and 33 nm, respectively. Notably, these values are in excellent agreement with the corresponding XRD results, confirming the consistency of the crystallite size measurements. The zeta potential analysis results of ZnO NPs and Al-doped ZnO NCs are shown in Figure 5. The zeta potential values of ZnO NPs, $Zn_{0.75}Al_{0.25}O$, $Zn_{0.5}Al_{0.5}O$, and $Zn_{0.25}Al_{0.75}O$ NCs dispersions were found to be +20, ±15, −10, and 13.7 mV, respectively. These results show the high stability of the prepared nanomaterials. These results are very similar to previously published articles [34–36].

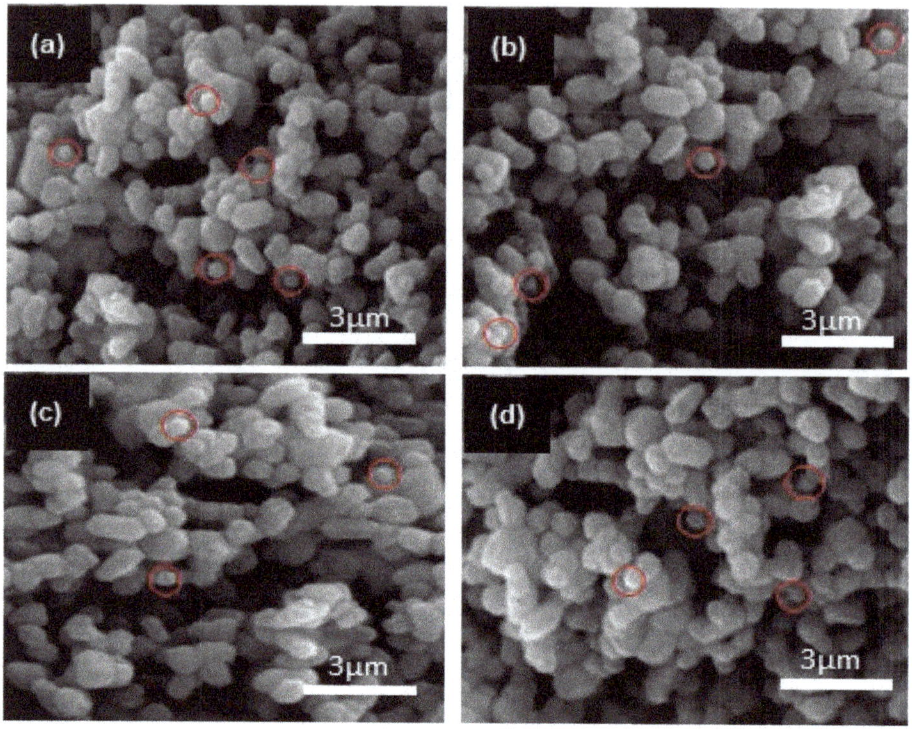

Figure 4. SEM results of (**a**) ZnO NPs and Al-doped ZnO NCs at varying concentrations (**b**) 0.25%, (**c**) 0.5%, and (**d**) 0.75%.

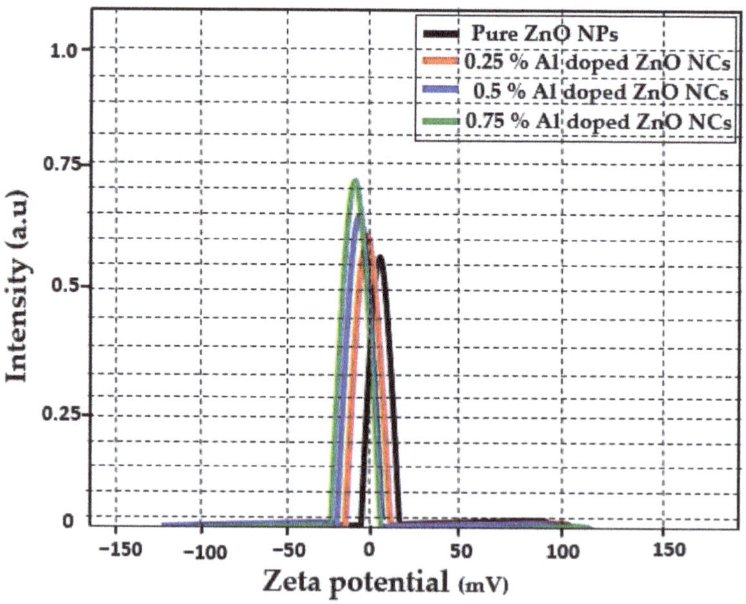

Figure 5. Zeta potential of ZnO NPs and Al-doped ZnO NCs at varying concentrations 0.25%, 0.5%, and 0.75%.

2.5. Energy Dispersive X-ray Spectroscopy

The EDX analysis was employed to measure the elemental composition of the prepared ZnO NPs and Al-doped ZnO NCs. The outcomes of the EDX analysis confirmed the presence of Zn and Al as primary elements and O as a secondary element in the prepared nanomaterials, as shown in Figure 6a–d. For example, Musleh et al. [37] and Aldalbahi et al. [38] stated comparable outcomes where zinc and aluminum elements were detected in the EDX spectrum, confirming the reduction in zinc and aluminum-doped zinc ions to Zn and Al elements.

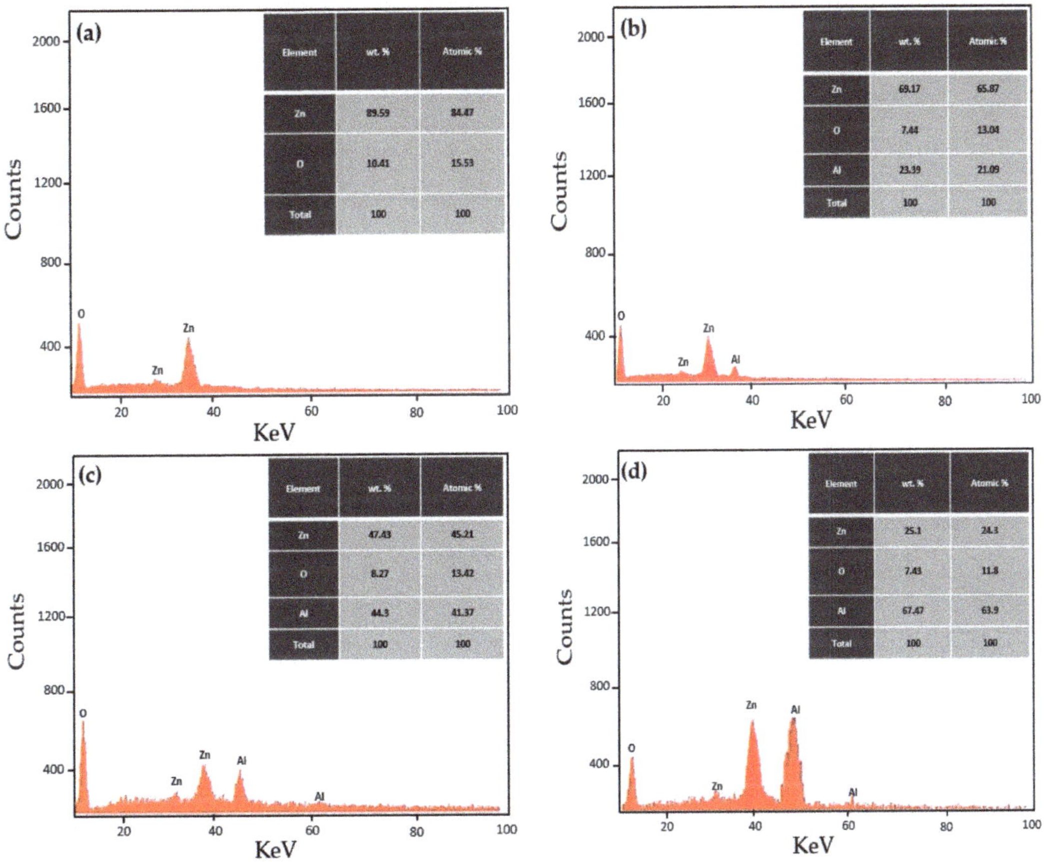

Figure 6. EDX spectrum of (**a**) ZnO NPs and Al-doped ZnO NCs at varying concentrations (**b**) 0.25%, (**c**) 0.5%, and (**d**) 0.75%.

2.6. Antimicrobial Activity

In this research work, the antimicrobial potential of ZnO NPs and Al-doped ZnO NCs at varying concentrations (0.25%, 0.5%, and 0.75%) was evaluated against four bacterial strains, including two Gram-positive strains, *Staphylococcus aureus* and *Bacillus cereus*, and two Gram-negative strains, *Escherichia coli*, and *Pasteurella multocida*. The results showed that ZnO NPs and Al-doped NCs exhibited the highest antimicrobial potential against the tested bacterial strains at a high-dose concentration of 40 mg/mL. Notably, the zones of inhibition were clear and significant, except for the $Zn_{0.75}Al_{0.25}O$ NCs sample, which showed very high inhibition against all bacterial strains (indicated by the symbol "#"), as shown in Figures 7 and 8.

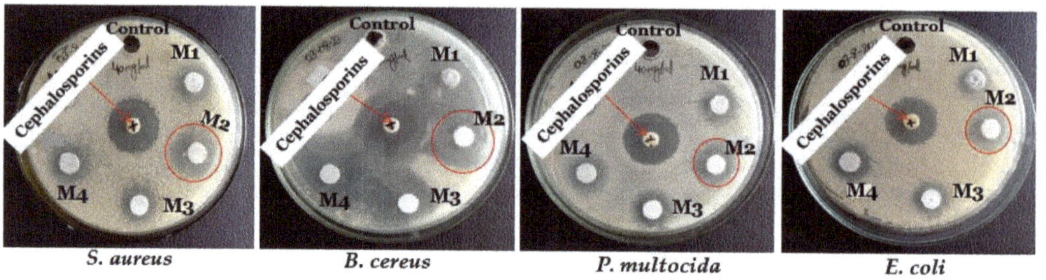

Figure 7. The antimicrobial activity of samples DMSO (control), standard antibiotic (cephalosporin), M1 (ZnO NPs), M2 ($Zn_{0.75}Al_{0.25}O$), M3 ($Zn_{0.5}Al_{0.5}O$), and M4 ($Zn_{0.25}Al_{0.75}O$ NCs) against *S. aureus*, *B. cereus*, *P. multocida*, and *E. coli*.

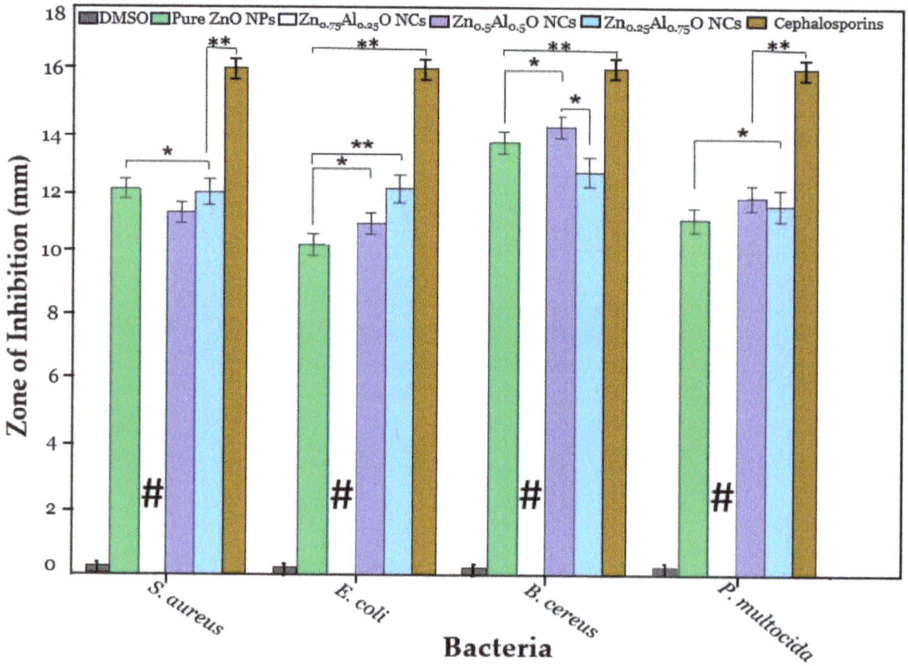

Figure 8. The antimicrobial activity of DMSO (control), standard antibiotic (cephalosporin), ZnO NPs, $Zn_{0.75}Al_{0.25}O$ NCs, $Zn_{0.5}Al_{0.5}O$ NCs, and $Zn_{0.25}Al_{0.75}O$ NCs at 40 mg/mL, where * is significance for $p \leq 0.05$ and ** is highly significance for $p \leq 0.01$.

Hence, the antimicrobial activity of $Zn_{0.75}Al_{0.25}O$ NCs was repeated at lower concentrations, and even at 150 µg/mL, it showed a significant zone of inhibition against each bacterial strain, which showed its highest antimicrobial potential against these bacterial strains. A standard antibiotic, cephalosporin, was used as a positive control, and dimethyl sulfoxide (DMSO) was used as a negative control and did not show any antimicrobial activity against any of the bacterial strains to determine the assay validity. ZnO NPs, $Zn_{0.75}Al_{0.25}O$, $Zn_{0.5}Al_{0.5}O$, and $Zn_{0.25}Al_{0.75}O$ NCs showed significant antimicrobial activity against all bacterial strains. Notably, $Zn_{0.75}Al_{0.25}O$ NCs exhibited the highest antibacterial efficiency, especially against *B. cereus*. Moreover, $Zn_{0.75}Al_{0.25}O$ NCs demonstrated the biggest inhibition zone against *S. aureus*, *B. cereus*, *E. coli*, and *P. multocida*, with 12 ± 0.09 mm, 15.5 ± 0.11 mm, 12.5 ± 0.07 mm, and 13.3 ± 0.11 mm, respectively, as shown in Figure 9 and Table 2. In this discussion, Yu et al. [23] described that Al-doped ZnO NCs showed great antibacterial activity against numerous bacterial strains, including *E. coli* and *P. aeruginosa*, especially against *S. aureus*. Also, it can decrease antimicrobial resistance (AMR). For instance, the antimicrobial activity of ZnO NPs can induce Zn^{2+} and ROS, such as hydroxyl radicals and superoxides, which can disrupt bacterial cell membrane integrity and DNA and inhibit enzyme activity, causing structural damage and leakage of cellular contents. These nanomaterials can prevent biofilm formation, which is important for bacterial survival and antibiotic resistance [38–41].

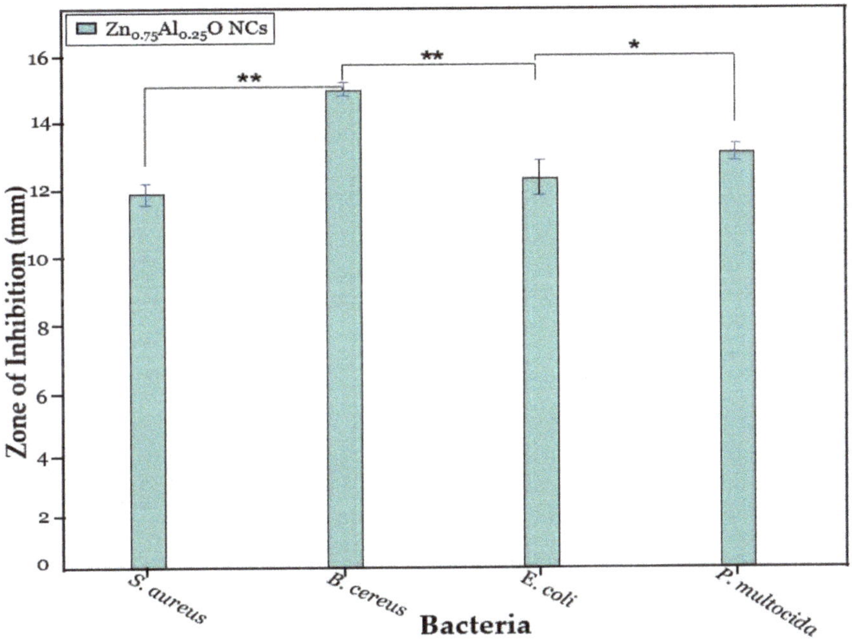

Figure 9. The antimicrobial activity of $Zn_{0.75}Al_{0.25}O$ NCs at 150 mg/mL, where * is significance for $p \leq 0.05$ and ** is highly significance for $p \leq 0.01$.

Table 2. Inhibition of ZnO NPs and Al-doped ZnO NCs at varying concentrations (0.25%, 0.5%, and 0.75%) against *S. aureus*, *E. coli*, *B. cereus*, and *P. multocida*.

Samples	*S. aureus*	*E. coli*	*B. cereus*	*P. multocida*
DMSO	0 ± 0.03 mm	0 ± 0.03 mm	0 ± 0.03 mm	0 ± 0.03 mm
M1	12 ± 0.05 mm	10.2 ± 0.07 mm	13.9 ± 0.08 mm	11.1 ± 0.08 mm
M2	12 ± 0.09 mm	12.5 ± 0.07 mm	15.5 ± 0.11 mm	13.3 ± 0.11 mm
M3	11.5 ± 0.07 mm	10.9 ± 0.09 mm	14.2 ± 0.08 mm	11.9 ± 0.09 mm
M4	12 ± 0.07 mm	12.2 ± 0.09 mm	13 ± 0.09 mm	11.8 ± 0.11 mm
Cephalosporins	16 ± 0.07 mm	16 ± 0.07 mm	16 ± 0.07 mm	16 ± 0.07 mm

2.7. Anticancer Activity

The 3-(4,5-dimethylthiazol-2-yl)-2,5-diphenyl-tetrazolium-bromide (MTT) assay was used to examine the cytotoxicity of ZnO NPs, $Zn_{0.75}Al_{0.25}O$, $Zn_{0.5}Al_{0.5}O$, and $Zn_{0.25}Al_{0.75}O$ NCs toward HepG-2, MCF-7, SKOV3, and normal cells. In the initial screening, the cell viability of ZnO NPs, $Zn_{0.75}Al_{0.25}O$, $Zn_{0.5}Al_{0.5}O$, and $Zn_{0.25}Al_{0.75}O$ NCs were tested against HepG-2, MCF-7, SKOV3, and normal cells at 100 µg/mL, and substantial cell viability was noticed for ZnO NPs and $Zn_{0.75}Al_{0.25}O$ NCs compared to $Zn_{0.5}Al_{0.5}O$ and $Zn_{0.25}Al_{0.75}O$ NCs (see Figure 10). Notably, $Zn_{0.75}Al_{0.25}O$ NCs exhibited > 70% cell viability toward HepG-2 cells. $Zn_{0.75}Al_{0.25}O$ NCs showed reasonably higher % cell viability. For further validation, the antiproliferative activity of $Zn_{0.75}Al_{0.25}O$ NCs was performed using a dose-dependent method (0, 3.125, 6.25, 12.5, 25, 50, and 100 μg/mL) toward all mentioned cell lines; then, dose-dependent curves were attained, as shown in Figure 11. The IC_{50} value was calculated as 20.47 μg/mL, indicating significant cell viability of $Zn_{0.75}Al_{0.25}O$ NCs. The cell viability results showed that $Zn_{0.75}Al_{0.25}O$ NCs not only enhanced HepG-2 cell destruction but also exhibited low cytotoxicity and high biocompatibility at low concentrations. Furthermore, $Zn_{0.75}Al_{0.25}O$ NCs have reduced toxicity toward normal liver cells compared to HepG-2, MCF-7, and SKOV3 cells. These results suggest that $Zn_{0.75}Al_{0.25}O$ NCs can serve as a promising preliminary point for further identification and isolation of compounds that inhibit HepG-2 cells or for the development of anticancer functional foods [34].

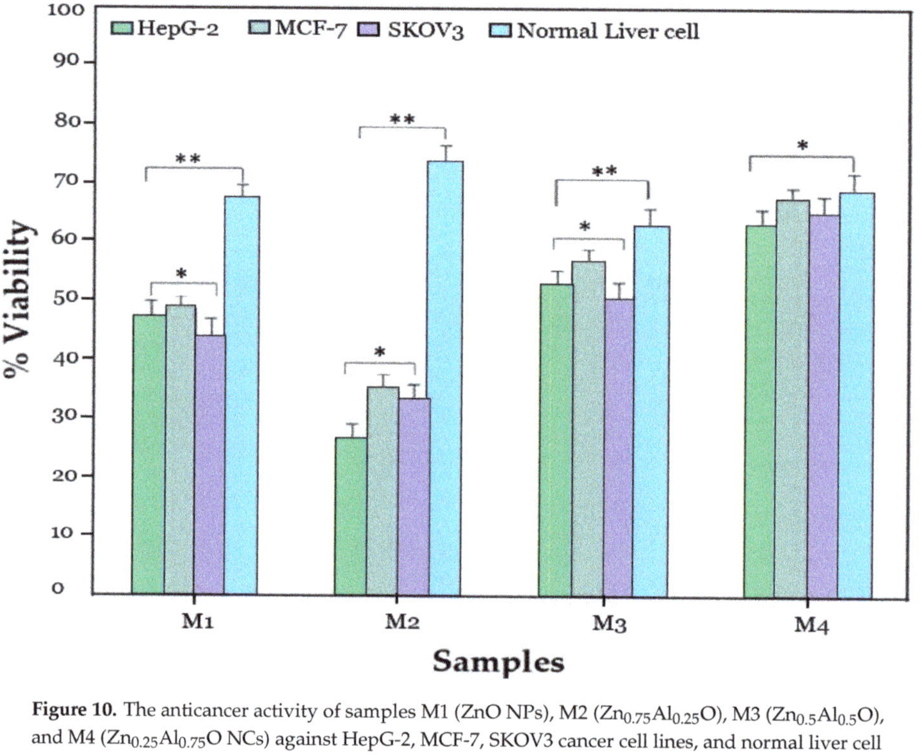

Figure 10. The anticancer activity of samples M1 (ZnO NPs), M2 ($Zn_{0.75}Al_{0.25}O$), M3 ($Zn_{0.5}Al_{0.5}O$), and M4 ($Zn_{0.25}Al_{0.75}O$ NCs) against HepG-2, MCF-7, SKOV3 cancer cell lines, and normal liver cell line, where * is significance for $p \leq 0.05$ and ** is highly significance for $p \leq 0.01$.

Allayeith et al. [41] described that ZnO NPs and their nanocomposites could be used for targeted cancer therapy, reducing side effects and improving the efficiency of cancer treatment. In these mechanisms, Al doping increases the surface area to volume ratio of ZnO and enhances the release of Zn^{2+} ions, improving interaction with cancer cells and bacterial strains (pathogens) and causing cell death and damage. Particularly, Al-doped ZnO NCs produce reactive oxygen species (ROS) and oxidative stress, leading to damage to cancer cell DNA, proteins, and lipids, as well as microbial damage. Anjum et al. [42] described that the anticancer activity of ZnO nanoparticles could induce programmed cell death (apoptosis) in cancer cells through mitochondrial dysfunction and activation of pro-apoptotic proteins. These nanoparticles can also inhibit cancer cell proliferation by arresting the cell cycle at specific stages (e.g., G2/M phase) and prevent the formation of new blood vessels, which is essential for cancer growth and metastasis. Notably, Al-doping can improve ROS generation, leading to enhanced antimicrobial and anticancer activity. Al-doping can facilitate the uptake of ZnO NPs by bacterial and cancer cells, enhancing their efficiency. Moreover, Al-doping can enhance the interaction between ZnO NPs and cell membranes, leading to superior structural damage.

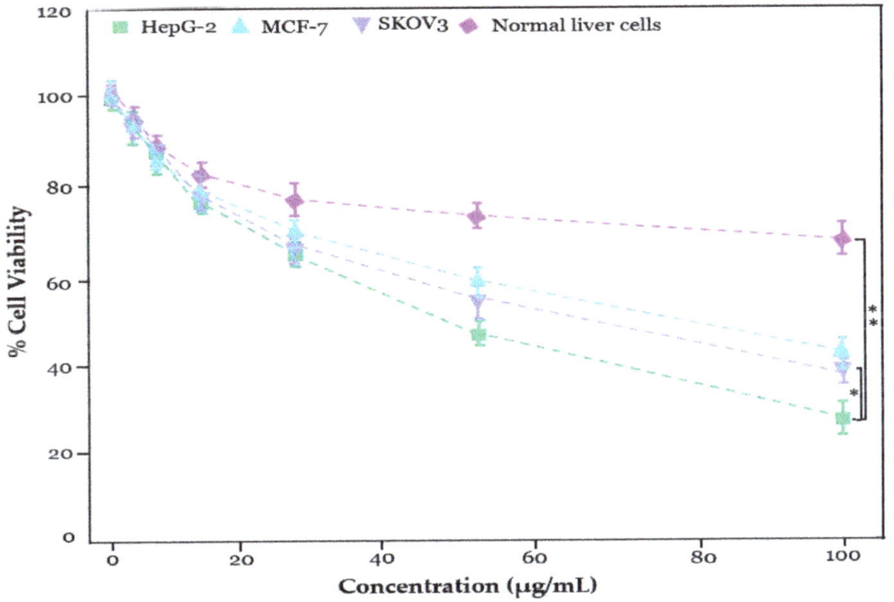

Figure 11. % cell viability of $Zn_{0.75}Al_{0.25}O$ NCs at varying concentration (µg/mL) against HepG-2, MCF-7, SKOV3, and normal liver cell line, where * is significance for $p \leq 0.05$ and ** is highly significance for $p \leq 0.01$.

3. Materials and Methods

3.1. Chemicals

Zinc nitrate [$Zn(NO_3)_2$, 98%], aluminum nitrate monohydrate [$Al(NO_3)_3$, 99.9%], sodium hydroxide (NaOH, 98%), MTT $\geq$ 98%, Dulbecco's Modified Eagles Medium (DMEM), dimethyl sulfoxide (DMSO), fetal bovine serum (FBS), and nutrient agar were used as chemical reagents without any further purification. All chemicals were purchased from the Punjab Scientific store, Sigma Aldrich (St. Louis, MO, USA).

3.2. Synthesis of ZnO NPs and Al-Doped ZnO Nanocomposites

A 0.5 M zinc nitrate solution was prepared using 100 mL of deionized water. In the next step, the solution was constantly stirred on a magnetic stirrer at 80 °C for 30 min. Furthermore, the pH was maintained between 10 and 11 by dropwise adding of NaOH solution. After 2 h of continuous stirring, a milky white precipitate formed. The precipitate was then centrifuged, washed, and dried. Finally, the ZnO NPs powder was annealed for 3 h at 300 °C in a muffle furnace. Similarly, Al-doped ZnO NCs were prepared by preparing a solution of 0.25% aluminum nitrate in 0.5 mM zinc nitrate. The solution was constantly stirred at 80 °C for 30 min. Finally, the same procedure was repeated twice with varying concentrations (0.5 and 0.75 wt%) [19,20]. The synthesis process of ZnO NPs and Al-doped ZnO NPs is shown in Scheme 1.

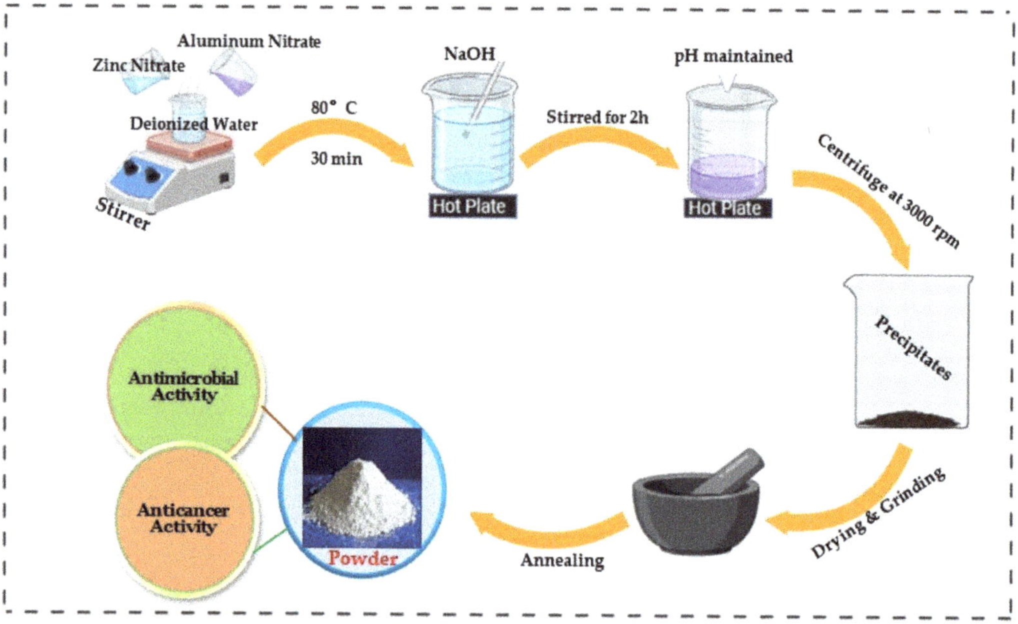

Scheme 1. Synthesis process of ZnO NPs and Al-doped ZnO NCs.

3.3. Characterization of Nanomaterials

SEM analysis was performed using the "Cube 10 Emcraft, Hanam, Republic of Korea" to analyze the surface morphology of the prepared nanomaterials. The UV-Vis spectrometer was employed to calculate the optical density on a "BK-UV1800PC double beam Biobase, Jinan, China". The XRD was performed using the "D8 advance Bruker, Karlsruhe, Germany" to examine the crystal structure using Cu Kα radiation (λ = 0.154). The FTIR analysis was performed on a "Spectrum 2, Perkin Elmer, Waltham, MA, USA" to analyze the functional groups. The EDX analysis was used to measure the elemental composition of the prepared nanomaterials on a "Zetasizer Nano ZS, Malvern, UK".

3.4. Well Diffusion Assay

The *Staphylococcus aureus* (*S. aureus*), *Bacillus cereus* (*B. cereus*), *Pasteurella multocida* (*P. multocida*), and *Escherichia coli* (*E. coli*) strains were obtained from the laboratory of Muhammad Ali, Department of Zoology, Government College University Faisalabad, Pakistan. The antimicrobial potential of ZnO NPs and Al-doped ZnO NCs was evaluated using a modified well diffusion method [21]. Nutrient agar powder (2.8 g) was dissolved in distilled water (100 mL). The media was then poured into sterile petri dishes in a biosafety cabinet and allowed to solidify for 2 hours. The plates were stacked and sealed, then incubated overnight at 37 °C to check for pre-experimental contamination. After 24 h, 100 µL of each strain (*E. coli, B. cereus, S. aureus*, and *P. multocida*) was inoculated onto agar plates at a concentration of 0.6 McFarland scale. Solutions of ZnO NPs and Al-doped ZnO NPs (40 mg/mL) were dispensed into separate plates in the agar medium and incubated at 37 °C for 24 h. Cephalosporin antibiotics were used as a positive control, while DMSO was used as a negative control to determine the validity of the assay. After 24 h, the inhibition zones were measured using a vernier caliper.

3.5. Cell Culturing

Liver cancer (HepG-2), breast cancer (MCF-7), ovarian cancer (SKOV3), and normal liver cell lines were gifted from the laboratory of Muhammad Ali, Department of Zoology,

Government College University Faisalabad (GCUF), Pakistan. HepG-2, MCF-7, SKOV3, and normal liver cells were cultured in a 96-well plate separately, using a culture medium supplemented with FBS (10%), DMEM (2 mM), penicillin and streptomycin (100 IU mL^{-1}), as well as HEPES (20 mM) as a buffering medium. The mixture was maintained under optimal conditions (37 °C, 5% CO_2) in a humidified atmosphere. The standard protocol was adopted as recommended by the Ethics Committee of GCUF. For more details, see our latest published paper [5,20].

3.6. MTT Assay

Using the MTT assay, the in vitro cytotoxicity of ZnO NPs and Al-doped ZnO NCs on the HepG-2, MCF-7, SKOV3, and normal liver cell lines was evaluated. HepG-2, MCF-7, SKOV3, and normal liver cells were separately cultured in a 96-well plate and incubated at 37 °C for 24 h. Then, the cells were incubated for 24 h in varying concentrations of nanocomposites for treatment. After 24 h, 10 microliters (μL) of MTT reagent was added, and the mixture was incubated for an additional 4 h. Subsequently, 150 microliters (μL) of DMSO was used to dissolve the formazan crystals, and the absorbance was determined using a microplate reader [28].

Equation (2) was used to calculate cell viability.

$$\%\ Cell\ Viability = \frac{Absorbance_C - Absorbance_T}{Absorbance_C} \times 100\ \% \tag{2}$$

3.7. Statistical Analysis

The statistical analysis was performed using GraphPad Prism (version 8.0). The mean ± standard deviation of each experiment is shown, with * $p < 0.05$ indicating statistical significance, and ** $p < 0.01$ indicating high statistical significance.

4. Conclusions

Zinc oxide nanoparticles and aluminum-doped zinc oxide nanocomposites were prepared using the coprecipitation method. The XRD outcomes revealed a hexagonal wurtzite phase structure. UV-Vis results determined maximum absorption spectra at 363 nm and ≥363, respectively, confirming the complete formation of ZnO NPs and Al-doped ZnO NCs. Zeta potential values of ZnO NPs, $Zn_{0.75}Al_{0.25}O$, $Zn_{0.5}Al_{0.5}O$, and $Zn_{0.25}Al_{0.75}O$ NCs were found to be +20, ±15, −10, and 13.7 mV, respectively. The SEM micrograph results revealed nonuniform, agglomerated, irregularly shaped particles with small and large spherical grains. With increasing Al^{3+} concentration, prominent changes in particle shape and size were observed. Additionally, an in vitro MTT assay of ZnO NPs and Al-doped ZnO NCs disclosed that they possessed significant anticancer activity toward HepG-2, MCF-7, SKOV3, and normal liver cells, respectively. The microbial activity of ZnO NPs and Al-doped ZnO NCs was also examined by well diffusion assay toward *S. aureus*, *B. cereus*, *E. coli*, and *P. multocida*. However, the $Zn_{0.75}Al_{0.25}O$ NCs exhibited the highest inhibition zone against *B. cereus*. Additionally, $Zn_{0.75}Al_{0.25}O$ NCs not only enhanced the destruction of cancerous cells but also showed low toxicity and high biocompatibility at low concentrations. These findings indicate that $Zn_{0.75}Al_{0.25}O$ NCs could serve as a new inhibitor for cytotoxic and antimicrobial agent, which suggests that it could be a promising candidate for in vivo anticancer and antimicrobial applications.

Author Contributions: Conceptualization, M.A. and M.F.-e.-A.; Writing—original draft, M.A.; Methodology, M.A. and M.T.; Software, M.A., M.T. and F.J.; Validation, M.F.-e.-A. and K.A.D.; Formal analysis, F.J., H.S. and J.R.; Investigation, M.A., M.F.-e.-A. and J.R.; Resources, M.T., H.S. and F.J.; Data curation, M.A.; Writing—review & editing, M.A., M.F.-e.-A., J.R. and K.A.D.; Visualization, M.A. and J.R.; Supervision, M.F.-e.-A. and K.A.D.; Project administration, M.F.-e.-A.; Funding acquisition, K.A.D. All authors have read and agreed to the published version of the manuscript.

Funding: This work was funded by researchers Supporting Project Number (RSP2024R388) King Saud University, Riyadh, Saudi Arabia.

Institutional Review Board Statement: Not applicable.

Informed Consent Statement: Not applicable.

Data Availability Statement: Data is contained within the article.

Acknowledgments: This work was funded by researchers Supporting Project Number (RSP2024R388) King Saud University, Riyadh, Saudi Arabia.

Conflicts of Interest: The authors declare no conflict of interest.

References

1. Chen, X.; Schluesener, H.J. Nanosilver: A nanoproduct in medical application. *Toxicol. Lett.* **2008**, *176*, 1–12. [CrossRef] [PubMed]
2. Asif, M.; Fakhar-e-Alam, M.; Hassan, M.; Sardar, H.; Zulqarnian, M.; Li, L.; Alothman, A.A.; Alangary, A.B.; Mohammad, S. Synergistic response of PEG coated manganese dioxide nanoparticles conjugated with doxorubicin for breast cancer treatment and MRI application. *Arab. J. Chem.* **2024**, *17*, 105958. [CrossRef]
3. Achrol, A.S.; Rennert, R.C.; Anders, C.; Soffietti, R.; Ahluwalia, M.S.; Nayak, L.; Peters, S.; Arvold, N.D.; Harsh, G.R.; Steeg, P.S.; et al. Brain metastases. *Nat. Rev. Dis. Primers* **2019**, *5*, 5. [CrossRef] [PubMed]
4. Asif, M.; Iqbal, W.; Fakhar-e-Alam, M.; Hussain, Z.; Saadullah, M.; Hassan, M.; Rehman, J.; Dahlous, K.A.; Al-Qahtani, N.H. Synthesis and Characterization of Chemically and Green-Synthesized Silver Oxide Particles for Evaluation of Antiviral and Anticancer Activity. *Pharmaceuticals* **2024**, *17*, 908. [CrossRef]
5. Tahir, M.; Fakhar-e-Alam, M.; Asif, M.; Iqbal, M.J.; Abbas, A.; Hassan, M.; Rehman, J.; Bhatti, Q.A.; Mustafa, G.; Alothman, A.A.; et al. Investigation of gadolinium doped manganese nano spinel ferrites via magnetic hypothermia therapy effect towards MCF-7 breast cancer. *Heliyon* **2024**, *10*, e24792. [CrossRef]
6. Hashem, A.H.; El-Sayyad, G.S. Antimicrobial and anticancer activities of biosynthesized bimetallic silver-zinc oxide nanoparticles (Ag-ZnO NPs) using pomegranate peel extract. *Biomass Conv. Bioref.* **2023**, *14*, 20345–20357. [CrossRef]
7. Hassan, S.E.; Fouda, A.; Radwan, A.A.; Salem, S.S.; Barghoth, M.G.; Awad, M.A.; Abdo, A.M.; El-Gamal, M.S. Endophytic Actinomycetes *Streptomyces* spp. mediated biosynthesis of copper oxide nanoparticles as a promising tool for biotechnological applications. *J. Biol. Inorg. Chem.* **2019**, *24*, 377–393. [CrossRef]
8. Menazea, A.A.; Awwad, N.S. Antibacterial activity of TiO_2 doped ZnO composite synthesized via laser ablation route for antimicrobial application. *J. Mater. Res. Technol.* **2020**, *9*, 9434–9441. [CrossRef]
9. Oves, M.; Aslam, M.; Rauf, M.A.; Qayyum, S.; Qari, H.A.; Khan, M.S.; Alam, M.Z.; Tabrez, S.; Pugazhendhi, A.; Ismail, I.M. Antimicrobial and anticancer activities of silver nanoparticles synthesized from the root hair extract of Phoenix dactylifera. *Mater. Sci. Eng. C* **2018**, *89*, 429–443. [CrossRef]
10. Lallo da Silva, B.; Abuçafy, M.P.; Berbel Manaia, E.; Oshiro Junior, J.A.; Chiari-Andréo, B.G.; Pietro, R.C.R.; Chiavacci, L.A. Relationship between structure and antimicrobial activity of zinc oxide nanoparticles: An overview. *Int. J. Nanomed.* **2019**, *14*, 9395–9410. [CrossRef]
11. Seshadri, V.D. Zinc oxide nanoparticles from Cassia auriculata flowers showed the potent antimicrobial and *in vitro* anticancer activity against the osteosarcoma MG-63 cells. *Saudi J. Biol. Sci.* **2021**, *28*, 4046–4054. [CrossRef] [PubMed]
12. Malik, A.R.; Sharif, S.; Shaheen, F.; Khalid, M.; Iqbal, Y.; Faisal, A.; Aziz, M.H.; Atif, M.; Ahmad, S.; Fakhar-e-Alam, M.; et al. Green synthesis of RGO-ZnO mediated Ocimum basilicum leaves extract nanocomposite for antioxidant, antibacterial, antidiabetic and photocatalytic activity. *J. Saudi Chem. Soc.* **2022**, *26*, 101438. [CrossRef]
13. Geffers, C.; Gastmeier, P. Nosocomial infections and multidrug-resistant organisms in Germany: Epidemiological data from KISS (the Hospital Infection Surveillance System). *Dtsch. Ärzteblatt Int.* **2011**, *108*, 87.
14. Waktole, G.; Chala, B. The Role of Biosynthesized Metallic and Metal Oxide Nanoparticles in Combating Anti-Microbial Drug Resilient Pathogens. *J. Biomater. Nanobiotechnol.* **2023**, *14*, 1–22. [CrossRef]
15. Sruthi, S.; Ashtami, J.; Mohanan, P.V. Biomedical application and hidden toxicity of Zinc oxide nanoparticles, Mater. *Today Chem.* **2018**, *10*, 175–186.
16. Baghdadi, N.; Salah, N.; Alshahrie, A.; Koumoto, K. Microwave irradiation to produce high performance thermoelectric material based on Al doped ZnO nanostructures. *Crystals* **2020**, *10*, 610. [CrossRef]
17. Ghotekar, S. Plant extract mediated biosynthesis of Al_2O_3 nanoparticles-a review on plant parts involved, characterization and applications. *Nanochemistry Res.* **2019**, *4*, 163–169.
18. Zhang, B.; Yang, K.; Zhang, K.; Wang, Q.; Wu, N. Migration transformation, prevention, and control of typical heavy metal lead in coal gangue: A review. *Int. J. Coal Sci. Technol.* **2023**, *10*, 85. [CrossRef]
19. Mahmood, A.; Munir, T.; Fakhar-e-Alam, M.; Atif, M.; Shahzad, K.; Alimgeer, K.S.; Gia, T.N.; Ahmad, H.; Ahmad, S. Analyses of structural and electrical properties of aluminium doped ZnO-NPs by experimental and mathematical approaches. *J. King Saud Univ.-Sci.* **2022**, *34*, 101796. [CrossRef]
20. Fakhar-e-Alam, M.; Amjad, I.; Saadullah, M.; Tahir, M.; Jawad, M.; Asif, M.; Atif, M.; Zara, S.; Rashad, M. Antitumor activity of zinc oxide nanoparticles fused with green extract of Nigella sativa. *J. Saudi Chem. Soc.* **2024**, *28*, 101814. [CrossRef]
21. Nanda, A.; Saravanan, M. Biosynthesis of silver nanoparticles from Staphylococcus aureus and its antimicrobial activity against MRSA and MRSE. *Nanomedicine* **2009**, *5*, 452–456. [CrossRef] [PubMed]

22. Rana, G.; Dhiman, P.; Sharma, A. Advances on ZnO hetro-structure as nanoadsorbant for heavy metal removals. In *ZnO and Their Hybrid Nano-Structures: Potential Candidates for Diverse Applications*; Materials Research Forum LLC: Millersville, PA, USA, 2023; Volume 146, pp. 173–201.
23. Yu, Q.; Wang, C.; Zhang, X.; Chen, H.; Wu, M.X.; Lu, M. Photochemical Strategies toward Precision Targeting against Multidrug-Resistant Bacterial Infections. *ACS Nano* **2024**, *18*, 14085–14122. [CrossRef] [PubMed]
24. Eddy, N.O.; Ukpe, R.A.; Garg, R.; Garg, R.; Odiongenyi, A.; Ameh, P.; Akpet, I.N.; Udo, S.E. Review of in-depth knowledge on the application of oxides nanoparticles and nanocomposites of Al, Si and Ca as photocatalyst and antimicrobial agents in the treatment of contaminants in water. *Clean Technol. Environ. Policy* **2023**, 1–32.
25. Guo, M.Y.; Ng, A.M.C.; Liu, F.; Djurisic, A.B.; Chan, W.K.; Su, H.; Wong, K.S. Effect of native defects on photocatalytic properties of ZnO. *J. Phys. Chem. C* **2011**, *115*, 11095–11101. [CrossRef]
26. Theerthagiri, J.; Salla, S.; Senthil, R.A.; Nithyadharseni, P.; Madankumar, A.; Arunachalam, P.; Maiyalagan, T.; Kim, H.S. A review on ZnO nanostructured materials: Energy, environmental and biological applications. *Nanotechnology* **2019**, *30*, 392001. [CrossRef]
27. Gudkov, S.V.; Burmistrov, D.E.; Smirnova, V.V.; Semenova, A.A.; Lisitsyn, A.B. A mini review of antibacterial properties of Al_2O_3 nanoparticles. *Nanomaterials* **2022**, *12*, 2635. [CrossRef]
28. Fatima, N.; ul Hassan, S.M.; Fakhar-e-Alam, M.; Asif, M.; Imtiaz, S.; Anwar, S.; Arooj, H.; Imran, M. Development of a novel fluorescence spectroscopy-based method using layered double hydroxides to study degradation of E. coli in water. *J. Mol. Struct.* **2024**, *1310*, 138248. [CrossRef]
29. Suresh, J.; Pradheesh, G.; Alexramani, V.; Sundrarajan, M.; Hong, S.I. Green synthesis and characterization of zinc oxide nanoparticle using insulin plant (Costus pictus D. Don) and investigation of its antimicrobial as well as anticancer activities. *Adv. Nat. Sci. Nanosci. Nanotechnol.* **2018**, *9*, 015008. [CrossRef]
30. Thongam, D.D.; Gupta, J.; Sahu, N.K.; Bahadur, D. Investigating the role of different reducing agents, molar ratios, and synthesis medium over the formation of ZnO nanostructures and their photo-catalytic activity. *J. Mater. Sci.* **2018**, *53*, 1110–1122. [CrossRef]
31. Srivastava, V.; Choubey, A.K. Kinetic and isothermal study of effect of transition metal doping on adsorptive property of zinc oxide nanoparticles synthesized via green route using Moringa oleifera leaf extract. *Mater. Res. Express* **2020**, *6*, 1250i7. [CrossRef]
32. Dhamodharan, P.; Manoharan, C.; Bououdina, M.; Venkadachalapathy, R.; Ramalingam, S. Al-doped ZnO thin films grown onto ITO substrates as photoanode in dye sensitized solar cell. *Sol. Energy* **2017**, *141*, 127–144. [CrossRef]
33. Sanguanprang, S.; Phuruangrat, A.; Thongtem, T.; Thongtem, S. Preparation of visible-light-driven Al-doped ZnO nanoparticles used for photodegradation of methylene blue. *J. Electron. Mater.* **2020**, *49*, 1841–1848. [CrossRef]
34. Sadiqa, A.; Rasul, A.; Hassan, M.; Sultana, S.; Jabeen, F. Identification of Novel Natural Inhibitors to Human 3-Phosphoglycerate Dehydrogenase (PHGDH) for Cancer Treatment. *Molecules* **2022**, *27*, 6108. [CrossRef] [PubMed]
35. Rani, V.; Verma, Y.; Rana, K.; Rana, S.V.S. Zinc oxide nanoparticles inhibit dimethylnitrosamine induced liver injury in rat. *Chem.-Biol. Interact.* **2018**, *295*, 84–92. [CrossRef] [PubMed]
36. Balaji, S.R.; Mandal, B.K. Synthesis, Characterization of ZnO and Al_2O_3 Nanoparticles and Its Application in Chromium Remediation Studies. *Asian J. Chem.* **2017**, *29*, 2459–2462. [CrossRef]
37. Musleh, H.; Zayed, H.; Shaat, S.; Al-Kahlout, A.; Tamous, H.; Issa, A.; Asad, J.; AlDahoudi, N. Enhancement of the performance of dye-sensitized solar cells using sensitized zinc oxide nanoparticles by rhodamine B dye. *Egypt. J. Chem.* **2019**, *62*, 111–123. [CrossRef]
38. Aldalbahi, A.; Alterary, S.; Ali Abdullrahman Almoghim, R.; Awad, M.A.; Aldosari, N.S.; Fahad Alghannam, S.; Alabdan, A.N.; Alharbi, S.; Alateeq, B.A.M.; Al Mohsen, A.A.; et al. Greener synthesis of zinc oxide nanoparticles: Characterization and multifaceted applications. *Molecules* **2020**, *25*, 4198. [CrossRef]
39. Krishnamoorthy, R.; Athinarayanan, J.; Periyasamy, V.S.; Alshuniaber, M.A.; Alshammari, G.; Hakeem, M.J.; Ahmed, M.A.; Alshatwi, A.A. Antibacterial mechanisms of zinc oxide nanoparticle against bacterial food pathogens resistant to beta-lactam antibiotics. *Molecules* **2022**, *27*, 2489. [CrossRef]
40. Ahmed, B.; Solanki, B.; Zaidi, A.; Khan, M.S.; Musarrat, J. Bacterial toxicity of biomimetic green zinc oxide nanoantibiotic: Insights into ZnONP uptake and nanocolloid–bacteria interface. *Toxicol. Res.* **2019**, *8*, 246–261. [CrossRef]
41. Allayeith, H.K. Zinc-Based Nanoparticles Prepared by a Top-Down Method Exhibit Extraordinary Antibacterial Activity Against Both Pseudomonas aeruginosa and Staphylococcus aureus. Ph.D. Thesis, Kent State University, Kent, OH, USA, 2020.
42. Anjum, S.; Hashim, M.; Malik, S.A.; Khan, M.; Lorenzo, J.M.; Abbasi, B.H.; Hano, C. Recent advances in zinc oxide nanoparticles (ZnO NPs) for cancer diagnosis, target drug delivery, and treatment. *Cancers* **2021**, *13*, 4570. [CrossRef]

pharmaceuticals

MDPI

Article

Antimicrobial and Oxidative Activities of Different Levels of Silver-Exchanged Zeolites X and ZSM-5 and Their Ecotoxicity

Elitsa L. Pavlova [1,*], Elena P. Nenova [2], Lyubomira D. Yocheva [3], Iliana A. Ivanova [2] and Peter A. Georgiev [1,*]

1 Faculty of Physics, Sofia University "St. Kliment Ohridski", 5 James Bourchier Blvd, 1164 Sofia, Bulgaria
2 Faculty of Biology, Sofia University "St. Kliment Ohridski", 8 Dragan Tsankov Blvd, 1164 Sofia, Bulgaria; elena_pnenova@abv.bg (E.P.N.); ilivanova@abv.bg (I.A.I.)
3 Faculty of Medicine, Sofia University "St. Kliment Ohridski, 1 Kozyak Str, 1407 Sofia, Bulgaria; lyubomirayocheva@abv.bg
* Correspondence: elli_pavlova@abv.bg (E.L.P.); pageorgiev@phys.uni-sofia.bg (P.A.G.); Tel.: +359-887982172 (E.L.P.); +359-28161470 (P.A.G.)

Abstract: Objectives: The antimicrobial, oxidative activities, and ecotoxicity of synthesized silver-loaded zeolites (X and ZSM-5(MFI), Si-to-Al ratios 12 and 25) were studied, linking antimicrobial properties to material structure and released active silver species. Methods: The materials were characterized by SEM, EDX, TEM, and XRPD. All materials, with a silver content of 1–3%wt for the Ss and about 35%wt for the X-zeolites, were tested against *Escherichia coli* and *Staphylococcus aureus*. Redox activity was studied in physiological (pH 7.4/37 °C) and optimal (pH 8.5/37 °C) conditions in chemiluminescent model systems. In the ecotoxicity tests, we used *Daphnia magna*. Results: A proportional correlation was observed between the bactericidal effect of and the silver content in the zeolites. AgX with a Si/Al ratio of ~1.23 and 35% silver showed a higher antimicrobial efficiency, particularly against Gram-negative *E. coli* versus Gram-positive *S. aureus*. The concentration thresholds were as follows: AgXas had a bactericidal effect at 0.003 g/L^{-1}, with an MIC at 0.0015 m/L^{-1} for *E. coli*; SA25-Ag, AgXcl, AgXrc had a bactericidal effect at 2.5 g/L^{-1}. The bacteria were more resilient than *Daphnia magna*, which showed a 90–100% lethality at Ag–zeolite concentrations of 0.00625 to 0.0125 g/L^{-1}. AgXas and AgXrc demonstrated strong reactive oxygen species generation at both the physiological and optimal pH, explaining their bactericidal effects. In general, the tested materials showed an inhibition of the generated reactive oxygen species depending on the model system and conditions. Conclusions: The silver species leached from the new materials explain their higher oxidation and bactericidal activity. While suitable for stringently controlled biological applications, their release into the environment, in concentrations higher than 0.01g/L^{-1}, should be avoided.

Keywords: Ag species; antimicrobial activity; ecotoxicity; luminescence; reactive oxygen species; zeolites

1. Introduction

Antimicrobial resistance (AMR) has been identified as one of the top public health and societal welfare threats from a global and long-term perspective [1]. The observed persistent increase in antibiotic resistance in pathogens and the substantial environmental pollution from antibiotics request an urgent scientific response by finding new efficient and environmentally friendly antimicrobial drugs. Well-known drugs with environmentally friendly potential include metallic silver and silver-containing compounds, for example, in the form of silver salts, silver-ion-releasing molecules, and charged or neutral clusters of Ag nanoparticles. Their mechanical, physical, and chemical interactions with cells are so fast and intensive that pathogens hardly ever develop resistance. Not surprisingly, various silver species have found applications in washing mixtures, cosmetics, nutrient supplements, surgical blades, socks, shoes, food containers, teeth prostheses, mobile phones, catheters, wound dressings, color coatings, water purification systems, and more [1–3]. Silver itself and silver salts, including nanoparticles and other sources of ionic silver species (e.g., surfaces of oxides, zeolites, and other minerals) have been used since ancient times. They have been applied to various medical conditions—curing wounds, diarrhea, and other illnesses. The Food and Drug Agency (FDA) of the United States approved silver for medical applications in the 1920s [2,3].

The silver concentration in foods, on the other hand, is regulated in many countries. The European Commission determined the concentration of silver ions of 50 ppb as being safe in foods, but the FDA limited the levels of zeolite silver to 5% in food packages. The dissolution of silver and its quantity in the environment depends on the ambient conditions and vehiculum. One possible natural carrier of different silver forms are microporous aluminosilicate materials, namely zeolites.

Zeolites are crystalline alumino-silicate frameworks of interconnected TO_4 (T = Si, Al) tetrahedra with a composition that can be given as $Si_{1-n}Al_nO_2xH_2O$. The addition of Al^{3+} to the structure results in a negatively charged skeleton. Positively charged, extra-framework ions, usually 1+ or 2+, are consequently included, to restore the neutrality of the structure. Such charge-compensating species could simply be protons, i.e., H^+, NH_4^+, alkali, alkaline earth or transition metal ions, heavier metal ions, or charged clusters of atoms. The Si/Al ratio of the framework can vary from 1, as in zeolite A, also known as Linde Type A (LTA), up to infinitely high numbers, as in pure siliceous zeolites like Silicalite-1 from the mobile five (MFI) type. As the Si/Al ratio increases, there are fewer charge-compensating sites available in the framework, and consequently, there is a lower loading capacity for extra framework ions. The most up-to-date atlas of zeolite skeleton types lists 133 structures, all with different topologies [4].

Some zeolites occur in nature, and 40 different structures of those are known. One of the most studied and widespread naturally occurring zeolites, for instance, is clinoptilolite, which has found various key environmental applications as a drug carrier, a heavy metals removal agent in aqueous solutions, and as a low-cost green composite magnetic dye adsorption and removal reagent [5–7]. On the other hand, many new zeolitic structures have only been synthesized in laboratory conditions, usually through hydrothermal processes, with or without the assistance of organic and/or inorganic templates [8]. In all of the microporous aluminosilicate zeolite structures, monovalent ions, e.g., Ag^+, are readily exchanged and the corresponding Ag forms are produced. Furthermore, different silver species, including charged clusters [9] and neutral nanoparticles [10], can be formed inside the carrier material's pores and surfaces under controlled conditions. Thus, Ag-exchanged zeolites, including some post-treatments, represent a convenient system to study the antimicrobial activity of silver as the active species in different forms and concentrations. Presumably, this activity must depend on the readiness of the material to release the active species in solutions where it interacts with the target. The ion release characteristics are known to vary with the host scaffold topology and the types of ions in the surrounding environment, e.g., in K^+-rich broth, Ag^+ has been found to escape more readily from clinoptilolite than from zeolite A or zeolite X/Y [11]. Silver species are polarized by the strong electric fields inside the framework, and this leads to a strong attraction between Ag^+ and the zeolite structure. In general, all zeolites show a high selectivity for Ag^+, and as the Si/Al ratio for a particular framework increases, the selectivity for Ag^+ tends to be higher [12].

In recent years, after a period of faded interest due to the wide use of organic medicines such as microbially produced antibiotics, the interest in silver-containing substances has attracted a constantly increasing amount of interest. Some excellent reviews and research works conducted in different labs are available, e.g., the one by Dutta and Wang (2019) [10]. When summarizing these studies in a few hundred works, consistent conclusions appeared to be difficult to draw. It seems there is an expected trend in the antimicrobial activity, increasing with the increase in the silver concentration, but exceptions also exist. Similarly, in various materials, either silver ions or silver nanoparticles have shown higher activities, while metallic silver particles are most often regarded as less active [10]. It should be mentioned that in many of the research works covered, a detailed and sound characterization of the state of silver was hard to achieve; moreover, it is well known that the relatively large size of monovalent silver ions, comparable to the size of the zeolite silicate rings, makes them quite mobile, thus experiencing a rather smooth potential surface inside the zeolite channels. This leads to a facile reduction and formation of either charged clusters or nanoparticles, under broad near-ambient conditions, during controlled treatments, e.g., vacuum and slightly elevated temperatures, reducing solutions. Accidental uncontrolled local overheating, however, may also occur during sample characterization procedures and/or storage, by irradiation with light, UV, higher-energy electromagnetic radiation e.g., X-rays, electron beams. For instance, Krishnani et al. (2012) [13] investigated the antimicrobial activity of Ag^+-exchanged zeolite A (39.4 wt% Ag, assuming the pure ionic form of the silver species) against *E. coli*, *Vibrio harveyi*, *V. cholerae*, and *V. parahaemolyticus*. The minimum inhibitory concentration (MIC) for *E. coli* and *V. harveyi* was 40 μg/mL, while for *V. cholerae* and *V. parahaemolyticus*, the MIC was higher, at 50 and 60 μg/mL, respectively (after 48 h contact, 10^9 CFU/mL). *V. cholerae* and *V. parahaemolyticus* have thicker cell walls and therefore require higher concentrations of Ag^+–zeolite. The presence of ammonia was found to increase the antimicrobial activity of the Ag^+–zeolite and was attributed to the toxicity of NH_3.

Kaali et al. (2013) studied the ion exchange isotherms of single, binary, and ternary mixtures of Ag^+, Cu^{2+}, and Zn^{2+} with zeolite A (Zeomic). Ag^+ shows a theoretical exchange of almost 100% [14]. This observation confirms the easy incorporation of silver into the zeolite material.

A lower Si-to-Al ratio often favors the auto reduction of silver and the formation of silver particles, which has been pointed out as the main reason for decreased antimicrobial activity as in the comparative study on Ag-X (Si/Al about 1.3) and Ag-Y (the same zeolitic topology but Si/Al about 2.5) [15]. This study found that Ag^+-ion-exchanged zeolite X (9.8 wt% Ag) and Ag^+-ion-exchanged zeolite Y (9.7 wt% Ag), both with micron-sized zeolite particles, show the following MIC values for *E. coli* and *Bacillus subtilis*: 300 μg/mL for AgX and 200 μg/mL for AgY (24 h exposure). For the yeasts *Saccharomyces cerevisiae* and *Candida albicans*, the MICs were 1000 μg/mL for both Ag-loaded zeolite X and Y (42 h exposure) [15]. The lower MIC value for bacteria treated with AgY, compared to AgX, was explained as arising from the metallic Ag (detected by spectroscopy) in AgX and the thinner cell wall. The thicker yeast cell envelope determines the higher concentrations of Ag^+ zeolite needed for growth inhibition [15]. Similar results were obtained by Kędziora et al. (2018) [16], which appear to support the view that metallic silver particles manifest their antimicrobial activity by releasing ionic silver species, and the minimal inhibition concentration of Ag^+ for Gram-negative bacteria is lower than that for Gram-positive bacteria because of their thicker cell walls.

In other scenarios, the cooperative effect of co-present counter ions like Zn^{2+} was studied by Ferreira et al. (2016) [17]. Their studies on bimetallic ZnAgY, tested for its activity against *E. coli* (CECT423), *B. subtilis* (4886), *C. albicans* (JGC 3456T), and *S. cerevisiae* (BY 4741), employed samples with different Ag/Zn loadings. ZnAgY with 3.03 wt% Zn and 6.04 wt% Ag were found to be most efficient, showing MICs of 100, 100, 300, and 300 μg/mL against *E. coli*, *B. subtilis*, *C. albicans*, and *S. cerevisiae*, respectively. For AgY with 9.70 wt% Ag loading, the comparable MIC values were higher, at 200 μg/mL for the bacteria and 1000 μg/mL for the yeast species, indicating enhanced activity with both ions present. The distribution of Zn^{2+} and Ag^+ in the zeolite was found to be uneven, suggesting that Ag^+ in particular is preferably found in the supercage sites, being large enough. A synergistic effect between Ag^+ and Zn^{2+}, however, was evident. Notably, and of great practical importance, the results of antimicrobial assays have been found to be reproducible even after two years of sample storage, which proves the stability of the zeolite [17].

Furthermore, the antimicrobial activity could be greatly influenced by the rate of the release of the silver species from the zeolite carrier. Parameters that could directly influence this property include the particle/crystallite size in which silver is accommodated, as well as possible surface morphology and/or the presence of particle structural hierarchy. The MIC and minimum bactericidal concentration (MBC) for Ag^+-exchanged zeolite X, with different morphologies, have been investigated by Chen et al. (2017) [18]. Submicron aggregates of 100–700 nm, containing ~24 nm primary particles, were compared to ~2 μm particles. The nanostructured zeolite had a hierarchical structure with both micro- and mesopores. The amount of Ag^+ loaded in both morphologies was similar, around 20–22 wt%. Not surprisingly, the Ag^+ release characteristics in a flow cell (with Na^+ eluent) showed a faster and larger amount of Ag^+ release from the nanozeolite. However, in a 24 h test, micron-sized and nanozeolites showed MIC and MBC values against *S. aureus* (*MRSA*) of 16 and 32 μg/mL, respectively. In short term experiments, e.g., 10 min, a faster Ag^+ release from the nanozeolite was evident, with 400 μg/mL of the zeolite killing *MRSA* (10^8 CFU/mL) in 3 min versus the 7 min needed for comparable concentrations of the micron-sized zeolite. Ag^+ nanozeolite was ineffective in the inhibition of the *MRSA* biofilm; it appeared to aid in the film's formation.

Cytotoxicity against human skin epithelial cells (WM-115) requires >128 μg/mL Ag^+-hierarchical zeolite, while human skin fibroblasts (Detroit 551) and monocytes (U-937) require concentrations of 64 μg/mL to significantly reduce the viability. These cytotoxic concentrations have been found to be significantly higher (2–4 times) than the MIC/MBC concentrations of *S. aureus* (*MRSA*) [19]. However, all other authors present a better sensitivity of the eukaryotes and a better cytotoxicity of the nanocrystalline silver forms.

Contrary to the above findings from Chen et al. (2017) [18], Youssef et al. (2017) [20], who compared the zeolites analcime, faujasite, and zeolite A at both the micron- and nano-scales (~200 nm), found that during Ag^+ exchange, the nanocrystalline zeolites degraded, but micron-sized zeolites were stable. Some very high levels of Ag^+ were found in faujasite (48 wt%) and analcime (50 wt%) from the micron-sized zeolites and zeolite A containing 24.6 wt%. In those studies, the agar plate diffusion method showed antimicrobial activity in the following order, analcime > faujasite > zeolite A, against *S. aureus*, *P. aeruginosa*, *C. albicans*, and *A. niger*, but no difference in the antimicrobial activity was observed between micron-sized and nano-sized zeolites [20].

Yet again, to compare the activities of silver ionic species vs. nanoparticles, Lv et al. (2009) [21] investigated the bactericidal activity of two types of titano-silicate: one loaded with silver ions and another with silver nanoparticles, with approximately equal weight percentages for the silver species in both materials. The average size of the silver nanoparticles was found to be below 5 nm. It turned out that the highly porous titano-silicate, loaded with silver ions released silver ions faster and in higher concentrations, but they were less active towards the bacteria than the nanoparticle-loaded titano-silicate. Consequently, they concluded that silver nanoparticles show a stronger bactericidal effect than ionic species [21].

The modification of zeolite A (NaA) by impregnation with different concentrations of silver ions (Ag^+) or silver nanoparticles (AgNPs) for antibacterial applications was also investigated by Jiraroj et al. (2014) [22]. The Ag^+-NaA composites were prepared by ion exchange with different concentrations of silver nitrate solution (25–200 mg/L), while AgNPs were subsequently prepared by reducing the impregnated Ag^+ with sodium borohydride. The NaA structure was not significantly affected by the addition of Ag^+ or AgNPs. The distinct nature of the silver species in the materials under study was manifested in the measured UV spectra as a broad absorption peak at 394 nm, characteristic of the AgNPs in the AgNP-NaA composites that were missing in the Ag^+-NaA material. The Ag^+-NaA and AgNPs-NaA composites both showed dose- and time-dependent antimicrobial activity against the Gram-positive *S. aureus* and Gram-negative *E. coli* bacteria. In these studies, the ionic silver species, i.e., Ag^+, showed a substantially higher antibacterial efficiency against *E. coli* than the AgNPs-NaA composites. Interestingly, for the same NaA zeolite materials with the highest silver loadings and at 3 h tests, the AgNPs-NaA composites were more effective than the Ag^+-exchanged NaA against *S. aureus* [22].

The antimicrobial activities of low-silver-content samples, presumably in their ionic form in a zeolite ZSM-5 carrier, which is a low-Al-content-type zeolite (i.e., high silica), were studied by Lalueza et al. (2010) [23]. The low Al content of the zeolite naturally resulted in low Ag^+ loadings, e.g., 0.2 wt%. Those Ag^+ -ZSM-5 samples showed lg4 suppression within 24 h for *S. aureus 9213* (10^9 CFU/mL); 25,000 ppm Ag^+ was released in a 300 g/L^{-1} zeolite culture medium with 0.23 wt% Ag within the first 4–6 h of exposure. During longer timeframes (24 h), although more Ag^+ was released, the biofilm formed (observed by SEM) around the zeolite hindered the migration of Ag^+, which reduced the antimicrobial activity. These studies indicate that not only is the amount of silver is important, but so is the active species release kinetics as well as the susceptibility towards biofilm formation. Biofilm formation has also been investigated in other studies, addressing the applicability of silver-loaded materials for marine applications, namely as antifouling agents. For instance, Yee et al. (2015) [24] reduced Ag^+-ZSM-5 (1–5 μm zeolite with Ag content 0.8–10 wt%) with citrate to obtain AgNPs (~1.48 nm). The adherent bacterial biomass of *H. pacifica*, a common marine-fouling organism, decreased by 81% in the presence of 10 wt% Ag^+. They observed that the formation of biofilm was greatly inhibited at higher Ag loadings of the zeolite, up to the highest achievable loadings of 10 wt%. Ag zeolite also inhibited the growth of other marine microalgae [24].

Singh et al. (2015) [25] noted that dispersed Ag metal on zeolite crystals can react with H_2O_2, producing O_2, which causes these crystals to move in a liquid medium. These Ag metal zeolites can be ion-exchanged with Ag^+ and have been found to exhibit antibacterial properties against *E. coli*. This self-induced movement resulting in increased contact with the bacteria is of interest in this study and could be used to understand the antimicrobial activity of the material, as the microorganisms form H_2O_2 during their metabolic activity [25].

As seen from the above retrospective review of previous works on the bactericidal activity of silver species, there is not yet agreement on the answers to the primary questions regarding which is the more active type of silver species, nor on the dependency of this activity on the silver concentration. No preferable type of active species carrier has been put forward either. Hence, clear recipes for preparing an efficient and reproducible antibacterial material based on silver or other metal ions or other metal species remain hard to devise.

The generation of reactive oxygen species (ROS) and prooxidant activity is an important biomarker of antibacterial activity and, on the other hand, is also demonstrative of the antioxidant effects of synthesized materials. It is very important to measure the kinetics of free radical generation in order to estimate the pro- and antioxidant properties of a material [26,27], keeping in mind that most materials change drastically their properties at the nanoscale or depending on the pH level. This information is very important to elucidate the mechanisms of action of new materials on micro- and macro-organisms.

The chemiluminescent assay is a fast and extremely sensitive method for use in such studies. It is applied to monitor the dynamics of free radical generation and to determine the prooxidant/antioxidant activity of various substances. According to this method, probes are applied that enhance the luminescent emission of these substances in order to achieve comparable signals. These reactions emit light within the range of 480–580 nm and can be harnessed to assess the quantum yield of the excited state products generated [27,28].

In the present work, we studied the bactericidal activity of the silver species originating from synthesized materials in their two main primary forms: silver nanoparticles with crystallite sizes in the range of 5–10 nm and monovalent silver ions, both contained in various microporous zeolite materials. The redox activity of these materials was tested in free radical oxidation reactions at pH 7.4 (physiological) and pH 8.5 (optimal) in chemiluminescent model systems. Finally, the environmental safety of all studied materials was tested via an acute toxicity test against young *Daphnia magna*.

2. Results and Discussion

The main physicochemical properties of the studied materials are summarized in Table 1. The silver chemical content was evaluated by the EDX analyses. The silicon-to-aluminum ratio is given as provided by the zeolite suppliers. The MFI and FAU topologies were verified by the XRPD analyses. The corresponding powder diffraction patterns are shown in Figures S1–S5, confirming that the main zeolitic structures remain intact after all treatments. However, a metallic silver FCC-phase was detected due to the presence of the corresponding (111) and (200) reflections in the PXRD patterns of both the reduced and reactivated silver-exchanged zeolite X materials, whereas the silver-exchanged ZSM-5 zeolites showed a pure MFI phase. Further inspection of the materials by SEM/EDX confirmed that some large metallic silver particles are present in the AgX materials (Figure 1).

Table 1. Materials' description, including sample batch notation, the parent zeolite Si-to-Al ratio, silver content, and the observed silver forms: isolated ionic species and metallic silver nanoparticles. The ivory color indicated for the zeolite–X materials comes from the binder material and is not specific to silver-exchanged X-zeolites, which are white.

Sample Material	Si/Al	Ag Cont. (EDX) %wt	Ag Form	Zeolite Type	Framework Type
AgXas	1.23	35	Ag^+, some Ag^0 (particles), ivory	X	FAU
AgXcl	1.23	35	mostly metallic Ag nanoparticles, black, some larger Ag metal clusters	X	FAU
AgXrc	1.23	35	Ag^+, some metallic nanoparticles, a few larger Ag particles	X	FAU
SA12-Ag	12	3.5	Ag^+, some Ag^0 (particles), white	ZSM-5	MFI
SA25-Ag	25	2.5	Ag^+, a few Ag^0 (particles), white	ZSM-5	MFI
SA25-$Ag_{0.33}$	25	1.2	Ag^+, white	ZSM-5	MFI

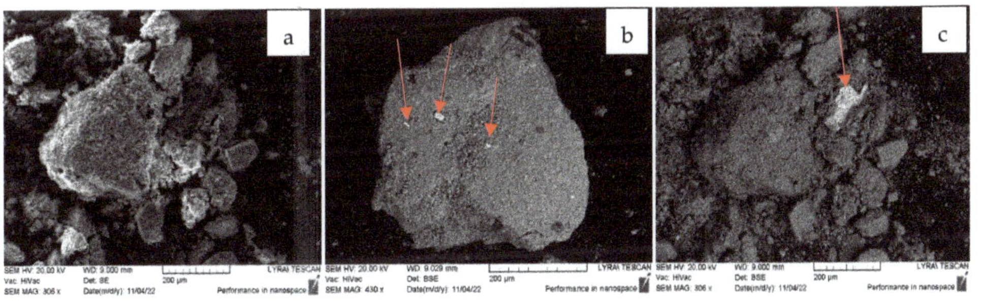

Figure 1. AgX and $Ag_{84}Na_2[(AlO_2)_{86}(SiO_2)_{106}] \times H_2O$, with a nominal 35%wt silver content, which was determined to be about 36% by EDX, and was up to 68% within the area of the large silver particle seen in the micrograph: (**a**) the as-received AgX material, sample AgXas; (**b**) the vacuum/heat reduced AgXcl; (**c**) the regenerated AgXrc material.

A deeper insight from TEM imaging (Figures 2–4) revealed that silver nanoparticles, 5–10 nm in diameter, are abundant in the reduced AgX material; some were still found in the regenerated material, but nanocrystalline particles of similar sizes were also found in the as-received AgX zeolite. A minor amount of hexagonal-phase silver could only be detected in some of the TEM images and only in the AgX zeolite sample (Figures 2 and 4), but not in the sample of the same zeolite material regenerated in air. We must also point out that the electron beam in microscopy experiments, under vacuum, locally heats up the samples under investigation, and this also might be sufficient to produce some reduced silver species, e.g., charged clusters and/or metallic nanoparticles.

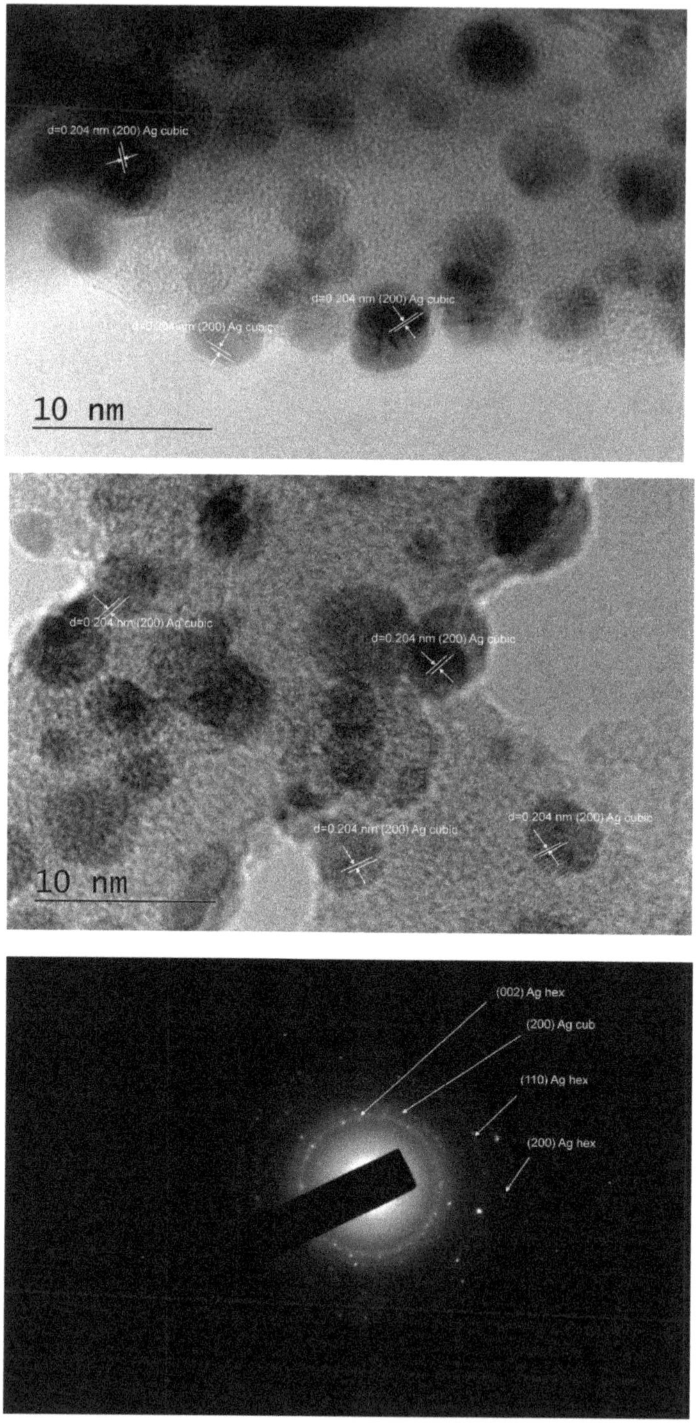

Figure 2. TEM images and analyses of the reduced Ag-exchanged zeolite X, sample AgXcl.

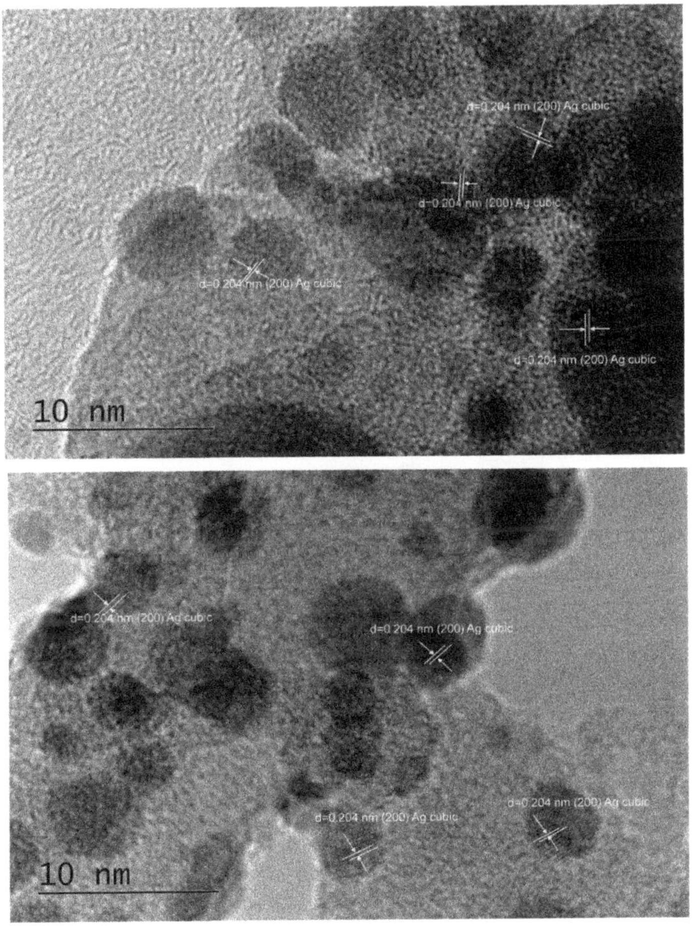

Figure 3. TEM images and analyses of the regenerated Ag-exchanged zeolite X, sample AgXrc.

Notably, during the PXRD experiments with the as-received AgX material, we observed a dark stripe on the sample after the measurements, exactly on the irradiated sample area, which is evidence that the intense X-ray beam causes silver reduction under ambient conditions. This should not be surprising, considering that silver exchange procedures are usually performed in the dark and samples, when stored for prolonged periods, are also kept in the dark. To avoid this effect, at least upon visual inspections, the X-ray tube's current was reduced down to 20 mA, and the reported diffraction patterns were measured for a total exposure time of about 1 h. However, one must generally bear in mind that, in zeolitic materials, isolated silver ionic species are relatively easily reduced to larger size species, such as metallic nanoparticles. Hence, experimental methods that involve the use of intense light beams directed to the sample and techniques involving uncontrolled local overheating of probed sample regions may not be considered truly nondestructive. These effects were not as pronounced for low silver contents, e.g., in high-silica zeolites like the ZSM-5 zeolites, in which we did not detect silver nanoparticles under equivalent characterization conditions. The corresponding PXRD patterns are shown in Figures S1–S5. However, some scarce silver nanoparticles were observed in the highest silver concentration, higher Al-content ZSM-5 zeolite sample SA12-Ag (Figure 5). The cause for the appearance of these metallic silver nanoparticles is unclear at present. It may be due to the electron beam in the TEM microscope, but may have also formed during the exchange procedure in the Ag-solution.

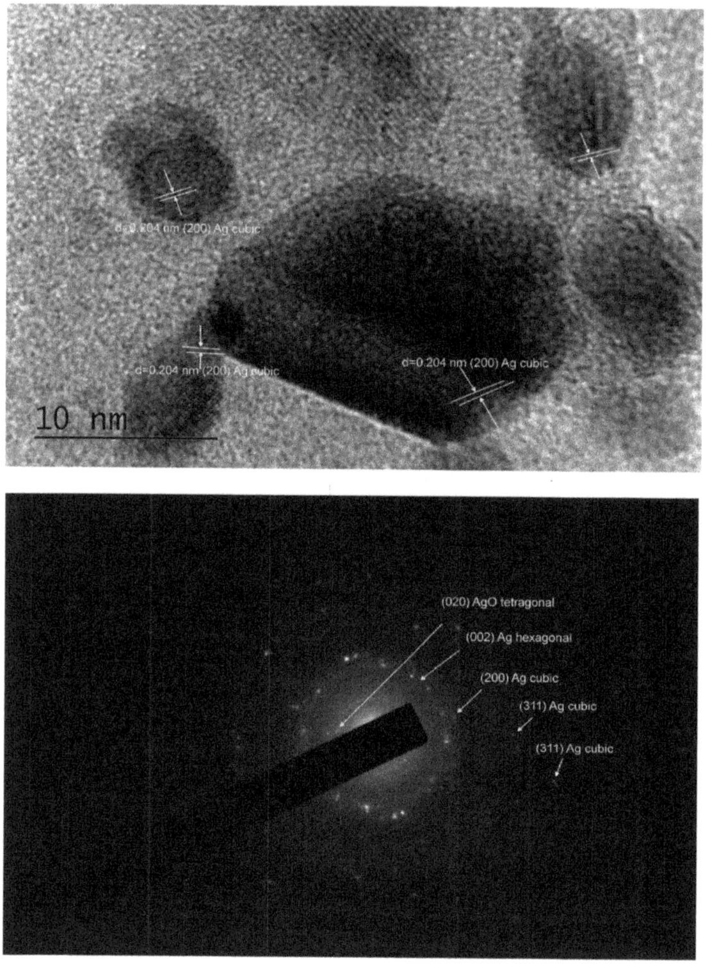

Figure 4. TEM images and analyses of the as-received Ag-exchanged zeolite X, sample AgXas, where some nanocrystalline silver particles could also be observed.

As seen in the TEM images above (Figures 2–4), all of the AgX samples contained metallic nanoparticles, which were quite abundant in the black AgXcl reduced material, with a particle size predominantly under 5 nm. As clearly indicated by the visually observed corresponding powders' colors (Figure 6), the as-received material must contain very few such metallic particles in contrast to the black AgXcl material (Figure 6). We spotted some in the synthesized material (Figure 4), of even larger than 5 nm size, e.g., about 10 nm. Although not abundant, such silver nanoparticles could even be found in the ZSM-5 material, with 3.5%wt silver-exchanged material. Evidently, the presence of silver nanoparticles in most silver exchanged materials is confirmed.

Figure 5. TEM images of SA12-Ag, 3.5%wt silver loading material, in which some scarce nanocrystalline silver was found too.

Figure 6. Visual appearance of the three AgX powder materials: from the left to the right AgXcl, AgXas, and AgXrc.

In our case, apart from the different zeolitic carrier material types, the investigated materials can also be distinguished by their silver content and form, as described in the third column of Table 1. Thus, we distinguish our materials mainly by the total silver content, once it had been controlled by the different Si-to-Al ratios of the X and MFI zeolite topologies, then by the different exchange protocols of the ZSM-5(MFI) materials, and third by the dominant presence of silver species: (i) mainly silver nanoparticles in the AgXcl material; (ii) mainly Ag^+ in the as received AgX material (AgXas) with a minority of silver nanoparticles; (iii) Ag^+ in the AgXr material, with abundant silver nanoparticles, including all three types with an equal total silver content; and (iv) Ag-exchanged ZSM-5 zeolites with different Ag contents, between 1 and 3.5%wt, that were an order of magnitude lower than that in the X-type zeolite samples, which led to a much larger separation of Ag ions within the framework and hence a much lower chance for silver reduction and nanoparticle formation. The corresponding PXRD patterns in Figures S3–S5 show only the pure MFI phases for all of the ZSM-5 materials, with no sign of metallic silver, in contrast to those for the AgX type of materials (Figures S1 and S2).

The chemiluminescent assay produced the following results:

- System I (Figure 7)

The Fenton's reaction between Fe^{2+} ions and H_2O_2 produces highly reactive, short-living $^{\cdot}OH$ and $^{\cdot}OOH$ radicals. Usually, the achieved chemiluminescent emission is much higher than the one from other mixtures:

(1) $Fe^{2+} + H_2O_2 \rightarrow Fe^{3+} + {}^{\cdot}OH + {}^{-}OH$

(2) $Fe^{3+} + H_2O_2 \rightarrow Fe^{2+} + {}^{\cdot}OOH + H^+$

The registered signals from this reaction, with the application of various synthesized materials, showed that, at pH 8.5 all reactions possess similar kinetics showing a plateau, which slightly increases with time. This is indicative of the silver species released in the solution with time. The sample containing control Ag nanoparticles (30–50 nm) emitted the highest signal (100%) as expected, with a maximum RLU of ~165,000. The AgXas sample showed about a 40% decrease in emission, followed by SA25-Ag at 70%, SA25-$Ag_{0.33}$ at 77%, SA12-Ag at almost 98%, and AgXcl at 99%. The AgXrc sample exhibited a 95% decreased emission in comparison to the control Ag nanoparticles.

At pH 7.4 (physiological), the kinetic curves were not smooth and had various fluctuations. AgXas demonstrated a prooxidant activity, that was 16% higher than the control Ag nanoparticles. All the other tested materials exhibited a suppressed emission in comparison to the control activity: SA25-Ag—20%; SA25-$Ag_{0.33}$—33%; SA12-Ag—35%; AgXcl—40%; AgXrc—33%.

The reactions are indicative of and comparative with the level of silver ions or nanoparticles released by the various samples and their effect on the free radical generation in comparison to the control with the pure Ag nanoparticles.

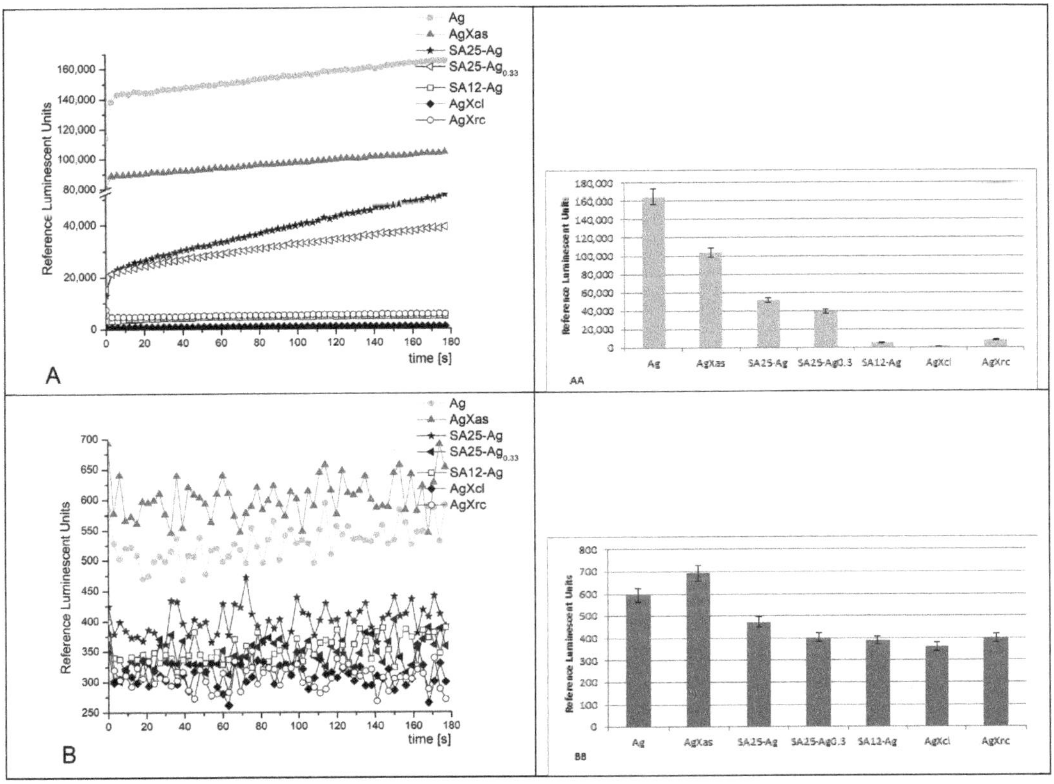

Figure 7. Fenton's chemiluminescence induced by ·OH and ·OOH radicals at pH 8.5 (**A**,**AA**) and pH 7.4 (**B**,**BB**), and the effect of the synthesized materials ($p \leq 0.05$); reagents: $FeSO_4$ (5×10^{-4} M) + H_2O_2 (1.5%) + lucigenin (10^{-4} M) at 37 °C.

- System II (Figure 8)

In this system, a strong oxidant, hydrogen peroxide, serves both as an oxidizing agent and an ROS. At alkaline pH, the kinetic curves were smooth, and the signal in all the reactions was increasing in time. The mixture, containing pure silver nanoparticles (control), emitted most light, as expected. All of the other tested samples demonstrated lower signals as follows: AgXas—10%, SA25-Ag—more than 60%, SA25-$Ag_{0.33}$—97%, SA12-Ag—97%, AgXcl—99%, and AgXrc—97%.

At pH 7.4, AgXas was as active as the control. All of the other tested samples presented a lower luminescent emission in comparison to the control with pure Ag nanoparticles, as follows: SA25-Ag—more than 93%; SA25-$Ag_{0.33}$—40%; SA12-Ag—70%; AgXcl—50%; AgXrc—78%. In general, the registered emissions very slightly increased with time, showing the release of reactive Ag species.

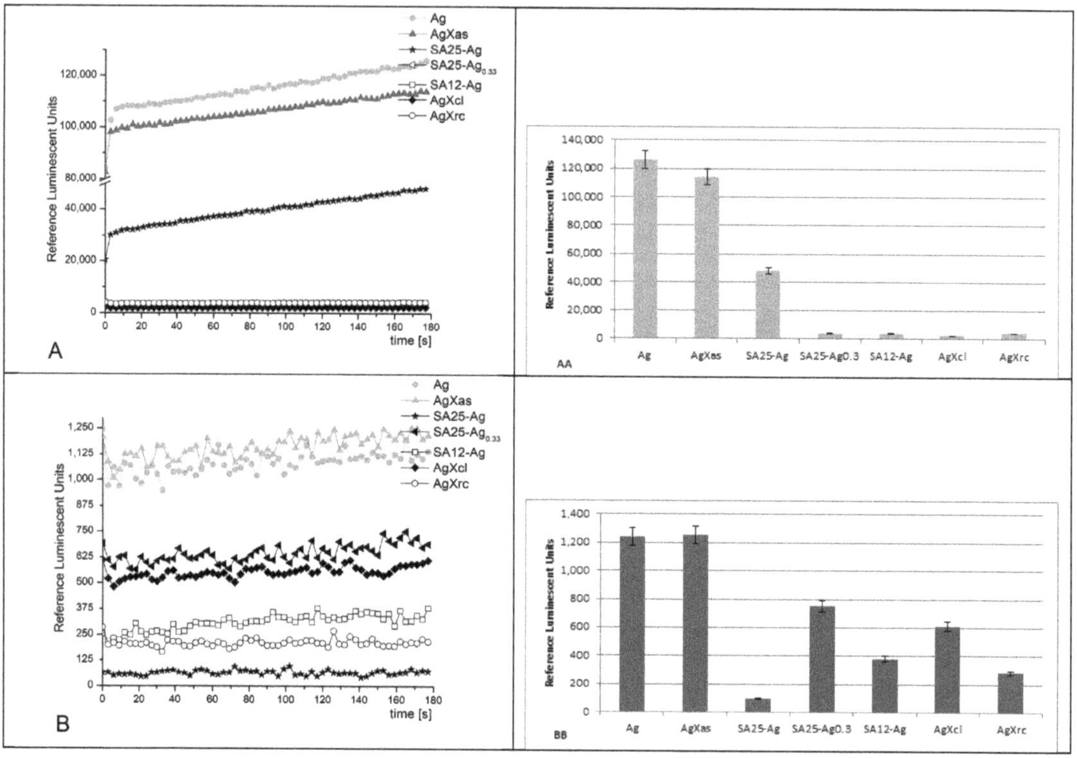

Figure 8. Chemiluminescence induced by H_2O_2 at pH 8.5 (**A**,**AA**) and pH 7.4 (**B**,**BB**) and the effect of the synthesized materials ($p \leq 0.05$); reagents: H_2O_2 (1.5%) + lucigenin (10^{-4} M) at 37 °C.

- System III (Figure 9)

The O_2^- generation in this system occurred according to the following chemical scheme [29,30]:

(1) PhMS + NAD.H + $H^+ \rightarrow$ PhMS.H_2 + NAD^+
(2) PhMS.H_2 + PhMS $\rightarrow$ 2PhMS.$H^\cdot$
(3) PhMS.$H^\cdot$ + $O_2 \rightarrow$ PhMS + $O_2{}^{-\cdot}$ + H^+

The superoxide radical produced is very short living and aggressive.

At pH 8.5, all the tested materials demonstrated an activity against that ROS lower than the control, as follows: AgXas—an 8% reduction; SA25-Ag—more than 40%; SA25-$Ag_{0.33}$—77%; SA12-Ag—more than 90%; AgXcl—80%; AgXrc—almost 70%. A higher level of oxidation was observed at the beginning of the reactions, and the emitted light decreased with time. The neutralization of that ROS was very effective for all the tested materials.

At pH 7.4, the observed properties were maintained for all materials except AgXrc. It was oxidized 25% more than the control consisting of pure Ag nanoparticles, indicating that it was more efficient in the release of silver species. All the other tested materials suppressed the generation of $O_2{}^{\cdot-}$ radicals: AgXas—46%; SA25-Ag—53%; SA25-$Ag_{0.33}$—54%; SA12-Ag—16%; AgXcl—more than 40%. Despite the significant fluctuations observed, the registered effects were maintained with time.

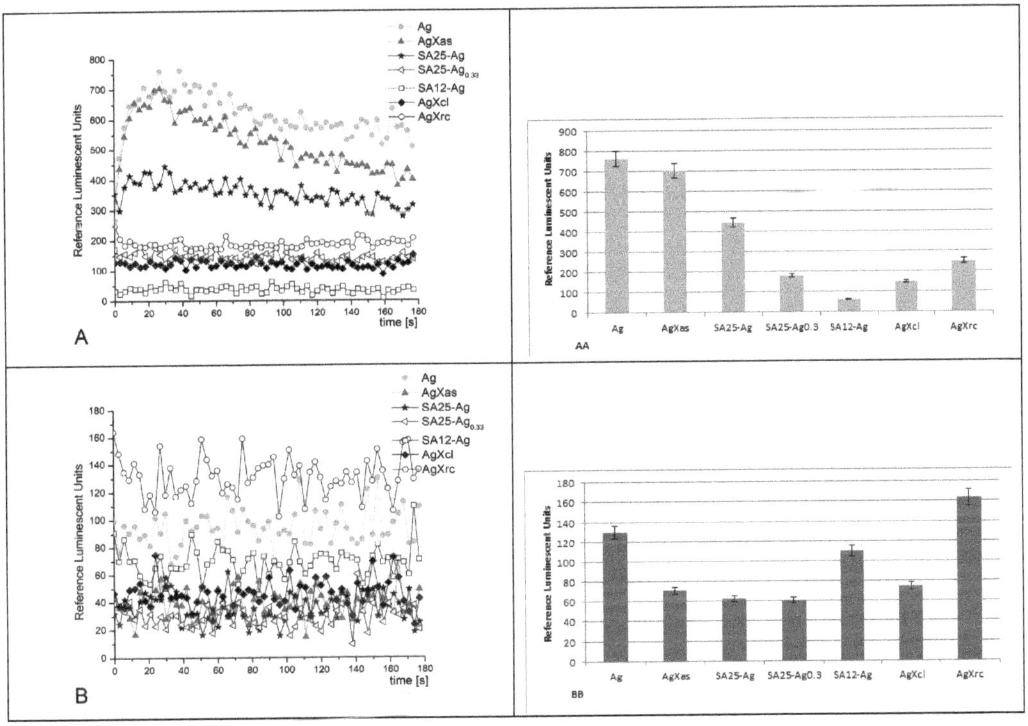

Figure 9. Chemiluminescence induced by $O_2{\cdot}^-$ radicals at pH 8.5 (**A**,**AA**) and pH 7.4 (**B**,**BB**) and the effect of the synthesized materials ($p \leq 0.05$); reagents: NADH (10^{-4} M) + PhMS (10^{-6} M) + lucigenin (10^{-4} M) at 37 °C.

In conclusion, AgXas and AgXrc appear to be the best candidates and rivals of the pure Ag nanoparticles in such ROS-generation reactions, especially within physiological conditions (pH 7.4/37 °C), showing a prooxidant activity. That result can explain their strong bactericidal effects.

The antibacterial function of the silver-doped materials was studied by monitoring the growth-inhibitory effect on *Staphylococcus aureus ATCC 25923* and *Escherichia coli ATCC 25922*. Figure 10 presents the results obtained from a screening using the most probable cell number method. It appeared that the well diffusion method was not suitable for studying zeolitic materials because agar hindered the diffusion of nanoparticles and the sterile zones around the wells were not clear. According to the well agar diffusion method, only the dispersion of AgXas diffused into the culture medium; around the 6 mm diameter well, a sterile zone with an average value of 12.5 mm was measured for *E. coli*. This indicated a bactericidal effect of the dispersion at a concentration of 5 g/L^{-1}. At lower concentrations, sterile zones were not detected. According to the spot test, with the same bacteria, a strong bactericidal effect on *E. coli* was observed, from 5 g/L^{-1} to the lowest concentration tested, 0.15 g/L^{-1}. The agar medium did not allow for the diffusion of zeolite dispersions. For that reason, the spot test (based on direct contact with bacterial cells) appeared to be more reliable for the demonstration of the bacterial toxicity (results are not presented).

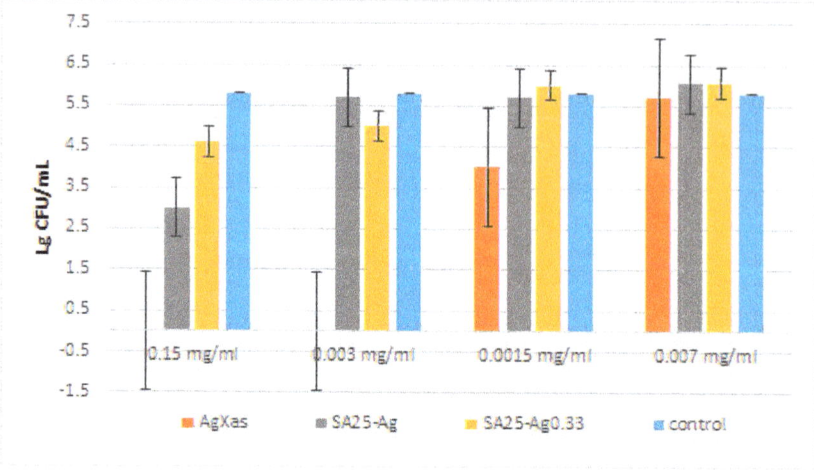

Figure 10. Results for *E. coli ATCC 25922* at the 4th h of impact of zeolites AgXas, SA25-Ag, and SA $Ag_{0.33}$.

The results obtained after counting the surviving cells on the inoculated plates after treatment with different types of Ag-loaded zeolites in liquid medium are presented in Figure 10 and Table 2.

Table 2. Results for *E. coli ATCC 25922* at the 4th h of impact with other zeolite materials, calculated in CFU/mL.

Material / Concentration	2.5 g/L^{-1}	0.25 mg/L^{-1}	0.025 g/L^{-1}	0.0025 g/L^{-1}
SA12-Ag fresh	0	>4 × 10^7	2 × 10^8	1 × 10^8
SA12-Ag old	0	>4 × 10^7	2 × 10^8	2 × 10^8
AgXcl fresh	0	>4 × 10^4	2 × 10^8	2 × 10^8
AgXcl old	0	>4 × 10^4	2 × 10^8	2 × 10^8
AgXrc fresh	0	2 × 10^8	2 × 10^8	2 × 10^8
AgXrc old	>4 × 10^4	2 × 10^8	2 × 10^8	2 × 10^8
control	1.14 × 10^8	1.14 × 10^8	1.14 × 10^8	1.14 × 10^8

As can be seen from Figure 10, AgXas exhibited the strongest effect on Gram-negative bacteria, as it killed all bacteria at concentrations of 0.003 g/L^{-1} and above. The MIC of AgXas against *E. coli* was 0.0015 g/L^{-1}. The other zeolites had almost no effect at the same concentration. The other strong bactericidal effect was observed for SA25-Ag. Perhaps the silver ions were more efficient. On the other hand, a higher Si-to-Al ratio seemed to be more beneficial, due to a higher release rate and concentration of silver ions in the media.

The other silver-loaded zeolites, SA25-Ag, AgXcl, and AgXrc, both fresh and old dispersions, manifested a bactericidal effect at a concentration of 2.5 g/L^{-1}. The zeolites with silver nanoparticles (those, originating from AgXcl or AgXrc, which also releases silver ions), possessed the strongest bactericidal effect in comparison to other zeolitic materials presented in Table 2. It demonstrated a 99% bactericidal effect on *E. coli* at a concentration of 0.25 g/L^{-1} too. This is more proof that the silver ions appear to be the most active silver species against bacteria.

As can be seen from Table 3, after 24 h the zeolite material AgXas manifested a 99.999% bactericidal effect on *E. coli ATCC 25922* at a concentration of 0.15 g/L^{-1}. The SA25-Ag material demonstrated the same bactericidal effect while the same carrier zeolite material, loaded with fewer silver ions, i.e., 1.2%wt vs. 2.5%wt, as shown in Table 1, showed only an inhibitory effect on the tested Gram-negative bacteria.

Table 3. Results for *E. coli ATCC 25922* at the 24th h of impact with zeolites AgXas, SA25-Ag, and SA $Ag_{0.33}$, calculated in CFU/mL.

Material \ Concentration	0.15 g/L^{-1}	0.003 g/L^{-1}	0.0015 g/L^{-1}	0.0007 g/L^{-1}
AgXas	0	3.85×10^8	$>4 \times 10^9$	$>4 \times 10^9$
SA25-Ag	1×10^1	$>4 \times 10^8$	$>4 \times 10^9$	1.15×10^9
SA25-$Ag_{0.33}$	$>4 \times 10^4$	$>1 \times 10^9$	$>1 \times 10^9$	1.15×10^9
control	1.11×10^9	1.11×10^9	1.11×10^9	1.11×10^9

As presented on Table 4, the freshly prepared dispersions of SA12-Ag, AgXcl as well as the old ones, that were stored in a refrigerator for 2 months, manifested a bactericidal effect at a concentration of 0.025 g/L^{-1}. SA12-Ag exhibited only a 10% inhibitory effect against *E. coli ATCC 25922*, at the same concentration. The SA12-Ag material demonstrated a bactericidal effect on *E. coli* at a concentration of 2.5 g/L^{-1}. The freshly prepared dispersion AgXrc produced a bactericidal effect at a concentration of 0.25 g/L^{-1}. After storage in a refrigerator for 2 months, the same dispersion lost its activity, with a bactericidal effect of 99.99% at a concentration of 2.5 g/L^{-1}. In addition to the dose/concentration-dependent effect, a time-dependent effect was also observed for samples SA12-Ag(I) fresh, AgXcl fresh, AgXcl old, and AgXrc fresh (when compared between the fourth and twenty-fourth hours on *E. coli ATCC 25922*).

Table 4. Results for *E. coli ATCC 25922* at the 24th h of impact with other zeolite materials, calculated in CFU/mL.

Material \ Concentration	2.5 g/L^{-1}	0.25 g/L^{-1}	0.025 g/L^{-1}	0.0025 g/L^{-1}
SA12-Ag fresh	0	0	0	1.9×10^9
SA-12Ag old	0	$>4 \times 10^7$	2×10^8	3.2×10^8
AgXcl fresh	0	0	0	1×10^7
AgXcl old	0	0	0	2×10^7
AgXrc fresh	0	0	2×10^8	2×10^8
AgXrc old	$>4 \times 10^4$	2×10^8	2×10^8	4×10^9
control	1.41×10^9	1.41×10^9	1.41×10^9	1.41×10^9

Clearly, the zeolites released both silver nanoparticles and silver ions, which leached from them quickly; their effect was strong during the first 4 h after interaction with the bacterial suspension. After 24 h, the bacterial cells presumably released proteins and other cell substances during their lysis, which decreased the effect of leached silver species from the zeolite materials.

Staphylococcus aureus is a Gram-positive bacterium with almost 80 layers of peptideglucan in its cell wall. The thickness of the cell wall determines the resistance of the bacteria to nanoparticles and other mechanical injuries. In our experiment, 0.15 g/L^{-1} of AgXas killed all living *S. aureus* cells, which was same as in the case with *E. coli* (Table 5). The results with other zeolitic materials loaded with silver ions were similar to those obtained with Gram-negative bacteria. The sample with less loaded silver, SA25-$Ag_{0.33}$, showed only four times fewer living bacterial cells in comparison to the control at the 4th h of treatment (Table 5). This result proved the importance of the concentration of eluted silver ions for the antibacterial effect of the material.

Table 5. Results for *S. aureus ATCC 25923*, at the 4th h of impact, calculated in CFU/mL.

Material \ Concentration	0.15 g/L^{-1}	0.003 g/L^{-1}	0.0015 g/L^{-1}	0.007 g/L^{-1}
AgXas	0	4×10^4	$>4 \times 10^8$	$>4 \times 10^8$
SA25-Ag	$>4 \times 10^4$	$>4 \times 10^8$	$>4 \times 10^8$	$>5 \times 10^8$
SA25-Ag$_{0.33}$	2×10^5	4×10^8	$>5 \times 10^8$	$>5 \times 10^8$
control	8.2×10^5	8.2×10^5	8.2×10^5	8.2×10^5

The results shown in Table 6 demonstrate the strongest bactericidal effect of the silver nanoparticle-containing zeolite AgXcl, at a concentration of 0.25 g/L^{-1}, on *S. aureus ATCC 25923* [31,32]. The most probable mechanism of action of the observed bactericidal effect is by the mechanical destruction of the cell wall of bacteria that are negatively charged by the leached nanoparticles; nanoparticles also interact with the water envelope around bacteria and the eluted silver ions that enter through the bacterial membrane in the cytoplasm are positive ions (I) [1–3]. The silver in these clusters may be the most concentrated material that releases more silver ions in comparison to other zeolite composites.

Table 6. Results for *S. aureus* 25923, at the 4th h of impact, calculated in CFU/mL.

Material \ Concentration	2.5 g/L^{-1}	0.25 g/L^{-1}	0.025 g/L^{-1}	0.0025 g/L^{-1}
SA12-Ag fresh	0	1.3×10^7	1×10^8	2×10^8
SA12-Ag old	0	2×10^7	2×10^8	2×10^8
AgXcl fresh	0	0	1×10^7	2×10^8
AgXcl old	0	1×10^7	2×10^8	2×10^9
AgXrc fresh	0	2×10^8	2×10^8	2×10^9
AgXrc old	$>4 \times 10^4$	2×10^8	2×10^9	2×10^9
control	1.21×10^9	1.21×10^9	1.21×10^9	1.21×10^9

A time-dependent antibacterial effect was also seen when the old water dispersions of zeolite materials lost their effect after storage in a refrigerator, despite sonication immediately before the second test. The results could be important when these materials are produced in large quantities, and the expiration date should be considered.

The findings about the effect of the synthesized zeolite-based materials on the lethality of *Daphnia magna*, evaluating their ecotoxicity, are shown on Figure 11.

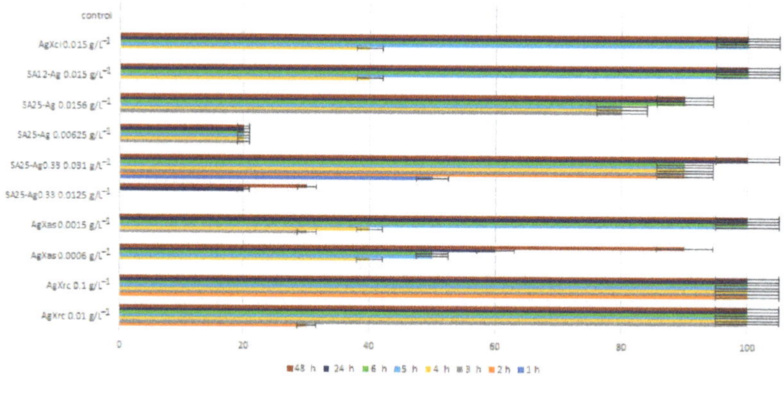

Figure 11. Interactions of Ag-loaded zeolites with *Daphnia magna*, given as lethality percentage.

Water dispersions from silver-loaded zeolite materials were tested immediately after sonication on young *Daphnia magna* individuals at four different concentrations (not all presented in the figure below).

Figure 11 shows that the pure zeolite did not cause lethality against the young *Daphnia* according to the acute tests. AgXas and SA12-Ag showed a 90% lethality at the end of 48th h. SA25-$Ag_{0.33}$, at a concentration of 0.0125 g/L^{-1}, had a 30% lethality. SA25-Ag, at a concentration of 0.00625 g/L^{-1}, had only a 20% lethality against *Daphnia magna* until the 48th h. AgXrc, AgXcl, and SA25-Ag were toxic and caused 100% lethality from the 2nd to the 4th h of the experiment. Generally, *Daphnia magna* were much more sensitive to the silver-loaded zeolite materials than the tested bacteria, presumably due to the slightly acidic pH of the water (6.5). These materials should not be released into the environment without control.

In general, silver nanoparticles exhibit strong antibacterial properties due to their small size. This small size increases the surface-to-volume ratio, enhancing the reactivity of the nanoparticles, which makes them effective against biofilm growth and microbial activity. They release silver ions over time, offering a sustained antimicrobial effect. However, the aggregation of nanoparticles can reduce their surface area, diminishing their reactivity and efficiency [16,33,34].

Silver ions (Ag^+) are also known for their strong antimicrobial effects, but their interaction with microbes tends to be more straightforward. They bind to the cellular components of bacteria, disrupting essential functions and starting the generation of ROS. However, they may have a lower efficiency compared to the silver nanoparticles due to their smaller size and predisposition to form complexes with other substances in water, such as chloride ions, which can reduce their availability and antimicrobial action [1–3,16,25,33–35].

In summary, the silver nanoparticles provide a more sustained and potentially more effective antimicrobial action compared to silver ions, but they are resource-intensive and pose concerns regarding aggregation and environmental impact. Silver ions, while effective, are limited by their tendency to form inactive compounds and may require higher concentrations to have the same level of efficacy as silver nanoparticles [16,31].

The release of Ag species at bactericidal concentrations into the environment should be avoided to prevent irreversible changes in the food chain and ecosystems. The synthesized silver-loaded zeolite materials could have a wide range of potential applications due to their unique properties, including antimicrobial activity, Ag species exchange capacity, high surface area, and possible catalytic properties. These results support their applications in air and water filtering and other similar systems, coatings for medical devices, wound dressings, fabrics with antimicrobial properties, soil additives, food packaging and preservation, construction materials, etc.

3. Materials and Methods

3.1. Materials

Ag-exchanged zeolite X (AgX) was purchased from Sigma-Aldrich, Steinheim, Germany, with a nominal composition of $Ag_{84}Na_2[(AlO_2)_{86}(SiO_2)_{106}] \times H_2O$, and a granular form with 35%wt silver content. The samples for antimicrobial and all other tests were prepared by gently grinding the material down to powder form. The as obtained powder was split into three samples: (1) AgXas—powder form of the as received material. (2) The rest of the powder was subjected to vacuum heat treatment in a vacuum oven up to 200 °C for 4 h and pressure down to about 1 mbar, measured at the exhaust line of the oven. This led to a visible change in the powder's appearance with a color change from ivory to black (sample AgXcl). (3) Part of this sample, AgXcl, was heated under a dry air flow in a furnace at 400 °C for 4 h, resulting in a color change back to ivory-reddish, designated as AgXrc. ZSM-5 materials, in their ammonium form, with silicon-to-aluminum ratios of 25 and about 12, were purchased from Zeolist International, Conshohocken, PA, USA (CBV2314G and CBV5524G). These were converted into their protonated forms by heating in air at 550 °C for about 2 h and labeled as SA12 (for Si-to-Al = 12) and SA25 (for Si-to-Al = 25) further in the text. The silver-exchanged samples were produced by wet exchange from 0.05 M silver nitrate water solutions at room temperature overnight in a dark room. The corresponding silver-exchanged powders were filtered out of the solutions, washed twice by stirring again in deionized water, and filtered. The resulting silver-exchanged materials were dried overnight in air at 80 °C. Ag-ZSM5 samples were white just like their original ammonia and H forms. Selected samples were characterized with SEM (LYRA I XMU, Tescan (Brno-Kohoutovice, Czech Republic)) and EDX (Quantax, Billerica, MA, USA) for agglomerate morphology and chemical composition. The zeolites and, eventually, the presence of silver metallic phases were checked by X-Ray Powder Diffraction (XRPD) using a URD-6 diffractometer, Freiberger Präzisionsmechanik GmbH, Rich. Seifert & Co. GmbH & Co. KG, Freiberg, Germany, using a Cu radiation source. Transmission Electron Microscopy (TEM) was performed with the materials and silver cluster/nanoparticle formation was anticipated to show the presence of metallic silver particles using a high-resolution, STEM JEOL JEM 2100 instrument with JEOL, Akishima, Japan. TEM images were recorded using a CCD camera (GATAN Inc., Pleasanton, CA, USA).

The following materials used for the chemiluminescent assay were purchased with a high purity. Those were iron sulfate (p. a.) (Merck, Darmstadt, Germany), hydrogen peroxide (30%) (Merck, Germany), phenazine methosulfate (PhMS) (N-methyldibenzopyrazine methyl sulfate salt) (p. a.) (Merck, Germany), lucigenin (bis-N-methylacridinium nitrate) (p. a.) (Sigma-Aldrich, St. Louis, MO, USA), β-nicotinamide adenine dinucleotide, reduced form (p. a.) (NAD.H, Boehringer, Ingelheim am Rhein, Germany), dimethyl sulfoxide (p. a.) (DMSO, Sigma-Aldrich), buffers at pH 8.5 and pH 7.4 (Sigma-Aldrich), and silver nanoparticles (99%, 30–50 nm) (Houston, TX, USA, US Research Nanomaterials Inc.). All chemicals were used as purchased.

3.2. Chemiluminescent Assay

We employed the luminescent method to study the effect of the synthesized silver materials on the kinetics of free radical oxidation reactions using activated chemiluminescence and a probe, lucigenin, in three model systems [35].

Usually, a medium with a higher alkalinity favors free radical generation and enables the achievement of reliable measurements and comparable differences. Two different media were tested: pH 7.4 and pH 8.5, physiological and alkaline, and optimal. Three ex vivo model systems were implemented [35].

The first model system is called Fenton's. It generates ·OOH and ·OH radicals: it contained 0.2 mol of sodium hydrogen phosphate buffer, with the selected pH, Fenton's reagent: $FeSO_4$ (5.10^{-4} mol)—H_2O_2 (1.5%), lucigenin (10^{-4} mol), and 0.025 g/L^{-1} of the tested material.

The second model system contained the strong oxidant hydrogen peroxide (H_2O_2): 0.2 mol sodium hydrogen phosphate buffer, with the chosen pH, H_2O_2 (1.5%), the chemiluminescent probe, lucigenin (10^{-4} mol), and 0.025 g/L^{-1} of the tested material.

The third model system generated superoxide radicals ($O_2{\cdot}^-$) in the reaction between NADH and PhMS [36]: it contained 0.2 mol of sodium hydrogen phosphate buffer with the chosen pH, NAD.H (10^{-4} mol), phenazine–metasulfate (10^{-6} mol), lucigenin (10^{-4} mol), and 0.025 g/L^{-1} of the tested material.

The control samples contained 30–50 nm silver nanoparticles (0.025 g/L^{-1}). The reactions were monitored for 3 min, every 3 s, at 37 °C (physiological). All tested materials were sonicated for 60 min before application. All experiments were performed in triplicate.

3.3. Microorganisms

The antimicrobial effect was tested on Gram-positive and Gram-negative bacteria, *Escherichia coli ATCC 25922* and *Staphylococcus aureus ATCC 25923*, which were both supplied by the National Bank of Industrial Microorganisms and Cell Cultures (NBIMCC), Sofia, Bulgaria. The test microorganisms were grown in nutrient broth (NB Conda, Madrid, Spain) at 37 °C and 180 rpm for 24 h with 2 sub-cultivations to reach the exponential growth phase. The microbial density of the cultures was adjusted to 0.5 according to the McFarland standard and used for the antimicrobial tests.

3.4. Antimicrobial Activity

Two classical methods were used to study the antibacterial effect of the synthesized nanoparticles: spot and well diffusion tests in agar medium. Aliquots of 100 μL microbial suspension in exponential phase with optical density 0.5 MacFarland were randomly spread on solid medium (Nutrient agar, Conda, Spain) for both tests.

Spot test: 10 μL drops of the tested nanomaterial were put on inoculated agar. The used concentrations for spots were 3.0 g/L^{-1}, 1.5 g/L^{-1}, 1.0 g/L^{-1}, and 0.75 g/L^{-1}.

Well diffusion test: Wells with a diameter of 6 mm were cut using a stopper borer and filled with 50 μL solutions of nanozeolites in concentrations of 1.0 g/L^{-1}, 0.5 g/L^{-1}, 0.3 g/L^{-1}, and 0.1 g/L^{-1}. For both methods, the plates were left for 2 h at 5 ± 1 °C to afford the diffusion of the dispersions, and after that cultivated for 24 h at 37 °C. The obtained sterile zones were measured with 0.5 mm accuracy. All experiments were performed with two replicates, and the results were presented as the mean value of two experiments.

The determination of minimum inhibitory (MIC) and minimum bactericidal (MBC) concentrations of the nanomaterials was defined after preparing consecutive two-fold dilutions in liquid medium, and after 24 h and 48 h of cultivation at 37 °C, aliquots of bacterial suspension were inoculated in agar plates to determine the amount of bacteria that survived. The colony-forming bacteria is calculated using the formula:

$$N = A \times 10^{n}/0.1$$

where A is the mean number of colonies from the two replicates counted on solid medium for a given dilution, 10^n is the corresponding dilution from which the inoculum was taken, and 0.1 is the amount of inoculum on one Petri dish.

In order to determine the stability and effectiveness of the zeolite dispersions in water, time-dependent experiments were performed. The dry materials were stored at room temperature in the dark, and water dispersions were stored in a refrigerator at 4–6 °C for 1–2 months and tested again. They were, respectively, labeled as "fresh" and "old" in the tables with results.

3.5. Environmental Toxicity by Testing the Lethality of Daphnia magna

Acute ecotoxicity tests were performed by using *Daphnia magna* for all types of materials that were applied in several concentrations, with three replicas of 10 young individuals for each substance and concentration, with the appropriate control. The concentrations varied from 0.0001 g/L^{-1} to 0.1 g/L^{-1}. The experiments were conducted at ambient temperature 22 ± 1 °C and pH level of the aquatic environment 6.5 ± 0.1. The observations were conducted every hour on the first day and every 12 h on the second day.

3.6. Statistics

The statistical analysis is performed by OriginPro 8 and Microsoft Office Excel.

4. Conclusions

We have successfully prepared a series of zeolite materials with two different topologies, namely FAU and MFI, containing different amounts of silver, spanning between a few % by wt and up to 35%wt, in either ionic or metallic nanoparticle forms. The observed effects could be explained by the release of Ag species as ions and nanoparticles, and also, the ions originating from them were as follows:

The material with the highest silver content of mostly ionic silver species, the AgXas material, presented a strong bactericidal activity, showing a bactericidal effect at a concentration of 0.15 g/L^{-1} for *E. coli ATCC 25922* and *S. aureus ATCC 25923*. Its measured MIC is 0.003 g/L^{-1} (99.999% growth inhibition) for both tested strains. At an even lower concentration of 0.0015 g/L^{-1}, it has an inhibitory effect on both bacteria too. SA25-Ag fresh and old dispersions, AgXcl fresh and old dispersions, and the AgXrc fresh dispersion had a bactericidal effect at a concentration of 2.5 g/L^{-1}, indicating that the high silica ZSM-5 zeolite performs with a similar efficiency at an order of magnitude-lower silver content compared to the Faujasites loaded with silver nanoparticles. An aging effect of the prepared silver zeolites test dispersions was observed for the first time by us. For instance, AgXrc old, with a high silver content (365 wt), lost its antibacterial properties after one month of storage at 4 °C. *E. coli ATCC 25922* was significantly more sensitive than *S. aureus ATCC 25923*, according to the microbiological results obtained.

AgXas and AgXrc presented a good prooxidant activity according to the control of pure Ag nanoparticles (30–50 nm) and appear to be the best candidates and rivals of Ag nanoparticles in ROS generation or oxidizing reactions, especially in physiological conditions (pH 7.4/37 °C). This result explains their bactericidal effects.

On the contrary, all the tested materials, including AgXas and AgXrc in some cases and conditions, exhibited strong inhibitory effects when tested in the selected ROS-generating systems. These results support their application for the prevention of free radical oxidation.

The demonstrated effects suggest the possible biological applications of the newly synthesized silver-loaded zeolitic materials, releasing Ag species. Their bactericidal or bacteriostatic effects would be gentle in comparison to silver ions originating from pure nanoparticles, as mentioned by Chen (2019) [19]. Most of these materials would not advance the accumulation of ROS and would not burden the metabolic processes connected with their generation, breaking the extremely sensitive oxidation equilibrium.

Daphnia magna was proved to be a very sensitive biomarker of ecotoxicity.

Supplementary Materials: The following supporting information can be downloaded at: https://www.mdpi.com/article/10.3390/ph17121586/s1, Figure S1. Powder diffraction pattern of the studied Ag-X materials compared to computed patterns for the FAU (Na-X) parent zeolite structure, FCC Ag-metal, hexagonal Ag-metal phase, and Ag2O; Figure S2. Same as on Figure S2, but around the high angle part of the pattern; Figure S3. PXRD of the lowest level silver loading ZSM-5 material with Si/Al=25, compared to the computed pattern of FCC metallic silver; Figure S4. PXRD patterns of all studied silver exchanged ZSM-5 zeolites, along with the computed pattern of the MFI prototype; Figure S5. Same as on Figure S4 but focused on the higher angles in the pattern, shown along with the computed pattern of the FCC phase metallic silver. As seen in both Figures S4 and S5, no extra phases are observed in the ZSM-5 Ag-exchanged materials.

Author Contributions: Conceptualization, P.A.G., I.A.I. and E.L.P.; methodology, P.A.G., I.A.I., E.L.P. and E.P.N.; samples characterization, P.A.G.; antibacterial properties, I.A.I. and L.D.Y.; chemiluminescence assay, E.L.P.; toxicity test with *Daphnia magna*, E.P.N.; writing and editing, P.A.G., E.L.P., I.A.I., L.D.Y. and E.P.N. All authors have read and agreed to the published version of the manuscript.

Funding: Project: MULTIFUNCTIONAL MAGNETIC NANOCERAMICS FOR ANTITUMOR AND ANTIBACTERIAL APPLICATIONS, Sofia University "St. Kliment Ohridski", Bulgaria, contract number: 80-10-18/10.05.2022 Project: BNSF co-finance grant KP-06-COST/5, 18/08/2021.

Data Availability Statement: Data are contained within the article.

Acknowledgments: E.P. is grateful to the project: CLEAN TECHNOLOGIES FOR SUSTAINABLE ENVIRONMENT—WATERS, WASTE, ENERGY FOR CIRCULAR ECONOMY, Ministry of Education and Science, Bulgaria, contract number: BG05M2OP001-1.002-0019. We would like to specially acknowledge the help and cooperation of D. Karashanova for the TEM and analyses.

Conflicts of Interest: The authors declare no conflicts of interest. The funders had no role in the design of the study; in the collection, analyses, or interpretation of data; in the writing of the manuscript; or in the decision to publish the results.

References

1. Antimicrobial Resistance Collaborators. Global burden of bacterial antimicrobial resistance in 2019: A systematic analysis. *Lancet* **2022**, *399*, 629–655. [CrossRef] [PubMed]
2. Alexander, J.W. History of the medical use of silver. *Surg. Infect.* **2009**, *10*, 289–292. [CrossRef] [PubMed]

3. Klasen, H.J. Historical review of the use of silver in the treatment of burns. I. Early uses. *Burns* **2000**, *26*, 117–130. [CrossRef] [PubMed]
4. Auerbach, S.M.; Carrado, K.A.; Dutta, P.K. *Handbook of Zeolite Science and Technology*, 1st ed.; CRC Press: Boca Raton, FL, USA, 2003; p. 1198. [CrossRef]
5. Nezamzadeh-Ejhieh, A.; Tavakoli-Ghinani, S. Effect of a nano-sized natural clinoptilolite modified by the hexadecyltrimethyl ammonium surfactant on cephalexin drug delivery. *Comptes Rendus Chim.* **2014**, *17*, 49–61. [CrossRef]
6. Nasiri-Ardali, M.; Nezamzadeh-Ejhieh, A. A comprehensive study on the kinetics and thermodynamic aspects of batch and column removal of Pb(II) by the clinoptilolite–glycine adsorbent. *Mat. Chem. Phys.* **2020**, *240*, 122142. [CrossRef]
7. Noori, M.; Tahmasebpoor, M.; Foroutan, R. Enhanced adsorption capacity of low-cost magnetic clinoptilolite powders/beads for the effective removal of methylene blue: Adsorption and desorption studies. *Mat. Chem. Phys.* **2022**, *278*, 125655. [CrossRef]
8. Król, M. Hydrothermal synthesis of zeolite aggregate with potential use as a sorbent of heavy metal cations. *J. Mol. Struct.* **2019**, *1183*, 353–359. [CrossRef]
9. Gellens, L.R.; Mortier, W.J.; Uytterhoeven, J.B. On the nature of the charged silver clusters in zeolites of type A, X and Y. *Zeolites* **1981**, *1*, 11–18. [CrossRef]
10. Dutta, P.; Wang, B. Zeolite-supported silver as antimicrobial agents. *Coord. Chem. Rev.* **2019**, *383*, 1–29. [CrossRef]
11. Colella, C. Ion exchange equilibria in zeolite minerals. *Miner. Depos.* **1996**, *31*, 554–562. [CrossRef]
12. Sherry, H.S. The ion-exchange properties of zeolites. I. Univalent ion exchange in synthetic faujasite. *J. Phys. Chem.* **1966**, *70*, 1158–1168. [CrossRef]
13. Krishnani, K.K.; Zhang, Y.; Xiong, L.; Yan, Y.; Boopathy, R.; Mulchandani, A. Bactericidal and ammonia removal activity of silver ion-exchanged zeolite. *Bioresour. Technol.* **2012**, *117*, 86–91. [CrossRef] [PubMed]
14. Kaali, P.; Czél, G. Single, binary and ternary ion exchanged zeolite as an effective bioactive filler for biomedical polymer composites. *Mater. Sci. Forum* **2013**, *729*, 234–239. [CrossRef]
15. Ferreira, L.; Fonseca, A.M.; Botelho, G.; Aguiar, C.A.; Neves, I.C. Antimicrobial activity of faujasite zeolites doped with silver. *Microporous Mesoporous Mater.* **2012**, *160*, 126–132. [CrossRef]
16. Kędziora, A.; Speruda, M.; Krzyżewska, E.; Rybka, J.; Łukowiak, A.; Bugla-Płoskońska, G. Similarities and differences between silver ions and silver in nanoforms as antibacterial agents. *Int. J. Mol. Sci.* **2018**, *19*, 444. [CrossRef]
17. Ferreira, L.; Guedes, J.F.; Almeida-Aguiar, C.; Fonseca, A.M.; Neves, I.C. Microbial growth inhibition caused by Zn/Ag-Y zeolite materials with different amounts of silver. *Colloids Surf. B Biointerfaces* **2016**, *142*, 141–147. [CrossRef]
18. Chen, S.; Popovich, J.; Iannuzo, N.; Haydel, S.E.; Seo, D.-K. Silver-ion-exchanged nanostructured zeolite X as antibacterial agent with superior ion release kinetics and efficacy against methicillin-resistant *Staphylococcus Aureus*. *ACS Appl. Mater. Interfaces* **2017**, *9*, 39271–39282. [CrossRef]
19. Chen, S. Synthesis and Applications of Nanostructured Zeolites from Geopolymer Chemistry. Ph.D. Thesis, Arizona State University, Tempe, AZ, USA, 2019; p. 189.
20. Youssef, H.F.; Abdel-Aziz, M.S.; Fouda, F.K. Evaluation of antimicrobial activity of different silver-exchanged nano and micronized zeolites prepared by microwave technique. *J. Porous Mater.* **2017**, *24*, 947–957. [CrossRef]
21. Lv, L.; Luo, Y.; Ng, W.J.; Zhao, X.S. Bactericidal activity of silver nanoparticles supported on microporous titanosilicate ETS-10. *Microporous Mesoporous Mater.* **2009**, *120*, 304–309. [CrossRef]
22. Jiraroj, D.; Tungasmita, S.; Tungasmita, D.N. Silver ions and silver nanoparticles in zeolite A composites for antibacterial activity. *Powder Technol.* **2014**, *264*, 418–422. [CrossRef]
23. Lalueza, P.; Monzón, M.; Arruebo, M.; Santamaria, J. Antibacterial action of Ag containing MFI zeolite at low Ag loadings. *Chem. Commun.* **2010**, *47*, 680–682. [CrossRef] [PubMed]
24. Yee, M.S.-L.; Khiew, P.S.; Tan, Y.F.; Chiu, W.S.; Kok, Y.-Y.; Leong, C.-O. Low temperature, rapid solution growth of antifouling silver-zeolite nanocomposite clusters. *Microporous Mesoporous Mater.* **2015**, *218*, 69–78. [CrossRef]
25. Singh, V.V.; Jurado-Sánchez, B.; Sattayasamitsathit, S.; Orozco, J.; Li, J.; Galarnyk, M.; Fedorak, Y.; Wang, J. Multifunctional silver-exchanged zeolite micromotors for catalytic detoxification of chemical and biological threats. *Adv. Funct. Mater.* **2015**, *25*, 2147–2155. [CrossRef]
26. Abebe, B.; Zereffa, E.A.; Tadesse, A.; Murthy, H.C.A. A review on enhancing the antibacterial activity of ZnO: Mechanisms and microscopic investigation. *Nanoscale Res. Lett.* **2020**, *15*, 190. [CrossRef] [PubMed]
27. Pavlova, E.L.; Toshkovska, R.D.; Doncheva, T.E.; Ivanova, I.A. Prooxidant and antimicrobic effects of iron and titanium oxide nanoparticles and thalicarpine. *Arch. Microbiol.* **2020**, *202*, 1873–1880. [CrossRef]
28. Pavlova, E.L.; Savov, V.M. Chemiluminescent in vitro estimation of the inhibitory constants of antioxidants ascorbic and uric acids in Fenton's reaction in urine. *Biochemistry* **2006**, *71*, 861–863. [CrossRef]
29. Hancock, J.T.; Desikan, R.; Neill, S.J. Role of reactive oxygen species in cell signaling pathways. *Biochem. Soc. Trans.* **2001**, *29*, 345–350. [CrossRef]
30. Nishikimi, M.; Rao, N.A.; Yagi, K. The occurrence of superoxide anion in the reaction of reduced phenazine methosulfate and molecular oxygen. *Biochem. Biophys. Res. Commun.* **1972**, *46*, 849–854. [CrossRef]
31. Sánchez, M.J.; Mauricio, J.E.; Paredes, A.R.; Gamero, P.; Cortés, D. Antimicrobial properties of ZSM-5 type zeolite functionalized with silver. *Mater. Lett.* **2017**, *191*, 65–68. [CrossRef]

32. Tosheva, L.; Belkhair, S.; Gackowski, M.; Malic, S.; Al-Shanti, N.; Verran, J. Rapid screening of the antimicrobial efficacy of Ag zeolites. *Colloids Surf. B Biointerfaces* **2017**, *157*, 254–260. [CrossRef]
33. Hong, F.; Chen, P.; Yu, X.; Chen, Q. The application of silver to decontaminate dental unit waterlines—A systematic review. *Biol. Trace Elem. Res.* **2022**, *200*, 4988–5002. [CrossRef] [PubMed]
34. Khan, I.; Saeed, K.; Khan, I. Nanoparticles: Properties, applications and toxicities. *Arab. J. Chem.* **2019**, *12*, 908–931. [CrossRef]
35. Pavlova, E.L.; Ivanova, I.A.; Staneva, A.D.; Kostadinova, A.S.; Kichukova, D.G.; Yocheva, L.D. Prooxidant, antioxidant and biological activity of nanocomposites of reduced graphene oxide, silver, copper and their combinations. *Chem. Pap.* **2022**, *76*, 6789–6800. [CrossRef]
36. Faulkner, K.; Fridovich, I. Luminol and lucigenin as detectors for $O_2^{\cdot -}$. *Free Radic. Biol. Med.* **1993**, *15*, 447–451. [CrossRef]

pharmaceuticals

Article

Synthesis, Cytotoxic and Genotoxic Evaluation of Drug-Loaded Silver Nanoparticles with Mebeverine and Its Analog

Mihaela Stoyanova [1], Miglena Milusheva [1,2], Milena Georgieva [3], Penyo Ivanov [3], George Miloshev [3], Natalia Krasteva [4], Kamelia Hristova-Panusheva [4], Mehran Feizi-Dehnayebi [5], Ghodsi Mohammadi Ziarani [5], Kirila Stojnova [6], Slava Tsoneva [7], Mina Todorova [1,*] and Stoyanka Nikolova [1,*]

1 Department of Organic Chemistry, Faculty of Chemistry, University of Plovdiv, 4000 Plovdiv, Bulgaria; stoyanova@uni-plovdiv.bg (M.S.); miglena.milusheva@uni-plovdiv.bg or miglena.milusheva@mu-plovdiv.bg (M.M.)
2 Department of Bioorganic Chemistry, Faculty of Pharmacy, Medical University of Plovdiv, 4002 Plovdiv, Bulgaria
3 Laboratory of Molecular Genetics, Epigenetics and Longevity, Institute of Molecular Biology "R. Tsanev", Bulgarian Academy of Sciences, 1113 Sofia, Bulgaria; milenageorgy@gmail.com (M.G.); p.ves.iv@gmail.com (P.I.); karamolbiol@gmail.com (G.M.)
4 Institute of Biophysics and Biomedical Engineering, Bulgarian Academy of Sciences, 1113 Sofia, Bulgaria; natalia.krasteva@yahoo.com (N.K.); kameliahristova@abv.bg (K.H.-P.)
5 Department of Organic Chemistry, Faculty of Chemistry, Alzahra University, Tehran P.O. Box 19938-93973, Iran; m.feizi@alzahra.ac.ir (M.F.-D.); gmohammadi@alzahra.ac.ir (G.M.Z.)
6 Department of General and Inorganic Chemistry with Methodology of Chemistry Education, Faculty of Chemistry, University of Plovdiv, 4000 Plovdiv, Bulgaria; stoynova@uni-plovdiv.bg
7 Department of Analytical Chemistry and Computer Chemistry, University of Plovdiv, 4000 Plovdiv, Bulgaria; slava.tsoneva@uni-plovdiv.bg
* Correspondence: minatodorova@uni-plovdiv.bg (M.T.); tanya@uni-plovdiv.bg (S.N.)

Abstract: Background: Irritable bowel syndrome (IBS) is a prevalent gastrointestinal disorder with a complex pathogenesis that necessitates innovative therapeutic approaches for effective management. Among the commonly used treatments, mebeverine (MBH), an antispasmodic, is widely prescribed to alleviate IBS symptoms. However, challenges in delivering the drug precisely to the colonic region often hinder its therapeutic effectiveness. To address this limitation, silver nanoparticles (AgNPs) have emerged as promising drug delivery systems, offering unique physicochemical properties that can enhance the precision and efficacy of IBS treatments. **Objectives:** This study aimed to synthesize AgNPs as drug delivery vehicles for MBH and a previously reported analog. The research focused on evaluating the cytotoxic and genotoxic effects of the AgNPs and forecasting their possibly harmful effects on future sustainable development. **Methods:** AgNPs were synthesized using a rapid method and functionalized with MBH and its analog. The nanoparticles were characterized using different techniques. Cytotoxicity and genotoxicity were evaluated in vitro. Additionally, in silico docking analyses were performed to explore their safety profile further. **Results:** In vitro assays revealed concentration-dependent cytotoxic effects and a lack of genotoxic effects with MBH-loaded AgNPs. A molecular docking simulation was performed to confirm this effect. **Conclusions:** The study underscores the potential of AgNPs as advanced drug delivery systems for safe and significant therapeutic implications for IBS. Future in vivo and preclinical investigations are essential to validate the safe range of exposure doses and evaluation standards for assessing AgNPs' safety in targeted and personalized medicine.

Keywords: silver nanoparticles; cytotoxicity; genotoxicity; in vitro; in silico; molecular docking; irritable bowel syndrome; mebeverine

 https://doi.org/10.3390/ph18030397

1. Introduction

Irritable bowel syndrome (IBS) is a prevalent gastrointestinal disorder characterized by symptoms such as abdominal pain, bloating, and altered bowel habits, including diarrhea and constipation. The exact cause of IBS remains unclear, but it is believed to involve a combination of factors, including gut–brain interactions, motility disturbances, and visceral hypersensitivity [1,2]. This illness affects 5% and 10% of the general population [3,4]. Management strategies typically focus on alleviating symptoms and improving patients' quality of life. Although the pathogenesis of irritable bowel syndrome remains unclear, it is known that abnormal gut–brain connection results in visceral hypersensitivity, abnormal CNS processing, and motility issues [5,6].

Mebeverine (MBH (**1**), Figure 1) is an antispasmodic commonly prescribed to manage IBS symptoms. It functions by directly relaxing the smooth muscles of the gastrointestinal tract, thereby reducing spasms and associated discomfort [7]. Unlike other antispasmodics, MBH does not exhibit typical anticholinergic side effects, such as dry mouth or blurred vision, making it a favorable option for many patients [8]. Consequently, MBH is widely used as a relaxant agent for treating gastrointestinal spasmodic disorders such as abdominal pain, intestinal disturbances, and bowel discomfort [9]. The main problem facing researchers in IBS treatment is targeting the medications directly to the colonic region [10]. Nanoparticles (NPs), on the other hand, have become ground-breaking instruments in various medical applications due to their unique qualities, such as increased surface reactivity, size control, and the capacity to be functionalized with therapeutic agents. Remarkably, they are valuable in drug delivery, diagnostics, cancer therapy, and gene therapy. To maximize safety and efficacy, NPs may enhance the stability and solubility of encapsulated cargos, facilitate transport across membranes, and extend circulation periods [11,12]. These factors have led to a large amount of NP research, which has produced encouraging outcomes in small animal models and in vitro [13].

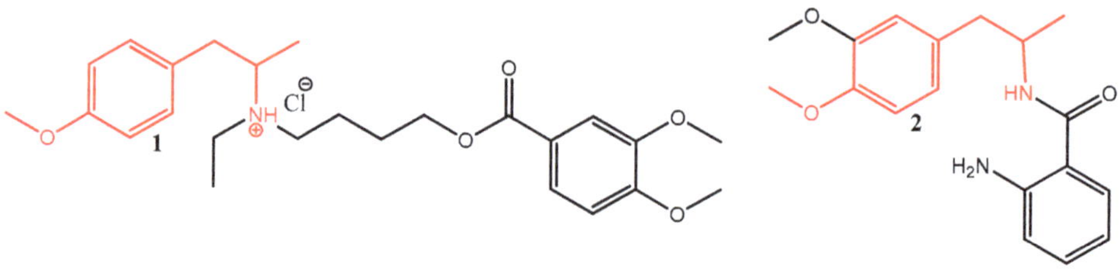

Figure 1. Structure of (**a**) MBH hydrochloride (**1**); (**b**) MBH analog (MA, **2**). Common structure in red.

Silver and gold nanoparticles (NPs) have garnered significant interest as drug delivery systems due to their unique physicochemical properties, such as high surface area-to-volume ratio, ease of functionalization, and excellent stability. In particular, silver nanoparticles (AgNPs) have demonstrated versatility in delivering a wide range of therapeutic agents. AgNPs enhance drug solubility, protect sensitive compounds from degradation, and provide controlled and sustained release at targeted sites. They have further been explored for the delivery of antibiotics, where they improve antimicrobial efficacy and help overcome resistance by targeting bacterial biofilms and disrupting cell membranes [14]. Additionally, AgNPs are used to deliver anti-inflammatory drugs, which enhance drug stability and provide targeted action at inflamed tissues, minimizing systemic side effects [15,16]. AgNPs have been incorporated into drug delivery systems in wound healing to promote tissue regeneration while simultaneously preventing infections, thanks to their inherent antimicrobial properties [17,18]. They are also employed for the topical delivery of corticosteroids and growth factors, improving the therapeutic outcomes for skin disorders and chronic wounds [19].

AgNPs have been utilized in gene therapy as carriers for nucleic acids, such as siRNA and plasmid DNA. By protecting these biomolecules from enzymatic degradation and facilitating their cellular uptake, AgNPs enable efficient gene silencing or expression in target cells [20]. They have additionally shown promise in the delivery of anti-diabetic drugs, where their functionalization with specific ligands enhances the targeting of pancreatic β-cells or improves glucose regulation [21]. Moreover, their role in delivering antiviral agents has been investigated, particularly for enhancing the stability and bioavailability of drugs targeting viral replication processes [22]. The ability to cross biological barriers, including the blood–brain barrier, makes this particular type of nanoparticle an attractive option for central nervous system (CNS) drug delivery, where it can enhance the bioavailability of therapeutic agents for treating neurological disorders [23]. These multifunctional capabilities of silver nanoparticles extend beyond their intrinsic biological activities, making them valuable tools for the precise and effective delivery of a diverse array of therapeutic agents across different medical applications.

On the other hand, size, surface chemistry, exposure techniques, and exposure duration are important factors that affect AgNP tissue distribution paths [24]. Individuals may be exposed to AgNPs by a variety of routes, such as blood circulation (intravenous injection), oral administration (gastrointestinal tract), inhalation (respiratory tract), and skin contact [25]. In vivo, injected AgNPs have a broad tissue distribution and a brief circulation duration through tail veins [16]. Although larger AgNPs may concentrate more in the spleen than in other organs, the liver is the primary target organ, followed by the spleen, lungs, and kidneys. It is interesting to note that blood levels of Ag stay elevated for up to six days following injection, after which they begin to decline. This shows that AgNPs may undergo time-dependent degradation and removal [26].

Additionally, a study on the toxicity of inhaled AgNPs found that their distribution to the liver and lungs causes inflammatory reactions like chronic alveolar inflammation and inflammatory cell infiltration [27]. Furthermore, migration to the gastrointestinal tract's acidic environment aids in dissolving AgNPs into silver ions when exposed orally. AgNPs also impact inflammation, apoptosis, and the expression of biochemical indicators of hepatotoxicity, including serum cholesterol and alkaline phosphatase [28–30]. Additionally, there were no discernible changes in the health condition of workers exposed to low levels of silver dust or soluble silver at threshold limit values [31]. A safe range of exposure levels and evaluation criteria for determining AgNP safety must be established. Assessing the biological impact is the primary goal of nanotoxicity, which is crucial for understanding mechanisms and predicting the potentially detrimental effects of AgNPs for their future development.

Here, we aim to address the challenges of targeted drug delivery for IBS by synthesizing AgNPs functionalized with MBH and MA. We evaluate these drug-loaded nanoparticles' cytotoxic and genotoxic profiles through in vitro assays and explore their molecular interactions using in silico docking analyses. This approach paves the way for sustainable and personalized medicine applications, offering innovative strategies to enhance therapeutic efficacy for IBS.

2. Results and Discussion

To explore the potential biomedical applications of silver nanoparticles (AgNPs), it is essential to develop synthesis and functionalization strategies that ensure efficacy, safety, and biocompatibility. Chemical reduction is a traditional procedure that entails dissolving a silver salt in a liquid phase to react with a reducing agent. To enhance biological activity while minimizing toxicity, we employed a chemical reduction as a one-pot synthesis method that leverages fructose's reducing and capping properties [16,32]. This approach facilitated the efficient formation of AgNPs, which were subsequently functionalized with MBH (**1**) and its analog MA (**2**) (Figure 1). AgNPs were selected due to their diverse biological activities, including antimicrobial, anti-inflammatory, and antioxidative effects [33–40]. Beyond these properties, their protective effect on the gastrointestinal tract is particularly significant [41,42]. Recent research highlights their potential to shield the gut lining by reducing oxidative stress, alleviating inflammation, and restoring gut microbiota balance [43]. This protective action extends to conditions such as inflammatory bowel disease (IBD) and gastric ulcers, where AgNPs may play a supportive therapeutic role. Additionally, AgNPs have been shown to mitigate damage caused by pathogens and chemical irritants, thus contributing to maintaining gastrointestinal health [15].

Using fructose as a reducing and capping agent was critical for achieving the desired properties of the synthesized AgNPs. At the boiling temperature of water (100 °C), Ag+ ions were rapidly reduced to their ground state, resulting in the formation of fructose-capped AgNPs (**3**) [44]. These nanoparticles were explicitly chosen for their lower toxicity than other capping agents [45,46], making them more biocompatible and suitable for medical research and practice applications.

2.1. *Particle Size and Characteristics of AgNPs with MBH (**4**) Compared to AgNPs with MA (**5**)*

To establish the size and shape of the AgNPs, TEM, DLS, and zeta potential were used. TEM images also indicate the shape and size of the nanoparticles and illustrate the individual nanoparticles. The TEM images confirmed the synthesis of smaller spherical particles with different sizes in the range of 3 to 10 and 90 to 95 nm for (**4**) (Figure 2a) and between 84 and 95 nm size for (**5**) (Figure 2b).

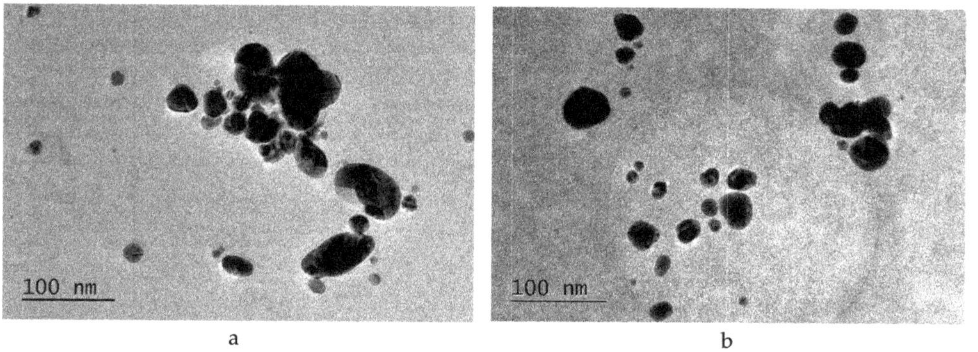

a b

Figure 2. TEM images of (**a**) AgNPs with MBH (**4**) and (**b**) AgNPs with MA (**5**).

AgNPs were obtained with controllable size due to carbohydrate assistance in the presence of $[Ag(NH_3)_2]^+$ [47]. It is well known that irrespective of the approach, biomolecules are involved in the reduction of silver nanoparticles and in an interaction with the upper face of silver, which is their initial connection of originating particles [48,49]. The carbohydrate coating of AgNPs decreases the agglomeration rate and size. Many groups are likely absorbed on the surface of silver due to covalent bonding to oxygen and nitrogen for the complex formation of silver [50]. The dynamic surface site of the AgNPs is determined based on the particle size, shape, and accumulation rate. The reaction conditions such as temperature, pH, extract volume, reactant concentration, and time describe the growing particles' size and gradation and influence the silver accumulates' shape and size [51].

The median average size of the obtained particles was confirmed by DLS (Figure 3). The histograms of both samples (**4**) and (**5**) displayed a bimodal particle size distribution, suggesting the presence of aggregates.

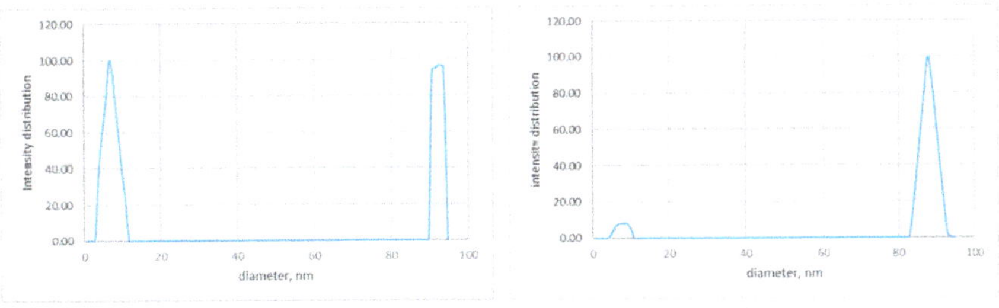

a b

Figure 3. Dynamic light scattering histograms of (**a**) AgNPs with MBH (**4**) and (**b**) AgNPs with MA (**5**).

The zeta potential of (**4**) was −19.19 mV, and that of (**5**) was −13.64 mV. Both samples' negative charge of the zeta potential refers to carboxylic groups of gluconic acid obtained when Ag^+ is reduced to Ag°. Carboxylic acids, obtained in the oxidation of sugars, provide a negative surface charge density to counteract the van der Waals forces responsible for particle coalescence. Self-assembled carboxylic acids ensure dense coating on the metal surfaces and stabilize them [52].

*2.2. UV-Vis and FTIR Spectra of AgNPs with MBH (**4**) and AgNPs with MA (**5**)*

The synthesis of the Ag NPs in an aqueous solution was monitored by recording the absorption spectra in the wavelength range of 190–600 nm (Figure 4). UV/Vis absorption spectroscopy is a crucial tool for investigating AgNPs in suspension. Each type of AgNP exhibits intriguing optical properties associated with surface plasmon resonance [53]. The time-dependent UV/Vis absorption spectra of AgNPs with MBH **4** (Figure 4a) and AgNPs with MA **5** (Figure 4b) are presented. Figure 4 shows that MBH (a) and MA (b) have absorption maxima at 229 nm and 225 nm, respectively, and fructose has an absorption peak at 189 nm. The indicator for the presence of AgNPs is the SPR characteristic band at 415 nm. The position of this band between 350 and 420 nm indicates the spherical shape of the particles [54].

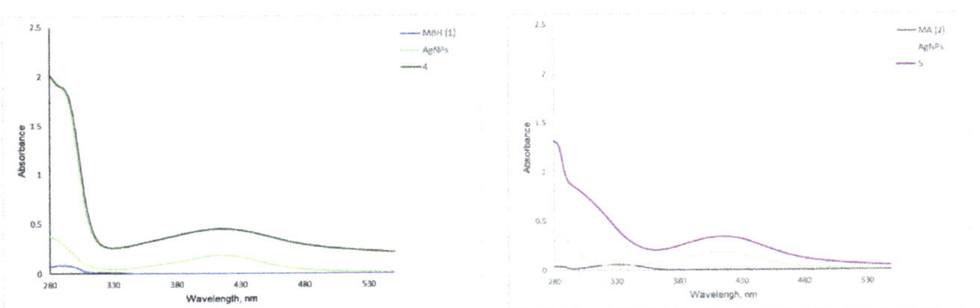

a b

Figure 4. Absorption spectra of (**4**) and (**5**): (**a**) MBH (**1**)—blue; AgNPs (**3**)—light green; AgNPs with MBH (**4**)—dark green; (**b**) MA (**2**)—grey; AgNPs (**3**)—light green; AgNPs with MA (**5**)—purple.

The SPR peak is broadening in both spectra (Figure 4a,b). The peak of (**4**) is symmetrical, which indicates the low degree of aggregation of the NPs [16,55]. A slight asymmetry is observed in the SPR peak of (5). The reasons for the asymmetry are different; it may be due to aggregation [56] or the existence of tautomeric forms [32]. The most likely explanation is the potential for intramolecular hydrogen bonding between the primary and secondary amino group in (**2**), which would result in forming a six-membered ring. We eliminate the potential for aggregation since the carbohydrate utilized to make the nanoparticles also helps stabilize them [57].

The deposition of (**1**) and (**2**) on AgNPs' surface, forming (**4**) and (**5**), has a negligible effect on the position of the SPR peak—422 nm (**4**) and 420 nm (**5**)—suggesting no significant alteration in their optical properties.

The majority of the distinctive bands of fructose are presented in the FTIR spectra of AgNPs with MBH (**4**) and AgNPs with MA (**5**) (Figure 5), along with a few extra and modified bands. In the FTIR spectrum of fructose (blue), the broad peak with a maximum at approximately 3417 cm^{-1} and its corresponding sharp band at 3524 cm^{-1} represent the stretching vibrations of the hydroxyl groups (-O-H), which are H-bonded and free hydrogen bonds, respectively. Additionally, the O-H stretching vibration mode linked to intramolecular and intermolecular hydrogen bonding was identified as the strong, broadband source at 3369–3368 cm^{-1} in (**4**) and (**5**), respectively.

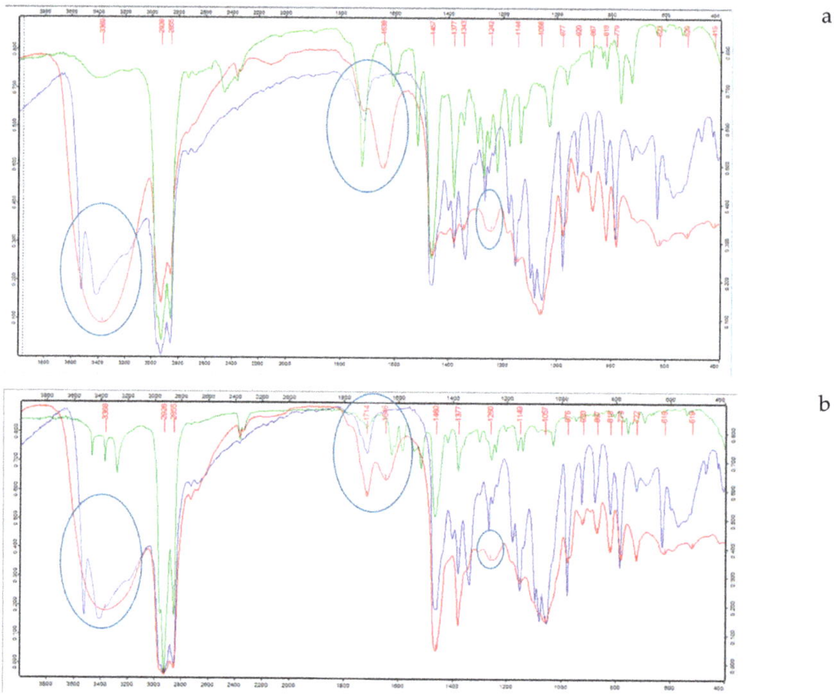

Figure 5. FTIR spectra of (**4**) and (**5**): (**a**) fructose—blue; AgNPs with MBH (**4**)—red; MBH (**1**)—green; (**b**) fructose—blue; AgNPs with MA (**5**)—red; MA (**2**)—green.

The 1058/1057 cm^{-1} peak can be assigned to C-O vibrations. The significant peak at 1713–1714 cm^{-1} is due to the C=O stretching of the -COOH group of the fructose unit. Compared to the FTIR spectrum of fructose, the corresponding peaks in the spectrum of (**4**) and (**5**) exhibit visible changes for the peaks of the COO– (around 1638 cm^{-1}) and O-H groups (around 3369 cm^{-1}). These can be considered an indication for the synthesis and stabilization of the AgNPs due to the responsible carboxyl and hydroxyl groups [32,44,58]. The weak broadband at 2104 cm^{-1} in both the (**4**) and (**5**) spectrum can be assigned to –CO-Ag linkages [59].

2.3. Effect of (1) to (5) on HepG2 Cell Morphology

The rapidly expanding application of nanoparticles in daily life raises concerns regarding their safety. AgNP risk assessments are intended to investigate biological consequences, potential causes, and practical strategies to reduce adverse effects. Less than 10 nm of Ag NPs may cause mutagenicity and is linked to genotoxicity due to oxidative stress-induced damage to DNA and chromosomes [60,61]. Elemental mapping of individual cells and transmission electron microscope pictures demonstrated that AgNPs can translocate to the nucleus and harm chromosomes and DNA [62–64]. Therefore, our primary goal was to assess the cytotoxic and genotoxic effects of synthesized AgNPs.

A sign of NPs' cytotoxicity is the alteration in normal cell morphology following treatment with NPs. Therefore, in this study, we assessed HepG2 cell morphology using phase-contrast microscopy at 24 h and 72 h after treatment with the tested substances. HepG2 cells were chosen for this investigation because they are widely recognized as excellent models for testing bioactivity. Their well-characterized metabolic and physiological properties closely mimic those of human hepatocytes, making them invaluable for studying the bioeffects and cytotoxicity of various substances [65]. At 24 h, only cells treated with substance (**3**) exhibited slight morphological changes, characterized by a more elongated cell shape. However, a more pronounced alteration in morphology was observed at 72 h. HepG2 cells treated with substances (**3**) and (**5**) displayed distinct apoptotic features. Compared to the untreated control cells, those treated with (**3**) and (**5**) exhibited irregular and fragmented morphology, with most cells detaching and dying. In contrast, untreated HepG2 cells maintained a healthy, polygonal shape with more regular dimensions and formed tightly packed monolayers (Figure 6).

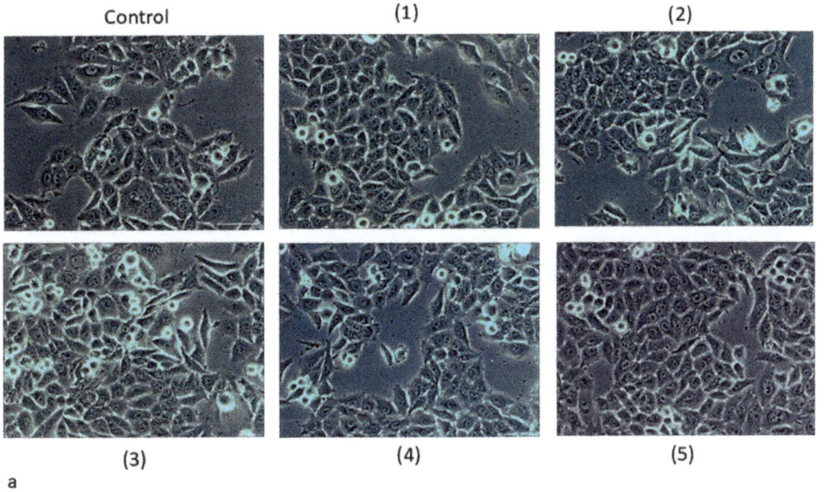

Figure 6. *Cont.*

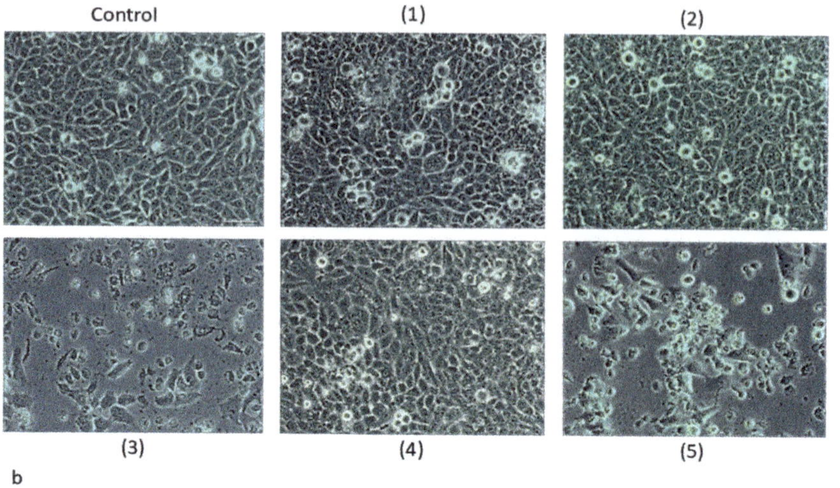

Figure 6. Phase-contrast micrographs of HepG2 cells treated for 24 h (**a**) and 72 h (**b**) with substances (**1**) to (**5**). Morphological changes in the cells were monitored to evaluate potential cytotoxic effects and alterations in cell structure. Non-treated HepG2 cells were used as the control group to establish baseline morphology and ensure the reliability of the observations. Images were captured at a magnification of 20×, with a scale bar of 50 μm for reference.

2.4. Effect of (1) to (5) on HepG2 Cell Viability

To further evaluate the cytotoxicity of the tested substances, HepG2 cells were treated with increasing concentrations (ranging from 10 μg/mL to 100 μg/mL) for 24 and 72 h, and the cell viability was assessed using the WST-1 assay. The results demonstrated a concentration-dependent decrease in cell viability at both time points, as depicted in Figure 7 and summarized in Table 1. At 24 h, substance (**3**) (AgNPs) exhibited the most significant cytotoxicity, reducing cell viability to 36.9% at the lowest concentration (10 μg/mL). This was followed by substance (**5**) (AgNPs with MA), which reduced viability by 57.7%, and substance (**4**) (AgNPs with MBH), with a reduction of 38.81%. Substance (**2**) (the analog) and substance (**1**) (MBH) showed the least cytotoxic effects, with reductions of 18.83% and 16.46%, respectively. All substances exhibited significant cytotoxicity at the highest concentration (100 μg/mL), reducing cell viability to 2.89–3.48%. The IC_{50} values calculated from these data confirmed the cytotoxicity order at 24 h as (**3**) > (**5**) > (**4**) > (**2**) > (**1**).

Table 1. IC_{50} values of the tested compounds **1–5** against HepG2 cells at the 24th h and 72nd h.

Samples	IC_{50} at 24th h [μg/mL]	IC_{50} at 72nd h [μg/mL]
1	17.16	27.50
2	24.69	29.39
3	6.35	17.79
4	14.16	17.94
5	8.88	13.93

Interestingly, at 72 h, a shift in the cytotoxicity profiles was observed. Substance (**1**) displayed a stimulatory effect on cell viability at the lowest concentration (10 μg/mL), indicating a potential adaptive cellular response. However, at higher concentrations (50 and 100 μg/mL), it became the most cytotoxic, showing a marked decrease in cell viability. Substance (**2**) alone continued to exhibit relatively low cytotoxicity, but when loaded onto AgNPs as substance (**5**), it became significantly more toxic, surpassing the cytotoxicity of (**3**) as plain AgNPs. At 72 h, the order of toxicity based on IC_{50} values shifted to (**5**) > (**3**) > (**2**) > (**4**) > (**1**).

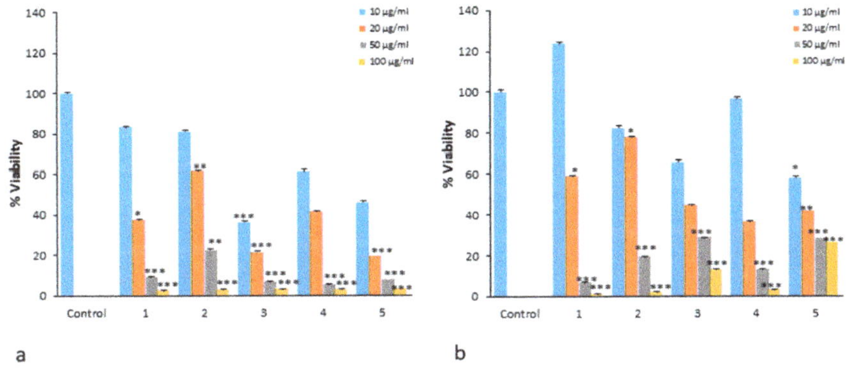

Figure 7. Analysis of cell viability using the WST-1 assay following treatment with substances (**1**) to (**5**): (**a**) at 24 h and (**b**) at 72 h. Non-treated HepG2 cells served as the control group for baseline comparison. Data are presented as mean ± SEM, with n = 3 independent experiments. Statistical significance is denoted by asterisks, with * $p < 0.05$, ** $p < 0.01$, and *** $p < 0.001$ indicating significant differences between the treated and control samples. These results provide insight into the cytotoxic effects of the tested substances over time.

These results seem to be supported by other authors who found that a significant reduction in cell survival was detected after 48 h of incubation following AgNP treatment [66].

These findings align with our hypothesis that AgNPs could be a viable delivery vehicle for MBH and MA, as they enhance the therapeutic agents' bioavailability while maintaining manageable cytotoxicity levels. The data also highlight the mitigated cytotoxicity of nanoparticle formulations compared to plain AgNPs, reinforcing their potential as effective and safe drug delivery systems. Notably, while AgNPs alone exhibited pronounced cytotoxic effects, their combination with MBH (**4**) or MA (**5**) effectively reduced the inherent toxicity of the nanoparticles, supporting their use in therapeutic applications.

The higher cytotoxicity observed in HepG2 cells treated with AgNPs (**3**) and AgNPs with MA (**5**) can be attributed to multiple factors. The small size and high surface area of AgNPs enhance cellular uptake, increasing the active agents' intracellular concentrations and amplifying their cytotoxic effects [67]. Additionally, combining AgNPs with MA in substance (**5**) likely results in a synergistic interaction, further potentiating its cytotoxicity. AgNPs also activate cellular stress pathways, including DNA damage and mitochondrial dysfunction, contributing to enhanced cell death. Moreover, the sustained release of MA from the nanoparticle matrix prolongs its bioactivity, maintaining a continuous cytotoxic effect [68–70]. These mechanisms collectively explain the pronounced cytotoxicity observed with substances (**3**) and (**5**), particularly at higher concentrations and extended exposure times. Despite these findings, the formulations demonstrate a promising balance of efficacy and safety, supporting their potential use as targeted therapeutic delivery systems.

2.5. Reactive Oxygen Species (ROS) Production

It has been established that the cytotoxicity and genotoxicity of NPs are determined by their absorption and the generation of ROS [71]. To explore whether AgNPs induce oxidative stress, potentially enhancing their effectiveness as a vehicle for MBH and MA by increasing cell susceptibility to therapeutic effects, we measured reactive oxygen species (ROS) levels 24 and 72 h after treatment with all tested substances. As shown in Figure 8, AgNPs did not significantly increase ROS production in HepG2 cells at 24 h. However, at the 72 h time point, ROS levels were significantly elevated compared to untreated control cells. Among the treatments, substances (**2**) and (**5**) induced the highest ROS production, while substance (**3**) (AgNPs alone) resulted in the lowest ROS induction. This pattern, however, does not directly align with the cytotoxicity results, where AgNPs exhibited the highest cytotoxic effects. This discrepancy suggests that the cytotoxic effects of AgNPs are not solely dependent on ROS generation. Instead, the potential of AgNPs as a drug delivery vehicle for MBH and MA may rely on additional mechanisms, such as enhancing the bioavailability of the loaded compounds and promoting their sustained release, which can ultimately improve therapeutic efficacy without primarily relying on oxidative stress pathways.

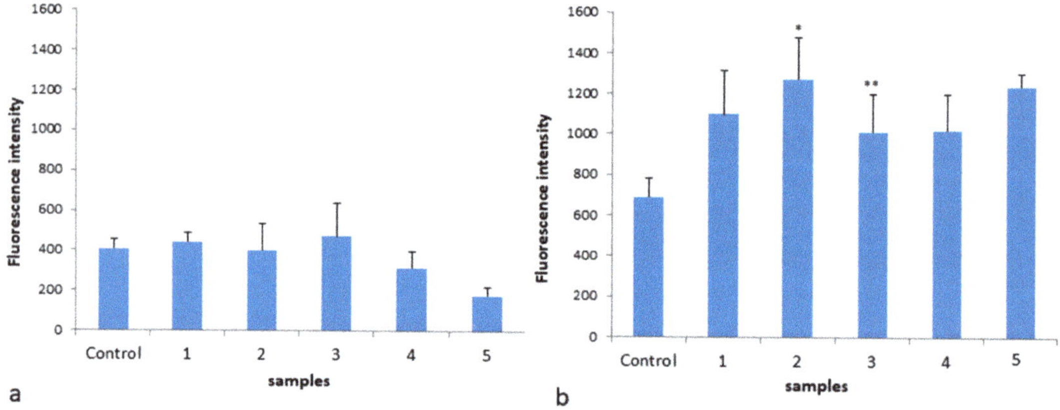

Figure 8. ROS production after treatment with (**1**) to (**5**) (**a**) at the 24th h and (**b**) at the 72nd h. Non-treated HepG2 cells were used as control. Bars are mean + SEM, n = 3. * $p < 0.05$; ** $p < 0.01$; asterisks indicate a significant difference between the sample and non-treated control.

2.6. Genotoxicity Testing of the Nanoformulations as Drug Delivery Vehicles for MBH and MA

The genotoxicity of compounds (**1**) to (**5**) was also evaluated. The results are presented in Figure 8 as box–whisker plots of the Olive Moment parameter. Following 24 h of treatment at IC_{50} concentrations (as detailed in Table 1), no genotoxic effects were detected in HepG2 cells (Figure 9a), indicating that DNA integrity remained intact. This suggests that neither the nanoparticles nor the tested compounds induced significant DNA damage during the initial exposure period.

In contrast, after the 72nd h of the treatment, a marked genotoxic effect was observed, particularly for compound (**5**) (Figure 9b). A slight increase in DNA strand breaks characterized this effect relative to the control group and other treatment conditions. The observed effect may stem from the synergistic interaction between AgNPs and the MA's structural components, which could potentiate genotoxicity over time. These findings underscore the importance of considering the chemical composition of nanoparticle-based systems and the temporal dynamics of their interaction with biological systems. Prolonged exposure to such materials may compromise genomic stability, raising potential safety concerns for extended treatment duration applications.

This evaluation provides a basis for understanding the safety parameters of the tested compounds and transitions logically to the next chapter, where molecular docking studies are employed to explore the biological interactions of compounds (**1**) to (**5**).

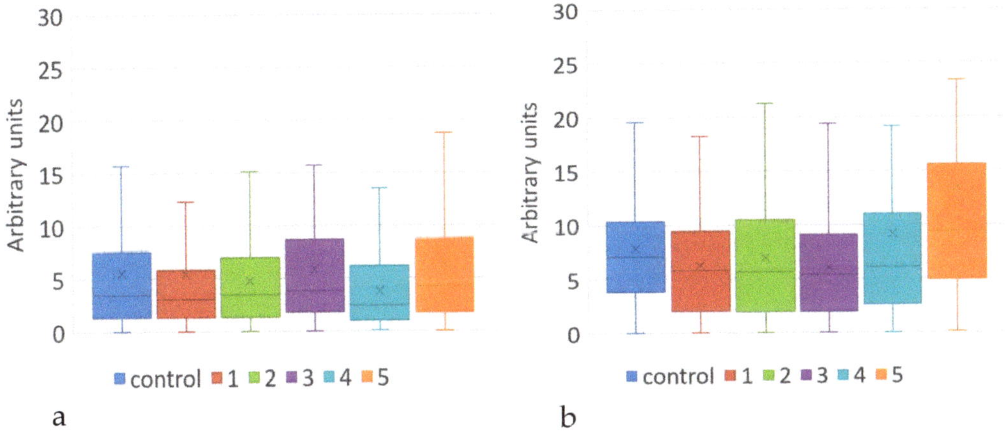

Figure 9. Genotoxic effects of the tested compounds **1** to **5** on HepG2 cells: DNA integrity across (**a**) 24th and (**b**) 72nd h. The genotoxic effects of compounds (**1**) to (**5**) were assessed using the neutral Comet Assay. HepG2 cells were cultured and treated at IC_{50} concentrations. DNA integrity was analyzed based on Olive Moment (OM) values, with data represented as (**a**) for 24 h treatments and (**b**) for 72 h treatments. Results are presented as box–whisker plots, reflecting three independent experiment repetitions.

2.7. Molecular Docking Simulations

The biological properties of (**1**) to (**5**) were further explored through molecular modeling. We limited the simulation of the nanoparticle structure to only a region consisting of two fructose molecules. Employing in silico docking simulations via AutoDock 4.2 and AutoDock Tools 1.5.6, the molecular interactions of these compounds with the VEGFR2 protein (PDB ID: 4AGD) were examined. The optimized molecular structures of the compounds were docked with VEGFR2 to evaluate their binding affinities, providing a comprehensive understanding of their potential biological effects. This computational approach was driven by the promising activity and safety potential of AgNPs demonstrated in experimental studies; by visualizing the key interaction patterns and identifying critical binding sites, the docking analysis aimed to validate and support the experimental findings. Such an integrative strategy offers more profound insights into the mechanism of action, enhancing the predictive accuracy of the therapeutic potential of these compounds.

The results of the docking analysis provide valuable insights into the binding interactions between the 4AGD protein and the tested compounds. These interactions' calculated binding free energies were −6.80, −5.91, −6.83, −7.47, and −6.59 kcal/mol, respectively. These results suggest that the trend of the inhibitory activity of compounds against the target protein is as follows: (**5**) > (**3**) > (**2**) > (**4**) > (**1**). Notably, this observed pattern is consistent with the cytotoxicity data, further supporting the correlation with binding affinity. Moreover, the docking analysis underscores the superior efficacy of (**5**) in inhibiting cellular growth compared to the other tested agents. This enhanced potency identifies it as a promising candidate for further exploration in therapeutic strategies. The studied compounds demonstrate significant interactions within the target protein's active site, as depicted in Figure 10.

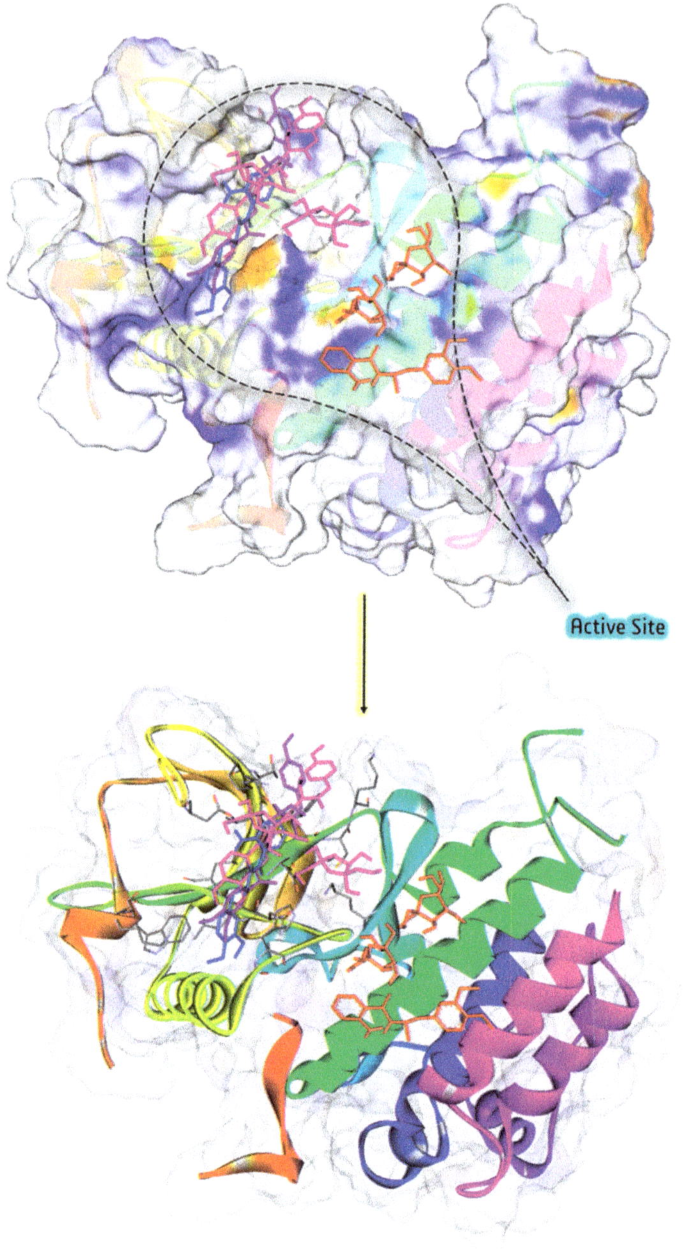

Figure 10. The docking perspective of (**1**) to (**5**) in the interaction with VEGFR2 protein.

A comprehensive visualization of these interactions is provided in Figure 11, which showcases detailed 3D interaction diagrams of the compounds during their binding process with the 4AGD protein.

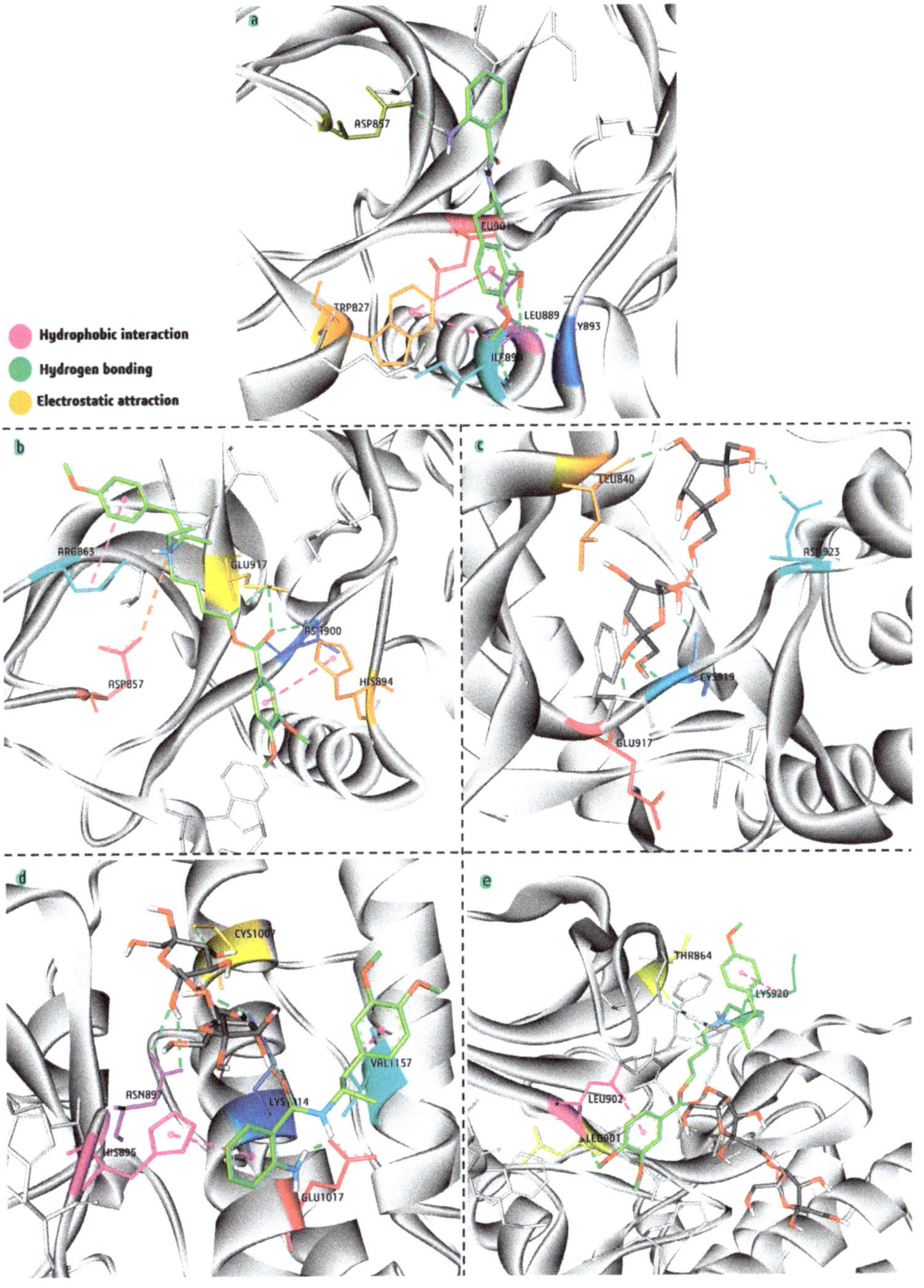

Figure 11. The 3D docking diagram of (**2**) (**a**), (**1**) (**b**), (**3**) (**c**), (**5**) (**d**), and (**4**) (**e**) in the interaction with VEGFR2 protein.

The findings reveal that AgNPs engage with the receptor primarily through hydrogen bonding, underscoring the critical role of these interactions in stabilizing the complex. In contrast, the binding behavior of (**1**) to (**5**) exhibits a more diverse interaction profile.

These systems rely predominantly on hydrophobic interactions in addition to hydrogen bonding. This dual-mode interaction strategy highlights the complex nature of ligand–receptor binding, where both polar and non-polar forces contribute to the overall stability and specificity of the molecular complex.

3. Materials and Methods

All solvents and reagents and MBH hydrochloride were purchased from Merck (Merck KGaA, Darmstadt, Germany). MA (**2**) was characterized using IR, ^{1}H-NMR, and ^{13}C-NMR. The melting point was measured on a Kruss M5000 melting point meter (A.Krüss Optronic GmbH, Hamburg, Germany). IR spectra were determined on a VERTEX 70 FT-IR spectrometer (Bruker Optics, Ettlingen, Germany). ^{1}H-NMR and ^{13}C-NMR spectra were recorded on a Bruker Avance III HD 500 spectrometer (Bruker, Billerica, MA, USA) at 500 MHz (^{1}H-NMR) and 125 MHz (^{13}C-NMR), respectively. Chemical shifts are given in relative ppm and were referenced to tetramethylsilane (TMS) (δ = 0.00 ppm) as an internal standard; the coupling constants are indicated in Hz. The NMR spectra were recorded at room temperature (ac. 295 K). The purity was determined by TLC using several solvent systems of different polarities. TLC was carried out on precoated 0.2 mm Fluka silica gel 60 plates (Merck KGaA, Darmstadt, Germany), using chloroform/n-hexane/methanol/acetone = 4:4:1:1 as a chromatographic system.

3.1. Synthetic Protocol

3.1.1. Synthesis of MA (**2**) (Scheme 1) [72]

A mixture of isatoic anhydride (**9**) (1.63 g, 10 mmol) and 1-(3,4-dimethoxyphenyl) propan-2-amine (**10**) (initially prepared from 1-(3,4-dimethoxyphenyl)propan-2-one) [73,74] (2.93 g, 15 mmol) dissolved in dichloromethane (30 mL) was stirred overnight at rt. The resulting solution was filtered on neutral Al_2O_3 and concentrated. Spectral data confirmed the structure of the obtained 2-amino-N-(1-(3,4-dimethoxyphenyl)propan-2-yl)benzamide (MA, **2**): m.p. 152.8 °C, ^{1}H NMR (500 MHz, $CDCl_3$) δ 1.14 (d, J = 6.7, 3H, CH_3), 2.70–2.80 (m, 2H, CH_2), 3.75 (s, 3H, OCH_3) 3.78 (s, 3H, OCH_3), 4.32 (dq, J = 13.4, 6.7 Hz, 1H, CH), 5.51 (broad s, 1H, NH_2), 5.86 (d, J = 7.9 Hz, 1H, NH), 6.56–6.59 (m, 1H, Ar), 6.66–6.68 (m, 3H, Ar), 6.74 (d, J = 5, 1H, Ar), 7.11–7.15 (m, 2H, Ar); ^{13}C NMR (126 MHz, $CDCl_3$) δ 168.49, 148.86, 147.72, 147.46, 132.25, 130.37, 126.98, 121.59, 117.91, 117.43, 117.07, 112.62, 111.18, 55.91, 55.84, 46.16, 41.66, 19.99; FT-IR, cm^{-1}: 3467 ν_{as} (-N-H, -NH_2), 3371 ν_s (-N-H, -NH_2), 3283 ν (-N-H, >NH-amide), 3067, 3055, 3033 ν (Csp^2-H, -Ph), 2998 ν_{as} (Csp^3-H, -OCH_3), 2972 ν_{as} (Csp^3-H, -CH_3), 2935, 2914 ν_{as} (Csp^3-H, -CH_2-), 2874 ν_s (Csp^3-H, -CH_3), 2838 ν_s (Csp^3-H, -OCH_3), ν_s (Csp^3-H, -CH_2-), 1626 ν (>C=O), secondary amide I, 1584 ν (C-C=C, -Ph1,2,4), 1539, 1514 δ (N-H) and ν (C-N), trans-secondary amide II, 1463 ν (C-C=C, -Ph$_{ortho}$), ν (C-C=C, -Ph1,2,4), δ_{as} (-CH_3), δ_{as} (-OCH_3), 1447 δ_s (-OCH_3), 1416 ν (C-C=C, -Ph1,2,4), 1358 δ_s (-CH_3), 1300 ν (C-C=C, -Ph$_{ortho}$), ν (C-N), secondary amide III, 1256 δ (-Csp^2-H, -Ph1,2,4), 1157 δ (-Csp^2-H, -Ph$_{ortho}$), ρ(-CH-)/ν (N-C) in –NH-CH-, 1139 δ (-Csp^2-H, -Ph1,2,4), 1029 δ (-Csp^2-H, -Ph$_{ortho}$); HRMS electrospray ionization (ESI) m/z calcd. for $[M+H]^+$ $C_{18}H_{23}O_3N_2^+$ = 315.17032, found 315.16956 (mass error Δm = −2.41 ppm).

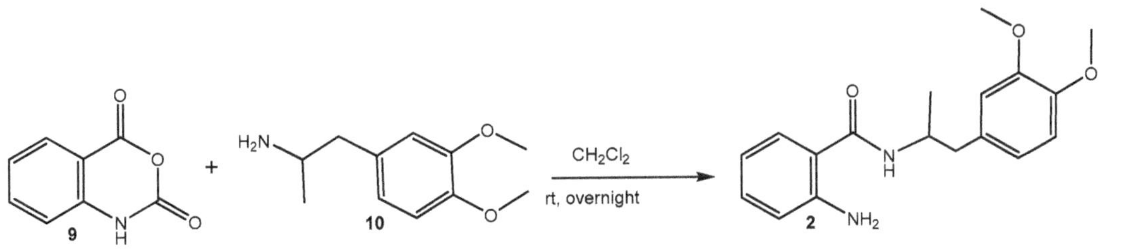

Scheme 1. Reaction pathway for the synthesis of MA (**2**).

3.1.2. Synthesis of Fructose-Assisted Ag NPs (**3**)

A total of 1.25 g (0.007 mol) fructose was dissolved in 25 mL water and refluxed for 5 min. Then, 0.63 mL of 0.01 M $AgNO_3$ solution was added, and the mixture was kept at boiling. In about 5 min, the solution turned pale yellow, indicating the formation of **6**.

3.1.3. Synthesis of MBH-Loaded AgNPs (**4**) and MA-Loaded Ag NPs (**5**)

A total of 1.25 g (0.007 mol) fructose was dissolved in 25 mL water and refluxed for 2 min; then, 0.63 mL of 0.01 M $AgNO_3$ solution was added. Then, MBH (**1**) or MA (**2**) was added in a concentration of 1 mg/mL. The solution's color changed to a light yellow after roughly 5 min of reflux, signifying the formation of (**4**) or (**5**).

3.2. Characterization of the Ag NPs: Analytical Techniques

After NP preparation, the solution was used for UV-Vis, transmission electron microscopy (TEM), dynamic light scattering (DLS), and zeta potential measurements.

3.2.1. TEM

The morphology, shape, and size of the nanoparticles were investigated by TEM (Talos F200C G2, Thermo Scientific, Landsmeer, The Netherlands). The nanoparticle suspension was added dropwise onto a formvar/carbon-coated copper grid, and then the TEM observation of the samples was performed at an operating voltage of 200 kV.

3.2.2. DLS and Zeta Potential

DLS measurements were performed on a Brookhaven BI-200 goniometer (Brookhaven Instrumetns Corporation, Holtsville, NY, USA) with vertically polarized incident light at a wavelength l = 632.8 nm supplied by a He–Ne laser operating at 35 mW and equipped with a Brookhaven BI-9000 AT digital autocorrelator (Brookhaven Instrumetns Corporation, Holtsville, NY, USA). The scattered light was measured for dilute aqueous dispersions in the concentration range 0.056–0.963 mg mL^{-1} at 25, 37, and 65 °C. Measurements were made at angles y in the range of 50–130°. The system allows measurements of ζ-potential in the range from −200 mV to +200 mV. All analyses were performed in triplicate at 25 °C.

3.2.3. UV-Vis Spectra

UV-Vis spectra were recorded between 320 and 800 nm using a Cary-60 UV-Vis spectrophotometer (Agilent Technologies, Santa Clara, CA, USA). Solution spectra were obtained by measuring the absorption of the prepared nanoparticle dispersions in a quartz cuvette with a 1 cm optical path.

3.2.4. FTIR Spectra

IR spectra were determined on a VERTEX 70 FT-IR spectrometer (Bruker Optics, Ettlingen, Germany). The spectra were collected from 600 cm^{-1} to 4000 cm^{-1} with a resolution of 4 nm and 32 scans. The instrument was equipped with a diamond attenuated total reflection (ATR) accessory. The IR spectra were analyzed with the OPUS-Spectroscopy Software, Bruker (Version 7.0, Bruker, Ettlingen, Germany).

3.3. Molecular Docking

An in silico docking analysis was carried out to explore the cytotoxic potential of the compounds further, complementing the results of in vitro experiments. The computational study utilized AutoDock 4.2 [75] and AutoDock Tools 1.5.6 [76], renowned molecular docking and binding analysis tools. The high-resolution crystallographic structure of the Vascular Endothelial Growth Factor Receptor 2 (VEGFR2) protein (PDB ID: 4AGD) was obtained from the RCSB Protein Data Bank (www.rcsb.org), accessed on 19 December 2024. Optimization procedures were undertaken to prepare the 4AGD structure for docking simulations. This process involved the removal of non-essential atoms, ligands, and water molecules that could interfere with the accuracy of the docking process. To further enhance the precision of the simulation, polar hydrogens were added, and Kollman charges were assigned to the structure. These modifications ensured a more accurate representation of the protein's binding environment, enabling a more reliable prediction of ligand–protein interactions. The structure of the compounds was obtained from the theoretical approach. This study employed computational techniques to construct and optimize the 3D structures of (**1**) and (**2**) and AgNPs functionalized with two fructose molecules and drugs. These molecular models were drawn using Gauss View 6.0 and optimized through the Gaussian 09W (Revision D.01, Gaussian, Inc.) software suite to facilitate docking simulations. The density functional theory (DFT) approach was utilized for geometry optimizations, explicitly employing the B3LYP method with the 6-311G(d,p) basis set for all atoms involved. To accurately identify the docking site, a grid box was established with a finely tuned spacing of 0.375 Å. The center of this grid was aligned with the active site at the coordinates corresponding to 4AGD, encompassing dimensions of 90 × 90 × 80. Docking simulations were conducted utilizing the Lamarckian Genetic Algorithm (LGA). Following the completion of the simulations, the resulting binding conformations were thoroughly examined using the Discovery Studio Visualizer 4.1 software to facilitate an in-depth analysis of the interactions. The docking procedure was meticulously carried out in alignment with the established protocols outlined in the published literature [77–79].

3.4. Cell Cultures

HepG2 cells, derived from a human liver carcinoma, were cultured in Dulbecco's modified Eagle's medium (DMEM) culture with 10% fetal bovine serum (FBS; Merck KGaA, Darmstadt, Germany). Cells grew in a humidified environment with 5% CO_2 and 95% humidity at 37 °C; 0.05% trypsin and 0.02% EDTA solution (Merck KGaA, Darmstadt, Germany) were used to detach cells for in vitro experiments. For cytotoxicity experiments (WST-1 assay, ROS, phase-contrast morphology), the cells were seeded in 24-well plates with a concentration of 2.5×10^4 cells/well. The cells were cultivated for 24 h and then treated with NPs.

3.5. Phase-Contrast Light Microscopy

Following a 24 and 72 h exposure to the studied NPs, phase-contrast observations were made to assess changes in cell morphology. Micrographs were taken at magnifications of 20× with a Zeiss microscope (Axiovert 25, Carl Zeiss, Jena, Germany) fitted with a digital camera.

3.6. WST-1 Assay

WST-1 was performed to evaluate cell viability and metabolic activity. At the end of incubation, the medium with NPs was removed, and the cells were washed with PBS.

Further, a 300 μL medium with 30 μL WST-1 (tetrazolium salt 4-(3-(4-iodophenyl)-2-(4-nitrophenyl)-2 h-5-tetrazolium)-1,3-benzene disulfonate) was added to all wells. Following a 1 h incubation period in the dark, an ELISA reader Thermo Scientific Multiskan Spectrum (Thermo Scientific, Tokyo, Japan) was employed to estimate the optical density of the samples at 450 nm.

3.7. DCFA-DA Analysis

As described before, intracellular reactive oxygen species (ROS) production was measured using 2,7-dichlorofluorescin diacetate (DCFH-DA, Sigma-Aldrich, St. Louis, MO, USA). DCFH-DA is a non-fluorescent compound that passively enters the cell, reacts with ROS, and forms the highly fluorescent compound dichlorofluorescin (DCF). In brief, HepG2 cells (2.5×10^4) were seeded in 24-well plates, allowing adherence. The next day, NPs with different concentrations were added to the cells and incubated for 24 h. Following respective exposure, cells were washed twice with PBS and incubated for 30 min in the dark in an FBS-free culture medium containing DCFH-DA (20 μM). Then, the DCFH-DA-containing medium was removed, the control (untreated) and treated cells were rinsed twice with PBS, and the fluorescence intensity of DCF was detected on a spectrofluorometer upon excitation at 485 nm and emission at 520 nm.

3.8. Genotoxicity Testing

The neutral Comet Assay evaluated the genotoxicity of (**1**) to (**5**). HepG2 cells were cultured in Dulbecco's modified Eagle's medium (DMEM) supplemented with 10% fetal bovine serum (FBS), 100 U/mL penicillin, and 10 μg/mL streptomycin, and maintained at 37 °C in a humidified atmosphere with 5% CO_2. Following treatment with the compounds for 24 and 72 h, the cells were collected by centrifugation at 2800× *g* for 5 min and washed with phosphate-buffered saline (PBS) to remove unbound substances. The cells were then resuspended in PBS and mixed with low-gelling agarose at a final concentration of 0.7% (*w*/*v*). This agarose–cell mixture was spread onto precoated microscopic slides and covered with a coverslip; after solidification at 4 °C for 10 min, the coverslip was removed. The slides containing the agarose-embedded cells were incubated in lysis solution (1 M NaCl; 50 mM EDTA, pH 8.0; 30 mM NaOH; 0.1% N-laurylsarcosine) for 60 min to disrupt cell membranes and remove proteins. Following lysis, the slides were submerged in alkaline electrophoresis buffer (10 mM EDTA, pH 8.0; 30 mM NaOH) for 30 min to unwind double-stranded DNA. Electrophoresis was conducted at a voltage of 0.45 V/cm for 20 min to allow DNA fragments to migrate towards the anode. After electrophoresis, the slides were rinsed with distilled water to neutralize the alkali in the gel and dehydrated by incubating in 75% and then 95% ethanol for 5 min each before air-drying. The DNA was stained using SYBR Green I (Molecular Probes) before observation under an epi-fluorescent microscope equipped with a 450–490 nm bandpass filter at a magnification of 250×. Comet Assay data quantitation was performed using CometScore software 2.0 (TriTek Corporation, Sumerduck, Virginia, USA), focusing on parameters such as Olive Moment (OM), which integrates both tail length and intensity to assess genotoxicity comprehensively.

3.9. Statistical Analysis

The obtained data were analyzed using Microsoft Excel software version 10 (Microsoft, Redmond, WA, USA). The presented bars illustrated mean values of the MDCK and HepG2 cell viability/metabolic activity ± STDV. Probability levels of * $p < 0.05$, ** $p < 0.01$, and *** $p < 0.005$ were considered statistically significant after the performance of Student's *t*-test.

4. Conclusions

MBH-loaded AgNPs and MA-loaded AgNPs were synthesized for the first time. The results of this study demonstrate the successful synthesis of AgNPs using fructose as a reducing and capping agent, leading to the formation of nanoparticles with controllable size and promising properties. The drug-loaded Ag NPs were characterized in vitro for their cytotoxic and genotoxic effect, compared to MBH, resp. MA, and plain AgNPs. Drug-loaded Ag NPs, on the other hand, showed better characteristics than MBH, or MA, itself.

Cytotoxicity testing revealed that AgNPs, particularly when combined with MA, exhibited concentration-dependent cytotoxic effects in HepG2 cells, with AgNPs alone showing the highest toxicity. Additionally, genotoxicity tests indicated no significant DNA damage at 24 h. Still, prolonged exposure to AgNPs in combination with MA resulted in slightly increased DNA strand breaks, highlighting the impact of treatment duration on genotoxicity.

Molecular docking studies provided further insights into the binding interactions between the compounds and the VEGFR2 protein, supporting the cytotoxicity data. The binding affinities suggested that the AgNPs loaded with MA (**5**) had more substantial interactions with the target protein, highlighting their potential as nanocarriers for mebeverine delivery. This computational analysis aligns with the experimental results and underscores the promising application of these formulations in enhancing drug delivery systems for IBS management.

Overall, the synthesized AgNPs show significant promise as drug delivery vehicles, with their combination with MBH and MA enhancing both bioavailability and efficacy. Their ability for targeted administration and low genotoxicity indicate that they may be developed into safe and effective medicines, warranting further investigation in vivo.

Author Contributions: Conceptualization, S.N. and M.T.; methodology, M.S., M.M., M.G., N.K., K.H.-P., M.T., G.M.Z. and K.S.; investigation, synthesis, and characterization, M.S., M.M., M.T., K.S., G.M.Z., S.T. and S.N.; genotoxicity testing, M.G., P.I. and G.M.; cell morphology and cytotoxicity testing, ROS, N.K. and K.H.-P.; DFT; software and molecular docking, M.F.-D.; writing—original draft preparation, S.N., M.T., N.K., K.H.-P., M.G., P.I., G.M. and M.F.-D.; writing—review and editing, S.N., N.K. and M.G.; visualization, S.N., N.K. and M.G.; supervision, S.N., M.G. and N.K.; project administration, S.N. All authors have read and agreed to the published version of the manuscript.

Funding: This research received no external funding.

Institutional Review Board Statement: Not applicable.

Informed Consent Statement: Not applicable.

Data Availability Statement: The original contributions presented in the study are included in the article. Further inquiries can be directed to the corresponding authors.

Acknowledgments: This study is supported by the Bulgarian Ministry of Education under the National Program "Young Scientists and Postdoctoral Students–2", Project № MUPD-HF-017.

Conflicts of Interest: The authors declare no conflicts of interest.

References

1. Buono, J.L.; Carson, R.T.; Flores, N.M. Health-related quality of life, work productivity, and indirect costs among patients with irritable bowel syndrome with diarrhea. *Health Qual. Life Outcomes* **2017**, *15*, 35. [CrossRef]
2. Frändemark, Å.; Törnblom, H.; Jakobsson, S.; Simrén, M. Work productivity and activity impairment in irritable bowel syndrome (IBS): A multifaceted problem. *Am. J. Gastroenterol.* **2018**, *113*, 1540–1549. [CrossRef]

3. Sperber, A.D.; Dumitrascu, D.; Fukudo, S.; Gerson, C.; Ghoshal, U.C.; Gwee, K.A.; Hungin, A.P.S.; Kang, J.Y.; Minhu, C.; Schmulson, M.; et al. The global prevalence of IBS in adults remains elusive due to the heterogeneity of studies: A Rome Foundation working team literature review. *Gut* **2017**, *66*, 1075–1082. [CrossRef]
4. Lacy, B.E.; Mearin, F.; Chang, L.; Chey, W.D.; Lembo, A.J.; Simren, M.; Spiller, R. Bowel disorders. *Gastroenterology* **2016**, *150*, 1393–1407. [CrossRef]
5. Raskov, H.; Burcharth, J.; Pommergaard, H.C.; Rosenberg, J. Irritable Bowel Syndrome, the Microbiota and the Gut-Brain Axis. *Gut Microbes* **2016**, *7*, 365–383. [CrossRef]
6. Shaikh, S.D.; Sun, N.; Canakis, A.; Park, W.Y.; Weber, H.C. Irritable Bowel Syndrome and the Gut Microbiome: A Comprehensive Review. *J. Clin. Med.* **2023**, *12*, 2558. [CrossRef]
7. Annahazi, A.; Roka, R.; Rosztoczy, A.; Wittmann, T. Role of Antispasmodics in the Treatment of Irritable Bowel Syndrome. *World J. Gastroenterol.* **2014**, *20*, 6031–6043. [CrossRef]
8. Daniluk, J.; Malecka-Wojciesko, E.; Skrzydlo-Radomanska, B.; Rydzewska, G. The Efficacy of Mebeverine in the Treatment of Irritable Bowel Syndrome—A Systematic Review. *J. Clin. Med.* **2022**, *11*, 1044. [CrossRef]
9. Chakraborty, D.S.; Hazra, A.; Sil, A.; Pain, S. Will Controlled Release Mebeverine Be Able to Surpass Placebo in Treatment of Diarrhoea Predominant Irritable Bowel Syndrome? *J. Fam. Med. Prim. Care* **2019**, *8*, 3173–3178.
10. Sowmya, C.; Reddy, C.S.; Priya, N.V.; Sandhya, R.; Keerthi, K. Colon-Specific Drug Delivery Systems: A Review on Pharmaceutical Approaches with Current Trends. *Int. Res. J. Pharm.* **2012**, *3*, 45–55.
11. Kou, L.; Bhutia, Y.D.; Yao, Q.; He, Z.; Sun, J.; Ganapathy, V. Transporter-Guided Delivery of Nanoparticles to Improve Drug Permeation across Cellular Barriers and Drug Exposure to Selective Cell Types. *Front. Pharmacol.* **2018**, *9*, 27. [CrossRef]
12. Blanco, E.; Shen, H.; Ferrari, M. Principles of Nanoparticle Design for Overcoming Biological Barriers to Drug Delivery. *Nat. Biotechnol.* **2015**, *33*, 941–951. [CrossRef]
13. Mitragotri, S.; Lammers, T.; Bae, Y.H.; Schwendeman, S.; De Smedt, S.; Leroux, J.C.; Peer, D.; Kwon, I.C.; Harashima, H.; Kikuchi, A.; et al. Drug Delivery Research for the Future: Expanding the Nano Horizons and Beyond. *J. Control. Release* **2017**, *246*, 183–184. [CrossRef]
14. Heidari, F.; Raoufi, Z.; Abdollahi, S.; Zare Asl, H. Antibiotic Delivery in the Presence of Green AgNPs Using Multifunctional Bilayer Carrageenan Nanofiber/Sodium Alginate Nanohydrogel for Rapid Control of Wound Infections. *Int. J. Biol. Macromol.* **2024**, *277*, 134109. [CrossRef]
15. Huang, Y.; Guo, X.; Wu, Y.; Chen, X.; Feng, L.; Xie, N.; Shen, G. Nanotechnology's Frontier in Combatting Infectious and Inflammatory Diseases: Prevention and Treatment. *Signal Transduct. Target. Ther.* **2024**, *9*, 34. [CrossRef]
16. Todorova, M.; Milusheva, M.; Kaynarova, L.; Georgieva, D.; Delchev, V.; Simeonova, S.; Pilicheva, B.; Nikolova, S. Drug-Loaded Silver Nanoparticles—A Tool for Delivery of a Mebeverine Precursor in Inflammatory Bowel Diseases Treatment. *Biomedicines* **2023**, *11*, 1593. [CrossRef]
17. Ahmad, N.; Bukhari, S.N.A.; Hussain, M.A.; Ejaz, H.; Munir, M.U.; Amjad, M.W. Nanoparticles Incorporated Hydrogels for Delivery of Antimicrobial Agents: Developments and Trends. *RSC Adv.* **2024**, *14*, 13535–13564. [CrossRef]
18. Krishnan, P.D.; Banas, D.; Durai, R.D.; Kabanov, D.; Hosnedlova, B.; Kepinska, M.; Fernandez, C.; Ruttkay-Nedecky, B.; Nguyen, H.V.; Farid, A.; et al. Silver Nanomaterials for Wound Dressing Applications. *Pharmaceutics* **2020**, *12*, 821. [CrossRef]
19. Kalvodová, A.; Dvořáková, K.; Petrová, E.; Michniak-Kohn, B.B.; Zbytovská, J. The Contest of Nanoparticles: Searching for the Most Effective Topical Delivery of Corticosteroids. *Pharmaceutics* **2023**, *15*, 513. [CrossRef]
20. Ren, Y.; Zhang, Y.; Li, X. Application of AgNPs in Biomedicine: An Overview and Current Trends. *Nanotechnol. Rev.* **2024**, *13*, 20240030. [CrossRef]
21. Kesharwani, P.; Gorain, B.; Low, S.Y.; Tan, S.A.; Ling, E.C.S.; Lim, Y.K.; Chin, C.M.; Lee, P.Y.; Lee, C.M.; Ooi, C.H.; et al. Nanotechnology-Based Approaches for Anti-Diabetic Drugs Delivery. *Diabetes Res. Clin. Pract.* **2018**, *136*, 52–77. [CrossRef]
22. Luceri, A.; Francese, R.; Lembo, D.; Ferraris, M.; Balagna, C. Silver Nanoparticles: Review of Antiviral Properties, Mechanism of Action and Applications. *Microorganisms* **2023**, *11*, 629. [CrossRef]
23. Hersh, A.M.; Alomari, S.; Tyler, B.M. Crossing the Blood-Brain Barrier: Advances in Nanoparticle Technology for Drug Delivery in Neuro-Oncology. *Int. J. Mol. Sci.* **2022**, *23*, 4153. [CrossRef]
24. Chen, X.; Schluesener, H.J. Nanosilver: A Nanoproduct in Medical Application. *Toxicol. Lett.* **2008**, *176*, 1–12. [CrossRef]
25. Ahamed, M.; Alsalhi, M.S.; Siddiqui, M.K. Silver Nanoparticle Applications and Human Health. *Clin. Chim. Acta* **2010**, *411*, 1841–1848. [CrossRef]
26. Lankveld, D.P.; Oomen, A.G.; Krystek, P.; Neigh, A.; Troost–de Jong, A.; Noorlander, C.W.; Van Eijkeren, J.C.H.; Geertsma, R.E.; De Jong, W.H. The Kinetics of the Tissue Distribution of Silver Nanoparticles of Different Sizes. *Biomaterials* **2010**, *31*, 8350–8361. [CrossRef]
27. Hyun, J.S.; Lee, B.S.; Ryu, H.Y.; Sung, J.H.; Chung, K.H.; Yu, I.J. Effects of Repeated Silver Nanoparticles Exposure on the Histological Structure and Mucins of Nasal Respiratory Mucosa in Rats. *Toxicol. Lett.* **2008**, *182*, 24–28. [CrossRef]

28. Kim, Y.S.; Kim, J.S.; Cho, H.S.; Rha, D.S.; Kim, J.M.; Park, J.D.; Choi, B.S.; Lim, R.; Chang, H.K.; Chung, Y.H.; et al. Twenty-Eight-Day Oral Toxicity, Genotoxicity, and Gender-Related Tissue Distribution of Silver Nanoparticles in Sprague-Dawley Rats. *Inhal. Toxicol.* **2008**, *20*, 575–583. [CrossRef]
29. Cha, K.; Hong, H.W.; Choi, Y.G.; Lee, M.J.; Park, J.H.; Chae, H.K.; Ryu, G.; Myung, H. Comparison of Acute Responses of Mice Livers to Short-Term Exposure to Nano-Sized or Micro-Sized Silver Particles. *Biotechnol. Lett.* **2008**, *30*, 1893–1899. [CrossRef]
30. Song, K.S.; Sung, J.H.; Ji, J.H.; Lee, J.H.; Lee, J.S.; Ryu, H.R.; Lee, J.K.; Chung, Y.H.; Park, H.M.; Shin, B.S.; et al. Recovery from Silver-Nanoparticle-Exposure-Induced Lung Inflammation and Lung Function Changes in Sprague Dawley Rats. *Nanotoxicology* **2013**, *7*, 169–180. [CrossRef]
31. Lee, J.H.; Mun, J.; Park, J.D.; Yu, I.J. A Health Surveillance Case Study on Workers Who Manufacture Silver Nanomaterials. *Nanotoxicology* **2012**, *6*, 667–669. [CrossRef]
32. Nikolova, S.; Milusheva, M.; Gledacheva, V.; Feizi-Dehnayebi, M.; Kaynarova, L.; Georgieva, D.; Delchev, V.; Stefanova, I.; Tumbarski, Y.; Mihaylova, R.; et al. Drug-Delivery Silver Nanoparticles: A New Perspective for Phenindione as an Anticoagulant. *Biomedicines* **2023**, *11*, 2201. [CrossRef]
33. Sukirtha, R.; Priyanka, K.; Antony, J.; Kamalakkannan, S.; Thangam, R.; Gunasekaran, P.; Krishnan, M.; Achiraman, S. Cytotoxic Effect of Green Synthesized Silver Nanoparticles Using *Melia azedarach* Against In Vitro HeLa Cell Lines and Lymphoma Mice Model. *Process Biochem.* **2012**, *47*, 273–279. [CrossRef]
34. Yesilot, S.; Aydin, C. Silver Nanoparticles: A New Hope in Cancer Therapy? *Eur. J. Med.* **2019**, *24*, 111–116. [CrossRef]
35. Khorrami, S.; Zarepour, A.; Zarrabi, A. Green Synthesis of Silver Nanoparticles at Low Temperature in a Fast Pace with Unique DPPH Radical Scavenging and Selective Cytotoxicity Against MCF-7 and BT-20 Tumor Cell Lines. *Biotechnol. Rep.* **2019**, *24*, e00393. [CrossRef]
36. Séguy, L.; Groo, A.-C.; Malzert-Fréon, A. How Nano-Engineered Delivery Systems Can Help Marketed and Repurposed Drugs in Alzheimer's Disease Treatment? *Drug Discov. Today* **2022**, *27*, 1575–1589. [CrossRef]
37. Muntimadugu, E.; Dhommati, R.; Jain, A.; Challa, V.G.S.; Shaheen, M.; Khan, W. Intranasal Delivery of Nanoparticle Encapsulated Tarenflurbil: A Potential Brain Targeting Strategy for Alzheimer's Disease. *Eur. J. Pharm. Sci.* **2016**, *92*, 224–234. [CrossRef]
38. Jojo, G.; Kuppusamy, G. Scope of New Formulation Approaches in the Repurposing of Pioglitazone for the Management of Alzheimer's Disease. *J. Clin. Pharm. Ther.* **2019**, *44*, 337–348. [CrossRef]
39. Jojo, G.; Kuppusamy, G.; De, A.; Karri, V.V.S.N.R. Formulation and Optimization of Intranasal Nanolipid Carriers of Pioglitazone for the Repurposing in Alzheimer's Disease Using Box-Behnken Design. *Drug Dev. Ind. Pharm.* **2019**, *45*, 1061–1072. [CrossRef]
40. Angelopoulou, E.; Piperi, C. DPP-4 Inhibitors: A Promising Therapeutic Approach Against Alzheimer's Disease. *Ann. Transl. Med.* **2018**, *6*, 255. [CrossRef]
41. Kose, O.; Béal, D.; Motellier, S.; Pelissier, N.; Collin-Faure, V.; Blosi, M.; Bengalli, R.; Costa, A.; Furxhi, I.; Mantecca, P.; et al. Physicochemical Transformations of Silver Nanoparticles in the Oro-Gastrointestinal Tract Mildly Affect Their Toxicity to Intestinal Cells In Vitro: An AOP-Oriented Testing Approach. *Toxics* **2023**, *11*, 199. [CrossRef]
42. Ibrahim, A.; Abbas, I.; Hussein, M.; Muter, S.; Abdulalah, M.; Morteta, A.; Suhayla, S.; Peshawa, A.; Nabaz, A.; Mahmood, A. Effect of nano silver on gastroprotective activity against ethanol-induced stomach ulcer in rats. *Biomed. Pharmacother.* **2022**, *154*, 113550. [CrossRef]
43. Patel, T.A.; Kevadiya, B.D.; Bajwa, N.; Singh, P.A.; Zheng, H.; Kirabo, A.; Li, Y.-L.; Patel, K.P. Role of Nanoparticle-Conjugates and Nanotheranostics in Abrogating Oxidative Stress and Ameliorating Neuroinflammation. *Antioxidants* **2023**, *12*, 1877. [CrossRef]
44. Filippo, E.; Serra, A.; Buccolieri, A.; Manno, D. Green synthesis of silver nanoparticles with sucrose and maltose: Morphological and structural characterization. *J. Non-Cryst. Solids* **2010**, *356*, 344–350. [CrossRef]
45. Kennedy, D.; Orts-Gil, G.; Lai, C.-H.; Müller, L.; Haase, A.; Luch, A.; Seeberger, P. Carbohydrate Functionalization of Silver Nanoparticles Modulates Cytotoxicity and Cellular Uptake. *J. Nanobiotechnol.* **2014**, *12*, 59. [CrossRef]
46. Pryshchepa, O.; Pomastowski, P.; Buszewski, B. Silver Nanoparticles: Synthesis, Investigation Techniques, and Properties. *Adv. Colloid Interface Sci.* **2020**, *284*, 102246. [CrossRef]
47. Iravani, S.; Korbekandi, H.; Mirmohammadi, S.; Zolfaghari, B. Synthesis of Silver Nanoparticles: Chemical, Physical, and Biological Methods. *Res. Pharm. Sci.* **2014**, *9*, 385–406.
48. Islam, N. Green Synthesis and Biological Activities of Gold Nanoparticles Functionalized with *Salix alba*. *Arab. J. Chem.* **2019**, *12*, 2914–2925. [CrossRef]
49. Tripathi, D. *Nanomaterials in Plants, Algae, and Microorganisms: Concepts and Controversies*; Academic Press: Cambridge, MA, USA, 2018; Volume 1.
50. Younas, M.; Ahmad, M.A.; Jannat, F.T.; Ashfaq, T.; Ahmad, A. 18—Role of Silver Nanoparticles in Multifunctional Drug Delivery. In *Micro and Nano Technologies, Nanomedicine Manufacturing and Applications*; Verpoort, F., Ahmad, I., Ahmad, A., Khan, A., Chee, C.Y., Eds.; Elsevier: Orlando, FL, USA, 2021; pp. 297–319. [CrossRef]
51. Ivask, A. Toxicity mechanisms in Escherichia coli vary for silver nanoparticles and differ from ionic silver. *ACS Nano* **2014**, *8*, 374–386. [CrossRef]

52. Mulvaney, P.; Liz-Marzan, L.; Giersig, M.; Ung, T.J. Silica encapsulation of quantum dots and metal clusters. *Mater. Chem.* **2000**, *10*, 1259. [CrossRef]
53. Park, M.; Sohn, Y.; Shin, W.G.; Lee, J.; Ko, S.H. Ultrason. *Sonochem.* **2015**, *22*, 35–40. [CrossRef] [PubMed]
54. Chowdhury, S.; Yusof, F.; Faruck, M.O.; Sulaiman, N. Process Optimization of Silver Nanoparticle Synthesis Using Response Surface Methodology. *Procedia Eng.* **2016**, *148*, 992–999. [CrossRef]
55. Huang, H.; Yang, X. Synthesis of Polysaccharide-Stabilized Gold and Silver Nanoparticles: A Green Method. *Carbohydr. Res.* **2004**, *339*, 2627–2631. [CrossRef] [PubMed]
56. Ezra, L.; O'Dell, Z.J.; Hui, J.; Riley, K.R. Emerging Investigator Series: Quantifying Silver Nanoparticle Aggregation Kinetics in Real-Time Using Particle Impact Voltammetry Coupled with UV-Vis Spectroscopy. *Environ. Sci. Nano* **2020**, *7*, 2509–2521. [CrossRef]
57. Rather, R.A.; Sarwar, R.K.; Das, N.; Pal, B. Impact of Reducing and Capping Agents on Carbohydrates for the Growth of Ag and Cu Nanostructures and Their Antibacterial Activities. *Particuology* **2019**, *43*, 219–226. [CrossRef]
58. Bhagyaraj, S.; Krupa, I. Alginate-Mediated Synthesis of Hetero-Shaped Silver Nanoparticles and Their Hydrogen Peroxide Sensing Ability. *Molecules* **2020**, *25*, 435. [CrossRef]
59. Mihaylov, M.; Chakarova, K.; Hadjiivanov, K.; Marie, O.; Daturi, M. FTIR Spectroscopy Study of CO Adsorption on Pt−Na−Mordenite. *Langmuir* **2005**, *21*, 11821–11828. [CrossRef]
60. Kim, Y.J.; Yang, S.I.; Ryu, J.C. Cytotoxicity and Genotoxicity of Nano-Silver in Mammalian Cell Lines. *Mol. Cell. Toxicol.* **2010**, *6*, 119–125. [CrossRef]
61. Zhang, T.; Wang, L.; Chen, Q.; Chen, C. Cytotoxic Potential of Silver Nanoparticles. *Yonsei Med. J.* **2014**, *55*, 283–291. [CrossRef]
62. Filipak Neto, F.; Cardoso da Silva, L.; Liebel, S.; Voigt, C.L.; Oliveira Ribeiro, C.A. de Responses of Human Hepatoma HepG2 Cells to Silver Nanoparticles and Polycyclic Aromatic Hydrocarbons. *Toxicol. Mech. Methods* **2018**, *28*, 69–78. [CrossRef]
63. Miranda, R.R.; Bezerra, A.G.; Oliveira Ribeiro, C.A.; Randi, M.A.F.; Voigt, C.L.; Skytte, L.; Rasmussen, K.L.; Kjeldsen, F.; Filipak Neto, F. Toxicological Interactions of Silver Nanoparticles and Non-Essential Metals in Human Hepatocarcinoma Cell Line. *Toxicol. Vitr.* **2017**, *40*, 134–143. [CrossRef] [PubMed]
64. Zuberek, M.; Wojciechowska, D.; Krzyzanowski, D.; Meczynska-Wielgosz, S.; Kruszewski, M.; Grzelak, A. Glucose Availability Determines Silver Nanoparticles Toxicity in HepG2. *J. Nanobiotechnol.* **2015**, *13*, 72. [CrossRef] [PubMed]
65. Ramos, A.A.; Prata-Sena, M.; Castro-Carvalho, B.; Dethoup, T.; Buttachon, S.; Kijjoa, A.; Rocha, E. Potential of Four Marine-Derived Fungi Extracts as Anti-Proliferative and Cell Death-Inducing Agents in Seven Human Cancer Cell Lines. *Asian Pac. J. Trop. Med.* **2015**, *8*, 798–806. [CrossRef] [PubMed]
66. Męczyńska-Wielgosz, S.; Wojewódzka, M.; Matysiak-Kucharek, M.; Czajka, M.; Jodłowska-Jędrych, B.; Kruszewski, M.; Kapka-Skrzypczak, L. Susceptibility of HepG2 Cells to Silver Nanoparticles in Combination with Other Metal/Metal Oxide Nanoparticles. *Materials* **2020**, *13*, 2221. [CrossRef]
67. Tran, Q.H.; Nguyen, V.Q.; Le, A.-T. Advances in Natural Sciences: Nanoscience and Nanotechnology. *Adv. Nat. Sci. Nanosci. Nanotechnol.* **2013**, *4*, 033001. [CrossRef]
68. Yusuf, A.; Almotairy, A.R.Z.; Henidi, H.; Alshehri, O.Y.; Aldughaim, M.S. Nanoparticles as Drug Delivery Systems: A Review of the Implication of Nanoparticles' Physicochemical Properties on Responses in Biological Systems. *Polymers* **2023**, *15*, 1596. [CrossRef]
69. Mitchell, M.J.; Billingsley, M.M.; Haley, R.M.; Wechsler, M.E.; Peppas, N.A.; Langer, R. Engineering Precision Nanoparticles for Drug Delivery. *Nat. Rev. Drug Discov.* **2021**, *20*, 101–124. [CrossRef]
70. Rivera, M.D.; Vazquez-Duhalt, R.; Castro-Longoria, E.; Juarez-Moreno, K. Synergistic Anticancer Effects and Reduced Genotoxicity of Silver Nanoparticles and Tamoxifen in Breast Cancer Cells. *J. Biochem. Mol. Toxicol.* **2024**, *38*, e23823. [CrossRef]
71. Kruszewski, M.; Brzoska, K.; Brunborg, G.; Asare, N.; Dobrzyńska, M.; Dušinská, M.; Fjellsbø, L.M.; Georgantzopoulou, A.; Gromadzka-Ostrowska, J.; Gutleb, A.C.; et al. Chapter Five—Toxicity of Silver Nanomaterials in Higher Eukaryotes. In *Advances in Molecular Toxicology*; Fishbein, J.C., Ed.; Elsevier: Amsterdam, The Netherlands, 2011; Volume 5, pp. 179–218.
72. Stoyanova, M.; Milusheva, M.; Gledacheva, V.; Stefanova, I.; Todorova, M.; Kircheva, N.; Angelova, S.; Pencheva, M.; Stojnova, K.; Tsoneva, S.; et al. Spasmolytic Activity and Anti-Inflammatory Effect of Novel Mebeverine Derivatives. *Biomedicines* **2024**, *12*, 2321. [CrossRef]
73. Milusheva, M.; Gledacheva, V.; Stefanova, I.; Pencheva, M.; Mihaylova, R.; Tumbarski, Y.; Nedialkov, P.; Cherneva, E.; Todorova, M.; Nikolova, S. In Silico, In Vitro, and Ex Vivo Biological Activity of Some Novel Mebeverine Precursors. *Biomedicines* **2023**, *11*, 605. [CrossRef]
74. Milusheva, M.; Gledacheva, V.; Batmazyan, M.; Nikolova, S.; Stefanova, I.; Dimitrova, D.; Saracheva, K.; Tomov, D.; Chaova-Gizdakova, V. Ex Vivo and In Vivo Study of Some Isoquinoline Precursors. *Sci. Pharm.* **2022**, *90*, 37. [CrossRef]
75. Garrett, M.; Huey, R.; Lindstrom, W.; Sanner, M.F.; Belew, R.K.; Goodsell, D.S.; Olson, A.J. AutoDock4 and AutoDock-Tools4: Automated Docking with Selective Receptor Flexibility. *J. Comp. Chem.* **2009**, *30*, 2785–2791.
76. MGL Tools. *1.5. 6 (ADT)/MGL Tools 1. 6*; The Scripps Research Institute: La Jolla, CA, USA, 2016.

77. Canakdag, M.; Feizi-Dehnayebi, M.; Kundu, S.; Sahin, D.; İlhan, İ.Ö.; Alhag, S.K.; Al-Shuraym, L.A.; Akkoc, S. Comprehensive evaluation of purine analogues: Cytotoxic and antioxidant activities, enzyme inhibition, DFT insights, and molecular docking analysis. *J. Mol. Struct.* **2025**, *1323*, 140798. [CrossRef]
78. Tahir, M.N.; Ashfaq, M.; Feizi-Dehnayebi, M.; Munawar, K.S.; Atalay, Ş.; Dege, N.; Guliyeva, N.; Sultan, A. Crystal structure, Hirshfeld surface analysis, computational study and molecular docking simulation of 4-aminoantipyrine derivative. *J. Mol. Struct.* **2025**, *1320*, 139747. [CrossRef]
79. Al-Farraj, E.S.; Qasem, H.A.; Aouad, M.R.; Al-Abdulkarim, H.A.; Alsaedi, W.H.; Khushaim, M.S.; Feizi-Dehnayebi, M.; Al-Ghamdi, K.; Abu-Dief, A.M. Synthesis, Structural Determination, DFT Calculation and Biological Evaluation Supported by Molecular Docking Approach of Some New Complexes Incorporating (E)-N′-(3, 5-di-Tert-Butyl-2-Hydroxybenzylidene) Isonicotino Hydrazide Ligand. *Appl. Organomet. Chem.* **2024**, *39*, e7768. [CrossRef]

pharmaceuticals

MDPI

Article

Green Synthesis and Characterization of Silver Nanoparticles Using *Azadirachta indica* Seeds Extract: In Vitro and In Vivo Evaluation of Anti-Diabetic Activity

Gauhar Rehman [1,*], Muhammad Umar [1], Nasrullah Shah [2], Muhammad Hamayun [3], Abid Ali [1], Waliullah Khan [2], Arif Khan [1], Sajjad Ahmad [1], Abdulwahed Fahad Alrefaei [4], Mikhlid H. Almutairi [4], Yong-Sun Moon [5] and Sajid Ali [5,*]

1. Department of Zoology, Abdul Wali Khan University Mardan, Mardan 23200, Pakistan; umarm5030@gmail.com (M.U.); drabid@awkum.edu.pk (A.A.); arifkhanpak83@gmail.com (A.K.); sajjadmohmand891@gmail.com (S.A.)
2. Department of Chemistry, Abdul Wali Khan University Mardan, Mardan 23200, Pakistan; nasrullah@awkum.edu.pk (N.S.); drwali@awkum.edu.pk (W.K.)
3. Department of Botany, Abdul Wali Khan University Mardan, Mardan 23200, Pakistan; hamayun@awkum.edu.pk
4. Department of Zoology, College of Science, King Saud University, Riyadh 11451, Saudi Arabia; afrefaei@ksu.edu.sa (A.F.A.); malmutari@ksu.edu.sa (M.H.A.)
5. Department of Horticulture and Life Science, Yeungnam University, Gyeongsan 38541, Republic of Korea; hangulmys@ynu.ac.kr

* Correspondence: gauhar@awkum.edu.pk (G.R.); sajidbioali@gmail.com (S.A.)

Abstract: Background: Diabetes mellitus (DM) is a non-communicable, life-threatening syndrome that is present all over the world. The use of eco-friendly, cost-effective, and green-synthesised nanoparticles as a medicinal therapy in the treatment of DM is an attractive option. Objective: In the present study, silver nanoparticles (AI-AgNPs) were biosynthesized through the green synthesis method using *Azadirachta indica* seed extract to evaluate their anti-diabetic potentials. Methods: These nanoparticles were characterized by using UV-visible spectroscopy, Fourier transform infrared spectrophotometers (FTIR), scanning electron microscopy (SEM), DLS, and X-ray diffraction (XRD). The biosynthesized AI-AgNPs and crude extracts of *Azadirachta indica* seeds were evaluated for anti-diabetic potentials using glucose adsorption assays, glucose uptake by yeast cells assays, and alpha-amylase inhibitory assays. Results: Al-AgNPs showed the highest activity (75 ± 1.528%), while crude extract showed (63 ± 2.5%) glucose uptake by yeast at 80 μg/mL. In the glucose adsorption assay, the highest activity of Al-AgNPs was 10.65 ± 1.58%, while crude extract showed 8.32 ± 0.258% at 30 mM, whereas in the alpha-amylase assay, Al-AgNPs exhibited the maximum activity of 73.85 ± 1.114% and crude extract 65.85 ± 2.101% at 100 μg/mL. The assay results of AI-AgNPs and crude showed substantial dose-dependent activities. Further, anti-diabetic potentials were also investigated in streptozotocin-induced diabetic mice. Mice were administered with AI-AgNPs (10 to 40 mg/kg b.w) for 30 days. Conclusions: The results showed a considerable drop in blood sugar levels, including pancreatic and liver cell regeneration, demonstrating that AI-AgNPs have strong anti-diabetic potential.

Keywords: green synthesis; AgNPs; *Azadirachta indica*; biosynthesize nanoparticles; diabetes; streptozotocin

1. Introduction

Diabetes mellitus (DM) is a chronic illness due to inadequate insulin production by pancreatic β cells. Diabetes could be hereditary or acquired. The insufficiency of insulin is a consequence of high blood glucose levels, which cause damage to various body systems, especially the circulatory and nervous systems [1,2]. As evident from the data of the International Diabetes Federation (IDF), the incidence of diabetes is rising internationally. In 2021, 536.6 million people had diabetes worldwide, and by the year 2045, it is projected to increase by 46% to 783.2 million [3]. As previously estimated by the IDF and other surveys, almost 50% of all patients with diabetes are oblivious to their illness [4]. Changing one's lifestyle to include increased physical exercise, consuming low-calorie foods, and avoiding inactive habits is necessary for DM prevention and control [5]. Although synthetic drugs like miglitol and acarbose have high inhibitory effects against alpha-amylase and alpha-glucosidase, they have consequences like causing diarrhea, nausea, vomiting, and intestinal swelling [6]. Therefore, plant extracts that are active in lowering serum glucose levels with slight or no side effects are used as hypoglycemic medications. Various compounds extracted from plants are used in combinational treatment for diabetes, such as *Azadirachta indica*, *T. indica*, and *Ceilba pentandrat, which* are well-known for their hypoglycemic properties. Therefore, the green synthesis methods are an attractive option. Green nanotechnology refers to the use of nanotechnology to enhance the environmental sustainability of processes producing negative externalities [7]. It includes making and using nano-products in support of sustainability. Biologically developed chemicals are used in these methods, which are not harmful for the environment [8]. Green nanotechnology has two goals: producing nanomaterials and products without harming the environment or human health, and producing nano-products that provide solutions to environmental problems [9]. Due to the wide range of applications of NPs, researchers, including biologists, chemists, physicists, and engineers, are working in this fascinating area [10]. Green synthesis has many advantages compared to chemical and physical methods; it is non-toxic, pollution-free, environmentally friendly, economical, and more sustainable [11,12]. Therefore, accessing green principles offers a high degree of safety, eco-friendliness, and cost-effectiveness. On green pathways, nanoparticles can be fabricated using natural compounds extracted from various biomass precursors, such as bacteria, fungi, biomolecules, and plant extracts [13]. The most important feature of the biogenic approach is the utilization of biologically reducing and capping agents to replace toxic chemicals. This alternation makes the biosynthetic method environmentally friendly, benign, and inexpensive [14].

The AgNPs in this study were developed under controlled conditions following parameters such as temperature, pressure, and reactant concentrations to obtain stable, spherical, and small-sized nanoparticles [15,16]. AgNPs have superiorities over other nanoparticles due to their outstanding properties such as small size, good conductivity, chemical stability, catalytic activity, optical, thermal, high electrical conductivity, and biological properties [17,18]. Moreover, the *Azadirachta indica* seed extract used in this study contains natural compounds such as Azadarachtin, Nimbin, and Nimbidin that have synergistic effects on AgNPs and enhance their anti-diabetic potentials [19,20]. A pH of 7 was the best to ensure the reduction of Ag+ to Ag0 during AgNPs production, and the greatest abundance of synthesized nanoparticles was obtained at pH 7–9. Several studies have shown that the production rate of AgNPs increases as the pH increases. Furthermore, AgNPs were almost spherical at higher pH values, and setting the pH at 8 substantially enhanced the reaction rate [21]. Temperature is one of the most important parameters that affects the size and morphology of biosynthesized AgNPs. Numerous studies have confirmed that the dimensions of AgNPs decrease as the reaction temperature increases, resulting in a change in their morphology [22].

The absorption intensity increased as the incubation time increased, owing to an increase in the amount of AgNPs produced. AgNPs have also been generated using Origanum vulgare L. extract, and the yield of nanoparticles increased with an increasing reaction time up to 3 h [23]. Song et al., 2022 [24] reported that absorbance increased when light intensity increased. Thus, it is expected that under sunlight, the reduction process of Ag+ ions can be completed within a few minutes, whereas the reaction requires a longer duration in the dark. Increasing the plant extract concentration in the reaction mixture can increase the absorbance intensity [25]. When using high extract concentrations, biomolecules act as reducing agents and cover the nanoparticle surfaces, preventing them from aggregating and increasing their stability. Due to their peculiar properties, they have been used for several applications, including as anti-bacterial agents, in industrial, household, and healthcare-related products, in consumer products, medical device coatings, optical sensors, and cosmetics, in the pharmaceutical industry, the food industry, in diagnostics, orthopedics, drug delivery, and as anti-cancer agents, and have ultimately enhanced the tumor-killing effects of anti-cancer drugs [26,27]. It has been shown that silver nanoparticles have anti-diabetic potential. Recent research in the field of plant-based nanomedicine has demonstrated that biosynthesized nanoparticles are more effective than crude extracts [28,29]. They have advantages, including increased surface area, solubility, and healing capacity. The synthesis of Ag nanoparticles using plants has several advantages compared to other biosynthesized nanoparticles, which is why the use of extract has received greater attention [30]. Moreover, the application of biomolecules as reducing, stabilizing, and capping agents rather than costly toxic chemicals makes the biosynthesis of nanoparticles an efficient process [31].

Several resources, such as pollens, polyoxometalates, irradiations, and polysaccharides, are employed for the environment-friendly biosynthesis of AgNPs [32]. The green synthesized silver nanoparticles have many applications, such as the breakdown of harmful pollutants, water purification, food preservation, and the production of nano-insecticides, nano-pesticides, and cosmetics [33]. Due to their bioactive nature, the scientific research community is now evaluating them as a new active mediator for the curing of diabetes mellitus [34]. Sharifi et al. synthesized the AgNPs through a green process and evaluated their anti-oxidant, anti-bacterial, and anti-inflammatory activities [35]. Jini and Sharmila have carried out the AgNPs synthesis through the green method and employed them as plant-mediated medicine for the management of diabetes [36]. Nagaraja et al. investigated the green synthesized AgNPs of leaf extract and stated that the particles act as an excellent anti-diabetic candidate [37]. Vinodhini et al. have biosynthesized AgNPs using *Allium fistulosum*, *Tabernaemontana divaricate*, and *Basella alba* extracts. The synthesized AgNPs exhibited high anti-oxidant and anti-diabetic activities [38]. By using *Acacia nilotica* extract, Zubair et al. sustainably created silver nanoparticles. They investigated the anticancer and anti-diabetic properties of the AgNPs [39]. Kaliammal et al. used an extract of *Zephyranthes candida* flower for the synthesis of AgNPs. Their results confirmed that the particles showed anti-diabetic, anti-inflammatory, anti-oxidant, and anti-cancer activities [40]. Badmus et al. synthesized the AgNPs using *Annona muricata* aqueous leaf extract. They found that the biosynthesized particles showed cytotoxicity in human keratinocyte cells (HaCaT) as well as in vitro anti-diabetic, antioxidant, lipid peroxidation inhibition, and anti-bacterial activities [41]. Das et al. (2021) biofabricated the AgNPs through green synthesis using *Dregea volubilis* flowers and established better anti-oxidant, anti-diabetic, and anti-bacterial activities [42]. Thirumal S. and Sivakumar used *Cassia auriculata* leaf extract for the green preparation of AgNPs and demonstrated their potent anti-diabetic activity [43]. Yarrappagaari et al. biosynthesized the AgNPs from the aqueous extract of *Cleome viscosa* and evaluated them for anti-bacterial, antioxidant, and anti-diabetic activities [44]. Sathiyaseelan et al. studied the fungus chitosan (FCS)-enclosed *Gynura procumbens* (GP) biosynthesized silver nanoparticles (GP-AgNPs) and found them to be excellent candidates for antibacterial and diabetic-associated enzyme inhibitory activities [45]. Similar studies reported AgNPs for their anti-diabetic, anti-cancer, anti-bacterial, and anti-inflammatory capacities [46]. The previous literature survey clearly indicates that various researchers have biosynthesized the AgNPs by taking different parts of the plant and evaluating their different biological activities.

According to previous literature [47], if the natural compounds coating the nanoparticles are themselves anti-diabetic in nature, a synergistic anti-diabetic potential of the final nanomaterial is observed [48], and *Azadirachta indica* seeds contain the most potent anti-diabetic compounds, such as Azadarachtin, Nimbin, Nimbidin, etc., so that must reveal high anti-diabetic potential; therefore, *Azadirachta indica* seeds extract-mediated

AgNPs are better than those AgNPs reported earlier in the literature [49,50]. Considering the vast potentiality of plants as sources, this work aims to investigate the use of *Azadirachta indica* seed extract for the biosynthesis of AgNPs, and the synthesized AgNPs were pragmatically characterized and investigated for their anti-diabetic potential. To the best of our knowledge, this research will represent the first reference to the use of *A. indica* seed extract for the green synthesis of silver nanoparticles and anti-diabetic activities.

2. Materials and Methods

2.1. Extract Preparation from Azadirachta indica Seeds

For the preparation of *A. indica* seed extract, the method of Hameed et al. [51] was followed. *Azadirachta indica* (Neem) seeds were purchased from an herbal store, identified, and specimens placed in the herbarium of the Department of Botany, Abdul Wali Khan University Mardan (AWKUM), with an accession number of AWKUM. Bot. 425.1.20. *Azadirachta indica* seeds were rinsed gently with clean water, dried, and crushed to powder using a plant grinder (Panasonic Model MX-AC210, Osaka, Japan). For the preparation of methanolic extract, 1 kg powder of *Azadirachta indica* was incubated in 2000 mL of 100% methanol (used as solvent) and stored at room temperature (25 ± 3 °C) for 5 days. After filtration, the solution was rotary evaporated at 48 °C under reduced pressure and used for the biosynthesis of AI-AgNPs.

2.2. Green Synthesizing of Azadirachta indica Seeds-Mediated Silver Nanoparticles (AI-AgNPs)

For the biosynthesis of AgNPs, the protocol of [52] was followed: a 100-mL solution of 1 mM $AgNO_3$ was prepared in a conical flask, and 20 mg of the *A. indica* seed extract was added with continuous stirring on a hot plate below the boiling point (70 °C) for 4 h. The AgNPs solution was then exposed to light for the process of reduction. The green synthesis of the AI-AgNPs was observed by the changing color of the solutions, which was recorded at regular intervals. The reaction was set to continue for 24 h, and then centrifugation was performed for 30 min at 10,000 rpm. The supernatant was discarded after centrifugation, and the pellet was washed three times with double-distilled water. This was followed by centrifugation to remove any remaining free compounds and obtain pure nanoparticles that were used for anti-diabetic activities. A total of 0.017 g of precursor ($AgNO_3$), which can theoretically yield 0.011 g of silver nanoparticles, and in this study, 0.015 g was obtained, which means the obtained efficiency is very good. A slight increase of 0.004 g in the weight of the obtained AgNPs is due to the coated organic layer on the surface of these particles. For further use, the precipitated AI-AgNPs were lyophilized and stored in a dry and cool place. Figure 1 represents the biosynthesis of AI-AgNPs.

Figure 1. Green Synthesis of AI-AgNPs.

2.3. Characterization of the Green Synthesized AI-AgNPs

The biosynthesized AI-AgNPs were examined for their physical, morphological, and chemical characteristics using different techniques, like UV-VIS spectroscopy, FTIR, SEM, and XRD.

2.3.1. Visible Observation

According to literature studies, silver nanoparticle solutions have a dark brown or dark reddish color [53]. The color of the *A. Indica* solution before the addition of $AgNO_3$ was yellowish, but after its treatment with $AgNO_3$, it changed to dark brown, which indicated the formation of AgNPs. (Figure 1) This color change is due to the property of quantum confinement, which is a size-dependent property of nanoparticles that affects the optical properties of the nanoparticles.

2.3.2. UV-Visible Spectral Analysis of AI-AgNPs

Using a UV-VIS spectrophotometer (Multiskan GO; Thermo Scientific, Waltham, MA, USA), the absorption spectra of the aqueous solution with dried 80 μg per mL silver nanoparticles were recorded at a 2-nm resolution between 200 and 800 nm to investigate how the light affected the green synthesis of AI-AgNPs. The experiment was repeated three times, and the absorption was measured at wavelengths between 200 and 800 nm with a 2-nm difference at a 1 h interval and recorded.

2.3.3. FTIR Analysis of AI-AgNPs

To identify distinct functional groups in the green synthesized AI-AgNPs, FTIR (IRTracer-100, Shimadzu, Kyoto, Japan) was used at wavelengths from 4000 to 500 cm^{-1} and the corresponding data were recorded.

2.3.4. X-ray Diffraction Technique (XRD) Analysis of AI-AgNPs

XRD was used to examine the purity and crystalline makeup of the biosynthesized AI-AgNPs following the method of [54] using an X-ray diffractometer (Model-D8 Advance, Bruker, Bremen, Germany).

2.3.5. Scanning Electron Microscopy (SEM)

A high-resolution image of the AI-AgNPs was produced by SEM (Hitachi, S-4300SE, Tokyo, Japan), which includes details about their size, shape, composition, electrical conductivity, topography, and other characteristics.

2.4. In Vitro Anti-Diabetic Potential

2.4.1. Assay for Uptake of Glucose by Yeast Cells

The glucose uptake by yeast cells experiment was conducted using a slightly modified method of Rehman et al. [55]. AI-AgNPs in various concentrations (10 μg/mL to 80 μg/mL) were combined with 1 mL of 5 mM glucose solutions, and the reaction solution was then incubated for 10 min at 37 °C. The final reaction was started by adding a 10%/V suspension of yeast. After being vortexed, the reaction mixture was kept for 60 min at 37 °C, then centrifuged for 5 min at 3800 rpm. The supernatant amount of glucose was determined using a spectrophotometer (UV 5100B spectrophotometer). First, the absorbance value of the control was found at 520 nm wavelength using the spectrophotometer, then the absorbance value of the test samples was determined, and the process was repeated in triplicate. The following formula was used to compute the uptake of glucose by yeast cells:

$$\%\ Glucose\ uptake = \frac{Absorbance\ Control - Absorbance\ Test\ sample}{Absorbance\ Control} \times 100$$

2.4.2. Glucose Adsorption Assay

The glucose adsorption method was used to evaluate the ability of AI-AgNPs to adsorb glucose by following the method of [56]. 1 g of the AI-AgNPs was combined with a 100 mL solution of glucose (5–30 mM glucose). The final mixture was incubated in a shaker at room temperature for up to 6 h before being centrifuged for 20 min at 4800 rpm. The absorbance values of G1 (glucose concentration before reaction, using a spectrophotometer), then the absorbance value and G6 (after 6 h concentration of glucose) were determined at 520 nm wavelength. Using the following formula, the % glucose adsorption was calculated.

$$\text{Bounded Glucose} = \frac{\text{G1} - \text{G6}}{\text{Sample weight}} \times \text{volume of sample}$$

G1—Glucose concentration before reaction.
G6—After 6 h concentration of glucose.

2.4.3. Alpha (α) Amylase Inhibition Assay

The alpha bonds of polysaccharides are hydrolyzed by the enzyme alpha-amylase to create glucose and maltose. This process of the alpha-amylase enzyme elevates blood sugar levels, and its inhibition will cause blood sugar levels to drop. According to the procedure described by [57], a reaction mixture containing porcine pancreatic amylase (500 µL) in phosphate buffer, a standard drug (250 µL), and AI-AgNPs was prepared and kept at 37 °C for 20 min. After that, 250 µL of PBS (100 mM) and 1% starch were added to the mixture and kept at 37 °C for 60 min. In the last step, 1 mL of a color reagent, di-nitro-salicylate was added to the solution and heated for 10 min. The optical density was determined at 540 nm. Using the formula below, the percentage of inhibition was calculated.

$$\%\ Inhibition = \frac{Control\ Absorbance - sample\ Absorbance}{Absorbance\ Control} \times 100$$

2.5. Analysis of In Vivo Antidiabetic Potentials

2.5.1. Experimental Animals and Conditions

Adult BALB/C mice weighing 25–30 g and 5 weeks old were bought from the Veterinary Research Institute Peshawar. The animals were kept in controlled environments (22 °C, 60% humidity, 12 h of light and darkness), provided with standard food, and provided with free access to water. All the experiments were carried out strictly in accordance with the guidelines of the National Research Council for the handling and use of lab animals. Before starting the experiments, approval of the protocol was obtained from the Ethical Committee of the Department of Zoology, Abdul Wali Khan University, Mardan, Pakistan.

Group I: Normal Control (0.1 M Citrate Buffer, 0.5 mL, and pH 4.5);
Group II: Diabetic Control (Streptozotocin 50 mg/kg body weight);
Group III: Diabetic mice administered with AI-AgNPs (10 mg/kg body weight);
Group IV: Diabetic mice administered with AI-AgNPs (20 mg/kg body weight);
Group IV: Diabetic mice administered with AI-AgNPs (30 mg/kg body weight);
Group V: Diabetic mice administered with AI-AgNPs (40 mg/kg body weight);

2.5.2. Induction of Diabetes in Mice by Streptozotocin

The mice were starved overnight before administering one dose (50 mg/kg body weight) of streptozotocin (STZ) was injected intraperitoneally into the diabetic groups, whereas citrate buffer alone was administered to the control group. The diabetes was confirmed by estimating the blood sugar after 72 h of STZ injection. Mice having a blood sugar of more than 120 mg/dl were classified as diabetic and employed in vivo experiments.

2.5.3. Blood Glucose Level of Experimental Mice

The experimental mice were starved for 12 to 16 h, blood samples were taken by the tail vein rupturing method, and the blood sugar was estimated using a one-touch glucometer (Life Scan, Inc., Milpitas, CA, USA). The glucose levels were observed on the 1st, 5th, 10th, 15th, 20th, 25th, and 30th days.

2.5.4. Histopathological Study of the Pancreas and Liver of Experimental Mice

The pancreas and liver were separated, and they were continually cleaned with phosphate-buffered saline. The tissue samples were subsequently fixed in 10% formalin and subjected to standard histopathology procedures. The sample tissues were immersed in wax and subsequently cut into 5µm-thick slices. Hematoxylin and eosin stain were used to stain the sections, and the morphology of the tissues was observed under the microscope.

2.5.5. Statistical Analysis

The results were expressed as mean ± SD using GraphPad Prism version 9. For statistical analysis of the data group, the mean was compared using a one-way analysis of variance (ANOVA). In addition, $p < 0.05$ was considered to be statistically significant.

3. Results

3.1. Visible Observation

The color of the A. Indica solution before the addition of $AgNO_3$ was yellowish, but after its treatment with $AgNO_3$, it changed to dark brown, which indicated the formation of AgNPs (Figure 2). This color change is due to the property of quantum confinement, which is a size-dependent property of nanoparticles that affects the optical properties of the nanoparticles.

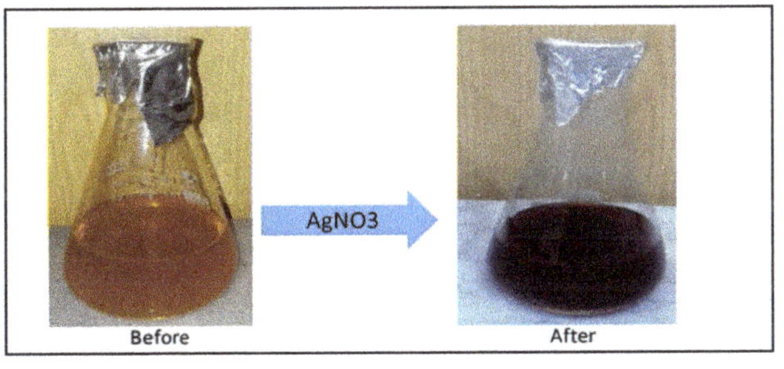

Figure 2. Color change of *A. indica* seed extract solution before and after the addition of $AgNO_3$.

3.2. UV-Visible and Bandgap Energy Analysis of AI-AgNPs

UV-visible analysis and band energy values for the synthesized nanoparticles are presented in Figure 3. Silver nanoparticles have a peak UV absorption wavelength between 400 and 450 nm. The synthesis of AI-AgNPs in the *A. indica* seed extract is indicated by the UV-visible density peaks of the *A. indica* UV-Vis spectra at about 441 nm shown in Figure 3a. The process of surface plasmon resonance, which results from stimulation of the surface plasmons that exist on the external surface of the AI-AgNPs and which is stimulated due to the applied electromagnetic field, is what causes the peak at 380 nm to appear [58]. The below Tauc plot equation was used to obtain the bandgap energy value for the produced nanoparticles shown in Figure 3b.

$$(\alpha h\nu)^{\gamma} = A(h\nu - Eg) \quad (1)$$

where as α is the coefficient of absorption, h is the planks constant, ν is the frequency of photons, A is the proportionality constant, γ = electronic transition, and Eg is bandgap energy depending on the transition and can have values of 2, 1/2, 2/3, or 1/3. Tauc plots for AgNPs are presented in Figure 3b. The plot of $(\alpha h\nu)^2$ against $(h\nu)$ results in a straight line, which explains that the edge of absorption is due to the direct transition (n = 1 for direct transition). The optical band gap (Eg) is indicated by the intercept of a straight line. The biosynthesized nanoparticles' direct bandgap energy is Eg = 2.43 eV.

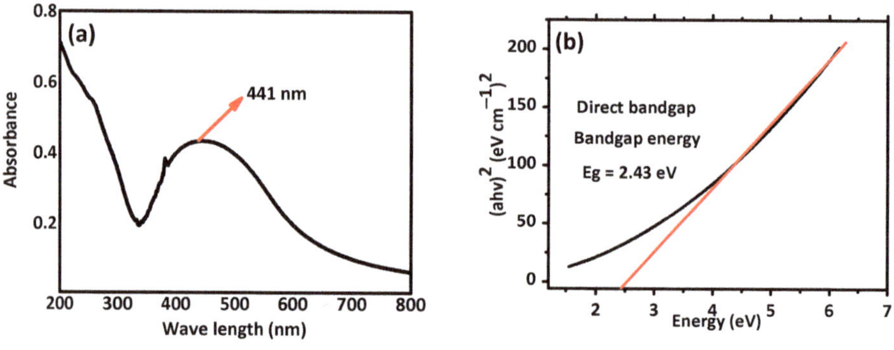

Figure 3. (**a**) UV-visible. (**b**) Tauc plots for bandgap energy analysis of AI-AgNPs. Plot of $(\alpha h\nu)^2$ against $(h\nu)$ results in a straight line, which explains that the edge of absorption is due to the direct transition (n = 1 for direct transition). The optical band gap (Eg) is indicated with the intercept of a straight line.

3.3. SEM Analysis of AI-AgNPs

To evaluate the morphology of the biosynthesized AI-AgNPs, the scanning electron microscope (SEM; JEOL JSM-7001F) was used. Figure 4A indicates from the SEM result of the biosynthesized nanoparticles that the particles are evenly and uniformly distributed and have a spherical shape at the nanoscale. Some agglomerations of the synthesized nanoparticles were seen because of the plant extract. However, there are also various plant extract components that function as stabilizing and capping agents, reducing the aggregation of the particles. Agglomeration led to collecting the nanoparticles in ordinary shapes, mostly by physical bond, due to the nature of the solvent used. We used biological sources with organic moieties on their surfaces and functional groups that interact with each other and cause agglomeration. There are many ways to de-agglomerate nanoparticles, such as sonication, ultrasound, isopropyl storage, heat, electrostatic charge, etc. We have overcome this issue by sonicating AgNPs before using them for biological activities. The coated layer of organic moieties on the silver nanoparticles gives them stability, and as a result, the nanoparticles are stable and mostly dispersed. The size distribution histogram of dynamic light scattering (DLS) indicates that the average size of these silver nanoparticles is 34.43 nm. Figure 4 shows the DLS pattern of the suspension of AI-AgNPs.

3.4. Fourier Transform Infrared Spectrophotometry (FTIR) and X-Ray Diffraction (XRD) Analysis

FTIR and XRD analysis were performed for structural identification, crystallinity, and phase shifts using an FTIR spectrophotometer (Spectrum Two TM FT-IR Spectrometer; PerkinElmer, Waltham, MA, USA) at different wavelengths ranging from 200 to 4000 cm^{-1}. Figure 5a,b shows the FTIR results of the pure *A. indica* and *A. indica*-mediated AgNPs. The extra band observed at 591 cm^{-1} corresponds to the Ag nanoparticles that appeared in *A. indica*-mediated AgNPs shown in Figure 5b. The peak at 1401.37 cm^{-1} may be due to the (C-O and C-H) bending vibrations of the *A. indica plant*. The stretching vibration of the C-O functional group of alcohol, ester, ether, or carboxylic acid is shown by the peaks at 1031.85 cm^{-1} and 1041 cm^{-1}, respectively. The peak at 1617.32 cm^{-1} and 1635 cm^{-1} might be caused by the C=O stretching vibration of alkenes, primary amines (N-H bending vibration), and amides (N-H bending and C=O stretching vibration), as well as the functional groups of aldehydes and ketones. The stretch at 2104 cm^{-1} is due to the C≡C of the alkene. The OH stretching vibration of the phenol causes the peaks at 3292 cm^{-1} and 3311.35 cm^{-1} respectively. Additionally, the presence of OH and C=O groups suggests that flavanones or terpenoids have been adsorbed on the surface of nanoparticles. Connections via π-electrons in the carbonyl groups may be liable for the reduction of Ag ions to AgNPs as well as for stability and as a capping agent. The presence of various functional groups in Figure 5b demonstrated the successful green synthesis of *A. indica*-mediated AgNPs.

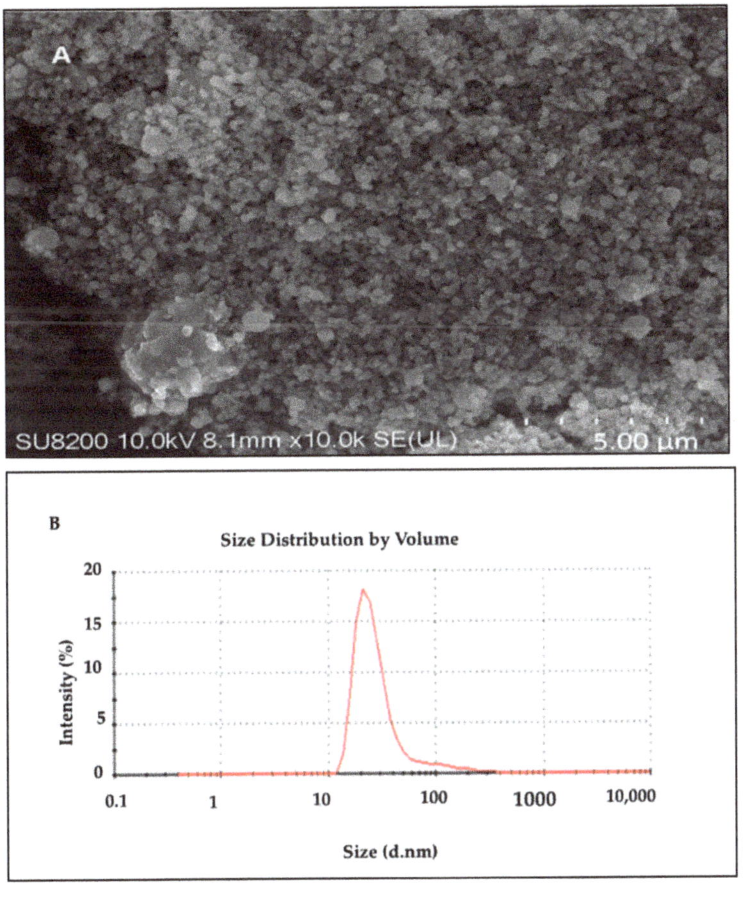

Figure 4. (**A**) SEM results of *A. indica*-mediated silver nanoparticles; (**B**) particle size analysis of AI-AgNPs using the DLS image system.

X-ray diffraction (XRD) is a popular analytical technique that has been used for the analysis of both molecular and crystal structures, qualitative identification of various compounds, quantitative resolution of chemical species, measuring the degree of crystallinity, isomorphous substitutions, particle sizes, etc. When X-ray light reflects on any crystal, it leads to the formation of many diffraction patterns, and the patterns reflect the physico-chemical characteristics of the crystal structures.

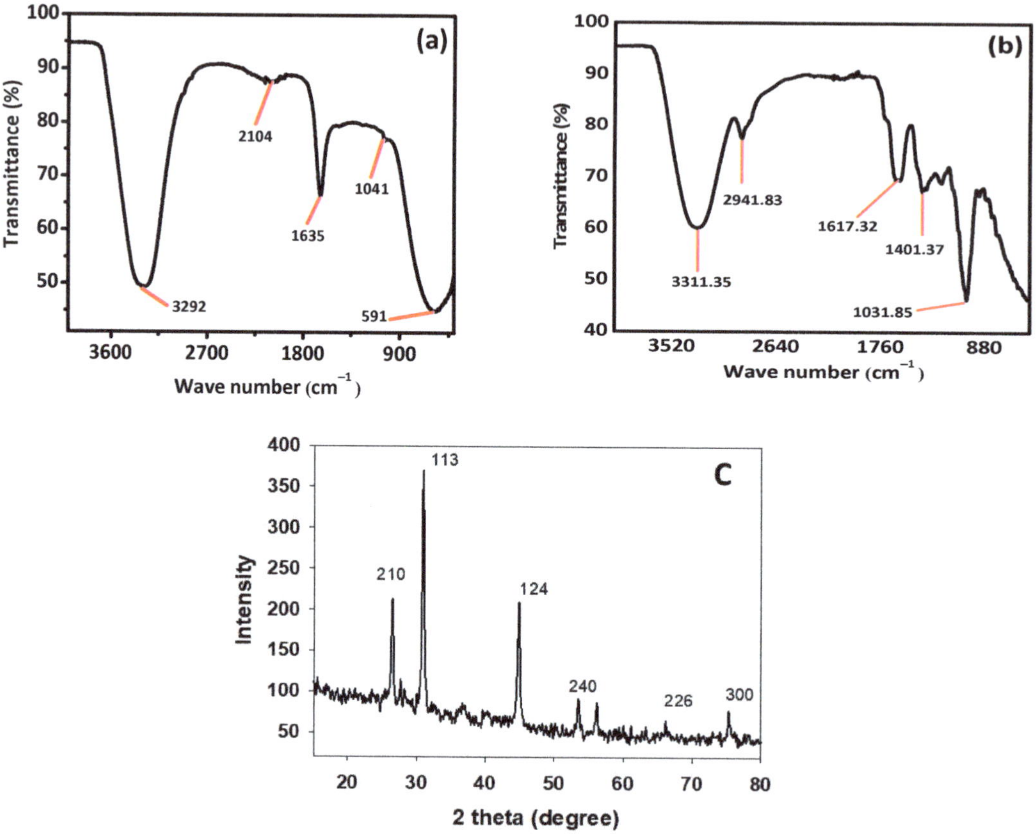

Figure 5. (**a**) FTIR of pure *A. indica*, (**b**) *A. indica*-mediated AgNPs, and (**c**) XRD spectrum of *A. indica*-mediated AgNPs.

The XRD result given in Figure 5c shows that there are four separate diffraction peaks in the 2θ range of 10° to 80°. The peaks seen at 2θ angles of 26.23°, 30.66°, 44.56°, 56.22°, 66.08°, and 75.35° correspond to the 210, 113, 124, 240, 226, and 300, planes correspondingly, which were indexed for a silver face-centered cube of (JCPDS file no. 04-0783).

The result clearly explains that the biosynthesized AI-AgNPs were crystalline. These findings support the presence of face-centered cubic AI-AgNPs. The unassigned peaks might have resulted from the plant extract-dependent capping agent involved in the stabilization of AgNPs, and the average size of the AgNPs was 34.43 nm.

3.5. Effects of AI-AgNPs and Crude Extract on the Uptake of Glucose by the Yeast Cells

Uptake of glucose by the yeast cells at different concentrations like 10, 20, 30, 40, 50, 60, 70, and 80 µg/mL of AI-AgNPs and crude extract were determined as AI-AgNPs 18 ± 2.64%, 23 ± 2.082%, 29 ± 1.692%, 37 ± 1.00%, 48 ± 2.82%, 57 ± 0.854%, 66 ± 1.311%, and 75 ± 1.528%, respectively, whereas crude extract was determined as 11.804 ± 1.0%, 16 ± 1.36%, 24 ± 2.03%, 31.63 ± 2.98%, 39.5 ± 1.36%, 48.467 ± 2.73%, 55.01 ± 2.015%, and 63 ± 2.5%, respectively, Acarbose was used as a standard drug in the same concentration as AI-AgNPs, and crude extract was taken. The results showed 21 ± 1%, 28 ± 1.528%, 34 ± 2.582%, 43 ± 1.206%, 52 ± 1.58%, 63 ± 1.637%, 70 ± 1.528%, and 80 ± 1.20% glucose uptake, respectively, sowing significant result (p-value ≤ 0.05) indicated in Figure 6A.

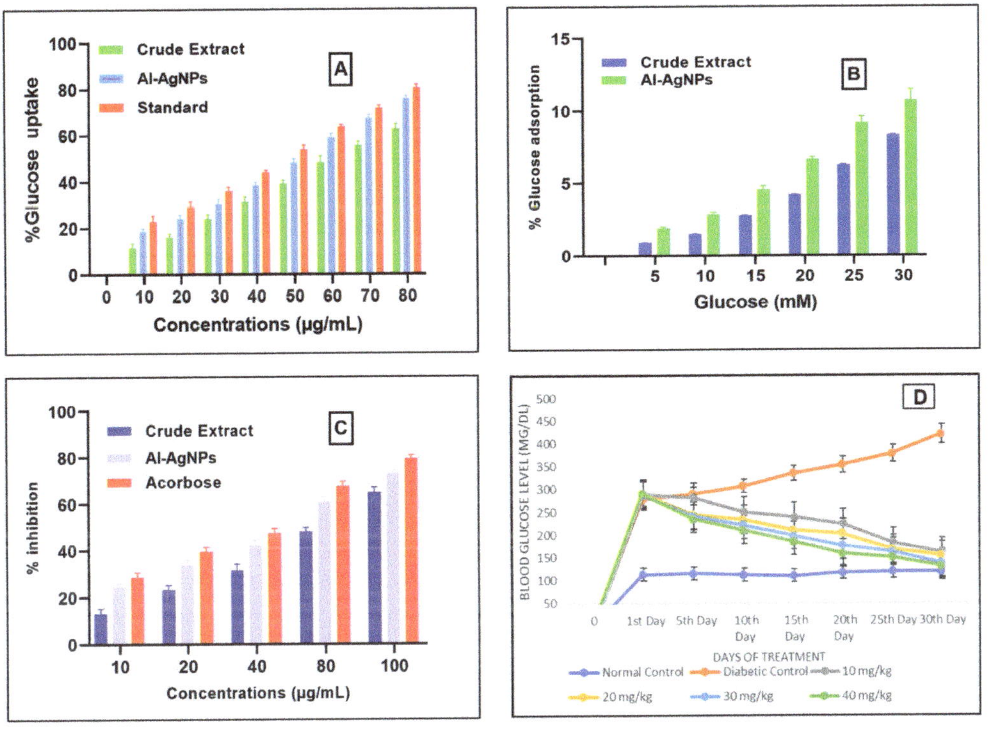

Figure 6. (**A**) Glucose uptake by yeast cells; (**B**) % Glucose adsorption; (**C**) inhibition of alpha-amylase; (**D**) blood glucose level of mice using *A. indica*-mediated AgNPs.

3.6. Analysis of Glucose Adsorption by A. indica Seed Extract-Mediated AgNPs and Crude Extract

Figure 6B shows the glucose adsorption activity of the green synthesized AI-AgNPs and crude extract at various concentrations of glucose that are 5, 10, 15, 20, 25, and 30 mM. AI-AgNPs showed 1.8 ± 0.62%, 2.88 ± 0.06%, 4.33 ± 0.153%, 6.56 ± 0.50%, 8.95 ± 1.00%, and 10.65 ± 1.58% adsorption of glucose, respectively, and crude extract showed 0.9 ± 0.02%, 1.5 ± 0.032%, 2.77 ± 0.015%, 4.200 ± 0.20%, 6.25 ± 0.04%, and 8.32 ± 0.258% adsorption of glucose, respectively. The results indicate that this effect is not just due to the adsorption of glucose but also to the uptake of glucose. Because the adsorption of glucose depends on the concentration of glucose, as the glucose concentration increases, the adsorption of glucose also increases, and glucose uptake in yeast cells depends on the concentration of AI-AgNPs and crude extract.

3.7. Impact of AI-AgNPs and Crude Extract on Inhibition of α-Amylase

The alpha-amylase inhibitory effect of AI-AgNPs (10–100 µg/Ml) was determined for the prepared AI-AgNPs and crude extract, as shown in Figure 6C. The percent inhibitory values of AI-AgNPs at 10, 20, 40, 80, 100µg/mL were noted as 23.7 ± 1.4%, 34 ± 1.99%, 42.13 ± 2.44%, 60.92 ± 2.00%, and 73.85 ± 1.114%, respectively, with an IC50 value of 48.26 µg/mL and that of crude extract. The percent inhibitory values of crude extract at 10, 20, 40, 80, and 100µg/mL were noted as 13.7 ± 2.15%, 23 ± 2.67%, 31.500 ± 2.78%, 48.62 ± 2.167%, and 65.85 ± 2.101%, respectively, with an IC50 value of 68.37 µg/mL. Whereas acarbose, taken as a standard drug, showed % inhibition at 10, 20, 40, 80, and 100 µg/mL as 27 ± 2.082%, 39.56 ± 1.913, 47.6 ± 1.97, 67.66 ± 2.086, 10 µg/mL, and 79.33 ± 1.528, respectively.

3.8. Analysis of Blood Glucose Level of Experimental Mice

Administration of STZ resulted in increased blood glucose levels, which were reversed by treating diabetic mice with 10 to 40 mg/kg body weight of AI-AgNPs for 30 days, resulting in a significant decrease in hyperglycemia as shown in Figure 6D. It is clear from this figure that there was a significant increase in blood glucose levels (277 $\pm$ 5.1 to 420 $\pm$ 22.03 mg/dL) in the diabetic control group. The normal control group did not show a significant increase in blood glucose level (113 $\pm$ 3.5 to 118 $\pm$ 4.72 mg/dL). Groups treated with AgNPs at doses of 10, 20, 30, and 40 mg/kg, respectively, showed a significant decrease ($p < 0.05$) in glucose level (288 $\pm$ 5.0 to 160 $\pm$ 5.54 mg/dL, 290 $\pm$ 2.5 to 154 $\pm$ 5.033 mg/dL, 287 $\pm$ 4.2 to 138 $\pm$ 7.50 mg/dL, 290 $\pm$ 8.3 to 131 $\pm$ 7.024 mg/dL), respectively.

3.9. Histological Study of Mice Pancreas

In the histological studies of pancreas, the normal control group displayed the pancreas with normal anatomy, which shows that the exocrine component of pancreas is well organized into small lobules and is densely packed with acinar cells, and the pancreatic lobules are divided by septa into healthy intralobular and interlobular connective tissue (Figure 7A). Pathological abnormalities in both the exocrine and endocrine systems were seen in the pancreas of the diabetic control group. Little vacuoles [shown by an arrow] in the enlarged acinar cell can be seen nearly in all the acinar cells. β-cells in the pancreas of STZ-treated mice are almost entirely damaged (Figure 7B). The diabetic group receiving a dose of 10 mg/kg b/w AI-AgNPs displayed general architectural deformation of the pancreas. Acinar injury manifested by cytoplasmic vacuolation and cell atrophy was seen in most exocrine acini [indicated by the arrow] of the pancreas (Figure 7C). Diabetic mice treated with a dose of 20 mg/kg b/w AI-AgNPs showed some restoration of islets of the pancreas. The basal region of acinar cells has medium-sized vacuoles (Figure 7D). The diabetic group treated with a dose of 30 m g/kg b/w AI-AgNPs revealed regeneration of islet cells. Moreover, the tiny vacuoles in the basal region of the acinar cells were significantly smaller (Figure 7E). The diabetic mice treated with a dose of 40 mg/kg b/w AI-AgNPs displayed the nearly normal shape of Islets of Langerhans [indicated by the arrow]. The moderate atrophic alteration occurred in the acinar cells, and the distinction between the exocrine and endocrine sections of the pancreas improved (Figure 7F).

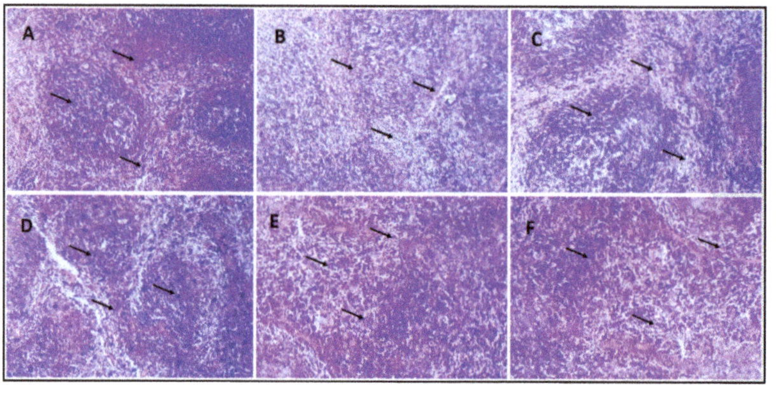

Figure 7. Histology section of the pancreas (**A**) Group–I: in normal control small lobules are densely packed with acinar cells (indicated by the arrow) (**B**) Group-II: in diabetic control β-cells of pancreas are almost entirely damaged (indicated by the arrow) (**C**) Group-III: treated with 10 mg/kg b/w AI-AgNPs; shown initial restoration of islets of the pancreas (indicated by the arrow) (**D**) Group-IV: treated with 20 mg/kg b/w AI-AgNPs; shows some restoration of islets of the pancreas (indicated by the arrow (**E**) Group-V: treated with 30 mg/kg b/w AI-AgNPs revealed moderate regeneration of islet cells (indicated by the arrow) (**F**) Group-VI: treated with 40 mg/kg b/w AI-AgNPs shown nearly normal shaped Islets of Langer-hans (indicated by the arrow).

3.10. Histological Analysis of Mice Liver

The histology of the normal control group showed a typical normal histological picture of the liver (Figure 8A). The hepatocytes of the diabetic control group showed significant deterioration and necrosis in the liver (Figure 8B). Differences in the size of vacuoles of hepatocytes can be seen in the livers of mice treated with a 10 mg/kg b/w dose (Figure 8C). In the livers of mice treated with doses of 20 mg/kg b/w and 30 mg/kg b/w, there was a noticeable decrease in liver deterioration and necrosis of hepatocytes (Figure 8D,E). The diabetic group of mice treated with 40 mg/kg b/w showed a histological appearance that was comparable to the normal control group (Figure 8F).

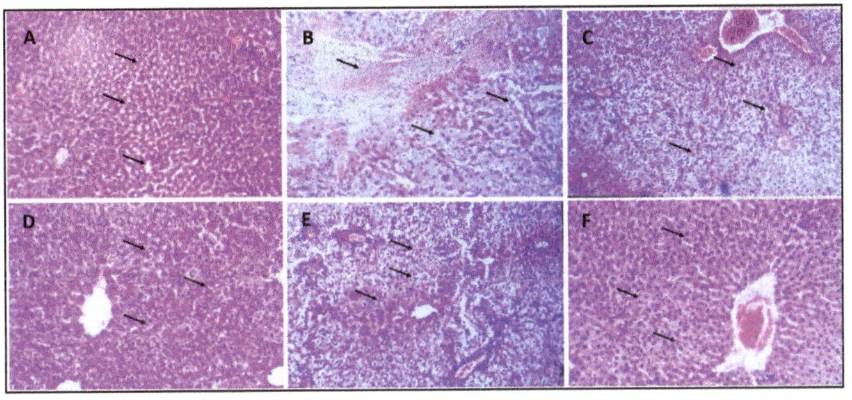

Figure 8. Histology sections of the liver. (**A**) Group–I: normal control; vacuoles and sinus are normal (indicated by arrow) (**B**) Group-II: diabetic control; Severe deterioration and necrosis are shown (indicated by arrow) (**C**) Group-III: treated with 10 mg/kg b/w AI-AgNPs; Some decrease occurred in deterioration and necrosis (indicated by arrow) (**D**) Group-IV: treated with 20 mg/kg b/w AI-AgNPs; moderate vacuoles and sinus regeneration are revealed (indicated by arrow) (**E**) Group-V: treated with 30 mg/kg b/w AI-AgNPs; Noticeable vacuoles and sinus regeneration occurred (indicated by arrow) (**F**) Group-VI: treated with 40 mg/kg b/w AI-AgNPs. Nearly normal vacuoles and sinus are shown (indicated by arrow).

4. Discussion

Nanotechnology has emerged as a leading technology in several sectors, with an ultimate application in agriculture, food, pharma, and biomedicine engineering. Nanoparticles, due to their small sizes, versatility, and readiness to couple with optical, textile, magnetic, electronic, mechanical, and chemical substances, are the candidates for novel applications in therapeutics, for example, anti-microbial, anti-oxidant, and cancer. In addition, nanoparticles have been extensively studied in the physical, chemical, and biological sciences [59]. Recently, several successful reports affirmed the production of these nanomaterials from natural sources such as plants and microbes. The biosynthesis of nanocomposites is a large-scale scientific domain with significant attention in biomedical applications due to their biocompatibility and multifunctional abilities [60].

Today's most popular oral anti-diabetic medications rarely provide long-lasting glycemic control. To fill the gap, several medicinal plant extracts are deemed effective in lowering blood glucose levels and are administered as anti-diabetic drugs [61]. Several research studies have described the role of metals and their oxides, such as silver, vanadium, chromium, magnesium, and zinc oxides, in carbohydrate metabolism and the control of DMT2 [62]. The preparation of Ag nanoparticles via the green synthesis approach is effective due to the use of non-toxic phytochemicals and the absence of hazardous elements as found in the chemical method [63]. There are three basic categories of these techniques, including biological, physical, and chemical methods. The best method for synthesizing nanoparticles is the biological method due to its simple, nontoxic, and cost-effective nature [64].

The biosynthesis of AgNPs from plants is an easy procedure that includes the interaction of silver nitrate ($AgNO_3$) with the biomolecule components of plant extracts [65]. Nanoparticles are formed primarily in three phases: an ion reduction reaction leads to cluster formation and then induces the growth of nanoparticles [66]. Each stage has unique characteristics depending on the reducing agent, its concentration, $AgNO_3$, and pH. The presence of hydroxyl groups (OH) in plant biomolecules, such as amino acids, proteins, alkaloids, flavonoids, polyphenols, enzymes, tannins, carbohydrates, and saponins, is associated with the stabilization and reduction of silver ions (Ag+) to Ag0. Its further reduction to Ag + leads to the formation of silver nuclei, resulting in the production of AgNPs [67]. Considering the vast potential of plants as sources of medicinal compounds, this work aims to biosynthesize AgNPs using *Azadirachta indica* seed extract. The synthesized AI-AgNPs were pragmatically characterized and investigated for their anti-diabetic potential for the first time. To the best of my knowledge, this research will represent the first reference to the use of *A. indica* seed extract for the green synthesis of silver nanoparticles in Pakistan. According to literature [68], *A. indica* is thought to have insulin-like effects, which could help lower blood glucose levels by improving insulin sensitivity or acting as an insulin mimetic. It may protect pancreatic beta cells from damage that is responsible for insulin production. Preserving these cells is crucial for maintaining proper insulin levels. Components in neem may contribute to the reduction of blood glucose levels, helping in the management of diabetes [69]. Furthermore, synergistic effects between *A. indica* and silver nanoparticles exist [70]. As reported in the previous literature, if the natural compounds coating the nanoparticles are themselves anti-diabetic in nature, a synergistic anti-diabetic potential of the final nanomaterial is observed, and *Azadirachta indica* seeds contain the most potent compounds, such as Azadarachtin, Nimbin, Nimbidin, etc., so that must reveal high anti-diabetic potential [71].

Silver nanoparticles can be a potential source of insulin sensitization as they increase the cytosolic calcium ion concentration and activate AMPK by phosphorylating it via the CAMKKβ pathway in SH-SY5Y cells and in rats [72]. AMPK activation enhances the sensitivity towards insulin, and it could mediate the insulin by increasing its action [73]. Insulin binds to its receptor and activates the phosphorylation cascade from IRS1, which induces the transport of glucose into the cells [57]. Studies have shown that animal models lacking IRS1 developed hyperglycaemia, or Type 2 diabetes mellitus; hence, increasing the protein levels of IRS1 will ultimately reduce the hyperglycemia complications [74]. Silver nanoparticles lead to a reduction in blood glucose levels by increasing the IRS1 and GLUT2 expression levels. In addition, silver nanoparticles elevate the expression levels of insulin and its secretion [75]. Natural compounds acting as high reducing agents (Azadirachtin, Vepinin, Limbocidin, etc.) present in the *Azadirachta indica* seed extract strongly affect the size and size distribution of nanoparticles; the stronger the reductant present in the extract, the higher the reaction rate, resulting in the synthesis of nanoparticles with a smaller size [76]. At the same time, the particle size distribution of nanoparticles remains narrow. Reducing agents are essential in the fabrication of nanoparticles to enhance their biomedical functionality by reducing their toxicity and enhancing their biocompatibility and bioavailability in living cells. They prevent clusters or aggregates of nanoparticles, enhance their colloidal stability, and prevent the uncontrolled growth of nanoparticles (especially metal and metal oxide nanoparticles) [77].

People are interested in nanotechnology in the fields of physics, chemistry, biology, nanomedicine, and electronics as a result of the contemporary scientific period. Nanotechnology may create a wide range of nanoscale materials with at least one dimension, ranging in size from 1 to 100 nm, known as nanoparticles (NPs) [78]. Because of their capacity to be modified at a scale where characteristics can be controlled, nanomaterials have opened new areas of scientific and industrial innovation.

In the current study, silver nanoparticles were synthesized through a biological method using the extract of *Azadirachta indica* seeds, which are well known for their strong biological potential. The present investigation indicated that post-treatment improved the hyperglycemic condition of STZ-induced mice with AI-AgNPs. These nanoparticles were characterized by visual examination (Figure 2), ultraviolet–visible (UV–vis) spectrophotometry (Figure 3), scanning electron microscopy (SEM) (Figure 4), and Fourier transform infrared (FTIR) (Figure 5a), which are in line with [41]. The formation of AI-AgNPs was predominantly detected by the variation in color of the reaction mixture from light brown to dark brown after treatment with a 1 mM silver salt ($AgNO_3$) solution. Several characterization techniques supported the successful production of NPs. The silver nanoparticles produced from *A. indica* showed a brown color change at an optimized time of 3 h. The maximum absorption was at 441 nm from the UV-visible analysis. Similar results were observed for silver nanoparticles in previous literature. Rajesh Kumar and Malarkodi [79] got an absorbance peak in the range of 400 to 450 nm when synthesized from different sources. The FTIR results of the pure *A. indica* and *A. indica*-mediated AgNPs. The extra band observed at 591 cm^{-1} corresponds to the Ag nanoparticles that appeared in *A. indica*-mediated AgNPs, as indicated in Figure 5b. The peak at 1401.37 cm^{-1} may be due to the (C-O and C-H) bending vibrations of the *A. indica plant*. The stretching vibration of the C-O functional group of alcohol, ester, ether, or carboxylic acid is shown by the peaks at 1031.85 cm^{-1} and 1041 cm^{-1}, respectively. The peak at 1617.32 cm^{-1} and 1635 cm^{-1} might be caused by the C=O stretching vibration of alkenes, primary amines (N-H bending vibration), and amides (N-H bending and C=O stretching vibration), as well as the functional groups of aldehydes and ketones. The stretch at 2104 cm^{-1} is due to the C≡C of the alkene. The OH stretching vibration of the phenol causes the peaks at 3292 cm^{-1} and 3311.35 cm^{-1}, respectively. Additionally, the presence of OH and C=O groups suggests that flavanones or terpenoids have been adsorbed on the surface of nanoparticles. Connections via π-electrons in the carbonyl groups may be liable for the reduction of Ag ions to AgNPs as well as for stability and as a capping agent. The presence of various functional groups in Figure 5b demonstrated the successful green synthesis of *A. indica*-mediated AgNPs [80,81]. The role of AgNPs in various metabolic diseases has been studied. In the present study, the AI-AgNPs were assessed for their anti-diabetic activity through in vitro studies using glucose uptake by yeast cells assays, glucose adsorption assays, and alpha-amylase inhibitory assays, as well as an in vivo study on STZ-induced diabetic mice. Glucose uptake by yeast cells at various concentrations of AI-AgNPs has determined good results (Figure 6A). These results were found to be good and in line with [77]. AI-AgNPs showed the highest activity (75 ± 1.528%), while crude extract showed (63 ± 2.5%) glucose uptake by yeast at 80 μg/mL. In the glucose adsorption assay, the highest activity of Al-AgNPs was 10.65 ± 1.58%, while crude extract showed 8.32 ± 0.258% at 30 mM, whereas in the alpha-amylase assay, Al-AgNPs exhibited the maximum activity of 73.85 ± 1.114% and crude extract 65.85 ± 2.101% at 100 μg/mL. These results were found to be similar to previous work [76].

Administration of STZ resulted in increased blood glucose levels; the treatment of AI-AgNPs has reversed the hyperglycemic condition of diabetic mice. AI-AgNPs (10 to 40 mg/kg b.w.) treatment for 30 days resulted in a significant decrease in blood glucose level (Figure 6D). It is clear from this figure that there was a significant increase in blood glucose levels (277–420 mg/dL) in the diabetic control group. The normal control group did not show a significant increase in blood glucose level (113–118 mg/dL). Groups treated with AgNPs at doses of 10, 20, 30, and 40 mg/kg, respectively, showed a significant decrease ($p < 0.05$) in glucose level (288–160, 290–154, 287–138, 290–131 mg/dL), respectively. Similar results were shown by [82,83].

In histopathological studies, the normal control group shows a normal architecture of the pancreas. The exocrine component forms a pancreas closely packed by acinar cells and arranged into small lobules. Pancreatic lobules are separated by intact intralocular and interlobular connective tissue septa, while the diabetic control group revealed pathological changes of both exocrine and endocrine components. The acinar cells were swollen, and small vacuoles were observed in almost all acinar cells. Islet β-cells are almost entirely lost in STZ-treated mice. Diabetic mice treated with different doses of AI-AgNPs revealed regeneration of islet cells. The small vacuoles in the basal area of acinar cells were also much smaller. The atrophic change of the acinar cells was less severe, and the border between the exocrine and endocrine portions became more distinct (Figure 7). These results are in line with [84–89].

Histological studies of the liver were carried out. Severe degeneration and necrosis in the hepatocytes were detected in the diabetic control group. Variations in the size of vacuoles were determined in the cytoplasm of hepatocytes in the 10 mg/kg b.w. AI-AgNPs-treated group. Degeneration and necrosis were found to be significantly reduced in the livers of mice in doses of 20 and 30 mg/kg b.w. of AI-AgNPs-treated groups. Similar histological appearance to the control group was found in the dose-40 mg/kg b.w. AI-AgNPs-treated group except for slight vacuolation and dilation of sinusoids (Figure 8). These results are in line with other mechanisms of interaction of nano Ag with cells that have been presented in the study [90], and various researchers have reported different approaches for diabetic medicine [91–97]. Considering the results obtained from anti-diabetic activity, green-synthesised AgNPs are good anti-diabetic agents. They may be crucial resources in the pharmacological and therapeutic domains for treating diabetes and other metabolic illnesses.

5. Conclusions

In this study, *Azadirachta indica*-conjugated silver nanoparticles were synthesized using their seed extract. The biosynthesized silver nanoparticles were proved to have excellent anti-diabetic potentials, which are due to the presence of natural compounds coating the nanoparticles, which are themselves anti-diabetic in nature. A synergistic anti-diabetic potential of the final nanomaterial was observed. Therefore, AgNPs produced by *A. indica* may be potentially utilized for the economical production of AgNPs for many pharmaceutical applications. To the best of my knowledge, this research will represent the first reference to the use of *A. indica* seed extract for the green synthesis of silver nanoparticles and their anti-diabetic activities in Pakistan.

Author Contributions: Conceptualization, G.R., M.U. and N.S.; methodology, M.U. and G.R.; software, M.H. and A.A.; validation, W.K., S.A. (Sajjad Ahmad) and Y.-S.M.; formal analysis, G.R. and M.U.; investigation, G.R.; resources, M.H. and A.F.A.; data curation, A.K., S.A. (Sajjad Ahmad) and G.R.; writing—original draft preparation, S.A. (Sajid Ali), M.H., A.K. and G.R.; writing—review and editing, M.H.A., S.A. (Sajid Ali), Y.-S.M. and G.R.; visualization, Y.-S.M., A.K., M.H., A.F.A., M.H.A. and S.A. (Sajid Ali); supervision, G.R.; project administration, G.R., A.F.A., S.A. (Sajid Ali) and M.H.A.; funding acquisition, A.F.A., S.A. (Sajid Ali), M.H.A., G.R. and M.H. All authors have read and agreed to the published version of the manuscript.

Funding: This research was funded by the Deputyship for Research and Innovation, "Ministry of Education" in Saudi Arabia (IFKSUOR3-185-3).

Institutional Review Board Statement: The experimentation ethics committee of the department of Zoology, Abdul Wali Khan University, Mardan, on the use of animal samples, approved the experimental protocol with access number AWKUM/ Zoo/2022/2790 on 11 April 2022. Moreover, it was carried out in strict compliance with the National Research Council guidelines on the care and use of the laboratory.

Informed Consent Statement: Not applicable.

Data Availability Statement: Data is contained within the article.

Conflicts of Interest: The authors declare no conflict of interest.

References

1. Mariadoss, A.V.A.; Saravanakumar, K.; Sathiyaseelan, A.; Karthikkumar, V.; Wang, M.H. Smart drug delivery of p-Coumaric acid loaded aptamer conjugated starch nanoparticles for effective triple-negative breast cancer therapy. *Int. J. Biol. Macromol.* **2022**, *195*, 22–29. [CrossRef]
2. Kumar, R.; Saha, P.; Kumar, Y.; Sahana, S.; Dubey, A.; Prakash, O. A Review on Diabetes Mellitus: Type 1 & Type 2. *World J. Pharm. Pharm. Sci.* **2020**, *9*, 838–850.
3. Defaei, M.; Taheri-Kafrani, A.; Miroliaei, M.; Yaghmaei, P. Improvement of stability and reusability of α-amylase immobilized on naringin functionalized magnetic nanoparticles: A robust nanobiocatalyst. *Int. J. Biol. Macromol.* **2018**, *113*, 354–360. [CrossRef] [PubMed]
4. Freiman, J.A.; Chalmers, T.C.; Smith, H.A.; Kuebler, R.R. The importance of beta, the type II error, and sample size in the design and interpretation of the randomized controlled trial: Survey of two sets of "negative" trials. In *Medical Uses of Statistics*; CRC Press: Boca Raton, FL, USA, 2019; pp. 357–389.
5. Abou Elmagd, M. Benefits, need and importance of daily exercise. *Int. J. Phys. Educ. Sports Health* **2016**, *3*, 22–27.
6. Padhi, S.; Nayak, A.K.; Behera, A. Type II diabetes mellitus: A review on recent drug based therapeutics. *Biomed. Pharmacother.* **2020**, *131*, 110708. [CrossRef] [PubMed]
7. Nasrollahzadeh, M.; Sajjadi, M.; Sajadi, S.M.; Issaabadi, Z. Green nanotechnology. In *Interface Science and Technology*; Elsevier: Amsterdam, The Netherlands, 2019; Volume 28, pp. 145–198.
8. Ijaz, I.; Gilani, E.; Nazir, A.; Bukhari, A. Detail review on chemical, physical and green synthesis, classification, characterizations and applications of nanoparticles. *Green Chem. Lett. Rev.* **2020**, *13*, 223–245. [CrossRef]
9. Manoj, K.; Rakesh, R.; Sinha, M.P.; Raipat, B.S. Different techniques utilized for characterization of metallic nanoparticles synthesized using biological agents: A review. *Balneo PRM Res. J.* **2023**, *14*, 534.
10. Noreen, S.; Tahir, M.B.; Hussain, A.; Nawaz, T.; Rehman, J.U.; Dahshan, A.; Alzaid, M.; Alrobei, H. Emerging 2D-Nanostructured materials for electrochemical and sensing Application—A review. *Int. J. Hydrogen Energy* **2022**, *47*, 1371. [CrossRef]
11. Gong, D.; Celi, N.; Zhang, D.; Cai, J. Magnetic biohybrid microrobot multimers based on chlorella cells for enhanced targeted drug delivery. *ACS Appl. Mater. Interfaces* **2022**, *14*, 6320–6330. [CrossRef]
12. Zhang, L.; Gong, C.; Bin, D. (Eds.) *Green Chemistry and Technologies*; Walter de Gruyter GmbH & Co KG: Berlin, Germany, 2018.
13. Afum, E.; Agyabeng-Mensah, Y.; Baah, C.; Agyapong, G.K.; Armas, J.A.L.; Al Farooque, O. Prioritizing zero-waste performance and green differentiation advantage through the Prism of circular principles adoption: A mediated approach. *J. Clean. Prod.* **2022**, *361*, 132182. [CrossRef]
14. Hussain, A.; Dar, B.A. Environmentally benign organic synthesis. In *Applications of Nanotechnology for Green Synthesis*; Springer: Berlin/Heidelberg, Germany, 2020; pp. 125–144.
15. Alsubhi, N.S.; Alharbi, N.S.; Felimban, A.I. Optimized Green Synthesis and Anticancer Potential of Silver Nanoparticles Using Juniperus procera Extract Against Lung Cancer Cells. *J. Biomed. Nanotechnol.* **2022**, *18*, 2249–2263. [CrossRef]
16. Khan, M.M.; Bhatti, Q.A.; Akhlaq, M.; Ishaq, M.; Ali, D.; Jalil, A.; Asghar, J.; Alarifi, S.; Elaissari, A. Assessment of Antimicrobial Potential of Plagiochasma rupestre Coupled with Healing Clay Bentonite and AGNPS. *BioMed Res. Int.* **2022**, *2022*, 4264466. [CrossRef] [PubMed]
17. Sajadi, S.M.; Kolo, K.; Hamad, S.M.; Mahmud, S.A.; Barzinjy, A.A.; Hussein, S.M. Green synthesis of the Ag/Bentonite nanocomposite UsingEuphorbia larica extract: A reusable catalyst for efficient reduction of nitro compounds and organic dyes. *ChemistrySelect* **2018**, *3*, 12274–12280. [CrossRef]
18. Talabani, R.F.; Hamad, S.M.; Barzinjy, A.A.; Demir, U. Biosynthesis of silver nanoparticles and their applications in harvesting sunlight for solar thermal generation. *Nanomaterials* **2021**, *11*, 2421. [CrossRef] [PubMed]
19. Mirzaei, Y.; Hamad, S.M.; Barzinjy, A.A.; Faris, V.M.; Karimpour, M.; Ahmed, M.H. In vitro effects of the green synthesized silver and nickel oxide nanoparticles on the motility and egg hatching ability of Marshallagia marshalli. *Emergent Mater.* **2022**, *5*, 1705–1716. [CrossRef]
20. Bergal, A.; Matar, G.H.; Andaç, M. Olive and green tea leaf extracts mediated green synthesis of silver nanoparticles (AgNPs): Comparison investigation on characterizations and antibacterial activity. *Bionanoscience* **2022**, *12*, 307–321. [CrossRef]
21. Asimuddin, M.; Shaik, M.R.; Adil, S.F.; Siddiqui, M.R.H.; Alwarthan, A.; Jamil, K.; Khan, M. Azadirachta indica based biosynthesis of silver nanoparticles and evaluation of their antibacterial and cytotoxic effects. *J. King Saud Univ. Sci.* **2020**, *32*, 648–656. [CrossRef]
22. Rakib-Uz-Zaman, S.M.; Hoque Apu, E.; Muntasir, M.N.; Mowna, S.A.; Khanom, M.G.; Jahan, S.S.; Akter, N.; Khan, M.A.R.; Shuborna, N.S.; Shams, S.M.; et al. Biosynthesis of silver nanoparticles from Cymbopogon citratus leaf extract and evaluation of their antimicrobial properties. *Challenges* **2022**, *13*, 18. [CrossRef]
23. Song, K.; Zhao, D.; Sun, H.; Gao, J.; Li, S.; Hu, T.; He, X. Green nanopriming: Responses of alfalfa (*Medicago sativa* L.) seedlings to alfalfa extracts capped and light-induced silver nanoparticles. *BMC Plant Biol.* **2022**, *22*, 323. [CrossRef]
24. Ahmed, J.; Mulla, M.Z.; Arfat, Y.A. Thermo-mechanical, structural characterization and antibacterial performance of solvent casted polylactide/cinnamon oil composite films. *Food Control* **2016**, *69*, 196–204. [CrossRef]
25. Haji, B.S.; Barzinjy, A.A. Jordan Journal of Physics. *Jordan J. Phys.* **2022**, *15*, 429–444.
26. Barzinjy, A.A.; Haji, B.S.; Fouad, H. Green Synthesis of Silver Nanoparticles Using *Citrullus colocynthis* Fruit Extract and the Eutectic-Based Ionic Liquid: Thin Film Application. *J. Nanoelectron. Optoelectron.* **2022**, *17*, 1328–1342. [CrossRef]

27. Rahman, A.; Rehman, G.; Shah, N.; Hamayun, M.; Ali, S.; Ali, A.; Shah, S.K.; Khan, W.; Shah, M.I.A.; Alrefaei, A.F. Biosynthesis and Characterization of Silver Nanoparticles Using *Tribulus terrestris* Seeds: Revealed Promising Antidiabetic Potentials. *Molecules* **2023**, *28*, 4203. [CrossRef] [PubMed]
28. Saqib, S.; Faryad, S.; Afridi, M.I.; Arshad, B.; Younas, M.; Naeem, M.; Zaman, W.; Ullah, F.; Nisar, M.; Ali, S.; et al. Bimetallic assembled silver nanoparticles impregnated in *Aspergillus fumigatus* extract damage the bacterial membrane surface and release cellular contents. *Coatings* **2022**, *12*, 1505. [CrossRef]
29. Saqib, S.; Ullah, F.; Naeem, M.; Younas, M.; Ayaz, A.; Ali, S.; Zaman, W. Mentha: Nutritional and Health Attributes to Treat Various Ailments Including Cardiovascular Diseases. *Molecules* **2022**, *27*, 6728. [CrossRef] [PubMed]
30. Hawar, S.N.; Al-Shmgani, H.S.; Al-Kubaisi, Z.A.; Sulaiman, G.M.; Dewir, Y.H.; Rikisahedew, J.J. Green synthesis of silver nanoparticles from *Alhagi graecorum* leaf extract and evaluation of their cytotoxicity and antifungal activity. *J. Nanomater.* **2022**, *2022*, 1058119. [CrossRef]
31. Akter, S.; Lee, S.Y.; Siddiqi, M.Z.; Balusamy, S.R.; Ashrafudoulla, M.; Rupa, E.J.; Huq, M.A. Ecofriendly synthesis of silver nanoparticles by *Terrabacter humi* sp. nov. and their antibacterial application against antibiotic-resistant pathogens. *Int. J. Mol. Sci.* **2020**, *21*, 9746. [CrossRef]
32. Asif, M.; Yasmin, R.; Asif, R.; Ambreen, A.; Mustafa, M.; Umbreen, S. Green Synthesis of Silver Nanoparticles (AgNPs), Structural Characterization, and their Antibacterial Potential. *Dose-Response* **2022**, *20*, 15593258221088709. [CrossRef]
33. Huq, M.A.; Ashrafudoulla, M.; Rahman, M.M.; Balusamy, S.R.; Akter, S. Green synthesis and potential antibacterial applications of bioactive silver nanoparticles: A review. *Polymers* **2022**, *14*, 742. [CrossRef]
34. Dinparvar, S.; Bagirova, M.; Allahverdiyev, A.M.; Abamor, E.S.; Safarov, T.; Aydogdu, M.; Aktas, D. A nanotechnology-based new approach in the treatment of breast cancer: Biosynthesized silver nanoparticles using *Cuminum cyminum* L. seed extract. *J. Photochem. Photobiol. B Biol.* **2020**, *208*, 111902. [CrossRef]
35. Sharifi-Rad, M.; Pohl, P.; Epifano, F.; Álvarez-Suarez, J.M. Green synthesis of silver nanoparticles using *Astragalus tribuloides* delile. root extract: Characterization, antioxidant, antibacterial, and anti-inflammatory activities. *Nanomaterials* **2020**, *10*, 2383. [CrossRef] [PubMed]
36. Jini, D.; Sharmila, S. Green synthesis of silver nanoparticles from *Allium cepa* and its in vitro antidiabetic activity. *Mater. Today Proc.* **2020**, *22*, 432–438. [CrossRef]
37. Nagaraja, S.; Ahmed, S.S.; DR, B.; Goudanavar, P.; Fattepur, S.; Meravanige, G.; Shariff, A.; Shiroorkar, P.N.; Habeebuddin, M.; Telsang, M. Green Synthesis and Characterization of Silver Nanoparticles of *Psidium guajava* Leaf Extract and Evaluation for Its Antidiabetic Activity. *Molecules* **2022**, *27*, 4336. [CrossRef] [PubMed]
38. Vinodhini, S.; Vithiya, B.S.M.; Prasad, T.A.A. Green synthesis of silver nanoparticles by employing the *Allium fistulosum*, Tabernaemontana divaricate and *Basella alba* leaf extracts for antimicrobial applications. *J. King Saud Univ. Sci.* **2022**, *34*, 101939. [CrossRef]
39. Zubair, M.; Azeem, M.; Mumtaz, R.; Younas, M.; Adrees, M.; Zubair, E.; Khalid, A.; Hafeez, F.; Rizwan, M.; Ali, S. Green synthesis and characterization of silver nanoparticles from *Acacia nilotica* and their anticancer, antidiabetic and antioxidant efficacy. *Environ. Pollut.* **2022**, *304*, 119249. [CrossRef]
40. Kaliammal, R.; Parvathy, G.; Maheshwaran, G.; Velsankar, K.; Devi, V.K.; Krishnakumar, M.; Sudhahar, S. Zephyranthes candida flower extract mediated green synthesis of silver nanoparticles for biological applications. *Adv. Powder Technol.* **2021**, *32*, 4408–4419. [CrossRef]
41. Badmus, J.A.; Oyemomi, S.A.; Adedosu, O.T.; Yekeen, T.A.; Azeez, M.A.; Adebayo, E.A.; Lateef, A.; Badeggi, U.M.; Botha, S.; Hussein, A.A.; et al. Photo-assisted bio-fabrication of silver nanoparticles using *Annona muricata* leaf extract: Exploring the antioxidant, anti-diabetic, antimicrobial, and cytotoxic activities. *Heliyon* **2020**, *6*, e05413. [CrossRef]
42. Das, B.; De, A.; Podder, S.; Das, S.; Ghosh, C.K. and Samanta, A. Green biosynthesis of silver nanoparticles using *Dregea volubilis* flowers: Characterization and evaluation of antioxidant, antidiabetic and antibacterial activity. *Inorg. Nano Met. Chem.* **2021**, *51*, 1066–1079. [CrossRef]
43. Thirumal, S.; Sivakumar, T. Synthesis of silver nanoparticles using *Cassia auriculata* leaves extracts and their potential antidiabetic activity. *Int. J. Botany Stud.* **2021**, *6*, 35–38.
44. Yarrappagaari, S.; Gutha, R.; Narayanaswamy, L.; Thopireddy, L.; Benne, L.; Mohiyuddin, S.S.; Vijayakumar, V.; Saddala, R.R. Eco-friendly synthesis of silver nanoparticles from the whole plant of *Cleome viscosa* and evaluation of their characterization, antibacterial, antioxidant and antidiabetic properties. *Saudi J. Biol. Sci.* **2020**, *27*, 3601–3614. [CrossRef]
45. Putra, I.M.W.A.; Fakhrudin, N.; Nurrochmad, A.; Wahyuono, S. Antidiabetic activity of *Coccinia grandis* (L.) Voigt: Bioactive constituents, mechanisms of action, and synergistic effects. *J. Appl. Pharm. Sci.* **2021**, *12*, 041–054.
46. Raja, B.D.; Sheela, D.S.; Priya, E.S.; Vanitha, A.; Kalimuthu, K.; Viswanathan, P. Phyto-mediated Synthesis of Silver Nanoparticles with Afrohybanthus travancoricus Leaf Aqueous Extract and Screening of their in vitro Antioxidant, Anti-Inflammatory, and Anti-diabetic Activities. *Pharmacogn. Res.* **2023**, *15*, 751–760. [CrossRef]
47. Barman, A.; Kotal, A.; Das, M. Synthesis of Metal Based Nano particles from Moringa Olifera and its Biomedical Applications: A Review. *Inorg. Chem. Commun.* **2023**, *158*, 111438. [CrossRef]
48. Sharma, D.; Radha, R.; Kumar, M.; Andrade-Cetto, A.; Puri, S.; Kumar, A.; Thakur, M.; Chandran, D.; Pundir, A.; Prakash, S.; et al. Chemical Diversity and Medicinal Potential of *Vitex negundo* L.: From Traditional Knowledge to Modern Clinical Trials. *Chem. Biodivers.* **2023**, *20*, e202301086. [CrossRef] [PubMed]

49. Ullah, H.; Ullah, I.; Rehman, G.; Hamayun, M.; Ali, S.; Rahman, A.; Lee, I.J. Magnesium and zinc oxide nanoparticles from datura alba improve cognitive impairment and blood brain barrier leakage. *Molecules* **2022**, *27*, 4753. [CrossRef]
50. Jyoti, K.; Baunthiyal, M.; Singh, A. Characterization of silver nanoparticles synthesized using *Urtica dioica* Linn. leaves and their synergistic effects with antibiotics. *J. Radiat. Res. Appl. Sci.* **2016**, *9*, 217–227. [CrossRef]
51. Sathiyaseelan, A.; Saravanakumar, K.; Mariadoss, A.V.A.; Wang, M.H. Biocompatible fungal chitosan encapsulated phytogenic silver nanoparticles enhanced antidiabetic, antioxidant and antibacterial activity. *Int. J. Biol. Macromol.* **2020**, *153*, 63–71. [CrossRef]
52. Nikolova, M.P.; Joshi, P.B.; Chavali, M.S. Updates on Biogenic Metallic and Metal Oxide Nanoparticles: Therapy, Drug Delivery and Cytotoxicity. *Pharmaceutics* **2023**, *15*, 1650. [CrossRef]
53. Rehman, G.; Hamayun, M.; Iqbal, A.; Ul Islam, S.; Arshad, S.; Zaman, K.; Ahmad, A.; Shehzad, A.; Hussain, A.; Lee, I. In vitro antidiabetic effects and antioxidant potential of *Cassia nemophila* pods. *BioMed Res. Int.* **2018**, *2018*, 1824790. [CrossRef]
54. Kim, W.H.; Song, H.O.; Jin, C.M.; Hur, J.M.; Lee, H.S.; Jin, H.Y.; Kim, S.Y.; Park, H. The Methanol Extract of Azadirachta indica A. Juss Leaf Protects Mice Against Lethal Endotoxemia and Sepsis. *Biomol. Ther.* **2012**, *20*, 96–103. [CrossRef]
55. Das, C.A.; Kumar, V.G.; Dhas, T.S.; Karthick, V.; Govindaraju, K.; Joselin, J.M.; Baalamurugan, J. Antibacterial activity of silver nanoparticles (biosynthesis): A short review on recent advances. *Biocatal. Agric. Biotechnol.* **2020**, *27*, 101593. [CrossRef]
56. Kaur, N.; Kumar, V.; Nayak, S.K.; Wadhwa, P.; Kaur, P.; Sahu, S.K. Alpha-amylase as molecular target for treatment of diabetes mellitus: A comprehensive review. *Chem. Biol. Drug Des.* **2021**, *98*, 539–560. [CrossRef] [PubMed]
57. Okka, E.Z.; Tongur, T.; Tarik Aytas, T.; Yilmaz, M.; Topel, Ö.; Sahin, R. Green Synthesis and the formation kinetics of silver nanoparticles in aqueous Inula Viscosa extract. *arXiv* **2022**, arXiv:2209.03022. [CrossRef]
58. Liu, Z.; Ma, S. Recent Advances in Synthetic α-Glucosidase Inhibitors. *ChemMedChem* **2017**, *12*, 819–829. [CrossRef] [PubMed]
59. Hall, B.M.; Balan, V.; Gleiberman, A.S.; Strom, E.; Krasnov, P.; Virtuoso, L.P.; Rydkina, E.; Vujcic, S.; Balan, K.; Gitlin, I.; et al. Aging of mice is associated with p16 (Ink4a)-and β-galactosidase-positive macrophage accumulation that can be induced in young mice by senescent cells. *Aging* **2016**, *8*, 1294. [CrossRef] [PubMed]
60. Giri, A.K.; Jena, B.; Biswal, B.; Pradhan, A.K.; Arakha, M.; Acharya, S.; Acharya, L. Green synthesis and characterization of silver nanoparticles using *Eugenia roxburghii* DC. extract and activity against biofilm-producing bacteria. *Sci. Rep.* **2022**, *12*, 8383. [CrossRef] [PubMed]
61. Uzair, B.; Liaqat, A.; Iqbal, H.; Menaa, B.; Razzaq, A.; Thiripuranathar, G.; Fatima Rana, N.; Menaa, F. Green and cost-effective synthesis of metallic nanoparticles by algae: Safe methods for translational medicine. *Bioengineering* **2020**, *7*, 129. [CrossRef]
62. Melkamu, W.W.; Bitew, L.T. Green synthesis of silver nanoparticles using *Hagenia abyssinica* (Bruce) JF Gmel plant leaf extract and their antibacterial and anti-oxidant activities. *Heliyon* **2021**, *7*, e08459. [CrossRef]
63. Yamamoto, K.; Imaoka, T.; Tanabe, M.; Kambe, T. New horizon of nanoparticle and cluster catalysis with dendrimers. *Chem. Rev.* **2019**, *120*, 1397–1437. [CrossRef]
64. Preety, R.; Anitha, R.; Rajeshkumar, S.; Lakshmi, T. Anti-diabetic activity of silver nanoparticles prepared from cumin oil using alpha amylase inhibitory assay. *Int. J. Res. Pharm. Sci.* **2020**, *11*, 1267–1269.
65. Košpić, K.; Biba, R.; Peharec Štefanić, P.; Cvjetko, P.; Tkalec, M.; Balen, B. Silver Nanoparticle Effects on Antioxidant Response in Tobacco Are Modulated by Surface Coating. *Plants* **2022**, *11*, 2402. [CrossRef] [PubMed]
66. Kurmi, U.G.; Zahir, A.; Musa, A.; Iyadunni, A.D.; Tomsu, U.A.; Patel, P.K.; Chukwuemeka, P.O. Review of nigerian medicinal plants used in the management of diabetes mellitus. *J. Clin. Med. Images Case Rep.* **2022**, *2*, 1–5. [CrossRef] [PubMed]
67. Chinnasamy, G.; Chandrasekharan, S.; Koh, T.W.; Bhatnagar, S. Synthesis, characterization, antibacterial and wound healing efficacy of silver nanoparticles from *Azadirachta indica*. *Front. Microbiol.* **2021**, *12*, 611560. [CrossRef] [PubMed]
68. Anwar, M.; Alghamdi, K.S.; Zulfiqar, S.; Warsi, M.F.; Waqas, M.; Hasan, M. Ag-decorated BiOCl anchored onto the g-C_3N_4 sheets for boosted photocatalytic and antimicrobial activities. *Opt. Mater.* **2023**, *135*, 113336. [CrossRef]
69. Khan, I.; Bawazeer, S.; Rauf, A.; Qureshi, M.N.; Muhammad, N.; Al-Awthan, Y.S.; Bahattab, O.; Maalik, A.; Rengasamy, K.R. Synthesis, biological investigation and catalytic application using the alcoholic extract of Black Cumin (*Bunium persicum*) seeds-based silver nanoparticles. *J. Nanostructure Chem.* **2021**, *12*, 59–77. [CrossRef]
70. Nguyen, N.T.T.; Nguyen, L.M.; Nguyen, T.T.T.; Nguyen, T.T.; Nguyen, D.T.C.; Tran, T.V. Formation, antimicrobial activity, and biomedical performance of plant-based nanoparticles: A review. *Environ. Chem. Lett.* **2022**, *20*, 2531–2571. [CrossRef] [PubMed]
71. Nagini, S.; Palrasu, M.; Bishayee, A. Limonoids from neem (*Azadirachta indica* A. Juss.) are potential anticancer drug candidates. *Med. Res. Rev.* **2023**, 1–40. [CrossRef]
72. Baby, A.R.; Freire, T.B.; Marques, G.D.A.; Rijo, P.; Lima, F.V.; Carvalho, J.C.M.D.; Rojas, J.; Magalhães, W.V.; Velasco, M.V.R.; Morocho-Jácome, A.L. *Azadirachta indica* (Neem) as a potential natural active for dermocosmetic and topical products: A narrative review. *Cosmetics* **2022**, *9*, 58. [CrossRef]
73. Li, L.; Li, L.; Zhou, X.; Yu, Y.; Li, Z.; Zuo, D.; Wu, Y. Silver nanoparticles induce protective autophagy via Ca^{2+}/CaMKKβ/AMPK/mTOR pathway in SH-SY5Y cells and rat brains. *Nanotoxicology* **2019**, *13*, 369–391. [CrossRef]
74. Zhang, R.; Qin, X.; Zhang, T.; Li, Q.; Zhang, J.; Zhao, J. Astragalus polysaccharide improves insulin sensitivity via AMPK activation in 3T3-L1 adipocytes. *Molecules* **2018**, *23*, 2711. [CrossRef]
75. Sudha, P.N.; Sangeetha, K.; Vijayalakshmi, K.; Barhoum, A. Nanomaterials history, classification, unique properties, production and market. In *Emerging Applications of Nanoparticles and Architecture Nanostructures*; Elsevier: Amsterdam, The Netherlands, 2018; pp. 341–384.
76. Lavin, D.P.; White, M.F.; Brazil, D.P. IRS proteins and diabetic complications. *Diabetologia* **2016**, *59*, 2280–2291. [CrossRef]

77. Alkaladi, A.; Abdelazim, A.M.; Afifi, M. Antidiabetic activity of zinc oxide and silver nanoparticles on streptozotocin-induced diabetic rats. *Int. J. Mol. Sci.* **2014**, *15*, 2015–2023. [CrossRef] [PubMed]
78. Jain, S.; Mehata, M.S. Medicinal plant leaf extract and pure flavonoid mediated green synthesis of silver nanoparticles and their enhanced antibacterial property. *Sci. Rep.* **2017**, *7*, 15867. [CrossRef] [PubMed]
79. Wang, H.; Tang, L.; Kong, Y.; Liu, W.; Zhu, X.; You, Y. Strategies for Reducing Toxicity and Enhancing Efficacy of Chimeric Antigen Receptor T cell Therapy in Hematological Malignancies. *Int. J. Mol. Sci.* **2023**, *24*, 9115. [CrossRef]
80. Keerthiga, N.; Anitha, R.; Rajeshkumar, S.; Lakshmi, T. Antioxidant activity of cumin oil mediated silver nanoparticles. *Pharmacogn. J.* **2019**, *11*, 787–789. [CrossRef]
81. Sano, T.; Ozaki, K.; Matsuura, T.; Narama, I. Giant mitochondria in pancreatic acinar cells of alloxan-induced diabetic rats. *Toxicol. Pathol.* **2010**, *38*, 658–665. [CrossRef] [PubMed]
82. Ara, C.; Andleeb, S.; Ali, S.; Majeed, B.; Iqbal, A.; Arshad, M.; Chaudhary, A.; Asmatullah; Muzamil, A. Protective potential of fresh orange juice against zinc oxide nanoparticles-induced trans-placental and trans-generational toxicity in mice. *Food Sci. Nutr.* **2023**, *11*, 5114–5128. [CrossRef] [PubMed]
83. Liu, H.; Zhang, M.; Meng, F.; Su, C.; Li, J. Polysaccharide-based gold nanomaterials: Synthesis mechanism, polysaccharide structure-effect, and anticancer activity. *Carbohydr. Polym.* **2023**, *321*, 121284. [CrossRef]
84. Zhang, J.; Wang, F.; Yalamarty, S.S.K.; Filipczak, N.; Jin, Y.; Li, X. Nano silver-induced toxicity and associated mechanisms. *Int. J. Nanomed.* **2022**, *17*, 1851–1864. [CrossRef]
85. Sun, S.J.; Deng, P.; Peng, C.E.; Ji, H.Y.; Mao, L.F.; Peng, L.Z. Extraction, Structure and Immunoregulatory Activity of Low Molecular Weight Polysaccharide from Dendrobium officinale. *Polymers* **2022**, *14*, 2899. [CrossRef]
86. Zhang, Y.; Zeng, M.; Li, B.; Zhang, B.; Cao, B.; Wu, Y.; Feng, W. Ephedra Herb extract ameliorates adriamycin-induced nephrotic syndrome in rats via the CAMKK2/AMPK/mTOR signaling pathway. *Chin. J. Nat. Med.* **2023**, *21*, 371–382. [CrossRef] [PubMed]
87. Zhang, Z.; Zhang, W.; Hou, Z.; Li, P.; Wang, L. Electrophilic Halospirocyclization of N-Benzylacrylamides to Access 4-Halomethyl-2-azaspiro[4.5]decanes. *J. Org. Chem.* **2023**, *88*, 13610–13621. [CrossRef] [PubMed]
88. Gao, Z.; Pan, X.; Shao, J.; Jiang, X.; Su, Z.; Jin, K.; Ye, J. Automatic interpretation and clinical evaluation for fundus fluorescein angiography images of diabetic retinopathy patients by deep learning. *Br. J. Ophthalmol.* **2022**, *107*, 1852–1858. [CrossRef] [PubMed]
89. Zhao, X.; Zhang, Y.; Yang, Y.; Pan, J. Diabetes-related avoidable hospitalisations and its relationship with primary healthcare resourcing in China: A cross-sectional study from Sichuan Province. *Health Soc. Care Community* **2022**, *30*, e1143–e1156. [CrossRef] [PubMed]
90. Sohrabi Kashani, A.; Packirisamy, M. Cancer-nano-interaction: From cellular uptake to mechanobiological responses. *Int. J. Mol. Sci.* **2021**, *22*, 9587. [CrossRef] [PubMed]
91. Yang, Y.Y.; Shi, L.X.; Li, J.H.; Yao, L.Y.; Xiang, D.X. Piperazine ferulate ameliorates the development of diabetic nephropathy by regulating endothelial nitric oxide synthase. *Mol. Med. Rep.* **2019**, *19*, 2245–2253. [CrossRef] [PubMed]
92. Chen, J.; Li, X.; Liu, H.; Zhong, D.; Yin, K.; Li, Y.; Wang, C. Bone marrow stromal cell-derived exosomal circular RNA improves diabetic foot ulcer wound healing by activating the nuclear factor erythroid 2-related factor 2 pathway and inhibiting ferroptosis. *Diabet. Med.* **2023**, *40*, e15031. [CrossRef]
93. Chen, Y.; Tan, S.; Liu, M.; Li, J. LncRNA TINCR is downregulated in diabetic cardiomyopathy and relates to cardiomyocyte apoptosis. *Scand. Cardiovasc. J.* **2018**, *52*, 335–339. [CrossRef]
94. Xiao, D.; Guo, Y.; Li, X.; Yin, J.; Zheng, W.; Qiu, X.; Veglio, F. The Impacts of SLC22A1 rs594709 and SLC47A1 rs2289669 Polymorphisms on Metformin Therapeutic Efficacy in Chinese Type 2 Diabetes Patients. *Int. J. Endocrinol.* **2016**, *2016*, 4350712. [CrossRef]
95. Su, M.; Hu, R.; Tang, T.; Tang, W.; Huang, C. Review of the correlation between Chinese medicine and intestinal microbiota on the efficacy of diabetes mellitus. *Front. Endocrinol.* **2023**, *13*, 1085092. [CrossRef]
96. Yu, T.; Xu, B.; Bao, M.; Gao, Y.; Zhang, Q.; Zhang, X.; Liu, R. Identification of potential biomarkers and pathways associated with carotid atherosclerotic plaques in type 2 diabetes mellitus: A transcriptomics study. *Front. Endocrinol.* **2022**, *13*, 981100. [CrossRef]
97. Yang, Y.; Chen, Z.; Yang, X.; Deng, R.; Shi, L.; Yao, L.; Xiang, D. Piperazine ferulate prevents high-glucose-induced filtration barrier injury of glomerular endothelial cells. *Exp. Ther. Med.* **2021**, *22*, 1175. [CrossRef]

pharmaceuticals

MDPI

Article

Gold Nanoparticles Synthesized with Triple-Negative Breast Cancer Cell Lysate Enhance Antitumoral Immunity: A Novel Synthesis Method

Raúl Rangel-López †, Moisés Ármides Franco-Molina, Cristina Rodríguez-Padilla and Diana Ginette Zárate-Triviño *,†

Universidad Autónoma de Nuevo León, Facultad de Ciencias Biológicas, Laboratorio de Inmunología y Virología, San Nicolás de los Garza 66455, Mexico; jrangelraul95@gmail.com (R.R.-L.); moyfranco@gmail.com (M.Á.F.-M.); cristina.rodriguezpd@uanl.edu.mx (C.R.-P.)

* Correspondence: diana.zaratetr@uanl.edu.mx

† These authors contributed equally to this work.

Abstract: Background: Gold nanoparticles enhance immunity, promotes antigen uptake by antigen-presenting cells (APCs), and boost the response against tumor antigens; therefore, they are a promising delivery vehicle. Tumor lysates have shown favorable responses as inductors of anti-cancer immunity, but the effectiveness of these treatments could be improved. Hybrid nanosystems gold nanoparticles with biomolecules have been show promising alternative on uptake, activation and response on immune system. **Objectives:** This study's objective was to develop a method of synthesizing gold nanoparticles employing a triple-negative breast cancer (4T1) cell lysate (AuLtNps) as a reducing agent to increase immunogenicity against breast cancer cells. **Methods:** Nanoparticle formation, size, and ζ potential were confirmed by surface plasmon resonance, dynamic light scattering, and transmission electron microscopy. Protein concentration was quantified using a Pierce BCA assay. The cytotoxic effects of treatments on murine macrophages were assessed, along with nanoparticle and tumor lysate uptake via epifluorescence microscopy. Using a murine model, cytokine secretion profiles were determined, and the efficacy in inhibiting the implantation of a 4T1 model was evaluated. **Results/Conclusions:** AuLtNps exhibited higher protein content than tumor lysate alone, leading to increased uptake and phagocytosis in murine macrophages, as confirmed by epifluorescence microscopy. Cytokine secretion analysis showed a proinflammatory response, with increased CD8+ and CD22+ lymphocytes and upregulation of APC markers (CD14, CD80, CD86, and MHC II+). Splenocytes demonstrated specific lysis of up to 40% against 4T1 tumor cells. In a murine model, AuLtNPs effectively inhibited tumor implantation, achieving an improved 90-days survival rate, highlighting their potential as an immunotherapy for triple-negative breast cancer.

Keywords: antigen presentation; breast cancer immunotherapy; gold nanoparticles; immune system activation; tumor cell lysates

1. Introduction

Cancer is a serious health problem, not only due to the disease itself but also because of the difficulty and high cost of treating it. One of the most relevant types of cancer is breast cancer; numerous therapies have been developed to reduce the size and aggression of such tumors [1,2]. One persistent concern is tumor regression or recurrence, which is the reappearance of the tumor after conventional treatments. In breast cancer, tumor recurrence is associated with the highest mortality rates [3].

To address this, new cancer treatments have been directed not only at combating the progression of the disease but also at preventing tumor recurrence [4]. Cancer immunotherapy has emerged as an alternative to treat various types of cancer [5]. Treatments include cellular therapy, which involves replacing damaged or non-functional patient cells with cells capable of fighting the disease [6]; monoclonal antibody therapy, which uses specific antibodies against molecules present in tumor cells to inhibit their development [7]; and cancer vaccines that aim to activate the antitumor immune system for long-term immunity [8]. Another approach, known as immunomodulation, involves the use of cytokines related to favorable prognosis [9].

Over the past 20 years, immunomodulation therapeutics have focused on enhancing the activity of cells responsible specifically or nonspecifically for antitumor immunity, such as natural killer cells or effector T lymphocytes [10]. However, challenges still exist in bridging the gap between innate and adaptive immunity, particularly in antigen presentation [11]. A successful immunotherapy must be able to achieve an adaptive immune response, which involves the recognition of tumor-associated antigens or neoantigens—proteins or peptides expressed exclusively by tumor cells [12].

Efforts have been made to develop vaccines against these antigens. Although they show efficacy in vivo and in vitro, many of these vaccines have not reached viable clinical applications [13]. Despite these challenges, significant progress has been made in recent years using pulsed dendritic cell therapies [14] and applying neoantigen vaccines against certain types of cancer. Notable successes include vaccines against human papillomavirus preventing cervical cancer and the recently approved vaccine against prostate cancer [15].

An emerging alternative proposes the use of autologous tumor lysates to generate specific immunity against the type of tumor from which the lysate was derived [16]; this is often combined with dendritic cell therapies to obtain a stronger response from the immune system [17]. Nevertheless, the efficiency of antigen presentation using these lysates could be improved. Studies using this therapeutic approach often report low effectiveness due to antigen presentation issues [18]. This happens because the same antigen-presenting cells responsible for presenting antigenic variants to T lymphocytes face difficulties in inducing specific antigen recognition and promoting antigen-specific cell lysis.

One of the initial methods developed to enhance the antigen capture processes from tumor cells and even whole tumor cells uses polymeric nanoparticles [19]. These nanoparticles encapsulate antigenic fragments derived from patient tumor cells, employing a polymer such as polylactic acid [20]; the encapsulation process includes antigens associated with autologous tumors. This approach put forward a therapy based on dendritic cells to stimulate autologous CD8+ T cells [21]. Through clinical trials, it was possible to increase the populations of antitumor CD8+ T lymphocytes, in addition to enhancing the secretion of IFN-γ and reducing the levels of IL-10 [22]. This effect was observed specifically in the group of patients treated with polymeric nanoparticles loaded with tumor lysate [23].

In recent years, nanotechnology has been used to deliver substances into biological systems, including tumor-associated antigens and neoantigens [24,25]. Gold nanoparticles, when attached to these molecules, can specifically activate the immune system, even generating antigen-specific antibodies [26,27].

This study aims to combine tumor lysates with gold nanoparticles, developing a one-step synthesis using a gold precursor salt and tumor lysates from a murine triple-negative breast cancer cell line (4T1). The goal is to enhance a specific antitumor immune response against this cell line using a murine model sensitive to tumor implantation.

2. Results

2.1. Formation of Gold Nanoparticles

When measured by UV-VIS spectroscopy, metal exhibits a characteristic band on a nanometric scale known as Surface Plasmon Resonance (SPR). In the case of gold, this band is within the range of 510 to 550 nm. Gold nanoparticles synthesized from tumor lysate derived from triple-negative murine breast cancer (AuLtNps) had a maximum absorbance at 537 nm, which remained consistent from day 1, with an intensity of 0.6 relative units (RU), until reaching 0.9 RU on day 30 (Figure 1a). AuLtNps showed a stable size measured by Dynamic Light Scattering (DLS), starting at 20 nm on day 1, increasing to 36.72 nm on day 7 through day 18, and reaching a size of 77.39 nm on day 31, indicating that the size increase is time-dependent (Figure 1b).

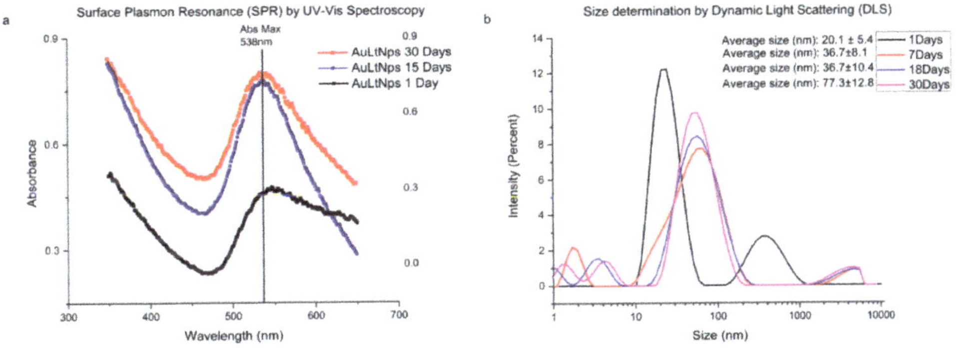

Figure 1. Characterization of gold nanoparticles synthesized from tumor lysate derived from triple-negative murine breast cancer (4T1). (**a**) Characterization through the identification of the surface plasmon resonance of AuLtNps up to 30 days. (**b**) Particle size determination by DLS.

2.2. Determination of ζ Potential

The electric superficial charge, known as ζ potential, was determined through DLS on days 1, 7, 18, and 31 to assess the particle stability in relation to the surrounding medium. A time-dependent increase in the surface charge of AuLtNps was identified, starting with a net charge of −18.9 mV on day 1, −20.1 mV on day 7, and −21.4 mV on day 18, eventually reaching −27.6 on day 31 (Table 1, Figure 2), indicating an increase in the repulsive force between particles.

Table 1. Size, ζ potential, and protein concentration in AuLtNPs.

Sampling Day	SPR Abs	Size nm	ζ Potential mV	Protein Concentration mg/mL
1	0.635	20.1 ± 5.4	−18.9	0.6126
15	0.865	36.7 ± 10.4	−21.4	0.754
30	1.21	77.3 ± 12.8	−27.6	0.902

2.3. Morphology Analysis of Gold Nanoparticles

Transmission Electron Microscopy (TEM) was used to observe the morphology and size of the nanoparticles on the first day of synthesis. The spherical morphology characteristic of metallic core–shell particles was observed (Figure 3a). Particle size analysis was conducted, revealing that most particles fell within the range of 9 to 16 nm (Figure 3b).

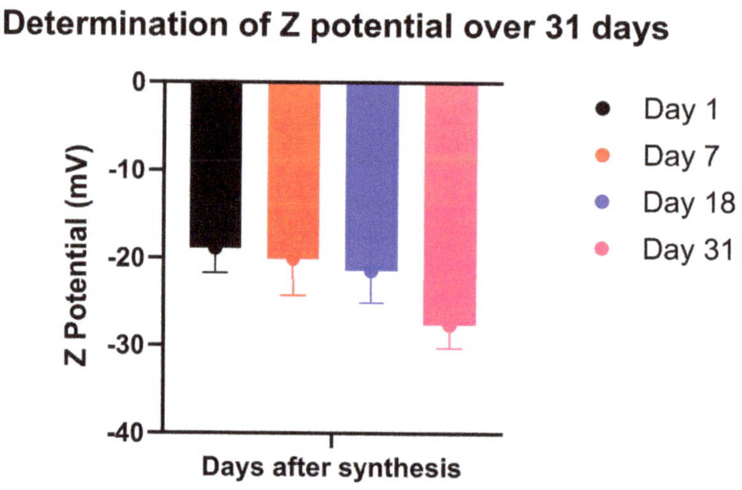

Figure 2. Determination of ζ potential up to 30 days post synthesis.

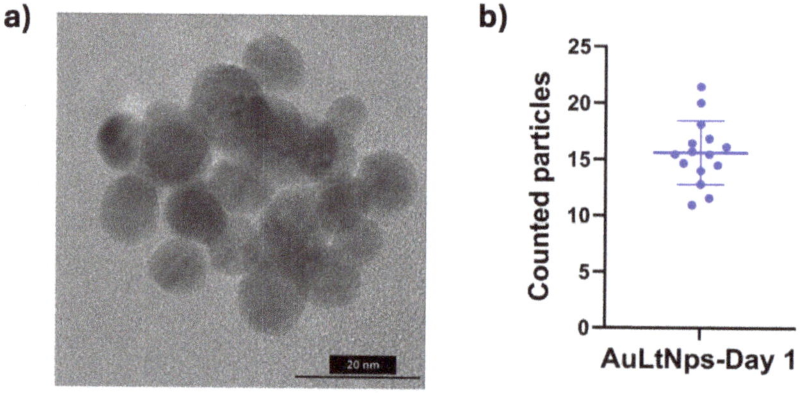

Figure 3. (**a**) Transmission electron microscopic image on the 1st day of AuLtNps synthesis. (**b**) Graphic representation of counted particles from the TEM image.

2.4. Determination of Protein Concentration in AuLtNPs

In order to confirm that the time-dependent growth of nanoparticles is associated with the total protein concentration, two nanoparticle solutions were selected with pH values of 8.5 and 9, using Tumor Lysate (LT) as control. A significant difference was observed between AuLTNps with pH 8.5 and AuLtNps with pH 9, which was more pronounced on the 31st day of evaluation compared to the control (Figure 4); this could be related to the degradation of tumor lysate and suggests that a protein is attached on the nanoparticles surface.

Polymeric nanoparticle systems have been successful in controlled molecular release [28]. Although, initially, the protein uptake phenomenon might be considered unfavorable, it proves to be advantageous in terms of antigen presentation. In comparison, gold nanoparticles have been proven to concentrate the antigen, allowing for the absorption of more antigenic variants confined within a nanoparticle [29] and a broader immune response against antigens.

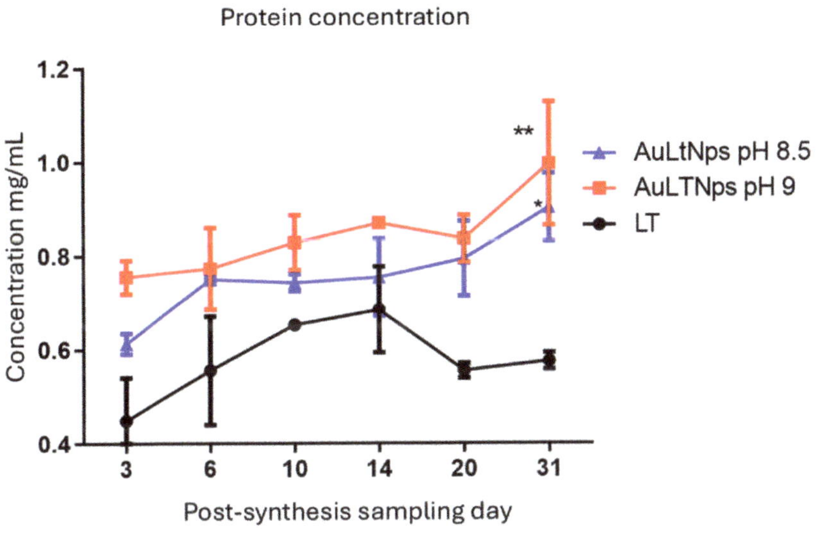

Figure 4. Determination of protein concentration in AuLtNPs. A significant difference was observed between AuLtNps pH 8.5 and AuLTNps pH 9, which was more pronounced at day 31 of evaluation compared to the control. * $p < 0.05$ and ** $p < 0.01$.

The biological activity of gold nanoparticles is significantly influenced by factors such as size and reducing agent. A novel method for gold nanoparticle synthesis was developed in [30], where tumor lysate was used as the reducing agent and $HAuCl_4$ was used as the metal precursor.

Repeated intraperitoneal administration of 12.5 nm AuNps did not elicit significant oxidative stress in mice. Given that nanoparticle size affects the biological effects in a model, smaller particles induced transient hepatic changes, whereas larger particles accumulated in the spleen. These findings highlight the importance of particle size in determining biodistribution and toxicity, indicating a dose-dependent accumulation pattern without associated physiological impairment [31,32].

It was determined that the size and negative surface charge increase over time, reaching 77 nm and −27.6 mV, respectively, after 30 days of synthesis, contributing to their stability. This increment in both size and surface charge is attributed to the protein corona effect. This phenomenon involves the net charge of the particle attracting proteins with an opposite charge. It has been observed that AuLtNps captures proteins from the surrounding solution, forming a hard and soft protein layer on the surface [33,34].

Although these interactions are not permanent, if the initial conditions are maintained, the charge present on the particle remains stable [35,36]. This phenomenon explains the increase in protein concentration over time, even without adding more protein to the sample; a similar phenomenon was reported on concentration of DNA or RNA with a cationic polymer [37,38].

2.5. Pearson Correlation Analysis

Pearson correlation of AuLtNps is a statistical method that measures the similarity or correlation between SPR intensity, particle size, surface charge, and protein concentration (Table 1) by comparing their attributes and calculating a score ranging from −1 to +1 (Table 2). A high score indicates high similarity and a high correlation between particle size and ζ potential. As time passes, the particle size and ζ potential increase, which indicates the stabilization of the components in the synthesis [39].

Table 2. Pearson correlation among PRS, size, ζ potential, and protein concentration in AuLtNPs.

Correlation Pearson r	SPR Abs	Size nm	ζ Potential mV	Protein Concentration mg/mL
SPR	1.000	0.993	−0.993	0.995
Size	0.993	1.000	−1.000	0.975
ζ potential	−0.993	−1.000	1.000	−0.976
Protein concentration	0.995	0.975	−0.976	1.000

2.6. Cytotoxicity Analysis

In order to determine the dose of AuLtNps and the controls, a dose-dependent cytotoxicity evaluation was performed on non-cancerous cell line NiH-3T3. On AuLtNps, the cell viability remained high, even at concentrations as high as 300 μM (Figure 5d).

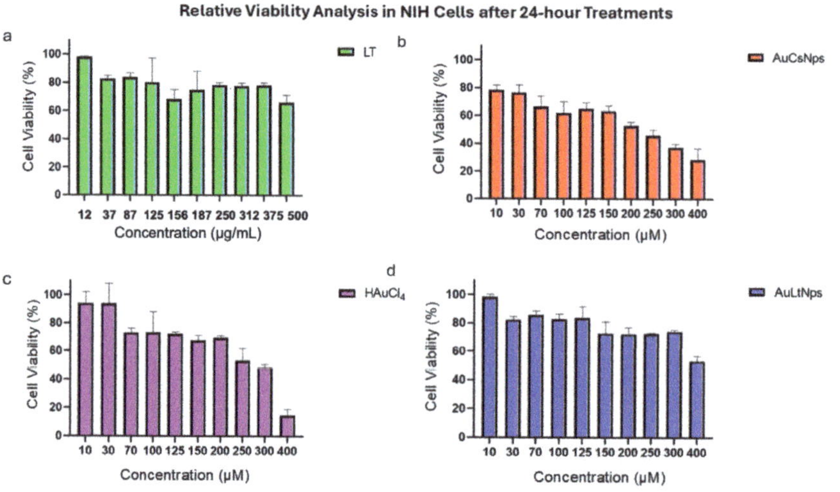

Figure 5. Relative viability analysis by MTS in NIH cells after 24 h. (**a**) LT is shown in green, (**b**) AuCsNps in red, (**c**) $HAuCl_4$ in purple, and (**d**) AuLtNps in blue. On AuLtNps, the cell viability remained at 80%, even at concentrations as high as 300 μM.

AuLtNps shows a minor cytotoxic effect on the fibroblast cell line in comparison to AuCsNps (Figure 5b) and $HAuCl_4$. This is due to the protein corona effect of proteins of the tumoral lysate on AuLtNps and the lower effect of ion release from gold atoms, in contrast to the AuCsNps and $HAuCl_4$, which lack protein in the synthesis [40].

Two treatments were administered at equivalent volumes (1:1 v/v), $HAuCl_4$ concentrations are expressed in micromolar units (μM), whereas those of LT are expressed in μg/mL; the difference in units reflects specific experimental considerations.

2.7. Macrophage Uptake

To determine whether AuLtNps were taken up by murine macrophages, the cells were extracted and cultured. Fluorescence was detected inside the macrophage, confirming the antigen uptake was more efficient for AuLtNps compared with the control LT (Figure 6).

The subcutaneous immunization route employed in this study is well-established, thanks to its minimal organ toxicity and bioaccumulation. Subcutaneously injected gold nanoparticles are efficiently phagocytosed by macrophages, which subsequently migrate to lymph nodes and blood vessels. This suggests that the reticuloendothelial system plays a pivotal role in AuNps clearance. Tumor lysate is normally taken up by intraperitoneal macrophages [41,42]; however, functionalization with gold nanoparticles potentiated the effect, as seen on the AuLtNps and AuCsNps samples (Figure 6a). The combination of CsNps+LT resulted in lower uptake compared to AuLtNps.

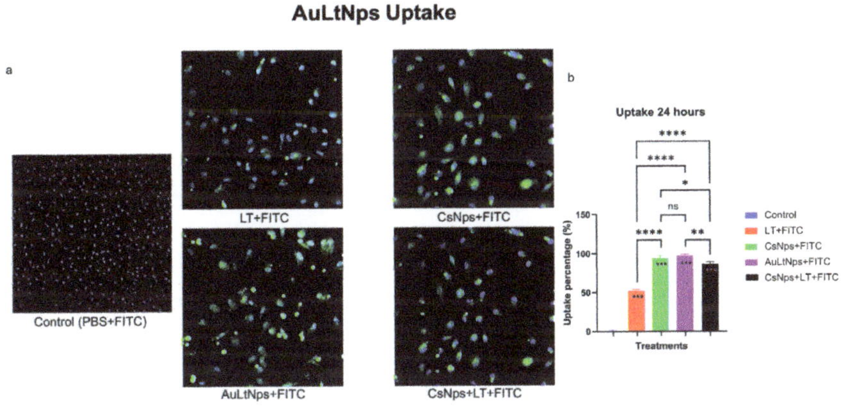

Figure 6. Determination of AuLtNps uptake in mouse intraperitoneal macrophages after 24 h. (**a**) Control group, tumor lysate (LT+FITC), AuCsNps (AuCsNps+FITC), AuLT (AuLtNps+FITC), and AuCsNps+LT(AuCsNps+LT+FITC). All cells were stained with blue DAPI to mark the nucleus and FITC as bright-green fluorescence. (**b**) Graph representing fluorescence detection. * p <0.05, ** p <0.01, and ***/**** p < 0.001, ns, not significant.

Fluorescence microscopy revealed interference within cells treated with AuLtNps. This interference did not occur in the case of AuCsNps but did with LT, suggesting distortion or interference as a signal that originates from inside the cell and not from the cell surface. Additionally, the overlap of DAPI staining for nuclei combined with the green fluorescence resulted in the cyan color observed in cells treated with LT and AuLtNps but not with CsNps, aligning with findings reported by Gao-Na Shi [19].

2.8. Cytokine Secretion Analysis

Mice immunized with AuLtNps and LT+Ady exhibited increased secretion of proinflammatory cytokines (IFN, TNF, IL-6, IL-10, and IL-12p70), whereas a high level of secretion of MCP-1 was observed in the group treated solely with AuLtNps (Figure 7). To determine whether prolonged exposure to different treatments induced chronic inflammation, sera were analyzed one week after the last immunization. Secretion of IL-6, IL-10, and IFN showed no variation compared to the control (Figure 8), whereas TNF, IL-12p70, and MCP-1 displayed variations relative to the control. Additionally, there was a noticeable increase in the secretion of IL-12p70 in all treatment groups except the control group.

Mice immunized with AuLtNps exhibited an increase in the secretion of proinflammatory cytokines such as IFN, IL-12p70, TNF, and MCP-1 after the second immunization compared to the other treatment groups. This demonstrates the potential of AuLtNps to activate the immune system. TNF secreted by macrophages is crucial for the formation and maintenance of granulomas, which play a critical role in the defense against intracellular organisms. TNF also participates in leukocyte trafficking and immune complex elimination [43].

IL-12p70 secretion, primarily produced by antigen-presenting cells (APCs), is crucial in both cellular and humoral immunity, as it induces interferon production and activates CD4+ T cells with a lymphocyte t helper 1 (Th1) phenotype. IL-12p70 also plays a significant role in T-cell trafficking, inducing functional adhesion molecules, such as selectin P and E ligand, which are recruited to sites requiring Th1-type immune responses [44].

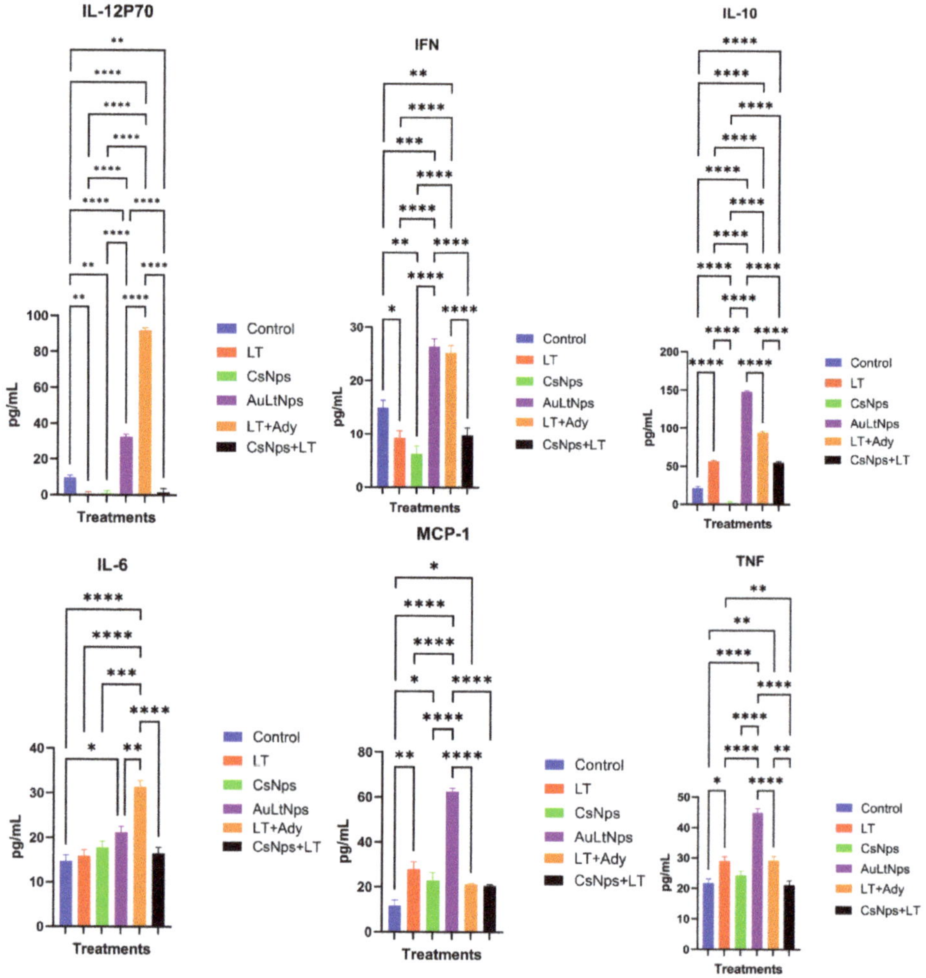

Figure 7. Determination of proinflammatory cytokines in mice sera 24 h after the second immunization. Experimental groups: control, tumor lysate, CsNps, tumor lysate + adjuvant [Al(OH)$_3$], and AuCs + tumor lysate. Mice treated with AuLtNps showed increments of IFN, IL-10, MCP-1, IL-6, and TNF. A statistical ANOVA with mean comparison was conducted to determine differences between groups. * p <0.05, ** p <0.01, and ***/**** $p < 0.001$.

A high level of MCP-1 secretion was observed in mice immunized with AuLtNps, which is key in regulating monocyte/macrophage migration and infiltration. This stimulation promoted monocyte migration to the bloodstream through the vascular endothelium, a fundamental process in immune surveillance. One week after the last immunization, IFN, IL-6, and IL-10 levels returned to normal compared to untreated controls. TNF, IL-12p70, and MCP-1 secretion showed no difference between groups, although they were higher than the control.

There was no chronic proinflammatory effect in the mice whose cytokine levels and secretion profiles were normalized after seven days compared to the untreated control. Rapid accumulation of intraperitoneally injected AuNps in the abdominal adipose tissue and liver of C57BL/6 mice was accompanied by reduced TNF-α and IL-6 mRNA. These findings suggest potential anti-inflammatory effects without overt organ damage [45]. MCP-

1 was the only cytokine with a significant elevation compared to the control, indicating ongoing immunogenic surveillance by macrophages and antigen-presenting cells, even after seven days [46].

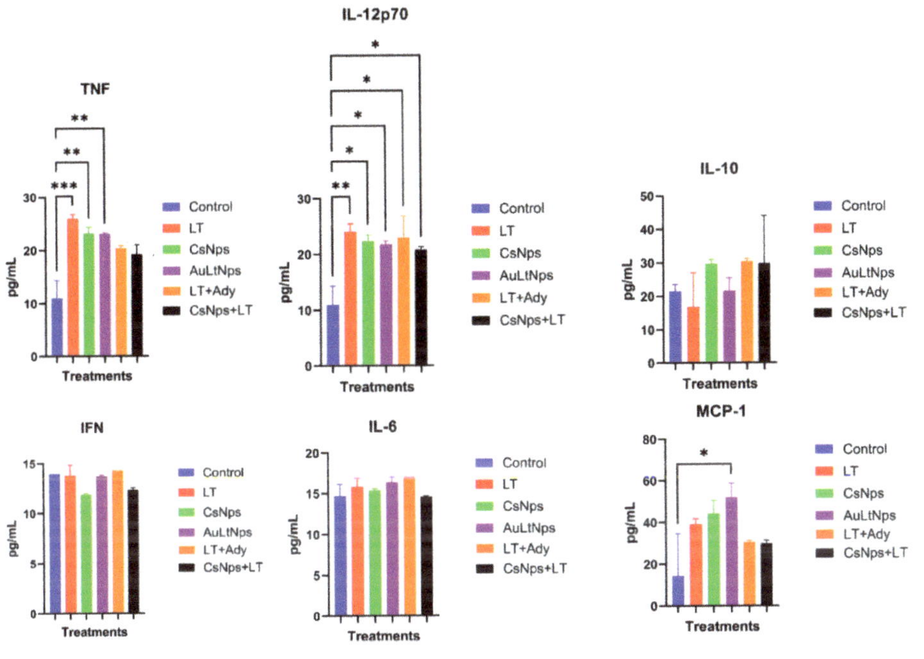

Figure 8. Determination of the presence of proinflammatory cytokines in mouse sera 7 days after immunization. Experimental groups: control, tumor lysate, CsNps, tumor lysate + adjuvant [$Al(OH)_3$], and AuCs + tumor lysate. The secretion of IL-10, IFN, and IL-6 was normalized; however TNF, Il12p70, and MCP-1 were elevated compared with the control in mice treated with AuLtNps, LT, and CsNps. A statistical ANOVA with mean comparison was conducted to determine differences between groups. * $p < 0.05$, ** $p < 0.01$, and *** $p < 0.001$.

2.9. *Lymphoid Lineage Activation and Antigen-Presenting Cell Markers*

Increases in the proportions of CD3, CD8, and CD22 cells were observed, whereas the proportion of CD34+ cells remained constant across treatments. A significant increase in CD3+ cells was observed for all treatments, except for the LT group.

The proportion of CD8+ cells showed a highly significant difference between AuLtNps and all other groups. This also occurred with CD22+, where the proportion of positive cells increased by up to three times when compared with the other treatments (Figure 9).

Markers for APCs are anti-CD80, CD11c, CD11b, MHC-II, and CD14. A highly significant increase in CD80 expression was identified in the CsNps, AuLtNps, LT+Ady, and CsNps+LT groups compared to the control. For CD86 cells, there was only a significant difference in the group treated with AuLtNps. In the case of CD11c+ and CD11b+ cell proportions, no significant differences were found among the treatments, although there was some variation across all groups.

Cells marked positive for MHC II and CD14+ showed a significant increase in the AuLtNps group in comparison with CsNps which displayed a notable decrease of this marker when compared to the control (Figure 10) [47].

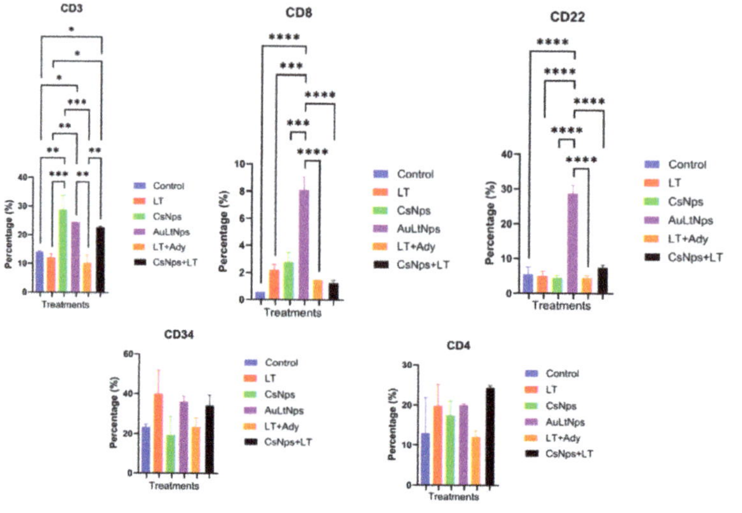

Figure 9. Determination of differentiation markers CD8, CD4, CD3, CD22, and CD34 in mouse spleen 7 days after immunization. Experimental groups: control, tumor lysate, CsNps, tumor lysate + adjuvant [$Al(OH)_3$], and AuCs + tumor lysate. Increments of CD3, CD4, and CD8 markers on cells were used on mice treated with AuLtNps, suggesting the presence of cytotoxic T lymphocytes and an increase in mature B lymphocytes. A statistical ANOVA was conducted to determine differences between the groups. * p <0.05, ** $p < 0.01$, and ***/**** $p < 0.001$.

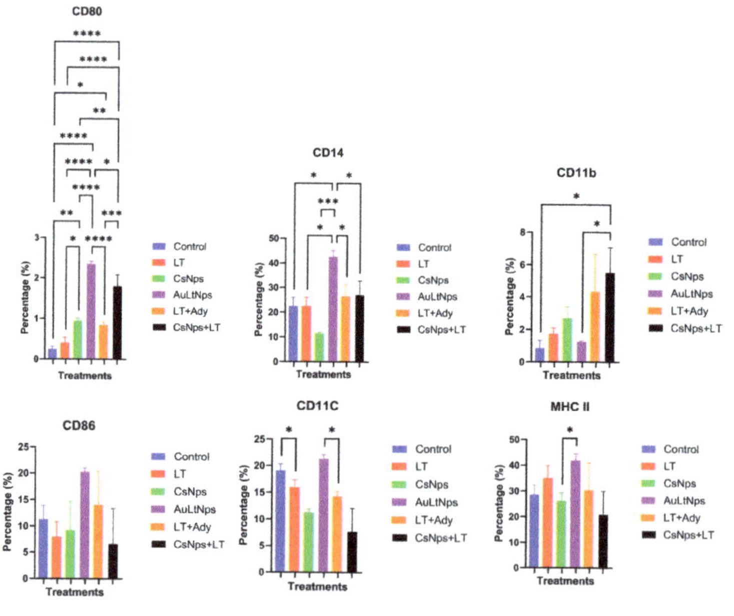

Figure 10. Determination of differentiation markers CD80, CD14, CD11c, CD11b, CD86, and MHC II in mice lymph nodes 7 days after immunization. Experimental groups were euthanized: Control, Tumor Lysate, CsNps, Tumor Lysate + Adjuvant [$Al(OH)_3$], and AuCs + Tumor Lysate. An increase of cells CD80, CD14, CD86, CD11c and MHC II in the mice treated with AuLtNps, which suggests a favorable maturation process for dendritic cells in lymph nodes. A statistical ANOVA was conducted to determine differences between the groups. * $p < 0.05$, ** $p < 0.01$, and ***/**** $p < 0.001$.

2.10. Specific Lysis Activity

Primed splenocytes of previously immunized mice exhibit an immune response against cancer cells due to their specific lytic capacity. To evaluate this, splenocytes extracted from immunized mice were cultured at ratios of 1:1, 1:10, 1:100, and 1:200 (4T1: splenocytes) and incubated for 24 h with viability marker 7-AAD, which is a DNA intercalator that fluoresces under flow cytometry only in dead or damaged cells. Only the AuLtNp group exhibited a significant difference, where the percentage of specific lysis reached up to 40% at ratios of 1:100 and 1:200 (Figure 11).

Splenocytes from immunized mice can specifically recognize 4T1 breast cancer cells, inducing an immunological defense. Splenocytes bind to cancer cells, activating the lytic capacity of the lymphocytes, which releases granzymes and perforins that produce pores in the cell membrane. Cell death is measured by the 7AAD marker, which enters dead cells. In controls, the 7AAD marker could not enter the impermeable cell membrane, indicating that AuLtNps showed a highly significant difference in comparison to all controls and that the immunization process of gold nanoparticle synthesis with tumoral lysates primes splenocytes to recognize 4T1 breast cancer cells [48].

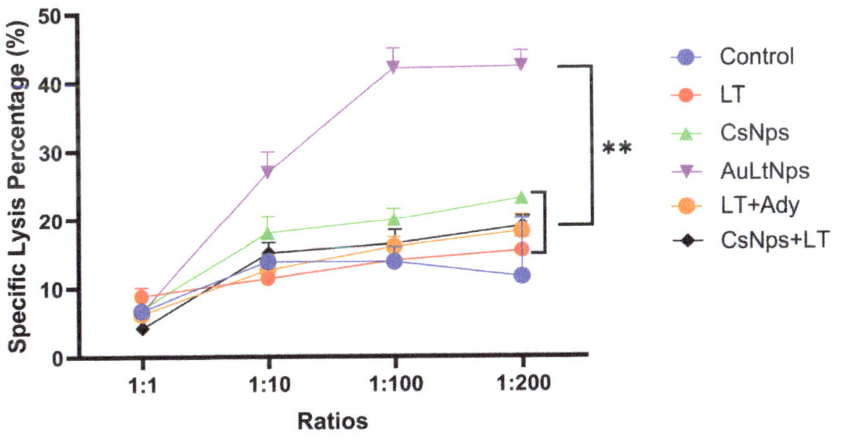

Figure 11. Assessment of prophylactic efficiency in a murine model of triple-negative breast cancer. Evaluation of specific cytotoxicity in lymphocytes extracted from the spleens of immunized mice at different ratios showed that cells treated with AuLtNps exhibited specific lysis of lymphocytes against 4T1 cells. A statistical ANOVA was performed to determine differences between the groups. ** $p < 0.01$.

2.11. Antitumor Efficacy of AuLtNps in Previously Immunized Balb/C Mice

A tumor challenge was conducted by inoculating 5×10^5 viable 4T1 cells into the mammary area of mice using an orthotopic model. The day when the tumor became palpable in the inoculated area was recorded as the tumor implantation day.

The AuLtNps group did not exhibit tumor implantation, with only slight inflammation at the injection site on the second day. In the LT group, three mice had tumor implantation on day 7, whereas one had it on day 10. The CsNps and LT+Ady groups showed tumor implantation on day 6, whereas the CsNps+LT group displayed tumor implantation on day 10. In the control group, tumor implantation occurred in all mice on day 7. All experimental groups were observed for a minimum of 21 days, without significant weight changes, but some groups experienced noticeable health deterioration (Figure 12).

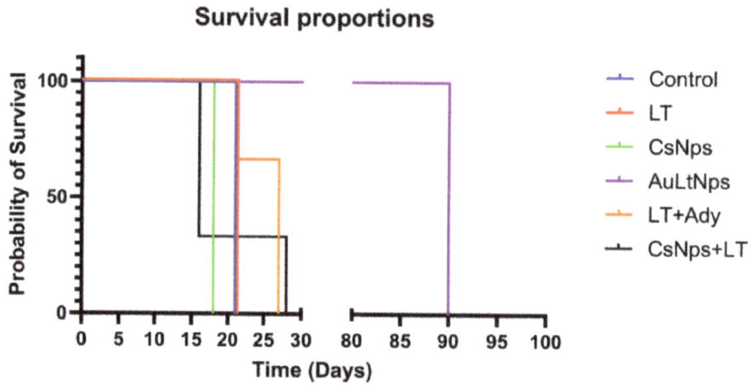

Figure 12. Comparative survival table for each experimental group extended up to a maximum of 90 days. Mice treated with CsNps had a maximum survival of 28 days, and with AuLtNps, it was possible to extend the survival rate to day 90.

On day 21, all individuals in the control group, as well as the LT group, were ethically euthanized due to the presence of metastases. In the CsNps group, one mouse died on day 16, and the remaining three were sacrificed on day 28. In the LT+Ady group, only two mice survived until day 27, whereas one died on day 5 and another on day 11. The CsNps+LT group was sacrificed on day 18 due overall deterioration. All four individuals treated with AuLtNps survived over 90 days without showing signs of tumor occurrence or metastasis.

The control group had an average lifespan of 21 days, whereas the LT group had an average lifespan of 20.5 days. The CsNps group had an average lifespan of 16 days, whereas the the CsNps+LT and LT+Ady groups reached 28 and 27 days, respectively. Finally, the AuLtNps group had an average lifespan of over 90 days (Table 3; Figure 12).

Table 3. Survival analysis of experimental groups.

Experimental Group	Survival Analysis Days (%)
Control	21 (100%)
Tumor cell lysate (LT)	21 (100%)
CsNps	18 (100%)
AuLtNps	90 (100%)
LT+Adjuvant	21 (33.3%), 8 (66.6%)
CsNps+LT	16 (66.6%), 28 (33.3%)

In vitro studies have shown the co-stimulatory effects of IL-10 with antigen immunization, promoting antigen-specific T-cell proliferation. An increase in IL-10 secretion is associated with proliferation of murine cytotoxic CD8+ T cells, and combined with IL-2, it enhances their cytolytic activity [49].

The anergy of CD4+ cells has been demonstrated between high IL-10 secretion and low expression of MHC II in APCs, which inhibits the MHC-CD28 stimulation pathway in T cells [50], as seen on cells treated with CsNps. In contrast, MHC II-positive cells in mice immunized with AuLtNps are increased by up to 40%, indicating activation of humoral immunity and active antigen presentation.

These results support the hypothesis that subcutaneous administration of AuNps is a safe and effective approach for inducing immune responses, as subcutaneous injection of AuNps induces localized macrophage recruitment and subsequent migration to regional lymph nodes. This mechanism likely contributes to the observed localization of nanoparticle-induced effects. Furthermore, repeated immunizations may promote a tolerogenic response, mitigating adverse effects.

Finally, mice immunized with AuLtNps showed a significant inhibition of 4T1 tumor cell implantation compared to all experimental groups, suggesting an antitumoral effect of the nanoparticles. An increase in antigen-presenting cells for the AuLtNps, AuCsNps, and LT+Ady groups correlates with a longer survival time, as does the inflammation pattern when secreting TNT, IL-12p70, and MCP-1 24 h and one week after the last immunization. AuLtNps exhibited a proinflammatory cytokine secretion profile with increases in CD3, CD8+ T, and CD4+ T cells, as well as an increase in the proportion of CD22+ cells, suggesting that nanoparticles obtained using tumor lysate as a reducing agent activate both humoral and cellular immunity.

3. Materials and Methods

3.1. Synthesis of Citrate-Reduced Gold Nanoparticles

Gold nanoparticles were synthesized based on the Turkevich method. The $HAuCl_4$ (Sigma Aldrich, St. Louis, MO, USA) solution was vigorously mixed with a 1% sodium citrate (Sigma Aldrich, St. Louis, MO, USA) solution at 2 mM in a 9:1 relation v/v. The mixture was exposed to UV light in a Crosslinker Ultraviolet (UVP, Upland, CA, USA) device at an intensity of 6000 $\mu J/cm^2$ for 30 min. Afterwards, the solution was allowed to rest at room temperature for 24 h, then refrigerated. These nanoparticles were employed as an experimental control [51].

3.2. Cell Lines

Murine breast cancer cell line 4T1 and murine fibroblast cell line NIH-3T3 were maintained under standard culturing conditions at 37 °C, 85% relative humidity, and 5% CO_2 in Dulbecco's Modified Eagle Medium (DMEM) (Gibco, New York, NY, USA) supplemented with 10% fetal bovine serum (FBS) (Gibco, New York, NY, USA) and 1% antimycotic antibiotic (Gibco, New York, NY, USA).

3.3. Synthesis of Tumor Lysate-Reduced Gold Nanoparticles

Gold nanoparticles reduced from a lysate of 4T1 cells were obtained from 1,200,000 viable cells lysed by sonication in three rounds of 10 min each. The lysate was mixed vigorously for with a 2 min with a 2 mM $HAuCl_4$ solution, then exposed to UV light using an Ultraviolet Crosslinker at an intensity of 6000 $\mu J/cm^2$ for 30 min. The solution was then allowed to cool in a refrigerator and stored for one week before use [52].

3.4. Characterization of Gold Nanoparticles

3.4.1. Surface Plasmon Resonance

Gold nanoparticles were characterized by Surface Plasmon Resonance (SPR) using a Nanodrop 2000c (Waltham, MA, USA) with UV-VIS spectroscopy. Measurements were taken every 3 days until day 30 [53].

3.4.2. Size and ζ Potential of Gold Nanoparticles

Nanoparticles were loaded in a capillary cell for spectrophotometry and analyzed at 25 °C; the size was determined by DLS. Laser Doppler electrophoresis was used to determine the ζ potential using an NS90 nanosizer instrument (Malvern Instruments, Malvern, UK) [54]. Measurements were obtained on days 1, 5, 12, 15, 24, and 30.

3.4.3. Size and Morphology Analysis via Transmission Electron Microscopy

Chitosan nanoparticles were mounted uniformly on carbon adhesive tape that was attached to a sample holder and analyzed by transmission electron microscopy (TEM). The resulting file was analyzed using Gwyddion software, version 3.0 to obtain size distribution histograms [55].

3.4.4. Protein Quantification Curve

The protein concentration was quantified using Pierce BCA Protein Assay Kits, (Thermo Scientific, Rockford, IL, USA) following the instructions of the manufacturer. Samples were analyzed via UV-VIS spectroscopy using a Nanodrop 2000c (Thermo Fisher, Walthman, MA, USA) and read at a wavelength of 560 nm. Measurements were performed in triplicate on days 3, 6, 10, 14, 20, and 31. The standard curve was made using albumin as a reference protein, at concentrations ranging from 0.2 to 2 mg/mL. After obtaining the data, mean variation analyses were performed using an ANOVA test for each measurement day [56].

3.5. Cytotoxicity Assay in NIH-3T3 Cells

Five thousand viable NIH-3T3 cells were treated for 24 h with tumor lysate-reduced gold nanoparticles (AuLtNps) at the following concentrations: 15, 35, 75, 120, 180, and 300 μM. Controls included tumor lysate, $HAuCl_4$, and sodium citrate nanoparticles (CsNps). Relative cell viability was evaluated using 3-(4,5-dimethylthiazol-2-yl)-2,5-diphenyltetrazolium bromide (MTT) (Cayman Chemical, Ann Arbor, MI, USA) [57]. The absorbance data were calculated using the following formula:

$$\text{Percentage of relative viability} = \frac{\text{Treatment absorbance (nm)}}{\text{Control absorbance (nm)}} \times 100 \quad (1)$$

3.6. Animal Model Immunization Schedule

Seven experimental groups were used: tumor lysate (LT), sodium citrate nanoparticles (CsNps), tumor lysate nanoparticles (AuLtNps), CsNps + LT, LT + [$Al(OH)_3$], and a control group using phosphate-buffered saline (PBS). Each group consisted of six female Balb/C mice aged 6 weeks. The immunization schedule followed the protocol described by Shi [19]; mice were injected once a week for three weeks (−21, −14, and −7), with a one-week break before subsequent tests. Immunization was performed without exceeding a maximum volume of 200 μL per mouse.

3.7. Tumor Challenge

One week after the final immunization, a subcutaneous inoculation was performed in the mammary gland using a 1 mL syringe and a 27 G needle, (Nipro, Mechelen, Belgium) injecting 500,000 viable 4T1 cells resuspended in 50 μL of Iscove's Modified Dulbecco's Medium (IMDM) medium supplemented with 10% SFB. Tumor implantation and development were closely monitored, with a maximum follow-up period of 21 days after inoculation [58]. If any mice exhibited excessive pain or difficulty in feeding or moving, they were euthanized following relevant guidelines.

3.8. Anesthesia and Euthanasia

All the in vivo experiments were performed following Mexican norm [59]. Mice were anesthetized using a mixture of ketamine (100 mg/kg) (Vetoquinol, Escobedo, México) and xylazine (10 mg/kg) (PiSA, Jálisco, México); a response test was conducted by gently pricking one of the mouse's paw pads. Once anesthesia was ensured, cardiac puncture was performed using a 1 mL syringe with a 25 G needle. The collected blood was stored in anticoagulant-containing tubes. After blood collection, cervical dislocation was carried out, and the mouse's spleen and lymphatic nodules were obtained [60].

3.9. Peritoneal Macrophage Antigen Uptake Assay

An incision was made in the skin covering the mouse's abdomen, and the skin was gently pulled back from the peritoneal membrane without piercing it. A total volume of 7 mL of DMEM was injected into the peritoneal cavity, gently agitated, then withdrawn from the mouse's peritoneal cavity and centrifuged at 1350 rmp for 10 min. The obtained cells were counted, and 100,000 cells were plated and incubated for 4 days with supplemented DMEM [61].

AuLtNPs, soluble LT, PBS, and CsNps treatments were conjugated with the fluorophore FITC. This was achieved through coupling by constant agitation for 3 h, using FITC at a final concentration of 1 mg/mL. After conjugation, the treatments were independently exposed to murine intraperitoneal macrophages for 24 h. Subsequently, the cell wells were washed to remove any remaining FITC that did not enter the cell, and DAPI staining was performed to identify the presence of cell nuclei. Samples were analyzed using an epifluorescence microscope to capture images of each treatment and count how many cells exhibited the characteristic green positive staining of the FITC marker. Only cells that presented both DAPI and FITC staining in or around the nucleus were considered positive. Finally, the percentage of cells that tested positive for FITC staining per treatment was plotted.

3.10. Determination of Proinflammatory Cytokines

After blood extraction, the serum was separated by centrifugation and analyzed for proinflammatory cytokines by flow cytometry using a BD CBA proinflammatory cytokine kit following the manufacturer's instructions (BD accuriTMC6, Piscataway, NJ, USA) [62].

3.11. Evaluation of Cellular Markers in Mouse Spleen After Immunization

After extraction of the spleen, perfusion with PBS was conducted to obtain lymphoid cells from the tissue. The cells were incubated with membrane markers CD4, CD3, CD8, CD22, and CD34 at a concentration of 1 μg/mL for 1 hour, then evaluated by flow cytometry [63].

3.12. Identification of Macrophage Differentiation Profile in Lymph Nodes of Immunized Mice

Three mice per group from all experimental groups were euthanized on day 1 of immunization, and peripheral lymph nodes from the axillary and inguinal were extracted. Perfusion with PBS was performed, passing at least 5 mL through each lymph node until a turbid solution was obtained. The lymph node cells were counted, and a minimum of 100,000 cells was used for each assay. Membrane markers CD14, CD11c, CD11b, CD80, CD86, and MHCII were employed at a concentration of 1 μg/mL for subsequent analysis by flow cytometry [64].

3.13. Lymphoid Lineage Activation and Antigen-Presenting Cell Markers

One week after the final immunization (day 1), three mice from each group were euthanized, and their spleens were extracted. Perfusion with PBS was performed, passing at least 5 mL through the spleen until a turbid solution was obtained. The extracted cells were disaggregated and marked with anti-CD3, CD4, CD8, CD22, and CD34 antibodies and analyzed by flow cytometry. The population of CD3+ cells was selected; then, the presence of CD4+ and CD8+ cells was determined [64].

3.14. Evaluation of Specific Cytotoxicity of T Lymphocytes

After the spleen was mechanically disaggregated, the lymphoid cells from the tissue were obtained by density gradient with Ficoll. These cells were cultured and subsequently exposed to viable 4T1 cells at ratios of 1:1, 1:10, 1:100, and 1:200. The cells were cultured for 24 h, and viability marker 7-Aminoactinomycin D (7-AAD) was used to identify dead cells. Gating was performed to determine the viability of 4T1 cells recollected from a co-culture by flow cytometry [65].

3.15. Antitumor Effect of AuLtNps in Previously Immunized Balb/C Mice

Previously immunized groups were inoculated subcutaneously with 5×10^5 viable 4T1 cells in the mammary gland. The appearance of the tumor was registered. The evaluation included implantation time, potential metastasis, and established survival [66].

3.16. Statistical Analysis

The data are represented as means ± standard deviation. Differences between the control group and experimental groups were analyzed using Student's t-test, whereas differences between groups were determined using a one-way ANOVA using GraphPad 10 software.

4. Conclusions

A novel synthesis method was developed to obtain gold nanoparticles using lysates derived from the 4T1 tumor cell line as a reducing agent, which showed immunostimulatory properties in vitro and in vivo. This method simplifies the production process and reduces synthesis costs, as it eliminates the need for a wide range of reducing and stabilizing agents.

The prophylactic effect of AuLtNps revealed a contrasting lack of tumor implantation in a murine model that was previously immunized and subsequently inoculated with viable 4T1 murine breast cancer cells.

In summary, reduced gold nanoparticles derived from tumor lysate enhance antigen presentation, maximizing the immune response from innate and adaptive immune systems and inhibiting tumor implantation in a triple-negative breast cancer murine model, preventing tumoral recurrence. Further studies should be performed with other cell lines to validate this nanosystem.

Author Contributions: Conceptualization, R.R.-L. and D.G.Z.-T.; Data curation, R.R.-L.; Formal analysis, R.R.-L., M.Á.F.-M. and D.G.Z.-T.; Funding acquisition, C.R.-P. and D.G.Z.-T.; Investigation, R.R.-L. and D.G.Z.-T.; Methodology, R.R.-L.; Project administration, M.Á.F.-M., C.R.-P. and D.G.Z.-T.; Resources, R.R.-L., M.Á.F.-M. and D.G.Z.-T.; Supervision, M.Á.F.-M., C.R.-P. and D.G.Z.-T.; Validation, R.R.-L., M.Á.F.-M. and D.G.Z.-T.; Visualization, R.R.-L.; Writing—original draft, R.R.-L.; Writing—review and editing, R.R.-L. and D.G.Z.-T. All authors have read and agreed to the published version of the manuscript.

Funding: This research received no external funding.

Institutional Review Board Statement: The animal study protocol was approved by the Institutional Review Board of Comité de Ética de Investigación y Bienestar Animal de la Facultad de Ciencias Biológicas de la Universidad Autónoma de Nuevo León (CEIBA-2019-002, 19 March 2019).

Informed Consent Statement: Not applicable.

Data Availability Statement: The raw data supporting the conclusions of this article will be made available by the authors on request.

Acknowledgments: We thank Laboratorio de Inmunología y Virología, Facultad de Ciencias Biológicas, Universidad Autónoma de Nuevo León, San Nicolás de los Garza, Mexico, as well as Alejandra Arreola-Triana, and Bruno Alejandro Valades Aguilar for their participation on the language review.

Conflicts of Interest: The authors declare no conflicts of interest.

References

1. Barrios, C. Global challenges in breast cancer detection and treatment. *Breast* **2022**, *62*, S3–S6. [CrossRef] [PubMed]
2. Anand, U.; Dey, A.; Chandel, A.; Sanyal, R.; Mishra, A.; Pandey, D.; De Falco, V.; Upadhyay, A.; Kandimalla, R.; Chaudhary, A.; et al. Cancer chemotherapy and beyond: Current status, drug candidates, associated risks and progress in targeted therapeutics. *Genes Dis.* **2022**, *10*, 1367–1401. [PubMed]
3. Hudson, B.; Lippman, M. Comment on The lingering mysteries of metastatic recurrence in breast cancer. *Br. J. Cancer* **2022**, *128*, 484–485. [CrossRef] [PubMed]
4. Houvenaeghel, G.; Boher, J.; Michel, V.; Bannier, M.; Minsat, M.; Tallet, A.; Lambaudie, E. Survival after breast cancer local recurrence according to therapeutic strategies. *Eur. J. Surg. Oncol.* **2017**, *43*, 1409–1414. [CrossRef] [PubMed]
5. Esfahani, K.; Roudaia, L.; Buhlaiga, N.; Del Rincon, S.; Papneja, N.; Miller, W. A review of cancer immunotherapy: From the past to the present, to the future. *Curr. Oncol.* **2020**, *27*, 87–97. [CrossRef] [PubMed]
6. Yang, Y.; Liu, J.; Lu, C.; Wei, J. CAR-T cell therapy for breast cancer: From basic research to clinical application. *Int. J. Biol. Sci.* **2022**, *18*, 2609. [CrossRef] [PubMed]
7. Zahavi, D.; Weiner, L. Monoclonal antibodies in cancer therapy. *Antibodies* **2020**, *9*, 34. [CrossRef] [PubMed]
8. Liao, J.; Zhang, S. Safety and efficacy of personalized cancer vaccines in combination with immune checkpoint inhibitors in cancer treatment. *Front. Oncol.* **2021**, *11*, 663264. [CrossRef] [PubMed]
9. Dranoff, G. Cytokines in cancer pathogenesis and cancer therapy. *Nat. Rev. Cancer* **2020**, *4*, 11–22. [CrossRef]
10. Sivori, S.; Pende, D.; Quatrini, L.; Pietra, G.; Della Chiesa, M.; Vacca, P.; Moretta, L. NK cells and ILCs in tumor immunotherapy. *Mol. Asp. Med.* **2021**, *80*, 100870. [CrossRef] [PubMed]
11. Morandi, F.; Yazdanifar, M.; Cocco, C.; Bertaina, A.; Airoldi, I. Engineering the bridge between innate and adaptive immunity for cancer immunotherapy: Focus on $\gamma\delta$ and NK cells. *Cells* **2020**, *9*, 1757. [CrossRef]
12. Blass, E.; Ott, P. Advances in the development of personalized neoantigen-based therapeutic cancer vaccines. *Nat. Rev. Clin. Oncol.* **2021**, *18*, 215–229. [CrossRef] [PubMed]
13. Morse, M.; Gwin, W., III; Mitchell, D. Vaccine therapies for cancer: Then and now. *Target. Oncol.* **2021**, *16*, 121–152. [CrossRef] [PubMed]
14. Laureano, R.; Sprooten, J.; Vanmeerbeerk, I.; Borras, D.; Govaerts, J.; Naulaerts, S.; Garg, A. Trial watch: Dendritic cell (DC)-based immunotherapy for cancer. *Oncoimmunology* **2022**, *11*, 121–152. [CrossRef] [PubMed]
15. Cheever, M.; Higano, C. PROVENGE (Sipuleucel-T) in prostate cancer: The first FDA-approved therapeutic cancer vaccine. *Clin. Cancer Res.* **2011**, *17*, 3520–3526. [CrossRef]
16. Shevchenko, J.; Khristin, A.; Kurilin, V.; Kuznetsova, M.; Blinova, D.; Starostina, N.; Sennikov, S. Autologous dendritic cells and activated cytotoxic T-cells as combination therapy for breast cancer. *Oncol. Rep.* **2020**, *43*, 671–680. [CrossRef] [PubMed]
17. Hatfield, P.; Merrick, A.; West, E.; O'Donnell, D.; Selby, P.; Vile, R.; Melcher, A. Optimization of dendritic cell loading with tumor cell lysates for cancer immunotherapy. *J. Immunother.* **2008**, *31*, 620–632. [CrossRef]
18. Foley, C.; Swan, S.; Swartz, M. Engineering challenges and opportunities in autologous cellular cancer immunotherapy. *J. Immunol.* **2024**, *212*, 188–189. [CrossRef]
19. Shi, G.; Zhang, C.; Xu, R.; Niu, J.; Song, H.; Zhang, X.; Kong, D. Enhanced antitumor immunity by targeting dendritic cells with tumor cell lysate-loaded chitosan nanoparticles vaccine. *Biomaterials* **2017**, *113*, 191–202. [CrossRef]
20. Thauvin, C.; Widmer, J.; Mottas, I.; Hocevar, S.; Allémann, E.; Bourquin, C.; Delie, F., Development of resiquimod-loaded modified PLA-based nanoparticles for cancer immunotherapy: A kinetic study. *Eur. J. Pharm. Biopharm.* **2019**, *139*, 253–261. [CrossRef] [PubMed]
21. Yang, W.; Zhu, G.; Wang, S.; Yu, G.; Yang, Z.; Lin, L.; Chen, X. In situ dendritic cell vaccine for effective cancer immunotherapy. *ACS Nano* **2019**, *11*, 3083–3094. [CrossRef] [PubMed]
22. Weller, S.; Li, X.; Petersen, L.; Kempen, P.; Clergeaud, G.; Andresen, T. Influence of different conjugation methods for activating antibodies on polymeric nanoparticles: Effects for polyclonal expansion of human CD8+ T cells. *Int. Immunopharmacol.* **2024**, *129*, 111643. [CrossRef]
23. Thangam, R.; Patel, K.; Kang, H.; Paulmurugan, R. Advances in engineered polymer nanoparticle tracking platforms towards cancer immunotherapy—Current status and future perspectives. *Vaccines* **2011**, *9*, 935. [CrossRef]
24. Chang, H.; Zou, Z.; Wang, Q.; Li, J.; Jin, H.; Yin, Q.; Xing, D. Targeting and specific activation of antigen-presenting cells by endogenous antigen-loaded nanoparticles elicits tumor-specific immunity. *Adv. Sci.* **2020**, *7*, 1900069. [CrossRef] [PubMed]

25. Trabbic, K.; Kleski, K.; Barchi, J., Jr. Stable gold-nanoparticle-based vaccine for the targeted delivery of tumor-associated glycopeptide antigens. *ACS Bio Med Chem Au* **2021**, *1*, 31–43. [CrossRef] [PubMed]
26. Dykman, L.; Khlebtsov, N. Immunological properties of gold nanoparticles. *Chem. Sci.* **2017**, *8*, 1719–1735. [CrossRef] [PubMed]
27. Kelly, H.; Kent, S.; Wheatley, A. Immunological basis for enhanced immunity of nanoparticle vaccines. *Expert Rev. Vaccines* **2019**, *18*, 269–280. [CrossRef] [PubMed]
28. Rizi, H.; Shin, D.; Rizi, S. Polymeric nanoparticles in cancer chemotherapy: A narrative review. *Iran. J. Public Health* **2022**, *51*, 226.
29. Shao, K.; Singha, S.; Clemente-Casares, X.; Tsai, S.; Yang, Y.; Santamaria, P. Nanoparticle-based immunotherapy for cancer. *ACS Nano* **2015**, *9*, 16–30. [CrossRef] [PubMed]
30. Niżnik, Ł.; Noga, M.; Kobylarz, D.; Frydrych, A.; Krośniak, A.; Kapka-Skrzypczak, L.; Jurowski, K. Gold Nanoparticles (AuNPs)-Toxicity, Safety and Green Synthesis: A Critical Review. *Int. J. Mol. Sci.* **2024**, *5*, 25. [CrossRef] [PubMed]
31. Ibrahim, K.E.; Al-Mutary, M.G.; Bakhiet, A.O.; Khan, H.A. Histopathology of the liver, kidney, and spleen of mice exposed to gold nanoparticles. *Molecules* **2018**, *23*, 1848. [CrossRef]
32. Lasagna-Reeves, C.; Gonzalez-Romero, D.; Barria, M. A.; Olmedo, I.; Clos, A.; Ramanujam, V. S.; Soto, C. Bioaccumulation and toxicity of gold nanoparticles after repeated administration in mice. *Biochem. Biophys. Res. Commun.* **2010**, *393*, 649–655. [CrossRef]
33. Amendola, V.; Pilot, R.; Frasconi, M.; Maragò, O. M.; Iatì, M. A. Surface plasmon resonance in gold nanoparticles: A review. *J. Phys. Condens. Matter* **2017**, *29*, 20. [CrossRef]
34. Doane, T.L.; Chuang, C.H.; Hill, R.J.; Burda, C. Nanoparticle ζ-potentials. *Accounts Chem. Res.* **2012**, *45*, 317–326. [CrossRef]
35. Lundqvist, M.; Stigler, J.; Cedervall, T.; Berggard, T.; Flanagan, M.; Lynch, I.; Dawson, K., The evolution of the protein corona around nanoparticles: A test study. *ACS Nano* **2011**, *5*, 7503–7509. [CrossRef]
36. Del Pino, P.; Pelaz, B.; Zhang, Q.; Maffre, P.; Nienhaus, G.; Parak, W. Protein corona formation around nanoparticles–from the past to the future. *Mater. Horizons* **2014**, *1*, 301–313. [CrossRef]
37. Devulapally, R.; Paulmurugan, R., Polymer nanoparticles for drug and small silencing RNA delivery to treat cancers of different phenotypes. *Wiley Interdiscip. Rev. Nanomed. Nanobiotechnol.* **2014**, *6*, 40–60. [CrossRef]
38. Kondinskaia, D.; Gurtovenko, A. Supramolecular complexes of DNA with cationic polymers: The effect of polymer concentration. *Polymer* **2018**, *142*, 277–284. [CrossRef]
39. Han, S.; Zal, T.; Sokolov, K.V. Fate of Antibody-Targeted Ultrasmall Gold Nanoparticles in Cancer Cells after Receptor-Mediated Uptake. *ACS Nano* **2021**, *22*, 9495–9508. [CrossRef]
40. Corbo, C.; Molinaro, R.; Parodi, A.; Toledano Furman, N.E.; Salvatore, F.; Tasciotti, E. The impact of nanoparticle protein corona on cytotoxicity, immunotoxicity and target drug delivery. *Nanomedicine* **2016**, *11*, 81–100. [CrossRef]
41. Reva, G.V.; Reva, I.V.; Yamamoto, T.; Usov, V.V.; Apanasevich, V.E.; Gul'kov, A.N.; Chaika, V.V. Reaction of dermal structures to subcutaneous injection of gold nanoparticles to CBA mice. *Bull. Exp. Biol. Med.* **2014**, *156*, 491–494. [CrossRef]
42. van Gelder, M.; Vanclée, A.; van Elssen, C.H.; Hupperets, P.; Wieten, L.; Bos, G.M. Bone marrow produces sufficient alloreactive natural killer (NK) cells in vivo to cure mice from subcutaneously and intravascularly injected 4T1 breast cancer. *Breast Cancer Res. Treat.* **2017**, *161*, 421–433. [CrossRef]
43. Klimp, A.; De Vries, E.; Scherphof, G.; Daemen, T. A potential role of macrophage activation in the treatment of cancer. *Crit. Rev. Oncol.* **2002**, *44*, 143–161. [CrossRef] [PubMed]
44. Hamza, T.; Barnett, J.; Li, B. Interleukin 12 a key immunoregulatory cytokine in infection applications. *Int. J. Mol. Sci.* **2010**, *11*, 789–806. [CrossRef]
45. Chen, H.; Dorrigan, A.; Saad, S.; Hare, D.J.; Cortie, M.B.; Valenzuela, S.M. In vivo study of spherical gold nanoparticles: Inflammatory effects and distribution in mice. *PLoS ONE* **2013**, *8*, e58208. [CrossRef] [PubMed]
46. Brown, C.; Vishwanath, R.; Aguilar, B.; Starr, R.; Najbauer, J.; Aboody, K.; Jensen, M. Tumor-derived chemokine MCP-1/CCL2 is sufficient for mediating tumor tropism of adoptively transferred T cells. *J. Immunol.* **2007**, *179*, 3332–3341. [CrossRef]
47. Michelini, S.; Barbero, F.; Prinelli, A.; Steiner, P.; Weiss, R.; Verwanger, T.; Horejs-Hoeck, J. Gold nanoparticles (AuNPs) impair LPS-driven immune responses by promoting a tolerogenic-like dendritic cell phenotype with altered endosomal structures. *Nanoscale* **2021**, *13*, 7648–7666. [CrossRef] [PubMed]
48. Frings, P.W.H.; Van Elssen, C.H.M.J.; Wieten, L. Matos, C.; Hupperets, P.; Schouten, H.; Bos, G.; van Gelder, M. Elimination of the chemotherapy resistant subpopulation of 4T1 mouse breast cancer by haploidentical NK cells cures the vast majority of mice. *Breast Cancer Res. Treat.* **2011**, *130*, 773–781. [CrossRef] [PubMed]
49. Santin, A.; Hermonat, P.; Ravaggi, A.; Bellone, S.; Pecorelli, S.; Roman, J.; Cannon, M. Interleukin-10 increases Th1 cytokine production and cytotoxic potential in human papillomavirus-specific CD8+ cytotoxic T lymphocytes. *J. Virol.* **2000**, *74*, 4729–4737. [CrossRef]
50. Sharma, S.; Stolina, M.; Lin, Y.; Gardner, B.; Miller, P.; Kronenberg, M.; Dubinett, S. T cell-derived IL-10 promotes lung cancer growth by suppressing both T cell and APC function. *J. Immunol.* **1999**, *163*, 5020–2018. [CrossRef] [PubMed]

51. Turkevich, J.; Stevenson, P.C.; Hillier, J. A study of the nucleation and growth processes in the synthesis of colloidal gold. *Discuss. Faraday Soc.* **1951**, *11*, 55–75. [CrossRef]
52. Dykman, L. A.; Staroverov, S. A.; Mezhenny, P. V.; Fomin, A. S.; Kozlov, S. V.; Volkov, A. A.; Shchyogolev, S. Y. Use of a synthetic foot-and-mouth disease virus peptide conjugated to gold nanoparticles for enhancing immunological response. *Gold Bull.* **2015**, *48*, 93–101. [CrossRef]
53. Catanzaro, L.; Scardaci, V.; Scuderi, M.; Condorelli, M.; D'Urso, L.; Compagnini, G. Surface plasmon resonance of gold nanoparticle aggregates induced by halide ions. *Mater. Chem. Phys.* **2023**, *308*, 128245. [CrossRef]
54. Oudih, S.; Tahtat, D.; Khodja, A.; Mahlous, M.; Hammache, Y.; Guittoum, A.; Gana, S. Chitosan nanoparticles with controlled size and zeta potential. *Polym. Eng. Sci.* **2023**, *53*, 1011–1021. [CrossRef]
55. Hinterwirth, H.; Wiedmer, S. K.; Moilanen, M.; Lehner, A.; Allmaier, G.; Waitz, T.; Lämmerhofer, M. Comparative method evaluation for size and size-distribution analysis of gold nanoparticles. *J. Sep. Sci.* **2013**, *36*, 2952–2961. [CrossRef]
56. Walker, J.M. The bicinchoninic acid (BCA) assay for protein quantitation. In *The Protein Protocols Handbook*; Springer: Berlin/Heidelberg, Germany, 2009; pp. 11–15.
57. Kumar, P.; Nagarajan, A.; Uchil, P. D. Analysis of cell viability by the MTT assay. *Cold Spring Harb. Protoc.* **2018**, *6*. [CrossRef]
58. Pulaski, B.; Ostrand-Rosenberg, S. Mouse 4T1 Breast Tumor Model. *Curr. Protoc. Immunol.* **2000**, *39*, 20–22. [CrossRef]
59. *NOM-062-ZOO-1999*; Especificaciones Técnicas para la Producción, Cuidado y uso de los Animales de Laboratorio. Diario Oficial de la Federación: Ciudad de México, México, 2001; pp. 107–165.
60. Arunachalam, K.; Sasidharan, S.P. *Bioassays in Experimental and Preclinical Pharmacology*; Humana Press: Totowa, NJ, USA, 2021.
61. Licht, R.; Jacobs, C.W.M.; Tax, W.J.M.; Berden, J.H.M. An assay for the quantitative measurement of in vitro phagocytosis of early apoptotic thymocytes by murine resident peritoneal macrophages. *J. Immunol. Methods* **1999**, *223*, 237–248. [CrossRef]
62. Morgan, E.; Varro, R.; Sepulveda, H.; Ember, J.A.; Apgar, J.; Wilson, J.; Gaur, A. Cytometric bead array: A multiplexed assay platform with applications in various areas of biology. *Clin. Immunol.* **2004**, *110*, 252–266. [CrossRef]
63. Czader, M. Clinical flow cytometry. In *Molecular Genetic Pathology*; Zhang, D., Cheng, L. Eds.; Humana Press: Totowa, NJ, USA, 2008; pp. 155–183.
64. Lambert, C.; Preijers, F.W.; Yanikkaya Demirel, G.; Sack, U. Monocytes and macrophages in flow: An ESCCA initiative on advanced analyses of monocyte lineage using flow cytometry. *Cytom. Part B Clin. Cytom.* **2017**, *92*, 180–188. [CrossRef]
65. Mousazadeh, R.; Hesaraki, S.; Bayat, M.; Jahandideh, A.; Hashemi, J. Anticancer tendency of aflatoxin B1 in 4T1 breast cancer cell line. *Gene Rep.* **2019**, *16*, 100442. [CrossRef]
66. Zhang, Y.; Zhang, G. L.; Sun, X.; Cao, K. X.; Ma, C.; Nan, N.; Wang, X. M. Establishment of a murine breast tumor model by subcutaneous or orthotopic implantation. *Oncol. Lett.* **2018**, *15*, 6233–6240. [CrossRef]

pharmaceuticals

MDPI

Article

Process Optimization for the Bioinspired Synthesis of Gold Nanoparticles Using *Cordyceps militaris*, Its Characterization, and Assessment of Enhanced Therapeutic Efficacy

Girish Gawas [1], Muniappan Ayyanar [2], Nilambari Gurav [3], Dinesh Hase [4], Vaishali Murade [5], Sameer Nadaf [6], Mohd Shahnawaz Khan [7], Rupesh Chikhale [8], Mohan Kalaskar [9] and Shailendra Gurav [1,*]

1 Department of Pharmacognosy, Goa College of Pharmacy, Goa University, Panaji 403001, India; girishgawas95@gmail.com
2 Department of Botany, A.V.V.M. Sri Pushpam College (Autonomous), Affiliated to Bharathidasan University, Poondi 613503, India; asmayyanar@yahoo.com
3 PES's Rajaram and Tarabai Bandekar College of Pharmacy, Goa University, Ponda 403401, India; nilagurav@rediffmail.com
4 Department of Pharmacognosy, Amrutwahini College of Pharmacy, Sangamner 422608, India; dinesh23787@gmail.com
5 Department of Chemistry and Research Centre, Padmashri Vikhe Patil College Pravaranagar, Loni 445001, India; vaishali.hase66@gmail.com
6 Bharati Vidyapeeth College of Pharmacy, Palus 416310, India; sam.nadaf@rediffmail.com
7 Department of Biochemistry, College of Science, King Saud University, Riyadh 11451, Saudi Arabia; moskhan@ksu.edu.sa
8 UCL School of Pharmacy, 29–39 Brunswick Square, London WC1N 1AX, UK; r.chikhale@ucl.ac.uk
9 R. C. Patel Institute of Pharmaceutical Education and Research, Shirpur 425405, India; kalaskar.mohan@gmail.com
* Correspondence: shailendra.gurav@nic.in; Tel.: +91-99-7083-5019

Abstract: The promising therapeutic implications of nanoparticles have spurred their development for biomedical applications. An eco-friendly methodology synthesizes gold nanoparticles using *Cordyceps militaris*, an edible mushroom (*Cord*-Au-NPs), using a quality-by-design approach (central composite design). UV-visible spectroscopy analysis revealed an absorption peak at 540–550 nm, thus confirming the synthesis of gold nanoparticles. *Cord*-Au-NPs have a crystalline structure, as evidenced by the diffraction peaks. The zeta potential value of −19.42 mV signifies the stability of *Cord*-Au-NPs. XRD study shows gold facets and EDX analysis revealed a strong peak of spherical nanoparticles in the gold region with a mean particle size of 7.18 nm and a polydispersity index of 0.096. The obtained peaks are closely associated with phenolic groups, lipids, and proteins, as examined by FTIR, suggesting that they function as the reducing agent. *Cord*-Au-NPs exhibited dose-dependent antioxidant, antidiabetic, and antibacterial activity. The method is eco-friendly, nontoxic, less time-consuming, and does not use synthetic materials, leading to higher capabilities in biomedical applications.

Keywords: green synthesis; central composite design; antioxidant activity; antidiabetic activity; process optimization

1. Introduction

Nanotechnology is an area of research and development concerned with designing environmentally friendly synthetic methods for material production [1]. It studies nanoscale materials with 1–100 nm diameters and their applications [2]. Researchers have been intrigued by the properties of nanostructures, such as nanoparticles, nanocapsules, nanotubes, and nanoflowers, for decades because of their applications in various fields, mainly applied medical sciences [3]. Nanoparticles with improved properties have become increasingly crucial in multiple health and environmental applications. Several synthesis methods, including chemical, physical, and biological ones, have been developed to create these nanoparticles [4,5]. However, one or more potentially harmful chemical species are adsorbed on the surface of materials synthesized using chemical and physical methods. Moreover, some toxic compounds may be used as reducing and stabilizing agents in chemically mediated nanoparticle synthesis to prevent aggregation, whereas the physical process involves high temperatures and expensive materials.

Gold is mainly a valuable, inert, and nontoxic metal, and it is employed to cure various diseases [6]. Nature provides a plethora of plant resources, including a purifying technique that is low-cost, highly reproducible, environmentally benign, and exact. There has been a rise in interest in environmentally friendly methods of manufacturing gold nanoparticles, which involve using plant extract, bacteria, and mushrooms as reducing agents and stabilizers. Gold nanoparticles (Au-NPs) have been successfully developed using greener substrates such as enzymes, fungi, and algae [2]. To overcome the limitations of conventional methods, the researchers developed eco-friendly and cost-effective biological methods at room temperature that produced highly stable nanoparticles that do not flocculate even without an external chemical stabilizing agent [7].

Cordyceps militaris is a mushroom fungus prevalent for its beneficial effects on sick and older people and those recovering from ailments. *C. militaris* is enriched with polysaccharides, cordycepin, proteins, adenosine, ergosterol, and myriocin. It has been used to treat various diseases, including lung function, kidney function, and the immune system [8]. Polysaccharides containing α-glucose, α-mannose, α-galactose, and α-arabinose with α-type glycosidic linkage have antioxidant, immunomodulatory, hypoglycemic, and anti-inflammatory properties [9,10].

Several researchers have reported the *C.-militaris*-mediated fabrication of metallic (silver, gold, and platinum) nanoparticles and their medicinal potential [11–13]. Li et al. [14] proposed photolytic degradation of methylene blue using *C.-militaris*-mediated zinc oxide nanoparticles. Suksiriworapong et al. [15] reported CD44-Targeted Lipid Polymer Hybrid Nanoparticles using *C. militaris* for their enhanced anti-breast-cancer effect. Rupa et al. [16] stated the antioxidant, antimicrobial, and anti-inflammatory activity of *C.-militaris*-mediated nanoemulsion.

With their proposed mechanism, we have evidenced the *C.-militaris*-mediated eco-synthesis of zinc oxide nanoparticles and their antibacterial, antidiabetic, and antioxidant potential in our research laboratory [17]. Despite the reports on *C.-militaris*-mediated synthesis of gold nanoparticles, no one has attempted to assess the effect of process variables in their eco-synthesis. The present study demonstrated the green synthesis employing quality by design approach (central composite design), characterization, and biological assessment of gold nanoparticles mediated through an aqueous extract of *C. militaris*, i.e., *Cord*-Au-NPs.

2. Results and Discussion

C. militaris acts as a stabilizing and reducing agent, while trichlorogold, i.e., gold chloride, is a gold precursor. The aqueous solution of *C. militaris* extract (CME) was added to the trichlorogold aqueous solution. The color change indicated the reduction of gold ions into gold nanoparticles, as shown in Figure S1 (Supplementary File). The dark pink color was formed at one hour of incubation; these color changes occur due to surface plasmon vibration being excited with the gold nanoparticle synthesis.

Nanoparticles of varying sizes, shapes, and morphologies are synthesized due to variations in the composition and concentration of reducing agents in CME. Reducing metal ions into metallic nanoparticles most likely involves oxidizing the hydroxyl group of polysaccharides to the carboxyl group. Reduced metal atoms are nucleated when metal ions are first activated from their monovalent or divalent oxidation states [18].

2.1. UV-Visible Spectroscopy Analysis

The use of UV-vis spectroscopy to assess the fabrication and stability of Au-NPs in an aqueous solution is a beneficial approach. UV-vis spectrum confirms the formation of gold nanoparticles from trichlorogold exhibiting a surface plasmon resonance (SPR) absorbance band around 540–550 nm (Figure 1A–C) [2,3]. The formation of Au-NPs in the sample was evident from the change in solution color from light yellow to ruby red. As light passes through the nanoparticles in the sample, the sample absorbs specific wavelengths, and the intensity of wavelengths is reduced in the beam emitted from the sample [19]. Monitoring the reaction kinetics using UV-vis spectroscopy confirmed the completion of the reaction in the sample after 3 min, as evident from the stability, with no further significant change beyond this time (Figure 1). The concentration of generated Au-NPs with the sample was determined spectrophotometrically using the Beer–Lambert law with an extinction coefficient of $1.8 \times 10^{10}\ M^{-1}\ cm^{-1}$ [5]. Furthermore, it gradually rises in intensity as a function of reaction time with a slight change in the peak wavelength. Therefore, it was possible to track the bioreduction of gold in an aqueous solution by diluting tiny aliquots of the sample solution with double-distilled water and subsequently measuring the solution's UV-vis spectrum.

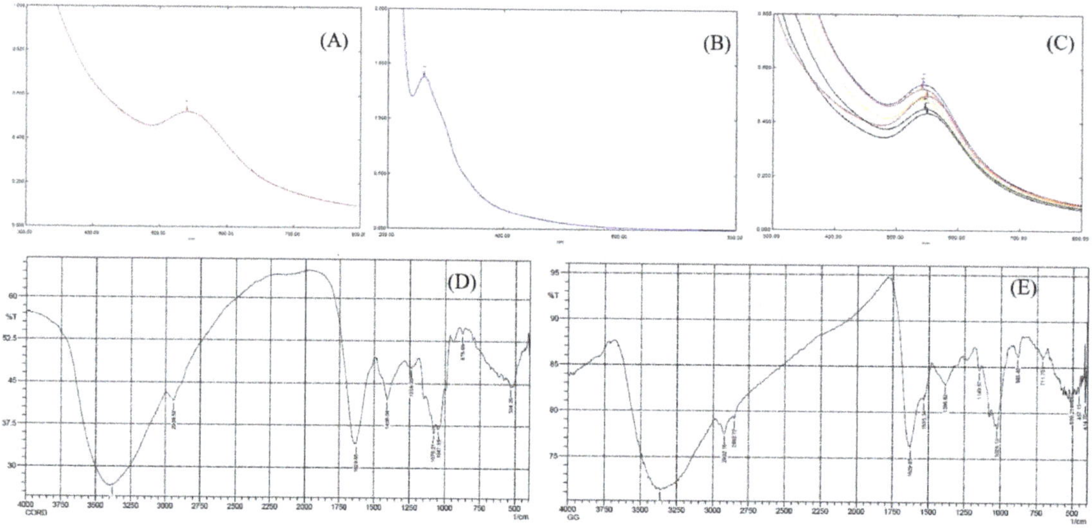

Figure 1. UV spectra of (**A**) Au-NPs, (**B**) CME, and (**C**) *Cord*-Au-NPs as a function of reaction time; FTIR spectra of CME (**D**) and *Cord*-Au-NPs (**E**).

2.2. Optimization by Factorial Design Approach

Effect of Variables on Absorbance, Particle Size, and Zeta Potential

Absorbance, particle size, and zeta potential values for *Cord*-Au-NPs batches ranged from 0.37 to 0.554, 4.22 to 16.68 nm, and −18.96 to −8.59 mV, respectively (Table 1). The linear model was the best fit for the response Y_1 (R^2: 0.9561 and PRESS: 0.0024), whereas the quadratic model was the best fit for the responses Y_2 (R^2: 0.9959 and PRESS: 2.37) and Y_3 (R^2: 0.9721 and PRESS: 14.30). This confirms that the suggested model can accurately predict the 95.61%, 99.59%, and 97.21% of variances in responses Y_1, Y_2, and Y_3, respectively. The models' significance and efficacy were assessed using ANOVA (Table 2).

Table 1. ANOVA results for responses Y_1 and Y_2.

Source	Response Y_1				Response Y_2				Response Y_3			
	Coeff.	Sum of Squares	F-Value	*p*-Value	Coeff.	Sum of Squares	F-Value	*p*-Value	Coeff.	Sum of Squares	F-Value	*p*-Value
Model	0.4689 [a]	0.0274	93.9	<0.0001 [b]	12.86 [a]	140.32	245.39	<0.0001 [b]	−13.69 [a]	90.47	34.89	0.0007 [b]
X_1 -	−0.0181	0.0039	26.88	0.0008	−0.9333	10.45	91.41	0.0002	−1.36	22.09	42.59	0.0013
X_2 -	0.0443	0.0235	160.93	<0.0001	0.5733	3.94	34.49	0.002	0.865	8.98	17.31	0.0088
X_1X_2					−2.24	19.98	174.72	<0.0001	−2.61	27.35	52.75	0.0008
X_1^2					0.4756	4.37	38.17	0.0016	0.581	6.52	12.57	0.0165
X_2^2					−1.92	70.92	620.17	<0.0001	−0.8252	13.14	25.35	0.004
Residual		0.0012				0.5718				2.59		
Lack of Fit		0.0009	1.33	0.4886 [c]		0.1552	0.2484	0.8586 [c]		1.6	1.07	0.5175 [c]
Pure Error		0.0002				0.4166				0.9978		
Cor Total		0.0286				140.89				93.06		

[a] Intercept; [b] significant; [c] non-significant.

Table 2. CCD matrix with coded and actual levels of independent and dependent variables.

Independent Variables	Levels				
	$-\alpha$	Low (1)	Medium (0)	High (+1)	$+\alpha$
X_1: CME conc. (% *w/v*)	4	6	8	10	12
X_2: $AuCl_3$ conc. (mM)	0.25	0.5	0.75	1	1.25
Dependent variables			Goal		
Y_1: Absorbance			Maximize		
Y_2: Particle size (nm)			Minimize		
Y_3: Zeta potential (mV)			Minimize		

Std	Run	X_1: CME Conc. (% *w/v*)	X_2: $AuCl_3$ Conc. (mM)	Y_1: Absorbance	Y_2: Particle size (nm)	Y_3: Zeta potential (mV)
11	1	8	0.75	0.461	13.26	−13.24
10	2	8	0.75	0.476	13.03	−14.65
9	3	8	0.75	0.482	12.38	−14.02
7	4	8	0.25	0.37	4.22	−18.96
4	5	10	1	0.507	8.96	−16.37
6	6	12	0.75	0.417	12.89	−14.56
1	7	6	0.5	0.442	9.32	−15.87
2	8	10	0.5	0.421	11.98	−12.81
5	9	4	0.75	0.509	16.68	−8.59
8	10	8	1.25	0.554	6.21	−15.44
3	11	6	1	0.519	15.24	−8.97

Small p (<0.05) and high F ratio values were statistically significant. The model's F-values were determined to be 93.90, 245.39, and 34.89 for the response variables Y_1, Y_2, and Y_3, respectively, highlighting the model's significance. The model's relevance is shown by the lack of fit values of 1.33 (Y_1), 0.2484 (Y_2), and 1.07 (Y_3), which show that the models adequately fit the experimental data and various parameters significantly affect the responses. The predicted R^2 values of 0.9158, 0.9832, and 0.8463 for Y_1, Y_2, and Y_3 agreed with the adjusted R^2 values of 0.9489, 0.9919, and 0.9443, respectively. The significance of the proposed models is illustrated by the plot of predicted against actual values (Figure S2).

The following polynomial equations illustrate the connection between independent and dependent variables:

$$Y_1\ (\text{Absorbance}) = +0.4689 - 0.0181\ X_1 + 0.0443\ X_2$$

$$Y_2\ (\text{Particle size}) = +12.86 - 0.9333\ X_1 + 0.5733\ X_2 - 2.24\ X_1X_2 + 0.4756\ X_1^2 - 1.92\ X_2^2$$

$$Y_3\ (\text{Zeta potential}) = -13.69 - 1.36\ X_1 + 0.8650\ X_2 - 2.61\ X_1X_2 + 0.5810\ X_1^2 - 0.8252\ X_2^2$$

Main terms X_1 and X_2 significantly affected response Y_1, whereas X_1, X_2, X_1X_2, X_1^2, and X_2^2 showed significant effects on responses Y_2 and Y_3. Factors X_1 and X_2 negatively and positively affected all the studied responses. These results were further validated by

2D contour and 3D response surface plots (Figure 2). Each variable's qualitative impact on each response parameter can be seen by carefully examining these charts. Interaction term X_1X_2 showed antagonistic effects on Y_2 and Y_3. The examination of interaction plots further supported this. Crossed lines in the plots (Figure S3) confirm interactions, while parallel lines indicate no interactions. Figure S4 shows model diagnosis plots, including a normal and residual versus run plot. A close look at the graphs reveals that the residuals typically follow a straight line, indicating that the errors are evenly distributed. The residuals vs. run plot also revealed a random dispersion.

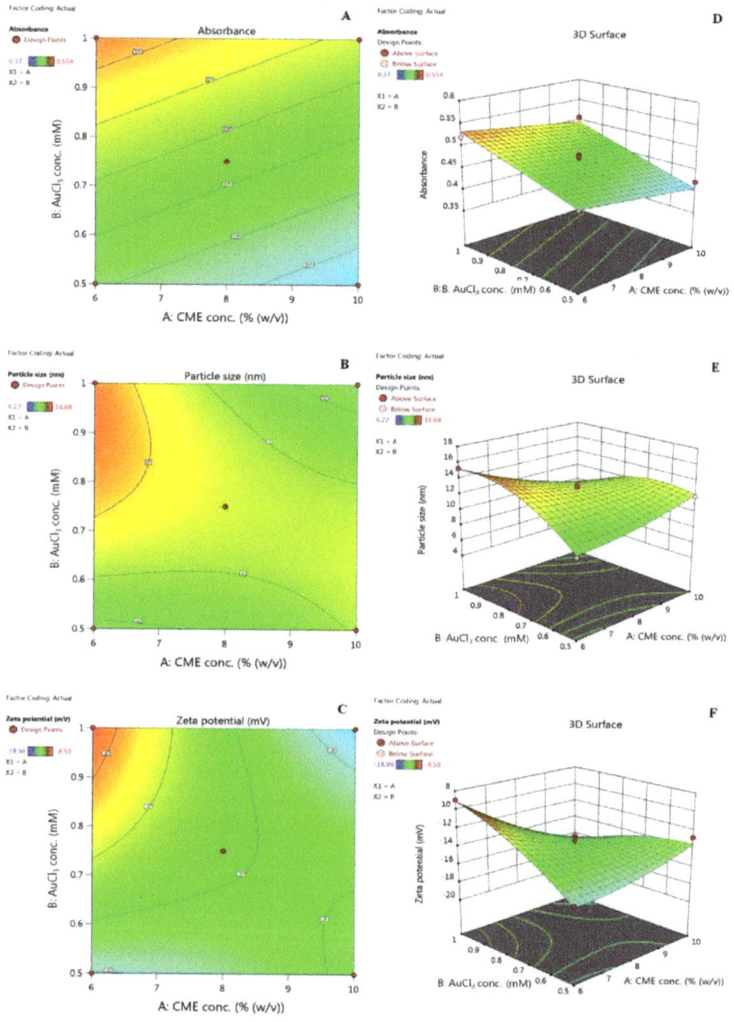

Figure 2. Two-dimensional contour plots (**A–C**) and three-dimensional response surface plots (**D–F**) show independent variables' effect on response variables.

2.3. Model Validation

The optimal range for the formulation parameters was determined using a graphical optimization strategy based on the desirability function approach and an overlaid contour plot. The optimal concentrations of CME and $AuCl_3$ for obtaining Au-NPs are displayed in the region marked in Figure S5. Optimal values of the synthesis variables under investigation were also determined using various optimization techniques. The results demonstrated that 10% (w/v) CME concentration and $AuCl_3$ at 1 mM concentration would offer Au-NPs with desired characteristics. Au-NPs were prepared using optimal values of independent variables and then examined concerning the investigated response variables. Predicted values for the absorbance, particle size, and zeta potential of the synthesized Au-NPs were 0.495, 8.82 nm, and −17.04 mV, respectively. The measured experimental values for the absorbance, particle size, and zeta potential were 0.512, 7.18 nm, and −19.42 mV, respectively. Predicted and actual experimental findings did not differ much, indicating the model's validity.

2.4. Fourier Transform Infrared (FTIR) Spectroscopy

FTIR analyzed the functional groups of CME and *Cord*-Au-NPs bioactives attached to the surface of the nanoparticles, where 4000–400 cm^{-1} was the scanning range. The biomolecules potentially accountable for stabilizing the gold nanoparticles were determined using FTIR analysis. The FTIR spectra of CME (Figure 1D) show a strong absorption band at 3381.21 cm^{-1} associated with the stretching vibration of -OH and -NH, related to the H_2O molecules and amino groups. Absorption bands at 2939.52 cm^{-1} and 1629.85 cm^{-1} correspond to -CH_2 groups in lipids and protein C=O stretching vibrations, respectively. Phenolic group vibrations were allocated to the 1408.04 cm^{-1} absorption band. The C-O-C and C-OH stretching vibrations of aromatic acid ester and phenolics are shown in the absorption band at 1238.30 cm^{-1}. At 1078, 1047, and 875 cm^{-1}, β-glucans were found to have β-glycosidic linkages that were characterized by absorption peaks [17].

The FTIR spectra of *Cord*-Au-NPs (Figure 1E) show absorption bands at 3361.88 cm^{-1}, 2922.16 cm^{-1}, 1619.85 cm^{-1}, 1149.57 cm^{-1}, and 1026.13 cm^{-1}, which can be allocated to -OH and N-H groups of water molecules and amino groups, -CH_2 groups in lipids, C=O stretching of proteins, C-O-C and C-OH groups of aromatic acid ester and phenolics, respectively, and β-glycosidic bonds of β-glucans. The decrease in intensity at the above absorption peaks in *Cord*-Au-NPs signifies the involvement of the phenolic group, lipids, and proteins in the reduction process. Furthermore, they essentially act as capping agents due to the presence of different groups of compounds, providing stability and preventing aggregation of Au-NPs. The *Cord*-Au-NPs exhibit a strong binding affinity to various functional groups, as reported in the CME, a layer covering Au-NPs.

According to the FTIR spectral analysis, the stabilization and reduction of the Au-NPs occur with the involvement of different functional groups of biomolecules reported in the extract used in the study. It is well known that phytochemicals act as stabilizing agents in the fabrication of metal nanoparticles [20]; further validation is essential to identify the active molecules that are responsible for Au-NP synthesis.

2.5. Morphological Analysis

The scanning electron microscope (SEM) images of the nano-sized *Cord*-Au-NPs describe the larger surface area. The SEM (Figure 3A) and high-resolution transmission electron microscope (HRTEM) (Figure 3B–D) images at different magnification levels were used to analyze the morphology of the *Cord*-Au-NPs. They revealed a cluster of predominantly spherical, irregular-shaped particles.

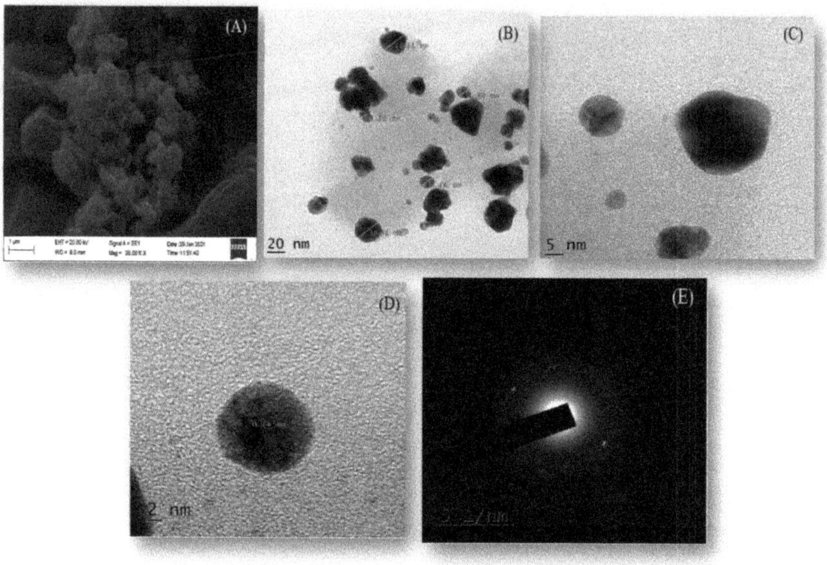

Figure 3. (**A**) SEM, (**B–D**) HRTEM, and (**E**) SAED pattern of *Cord*-Au-NPs.

In addition, the selected area electron diffraction (SAED) (Figure 3E) pattern specified the polycrystalline nature of the *Cord*-Au-NPs [17]. Figure 3A illustrates the *Cord*-Au-NPs synthesized as standard protocol and shows spherical morphology with low distribution.

2.6. Particle Size and Zeta Potential Analysis

The size of *Cord*-Au-NPs varies from 1 to 100 nm, with an average particle size of 7.18 nm and polydispersity index (PDI) of 0.096 (Figure 4A). Nanoparticles' net surface charge determines their zeta potential. Nanoparticles resist each other in solution, causing a Coulomb explosion; thus, they do not agglomerate. Nanoparticles with zeta potential scaling from +30 mV to −30 mV are considered stable [18,21]. *Cord*-Au-NPs had a zeta potential of −19.42 mV (Figure 4B). Zeta potential values are negative, indicating that the capping agents effectively stabilize nanoparticles by preventing agglomeration through intense negative charges. Zeta potential values are used as a hallmark indication of the stability of colloidal particles, and absolute values replicate the net electrical charge on the particles' external surface that arises from the surface functional groups [5].

2.7. X-ray Diffraction (XRD) Analysis

XRD test is used to determine the size and evaluate the crystalline structure of the synthesized1 samples and provides information about the crystallinity of the NPs [19]. It is clear from the XRD pattern (Figure 4C,D) that the gold facets (111), (200), (220), and (311) are all represented by peaks with 2θ values between 38.26, 44.23, 64.75, and 77.67. Au-NPs synthesized using *Turbinaria conoides* reported similar observations [7]. As can be seen from the XRD pattern, the gold nanoparticles synthesized are crystalline. Debye–Scherrer equation, i.e., $d = k\lambda/(\beta\ \mathrm{Cos}\theta)$, was used and revealed an average crystalline size of 6.72 nm. The reported peak values of Cord-Au-NPs also matched the planes and face-centered cubic structures of Au-NPs prepared by other green synthesis methods by various researchers [22–24].

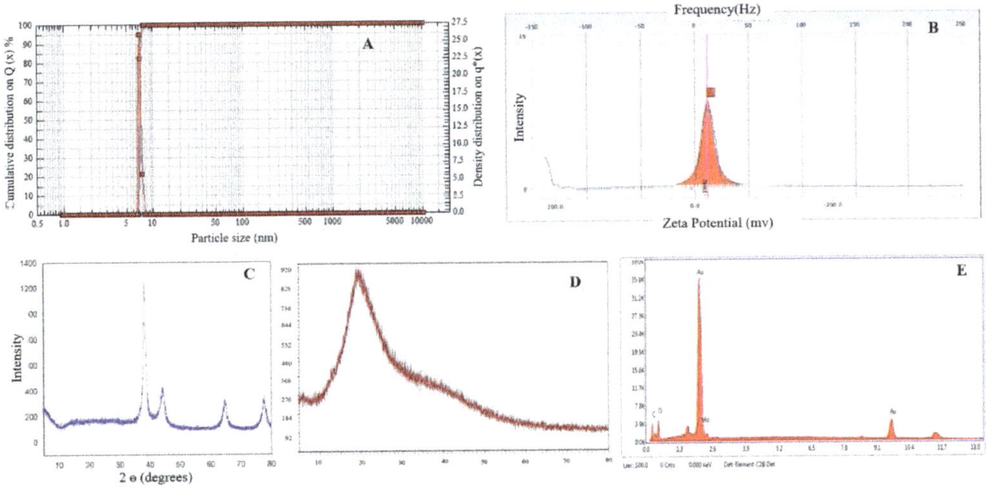

Figure 4. Particle size (**A**), zeta potential (**B**), XRD (**C**,**D**), and EDX (**E**) analysis of *Cord*-Au-NPs.

2.8. Energy-Dispersive X-ray Analysis

Due to SPR, metallic NPs have a typical optical absorption peak of around 2.15 keV [17,25,26]. The energy-dispersive X-ray (EDX) analysis (Figure 4E) of *Cord*-Au-NPs illustrated a strong peak in the gold region (around 2.15 keV) that validated the synthesis of Au-NPs. All additional signals may have been caused by elements present in CME, as shown by the EDX data's carbon grid signal.

2.9. Brine Shrimp Lethality Assay

The brine shrimp lethality experiment was used to test the cytotoxicity of biogenetically synthesized *Cord*-Au-NPs at varying doses for 24 h. Brine shrimps were highly vulnerable to *Cord*-Au-NPs, even at relatively low concentrations (Table S2) (Supplementary File). *Cord*-Au-NPs exhibited an LC_{50} value of 192.23 ± 57.34 μg/mL; given that the LC_{50} is between 100 and 1000 ppm, they may have some biological activity. After 24 h of exposure to *Cord*-Au-NPs, concentration-dependent death in brine shrimp was observed. Hitherto, the green synthesized *Cord*-ZnO-NPs using *C. militaris* and rutin-conjugated titanium dioxide nanoparticles (Rut-TiO_2NPs) have evidenced cytotoxic effects on brine shrimps [25].

2.10. In Vitro Antidiabetic Activity

The α-amylase and α-glucosidase inhibitory activity test results showed that *Cord*-Au-NPs exhibited a higher inhibitory effect than CME (Figure 5A) and gave equivalent results to the standard acarbose used in the study.

Significant α-glucosidase enzyme inhibition (53.27 ± 0.07 μg/mL) was observed in *Cord*-Au-NPs compared to the standard acarbose (56.40 ± 0.09 μg/mL). Likewise, *Cord*-Au-NPs showed a notable α-amylase inhibitory effect (55.78 ± 0.33 μg/mL), which is comparable to the results of standard (54.74 ± 0.80). Previously, Nigella-sativa-mediated Au-NPs showed effective antidiabetic activity by inhibiting α-amylase (81% at 500 μg/mL) and α-glucosidase (82.3% at 500 μg/mL) enzymes [27]. Furthermore, oral administration of Fritillaria-cirrhosis-mediated Au-NPs possessed good antidiabetic potential in STZ-induced diabetic rats with appreciable restoration of serum, hepatic, and renal markers to normal by Au-NPs treatment [28]. *C. militaris*-mediated ZnO nanoparticles also exhibited a remarkable inhibitory effect against the enzymes α-amylase (IC_{50} of 46.29 ± 0.49 μg/mL) and α-glucosidase (47.21 ± 0.19 μg/mL), with *C. militaris* having a moderate enzyme inhibitory effect [17,25].

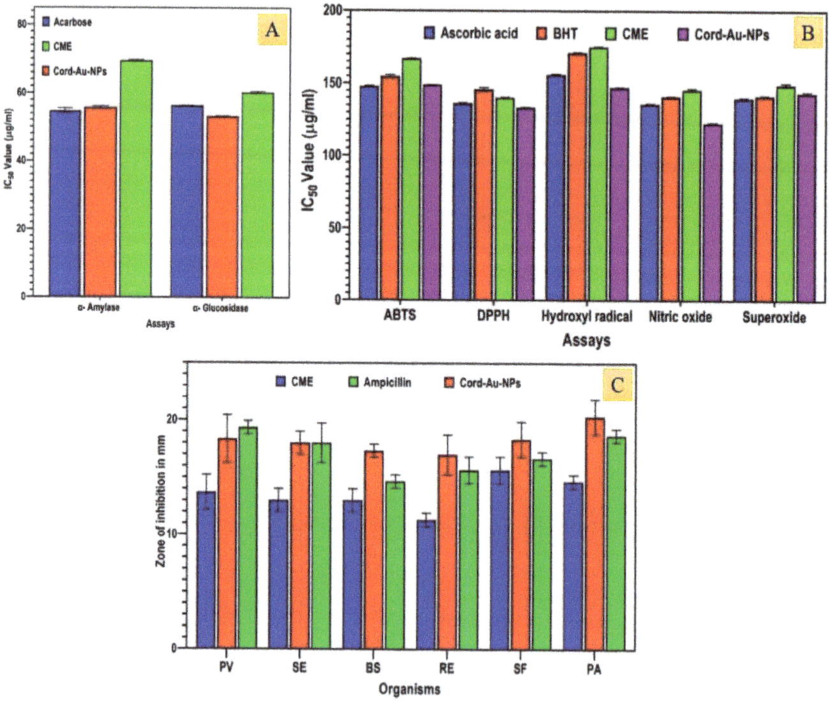

Figure 5. Antidiabetic ((**A**) α-amylase and α-glucosidase enzyme inhibition assays), antioxidant ((**B**) DPPH, ABTS, superoxide, nitric oxide, and hydroxyl radical scavenging assays), and antibacterial activity of CME and *Cord*-Au-NPs (PV—*Proteus vulgaris*, SE—*Staphylococcus epidermidis*, BS—*Bacillus subtilis*, RE—*Rhodococcus equi*, SF—*Shigella flexneri*, PA—*Pseudomonas aeruginosa*).

2.11. In Vitro Antioxidant Activity

The antioxidant potential of CME and Cord-Au-NPs was determined by DPPH, ABTS, superoxide, NO, and hydroxyl radical assays. All the tested samples revealed dose-dependent radical scavenging activity (Figure 5B). *Cord*-Au-NPs exhibited the highest ABTS radical scavenging activity (148.90 ± 0.05 µg/mL). A maximum hydroxyl radical scavenging effect was observed in *Cord*-Au-NPs with an IC_{50} of 147.27 ± 0.08 µg/mL. The maximum DPPH radical scavenging activity was observed in *Cord*-Au-NPs (133.29 ± 0.31 µg/mL), lower than the standard ascorbic acid and higher than the BHT. At the same time, moderate DPPH scavenging activity was recorded in CME (140.15 ± 0.29 µg/mL). According to Milanezi et al. [29], quercetin-capped Au-NPs showed a higher antioxidant potential on DPPH, ABTS, and NO radical scavenging assays. Similar findings were also reported by Zayadi et al. [30]. They observed significant antioxidant (DPPH free radical scavenging potential of 55.1% and hydroxyl radical scavenging of 50.0%) activity in *Zingiber officinale*-mediated synthesis of colloidal biogenic Au-NPs.

For the superoxide free radical scavenging assay, CME (149.19 ± 1.03 µg/mL) exhibited moderate effect and *Cord*-Au-NPs (143.56 ± 0.65 µg/mL) showed excellent free radical scavenging activity when compared to the standard, ascorbic acid (140.01 ± 0.29 µg/mL), and BHT (141.56 ± 0.22 µg/mL) (Figure 5B). Nitric oxide scavenging effect of the CME was found to be 145.85 ± 0.81 µg/mL. In contrast, the scavenging value was determined to be superior in *Cord*-Au-NPs, with an IC_{50} of 122.94 ± 0.63 µg/mL, which was recorded as a better activity than the ascorbic acid (136.02 ± 0.19 µg/mL) and BHT (141.06 ± 0.23 µg/mL). Similarly, Sathishkumar et al. [31] have demonstrated that the biosynthesis of Au-NPs with *Couroupita guianensis* fruits showed potential free radical scavenging activity on DPPH, reducing power assay, hydroxyl radical scavenging, and superoxide radical scavenging assays. Lydia et al. (2020) [32] reported that Au-NPs synthesized from *Punica granatum* revealed significant DPPH (23.6 ± 1.5 to 62.5 ± 1.8%) and hydrogen peroxide (21.6 ± 1.3 to 62.8 ± 1.8%) radical scavenging activity. Similarly, the calcinated form of the titanium oxide nanoparticles exhibited considerable free radical scavenging abilities using assays such as DPPH (IC_{50} of 95.32 ± 0.43), NO (IC_{50} of 110.03 ± 0.99 μg/mL), superoxide (IC_{50} of 136.27 ± 0.63 μg/mL), and ABTS method (IC_{50} of 119.11 ± 0.94 μg/mL) [17,25].

2.12. In Vitro Antibacterial Activity

Cord-Au-NPs showed a strong antibacterial effect by inhibiting the growth of studied bacteria at 300 μg/mL concentration (Figure 5C). *Cord*-Au-NPs exhibited a higher inhibitory effect against the growth of *Pseudomonas aeruginosa* MTCC 1748 (20.33 ± 1.52 mm), *Proteus vulgaris* MTCC 426 (18.33 ± 2.08 mm), *Staphylococcus epidermis* MTCC 435 (17.33 ± 2.08 mm), and *Shigella flexneri* MTCC 1457 (16.66 ± 0.57 mm) (Figure 6), whereas *Bacillus subtilis* MTCC 441 and *Rhodococcus equi* MTCC 2558 were most sensitive for CME. *Cord*-Au-NPs were observed to have the potential to pierce bacterial cell walls and substantially destroy cell membranes.

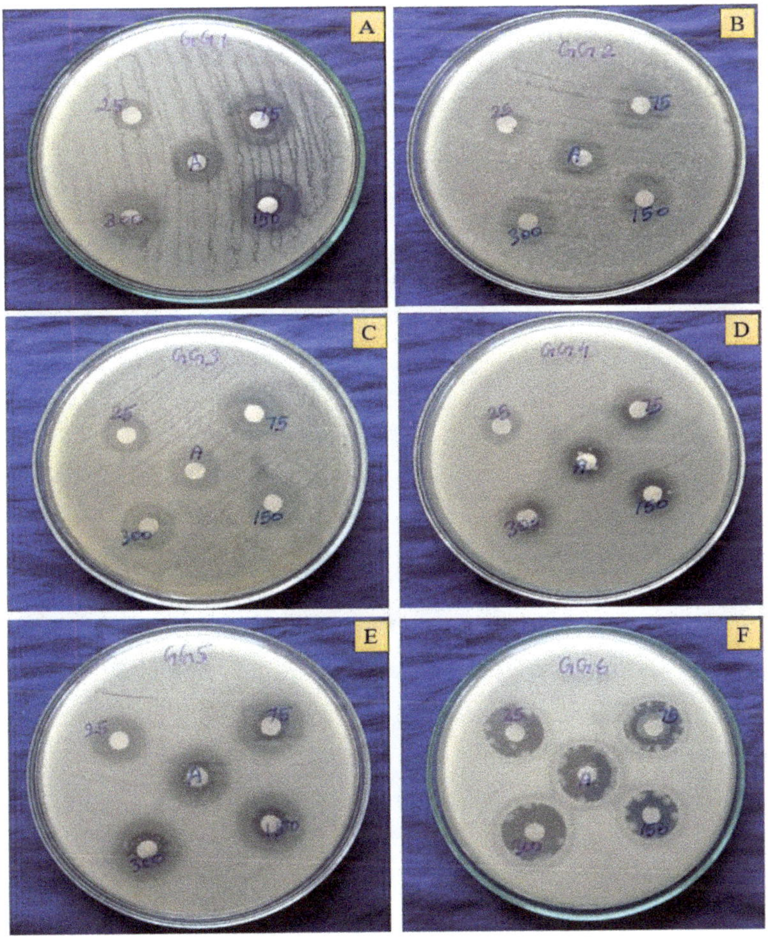

Figure 6. Zone of inhibition results for the *Cord*-Au-NPs. (**A**)—*Proteus vulgaris*, (**B**)—*Staphylococcus epidermidis*, (**C**)—*Bacillus subtilis*, (**D**)—*Rhodococcus equi*, (**E**)—*Shigella flexneri*, (**F**)—*Pseudomonas aeruginosa*.

In agreement with our results, *Alternanthera bettzickiana* leaf-extract-mediated biocompatible Au-NPs exhibited significant antimicrobial activity against *B. subtilis* (14 ± 0.15 mm), *S. aureus* (16 ± 0.88 mm), *S. typhi* (16 ± 0.44 mm), *P. aeruginosa* (14 ± 0.58 mm), *Micrococcus luteus* (22 ± 0.44 mm), and *Enterobacter aerogenes* (06 ± 0.15 mm) [33]. Likewise, *Halymenia dilatata*-mediated biogenic synthesized Au-NPs exhibited significant antibacterial activity, with the highest zone of inhibition (21 mm) at 100 µg/mL against the selected pathogenic bacteria [34]. Bacterial drug susceptibility was assessed using the MIC value, which was also vital for testing the efficacy of antimicrobial compounds [35]. In the present investigation, *Cord*-Au-NPs showed a good inhibitory effect against the growth of *B. subtilis* and *P. vulgaris*, with an MIC of 23.44 µg/mL, followed by *P. aeruginosa* and *S. flexneri* (46.88 µg/mL) (Table S1). A moderate inhibitory effect was observed against *S. epidermis* and *S. flexneri* growth, with an MIC of 93.74 µg/mL. Likewise, CME showed potential inhibition against *P. vulgaris*, with an MIC of 46.80 µg/mL, followed by *R. equi* and *P. aeruginosa* (93.75 µg/mL). Similar results were reported by Nagalingam et al. [33] in the pathogenic bacterial strains *B. subtilis*, *S. aureus*, *M. luteus*, *E. aerogenes*, *S. typhi*, and *P. aeroginosa*, with the MIC of 10–40 µg/mL with Au-NPs synthesized from *A. bettzickiana* leaf extracts.

2.13. Cytotoxicity of Cord-Au-NPs

Cytotoxicity assay of *Cord*-Au-NPs inhibited 81.1% and 76.4% of the growth of Human Breast Cancer (MDA-MB-31) and Human Colon Adenocarcinoma (HT-29) cell lines at 90 µg/mL.

Figure 7 depicts the significant growth inhibition of *Cord*-Au-NPs (Figure 7C,F) against selected cell lines compared to control (Figure 7A,D) and Adriamycin (ADR) (Figure 7B,E). Moreover, Hosny et al. [36] revealed the cytotoxicity effect of Au-NPs using MTT assay with higher efficiency in inhibiting the growth and proliferation of human breast cancer cells.

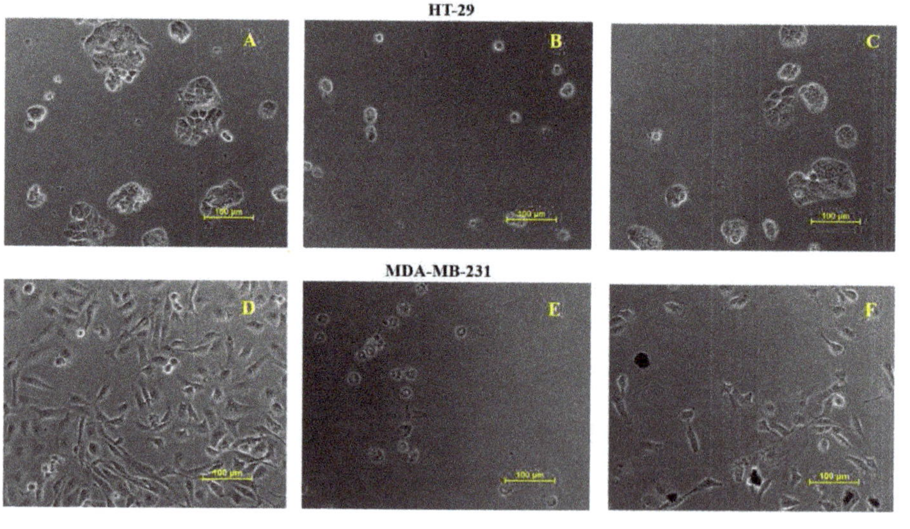

Figure 7. Cytotoxicity studies of test drugs ((**A**)—HT-29-Control, (**B**)—ADR, (**C**)—*Cord*-Au-NPs, (**D**)—MDA-MB-231 Control, (**E**)—ADR and (**F**)—*Cord*-Au-NPs).

3. Materials and Methods

3.1. Chemicals

Analytical-grade Trichlorogold ($AuCl_3$, 99%) was purchased from Sigma-Aldrich, St. Louis, MO, USA. All other chemicals used were of analytical grade.

3.2. Preparation of Test Material

As per our earlier reports, the fruiting bodies of *C. militaris* were collected from the Department of Biotechnology—Technology Incubation Centre, Bodoland University, Assam, India, followed by their microscopic and macromorphological examinations to aid their identification and confirmation [17]. A total of 10 g of freeze-dried *C. militaris* powder was added to 50 mL of double distilled water and boiled at 60 °C for 15 min. The obtained *C. militaris* extract was filtered through Whatman No.1 filter paper and stored at 4 °C until further use.

3.3. Cord-Au-NPs Synthesis

Cord-Au-NPs were synthesized following a reported method [2]. Briefly, 10 mL of CME was added to 100 mL of 1 mM gold chloride solution in a conical flask and kept in the dark at room temperature. The color change was visually observed, and the time taken for the color to change was noted. Samples were drawn routinely and analyzed spectrophotometrically using a UV-visible Spectrophotometer (Shimadzu, UV-2700, Kyoto, Japan) to verify the synthesis of *Cord*-Au-NPs.

3.3.1. Optimization by Factorial Design Approach

The synthesis technique for *Cord*-Au-NPs was optimized utilizing a central composite design (CCD) (Design-Expert software V13.0; Stat-Ease Inc., Minneapolis, MN, USA). A total of 11 runs (R1–R11) were produced using software and the effects of independent variables were evaluated, namely, X_1: CME concentration (% (w/v)) and X_2: $AuCl_3$ concentration (mM) on Y_1: absorbance, Y_2: particle size (nm), and Y_3: zeta potential of prepared Au-NPs. Table 2 shows all independent and response variables and their coded and actual levels. As shown in Table 2, 11 experimental runs were generated with five distinct levels for both independent variables.

The experimental runs' center point (X_1 = 8% w/v and X_2 = 0.75 mM) was analyzed in triplicate to reduce pure error. Model suitability was decided based on the coefficient of determination (R^2), predicted error sum of squares (PRESS), and lack of fit analysis. The significance of the model was also tested with analysis of variance (ANOVA). Effects of independent variables on responses were visualized using two contour plots and 3D surface plots.

3.3.2. Model Validation

Graphical optimization was utilized to achieve the optimized values for the CME concentration and X_2: $AuCl_3$ concentration that can offer Au-NPs of desired attributes, i.e., minimum particle size, zeta potential, and maximum absorbance. Finally, checkpoint analysis was performed at ideal synthesis conditions to validate the model and obtain an optimized experimental procedure.

3.4. Physicochemical Characterization

The synthesized *Cord*-Au-NPs were subjected to physicochemical characterization [2,17,20] using Fourier transform infrared (FT-IR) spectrophotometer (Make: Shimadzu, IR Affinity-1), an X-ray diffractometer (XRD) (Rigaku, Ultima IV, Tokyo, Japan), a scanning electron microscope with energy-dispersive X-ray (EDX) (Carl ZEISS EVO-18) analyzer, high-resolution transmission electron microscope (HRTEM) (Make: Joel, Model: JEM 2100), and particle size and zeta potential analyzer (Nanophox, NX0088—Sympatec, Clausthal-Zellerfeld, Germany). The detailed procedures for the physicochemical evaluation of green synthesized *Cord*-Au-NPs are given in the Supplementary File.

3.5. Therapeutic Investigation

3.5.1. Brine Shrimp Lethality Bioassay

Cord-Au-NPs, at various concentrations, were tested for cytotoxicity using the Brine Shrimp Lethality Bioassay with *Artemia salina* eggs, as described in the literature [17,20].

3.5.2. Antidiabetic Activity

The antidiabetic potential of *Cord*-Au-NPs was ascertained using in vitro enzyme (α-amylase and α-glucosidase) inhibitory assays [17,20]. Starch solution (1%) was used as a substrate for α-amylase and p-nitrophenyl-α-D-glucopyranoside (PNPG) for α-glucosidase inhibitory assays, with acarbose as an antidiabetic reference.

For the α-amylase inhibitory assay, 100 μL of the different concentrations of CME and *Cord*-Au-NPs (3.125, 6.25, 12.5, 25, 50, 100, and 200 μg/mL) were added with 1% starch solution in 20 mM phosphate buffer as the substrate solution. The mixture was incubated at 25 °C for 10 min, and then 100 μL of porcine pancreatic α-amylase enzyme (0.5 mg/mL) was added, followed by incubating the mixture for 10 min at 25 °C. To this, 200 μL of dinitro salicylic acid reagent was added to terminate the reaction. Then, the mixture was again incubated at 100 °C for 5 min. After cooling, the absorbance of samples was measured at 540 nm [10].

For the α-glucosidase inhibitory assay, 50 μL of CME and *Cord*-Au-NPs in different concentrations (3.125, 6.25, 12.5, 25, 50, 100, and 200 μg/mL) was added with 100 mM sodium phosphate buffer and 50 μL of 5 mM p-nitrophenyl-α-d-glucopyranoside solution and the mixture was incubated at 37 °C for 5 min. To this mixture, 100 μL of phosphate buffer with 0.1 U/mL α-glucosidase enzyme was added, and after 30 min of incubation, absorbance was measured at 405 nm.

The α-amylase and α-glucosidase enzyme inhibitory effects of CME and *Cord*-Au-NPs were calculated using the following formula:

$$\% \text{ inhibition} = [(\text{Control Abs} - \text{Sample Abs})/\text{Control Abs}] \times 100$$

3.5.3. Antioxidant Activity

Antioxidants play a vital role in the treatment and management of several diseases. Hence, the antioxidant activity of *Cord*-Au-NPs was evaluated using different in vitro assays (2, 2-diphenyl-1-picrylhydrazyl (DPPH) radical, nitric oxide (NO) radical, superoxide radical, hydroxyl radical, and 2, 2′-azino-bis-(3-ethylbenzo-thiozoline-6-sulfonic acid) diammonium salt (ABTS) scavenging assay). These methods assess free radical inhibition based on the endpoints, radical generation, and reproducibility. In all these assays, CME and *Cord*-Au-NPs were prepared using 31.25, 62.5, 125, 250, and 500 μg/mL concentrations by adding respective reagents followed by the standard methodologies [17,20].

3.5.4. Antibacterial Activity

Antibacterial activity of CME and *Cord*-Au-NPs were studied against the Gram-positive and Gram-negative bacterial strains *Proteus vulgaris* MTCC 426, *Staphylococcus epidermis* MTCC 435, *Bacillus subtilis* MTCC 441, *Rhodococcus equi* MTCC 2558, *Shigella flexneri* MTCC 1457, and *Pseudomonas aeruginosa* MTCC 1748 by agar disc diffusion and microdilution methods [17,20]. Different concentrations (25, 75, 150, and 300 μg/mL) were used in the agar disc diffusion assay with amoxicillin as standard (1 mg/mL). The Mueller–Hinton (MH) agar medium inoculated with a loop of each tested bacterium was used as a working culture. The prepared discs were soaked in test solutions and placed on the plates with inoculated agar and bacterial strains before incubating at 37 °C for 24 h. Then, the diameter (mm) of the inhibition zone (growth zone of bacterial strains around the discs) was measured to analyze the antibacterial effect of CME and *Cord*-Au-NPs.

In the microdilution assay, the minimal concentration at which the growth of bacteria was inhibited is considered as minimum inhibitory concentration (MIC). The bacterial strains in the above assay were used to determine the MIC in 96-well microtiter plates. Briefly, 50 μL of CME and *Cord*-Au-NPs were poured into each well at different concentrations (500–0.488 mg/mL in two-fold serial dilution) with 50 μL of MH broth. A total of 50 μL of bacterial inoculum at 106 CFU/mL was added to each well and then incubated at 37 °C for 24 h. Then, 20 μL of *p*-iodonitro tetrazolium dye (INT) at a concentration of 0.5 mg/mL was added to each well and again incubated at 37 °C for 30 min with amoxicillin

(30–0.029 mg/mL) as standard. The change in color after adding the INT dye embodies the MIC of CME and *Cord*-Au-NPs.

3.5.5. Anticancer Activity

Cell lines, i.e., Human Breast Cancer (MDA-MB-31) and Human Colon Adenocarcinoma (HT-29), were procured from the National Centre for Cell Science (NCCS), Pune, India. Anticancer activity of *Cord*-Au-NPs was appraised against selected cell lines using Sulforhodamine B (SRB) assay at various concentrations, i.e., 10 μg/mL, 20 μg/mL, 40 μg/mL, and 80 μg/mL [24]. The percent growth inhibition was calculated using the following formula:

$$[Ti/C] \times 100\%$$

where Ti—test drug concentration; C—control growth.

4. Conclusions

The biosynthesis of gold nanoparticles was carried out using an aqueous extract of *C. militaris*. UV-vis spectral analysis was used to validate the formation and stability of nanogold. The synthesized nanoparticles were used to characterize TEM, SEM, PSA, zeta potential analysis, ED-X-ray studies, XRD, and FTIR analysis. The gold nanoparticles possess significant antidiabetic, antioxidant, and antibacterial activities. Cytotoxicity study also confirmed the nontoxicity of *Cord*-Au-NPs. The current approach of edible-mushroom-mediated synthesis of nanoparticles deliberates the improved therapeutic potential and *Cord*-Au-NPs as a potential source of biomedical applications. This adapted synthesis method is simple, effective, environmentally friendly, nontoxic, and labor-saving.

Supplementary Materials: The following supporting information can be downloaded at: https://www.mdpi.com/article/10.3390/ph16091311/s1, Figure S1: Green synthesis of gold nanoparticles (a) initial color (b) color change after 1 h of incubation; Figure S2: Actual *vs* predicted plot for responses (A) Y_1, (B) Y_2 and (C) Y_3; Figure S3: Interaction plot for responses (A) Y_1, (B) Y_2 and (C) Y_3; Figure S4: Normal plots of residuals (A–C) and residual *vs* run plots (D–F) for responses Y_1 (A and D), Y_2 (B and E) and Y_3 (C and F); Figure S5: Overlay contour plot; Table S1: MICs of CME and *Cord*-Au-NPs against the studied bacterial strains; Table S2: Brine shrimp lethality results of *Cor*-Au-NPs.

Author Contributions: Conceptualization, S.G.; methodology, G.G. and M.A.; software, N.G. and M.K.; validation, D.H., V.M., and S.N.; formal analysis, N.G.; investigation, G.G.; resources, R.C. and M.S.K.; data curation, G.G. and M.K.; writing—original draft preparation, G.G. and M.A.; writing—review and editing, V.M.; visualization, D.H.; supervision, S.G.; project administration, S.G.; funding acquisition, M.S.K. and R.C. All authors have read and agreed to the published version of the manuscript.

Funding: The authors acknowledge the generous support from the Research Supporting Project (RSP2023R352) by King Saud University, Riyadh, Kingdom of Saudi Arabia.

Institutional Review Board Statement: Not applicable.

Informed Consent Statement: Not applicable.

Data Availability Statement: Data is contained within the article and supplementary material.

Conflicts of Interest: The authors declare no conflict of interest. The funders had no role in the design of the study; in the collection, analyses, or interpretation of data; in the writing of the manuscript; or in the decision to publish the results.

References

1. Sharma, B.; Purkayastha, D.D.; Hazra, S.; Gogoi, L.; Bhattacharjee, C.R.; Ghosh, N.N.; Rout, J. Biosynthesis of gold nanoparticles using a freshwater green alga, Prasiola crispa. *Mater. Lett.* **2014**, *116*, 94–97. [CrossRef]
2. Lee, K.X.; Shameli, K.; Miyake, M.; Kuwano, N.; Khairudin, N.B.B.A.; Mohamad, S.E.B.; Yew, Y.P. Green Synthesis of Gold Nanoparticles Using Aqueous Extract of *Garcinia mangostana* Fruit Peels. *J. Nanomater.* **2016**, *2016*, 8489094. [CrossRef]

3. Shankar, P.D.; Shobana, S.; Karuppusamy, I.; Pugazhendhi, A.; Ramkumar, V.S.; Arvindnarayan, S.; Kumar, G. A review on the biosynthesis of metallic nanoparticles (gold and silver) using bio-components of microalgae: Formation mechanism and applications. *Enzym. Microb. Technol.* **2016**, *95*, 28–44. [CrossRef] [PubMed]
4. Nayak, S.; Sajankila, S.P.; Goveas, L.C.; Rao, V.C.; Mutalik, S.; Shreya, B.A. Two fold increase in synthesis of gold nanoparticles assisted by proteins and phenolic compounds in Pongamia seed cake extract: Response surface methodology approach. *SN Appl. Sci.* **2020**, *4*, 634. [CrossRef]
5. Aljabali, A.A.A.; Akkam, Y.; Al Zoubi, M.S.; Al-Batayneh, K.M.; Al-Trad, B.; Alrob, O.A.; Alkilany, A.M.; Benamara, M.; Evans, D.J. Synthesis of Gold Nanoparticles Using Leaf Extract of *Ziziphus zizyphus* and their Antimicrobial Activity. *Nanomaterials* **2018**, *8*, 174. [CrossRef]
6. Rajeshkumar, S.; Malarkodi, C.; Gnanajobitha, G.; Paulkumar, K.; Vanaja, M.; Kannan, C.; Annadurai, G. Seaweed-mediated synthesis of gold nanoparticles using Turbinaria conoides and its characterization. *J. Nanostructure Chem.* **2013**, *3*, 44. [CrossRef]
7. Senapati, S.; Syed, A.; Moeez, S.; Kumar, A.; Ahmad, A. Intracellular synthesis of gold nanoparticles using alga Tetraselmis kochinensis. *Mater. Lett.* **2012**, *79*, 116–118. [CrossRef]
8. Zhang, G.; Huang, Y.; Bian, Y.; Wong, J.H.; Ng, T.B.; Wang, H. Hypoglycemic activity of the fungi Cordyceps militaris, Cordyceps sinensis, Tricholoma mongolicum, and Omphalia lapidescens in streptozotocin-induced diabetic rats. *Appl. Microbiol. Biotechnol.* **2006**, *72*, 1152–1156. [CrossRef]
9. Chen, X.; Wu, G.; Huang, Z. Structural analysis and antioxidant activities of polysaccharides from cultured Cordyceps militaris. *Int. J. Biol. Macromol.* **2013**, *58*, 18–22. [CrossRef]
10. Jędrejko, K.J.; Lazur, J.; Muszyńska, B. *Cordyceps militaris*: An Overview of Its Chemical Constituents in Relation to Biological Activity. *Foods* **2021**, *10*, 2634. [CrossRef]
11. Ji, Y.; Cao, Y.; Song, Y. Green synthesis of gold nanoparticles using a *Cordyceps militaris* extract and their antiproliferative effect in liver cancer cells (HepG2). *Artif. Cells Nanomed. Biotechnol.* **2019**, *47*, 2737–2745. [CrossRef] [PubMed]
12. Liu, L.; Jing, Y.; Guo, A.; Li, X.; Li, Q.; Liu, W.; Zhang, X. Biosynthesis of Platinum Nanoparticles with Cordyceps Flower Extract: Characterization, Antioxidant Activity and Antibacterial Activity. *Nanomaterials* **2022**, *12*, 1904. [CrossRef] [PubMed]
13. Wang, L.; Liu, C.-C.; Wang, Y.-Y.; Xu, H.; Su, H.; Cheng, X. Antibacterial activities of the novel silver nanoparticles biosynthesized using Cordyceps militaris extract. *Curr. Appl. Phys.* **2016**, *16*, 969–973. [CrossRef]
14. Li, J.F.; Rupa, E.J.; Hurh, J.; Huo, Y.; Chen, L.; Han, Y.; Ahn, J.C.; Park, J.K.; Lee, H.A.; Mathiyalagan, R.; et al. Cordyceps militaris fungus mediated Zinc Oxide nanoparticles for the photocatalytic degradation of Methylene blue dye. *Optik* **2019**, *183*, 691–697. [CrossRef]
15. Suksiriworapong, J.; Pongprasert, N.; Bunsupa, S.; Taresco, V.; Crucitti, V.C.; Janurai, T.; Phruttiwanichakun, P.; Sakchaisri, K.; Wongrakpanich, A. CD44-Targeted Lipid Polymer Hybrid Nanoparticles Enhance Anti-Breast Cancer Effect of *Cordyceps militaris* Extracts. *Pharmaceutics* **2023**, *15*, 1771. [CrossRef]
16. Rupa, E.J.; Li, J.F.; Arif, M.H.; Yaxi, H.; Puja, A.M.; Chan, A.J.; Hoang, V.-A.; Kaliraj, L.; Yang, D.C.; Kang, S.C. *Cordyceps militaris* Fungus Extracts-Mediated Nanoemulsion for Improvement Antioxidant, Antimicrobial, and Anti-Inflammatory Activities. *Molecules* **2020**, *25*, 5733. [CrossRef]
17. Dias, C.; Ayyanar, M.; Amalraj, S.; Khanal, P.; Subramaniyan, V.; Das, S.; Gandhale, P.; Biswa, V.; Ali, R.; Gurav, N.; et al. Biogenic synthesis of zinc oxide nanoparticles using mushroom fungus *Cordyceps militaris*: Characterization and mechanistic insights of therapeutic investigation. *J. Drug Deliv. Sci. Technol.* **2022**, *73*, 103444. [CrossRef]
18. Saoji, S.D.; Rarokar, N.R.; Dhore, P.W.; Dube, S.O.; Gurav, N.S.; Gurav, S.S.; Raut, N.A. Phospholipid Based Colloidal Nanocarriers for Enhanced Solubility and Therapeutic Efficacy of Withanolides. *J. Drug Deliv. Sci. Technol.* **2022**, *70*, 103251. [CrossRef]
19. Moradpoor, H.; Safaei, M.; Golshah, A.; Mozaffari, H.R.; Sharifi, R.; Imani, M.M.; Mobarakeh, M.S. Green synthesis and antifungal effect of titanium dioxide nanoparticles on oral Candida albicans pathogen. *Inorg. Chem. Commun.* **2021**, *130*, 108748. [CrossRef]
20. Fan, D.; Li, L.; Li, Z.; Zhang, Y.; Ma, X.; Wu, L.; Zhang, H.; Guo, F. Biosynthesis of selenium nanoparticles and their protective, antioxidative effects in streptozotocin induced diabetic rats. *Sci. Technol. Adv. Mater.* **2020**, *21*, 505–514. [CrossRef]
21. Rodrigues, K.; Gurav, S.; Joshi, A.; Krishna, M.; Bhandarkar, A. Porous polymeric carrier system for modified drug release of boswellic acid. *Chem. Sci. J.* **2020**, *11*, 1–12. [CrossRef]
22. Karuppiah, C.; Palanisamy, S.; Chen, S.-M.; Emmanuel, R.; Muthupandi, K.; Prakash, P. Green synthesis of gold nanoparticles and its application for the trace level determination of painter's colic. *RSC Adv.* **2015**, *5*, 16284–16291. [CrossRef]
23. Yuan, C.G.; Huo, C.; Gui, B.; Cao, W.P. Green synthesis of gold nanoparticles using Citrus maxima peel extract and their catalytic/antibacterial activities. *IET Nanobiotechnol.* **2017**, *11*, 523–530. [CrossRef] [PubMed]
24. Uthaman, S.; Kim, H.S.; Revuri, V.; Min, J.-J.; Lee, Y.-K.; Huh, K.M.; Park, I.-K. Green synthesis of bioactive polysaccharide-capped gold nanoparticles for lymph node CT imaging. *Carbohydr. Polym.* **2018**, *181*, 27–33. [CrossRef] [PubMed]
25. Dessai, S.; Ayyanar, M.; Amalraj, S.; Khanal, P.; Vijayakumar, S.; Gurav, N.; Rarokar, N.; Kalaskar, M.; Nadaf, S.; Gurav, S. Bioflavonoid mediated synthesis of TiO_2 nanoparticles: Characterization and their biomedical applications. *Mater. Lett.* **2022**, *311*, 131639. [CrossRef]
26. Singh, P.; Kim, Y.J.; Yang, D.C. A strategic approach for rapid synthesis of gold and silver nanoparticles by Panax ginseng leaves. *Artif. Cells Nanomed. Biotechnol.* **2016**, *44*, 1949–1957. [CrossRef]

27. Veeramani, S.; Narayanan, A.P.; Yuvaraj, K.; Sivaramakrishnan, R.; Pugazhendhi, A.; Rishivarathan, I.; Jose, S.P.; Ilangovan, R. Nigella sativa flavonoids surface coated gold NPs (Au-NPs) enhancing antioxidant and anti-diabetic activity. *Process. Biochem.* **2021**, *114*, 193–202. [CrossRef]
28. Guo, Y.; Jiang, N.; Zhang, L.; Yin, M. Green synthesis of gold nanoparticles from Fritillaria cirrhosa and its anti-diabetic activity on Streptozotocin induced rats. *Arab. J. Chem.* **2020**, *13*, 5096–5106. [CrossRef]
29. Milanezi, F.G.; Meireles, L.M.; de Christo Scherer, M.M.; de Oliveira, J.P.; da Silva, A.R.; de Araujo, M.L.; Endringer, D.C.; Fronza, M.; Guimarães, M.C.C.; Scherer, R. Antioxidant, antimicrobial and cytotoxic activities of gold nanoparticles capped with quercetin. *Saudi Pharm. J.* **2019**, *27*, 968–974. [CrossRef]
30. Zayadi, R.A.; Abu Bakar, F.; Ahmad, M.K. Elucidation of synergistic effect of eucalyptus globulus honey and Zingiber officinale in the synthesis of colloidal biogenic gold nanoparticles with antioxidant and catalytic properties. *Sustain. Chem. Pharm.* **2019**, *13*, 100156. [CrossRef]
31. Sathishkumar, G.; Pradeep, K.; Vignesh, V.; Rajkuberan, C.; Jeyaraj, M.; Selvakumar, M.; Rakhi, J.; Sivaramakrishnan, S. Cannonball fruit (*Couroupita guianensis*, Aubl.) extract mediated synthesis of gold nanoparticles and evaluation of its antioxidant activity. *J. Mol. Liq.* **2016**, *215*, 229–236.
32. Lydia, D.E.; Khusro, A.; Immanuel, P.; Esmail, G.A.; Al-Dhabi, N.A.; Arasu, M.V. Photo-activated synthesis and characterization of gold nanoparticles from Punica granatum L. seed oil: An assessment on antioxidant and anticancer properties for functional yoghurt nutraceuticals. *J. Photochem. Photobiol. B Biol.* **2020**, *206*, 111868. [CrossRef]
33. Nagalingam, M.; Kalpana, V.; Devi Rajeswari, V.; Panneerselvam, A. Biosynthesis, characterization, and evaluation of bioactivities of leaf extract-mediated biocompatible gold nanoparticles from *Alternanthera bettzickiana*. *Biotechnol. Rep.* **2018**, *19*, e00268.
34. Vinosha, M.; Palanisamy, S.; Muthukrishnan, R.; Selvam, S.; Kannapiran, E.; You, S.; Prabhu, N.M. Biogenic synthesis of gold nanoparticles from Halymenia dilatata for pharmaceutical applications: Antioxidant, anti-cancer and antibacterial activities. *Process. Biochem.* **2019**, *85*, 219–229. [CrossRef]
35. Veiga, A.; Toledo, M.d.G.T.; Rossa, L.S.; Mengarda, M.; Stofella, N.C.; Oliveira, L.J.; Gonçalves, A.G.; Murakami, F.S. Colorimetric microdilution assay: Validation of a standard method for determination of MIC, IC50%, and IC90% of antimicrobial compounds. *J. Microbiol. Methods* **2019**, *162*, 50–61. [CrossRef] [PubMed]
36. Hosny, M.; Fawzy, M.; Abdelfatah, A.M.; Fawzy, E.E.; Eltaweil, A.S. Comparative study on the potentialities of two halophytic species in the green synthesis of gold nanoparticles and their anticancer, antioxidant and catalytic efficiencies. *Adv. Powder Technol.* **2021**, *32*, 3220–3233. [CrossRef]

pharmaceuticals

MDPI

Review

A Review on Biomedical Applications of Plant Extract-Mediated Metallic Ag, Au, and ZnO Nanoparticles and Future Prospects for Their Combination with Graphitic Carbon Nitride

Priyanka Panchal [1,*], Protima Rauwel [2], Satya Pal Nehra [3], Priyanka Singh [4], Mamta Karla [5], Glemarie Hermosa [1] and Erwan Rauwel [1,*]

1 Institute of Veterinary Medicine and Animal Sciences, Estonian University of Life Sciences, 51006 Tartu, Estonia; glemarie.hermosa@emu.ee

2 Department of Aeronautical Engineering, Estonian Aviation Academy, 61707 Tartu, Estonia; protima.rauwel@eava.ee

3 Center of Excellence for Energy and Environmental Studies, Deenbandhu Chhotu Ram University of Science and Technology, Sonipat 131039, India; spnehra.energy@dcrustm.org

4 Faculty of Management Studies, University of Delhi, Delhi 110007, India; drpriyancka@gmail.com

5 Computer Science and Engineering Department, Deenbandhu Chhotu Ram University of Science and Technology, Sonipat 131039, India; mamtakarla21@gmail.com

* Correspondence: priyanka.panchal@emu.ee (P.P.); erwan.rauwel@emu.ee (E.R.)

Abstract: Since the publication of the 12 principles of green chemistry in 1998 by Paul Anastas and John Warner, the green synthesis of metal and metal oxide nanoparticles has emerged as an eco-friendly and sustainable alternative to conventional chemical methods. Plant-based synthesis utilizes natural extracts as reducing and stabilizing agents, minimizing harmful chemicals and toxic by-products. Ag nanoparticles (Ag-NPs) exhibit strong antibacterial activity; Au nanoparticles (Au-NPs) are seen as a promising carrier for drug delivery and diagnostics because of their easy functionalization and biocompatibility; and ZnO nanoparticles (ZnO-NPs), on the other hand, produce reactive oxygen species (ROS) that kill microorganisms effectively. These nanoparticles also demonstrate antioxidant properties by scavenging free radicals, reducing oxidative stress, and preventing degenerative diseases. Green syntheses based on plant extracts enhance biocompatibility and therapeutic efficacy, making them suitable for antimicrobial, anticancer, and antioxidant applications. Applying a similar "green synthesis" for advanced nanostructures like graphitic carbon nitride (GCN) is an environmentally friendly alternative to the traditional ways of doing things. GCN exhibits exceptional photocatalytic activity, pollutant degradation efficiency, and electronic properties, with applications in environmental remediation, energy storage, and biomedicine. This review highlights the potential of green-synthesized hybrid nanocomposites combining nanoparticles and GCN as sustainable solutions for biomedical and environmental challenges. The review also highlights the need for the creation of a database using a machine learning process that will enable providing a clear vision of all the progress accomplished till now and identify the most promising plant extracts that should be used for targeted applications.

Keywords: green synthesis; silver nanoparticles; gold nanoparticles; zinc oxide nanoparticles; graphitic carbon nitrides; nanocomposite; antimicrobial activity; antioxidant activity; cytotoxicity

1. Introduction

In contrast to conventional chemical and physical synthesis techniques, the green synthesis of nanoparticles (NPs), particularly silver (Ag-NPs) [1–4], gold (Au-NPs) [5–9], and zinc oxide (ZnO-NPs) [10–14], has become an important and sustainable alternative [15–17]. The production of conventional NPs usually entails the use of hazardous chemicals and high temperatures. These processes not only take a lot of energy but also produce toxic by-products that may pollute the environment [18–22]. The primary advantages of green synthesis are the use of renewal resources and its capacity to generate nanoparticles with enhanced properties [23–26]. Bioactive substances in plant extracts, such as polyphenols, flavonoids, terpenoids, and alkaloids, function as reducing agents and stabilizers, facilitating the regulated and quick synthesis of nanoparticles without the need of harmful chemicals [27–30]. These phytochemical compounds improve the stability, solubility, and dispersibility of the nanoparticles, which are crucial for their properties and efficacy in many applications [31–35].

Moreover, the antimicrobial properties of green-synthesized nanoparticles are of interest because these nanoparticles have significant antimicrobial efficacy against a wide range of pathogens, including bacteria, fungi, and viruses [36–40]. Their mechanisms of action often entail the production of reactive oxygen species (ROS), resulting in oxidative damage to bacterial cell membranes, proteins, and DNA, ultimately causing cell death [38–40]. By minimizing dependence on synthetic chemicals, green synthesis guarantees that these nanoparticles are safer for environmental and human health. Beyond their antibacterial characteristics, Ag-NPs, Au-NPs, and ZnO-NPs also exhibit antioxidant activity. For example, Au-NPs have shown the ability to promote the production of antioxidant enzymes, including superoxide dismutase (SOD) and catalase, which mitigate detrimental free radicals, thereby safeguarding cells from oxidative injury. These qualities are exceedingly advantageous for preventive healthcare, underscoring the significance of the plant-based synthesis in enhancing human health [8,9,26,41–47]. Another notable benefit of green-synthesized Ag-NPs, Au-NPs, and ZnO-NPs is their potential for targeted cancer therapy, attributed to their nano size, extensive surface area, and facile functionalization with biomolecules, rendering them exemplary candidates for selective interactions with cancer cells [7,8,14,19,48–53].

Research indicates that these nanoparticles may trigger death in cancer cells by mechanisms including DNA damage, mitochondrial malfunction, and reactive oxygen species production [14,19,31]. The capacity to functionalize these nanoparticles with particular ligands such as Folic Acid, Hyaluronic Acid, Galactose, and Cholesterol-based ligands, which target cancer cells overexpressing folate receptors, facilitates targeted medication delivery, minimizing adverse effects and enhancing therapeutic success. This focused methodology is facilitated by green synthesis that can alter the surface of nanoparticles to improve their specificity and selectivity in biological applications [54–56]. Moreover, clinical trials have already shown that plant extract-mediated nanoparticle synthesis is a viable approach in the field of nanomedicine [57,58]. Currently, most nanomaterials used in nanomedicine are synthesized using conventional methods. However, these traditional synthesis processes are often complex and/or expensive [58,59]. In contrast, plant-based synthesis methods offer a promising, cost-effective alternative with the potential to enhance nanoparticle properties. Therefore, it is crucial to identify the most suitable plant extracts and synthesis protocols. This can be achieved through the application of machine learning tools to develop a comprehensive database. Such a database could provide insights into optimal synthesis conditions or highlight areas for improvement to ensure that plant-mediated nanoparticles match the performance of those produced by conventional methods [60–62].

In a similar way, graphitic carbon nitride (GCN) has become a particularly appealing material in recent years owing to its many possible applications in photocatalysis, energy storage, environmental remediation, and biomedicine. Its distinctive structure, superior thermal stability, and exceptional optical qualities make it a material of considerable worldwide interest [63–67]. Currently, GCN is primarily synthesized using nitrogen-rich chemical reagents such as urea, melamine, cyanamide, and dicyandiamide. In contrast, green synthesis emphasizes the use of nitrogen-rich plant-based precursors, aligning with the principles of sustainable and eco-friendly chemistry while reducing environmental impact.

2. Different Synthesis Methods Used for Nanoparticles Investigation

Top–Down Approach: The top–down approach to NPs synthesis involves the use of physical methods to break down bulk materials into nanoscale particles. Physical methods include ball milling, evaporation–condensation [68–72], arc discharge [70,73], laser ablation [70], and spray pyrolysis [71,72]. These methods offer the advantage of producing NPs with minimal chemical contamination, ensuring high-quality end products. However, these physical methods are often associated with significant drawbacks. They are typically energy-intensive and expensive due to the need for sophisticated equipment and high operational costs.

Bottom–Up approach: The chemical synthesis process for NPs synthesis involves the reduction of metal ions into zero-valent atoms and leads to the formation of NPs typically ranging from 1 to 100 nm. In this method, chemical reagents play a crucial role in reducing metal ions and promoting the growth of crystalline structures. However, it is often necessary to incorporate additional stabilizers to uphold the stability of the nanoparticles, preventing aggregation and ensuring a controlled particle size and shape, which is crucial for attaining the desired physiological properties [73–76]. The chemical synthesis process, while versatile, presents several limitations. Common chemical methods encompass conventional reduction, sol-gel processes [77], chemical vapor deposition [78], reverse micelle synthesis, co-precipitation, electrochemical reduction, and solvothermal methods [79]. Conversely, green synthesis utilizes living organisms and bio-derived compounds, including plant extracts, microorganisms (such as bacteria, fungi, and viruses), and enzymes to generate nanoparticles [80].

This approach presents notable benefits compared to conventional chemical methods, as it eliminates the need for hazardous substances, promotes more eco-friendly processes, and can result in nanoparticles that exhibit enhanced biocompatibility and stability [81–83]. The approach provides streamlined, secure, and eco-conscious methods while preserving oversight of nanoparticle properties, leading to its growing preference in biomedical applications and various other sectors [84–91], as shown in Figure 1 and Table 1.

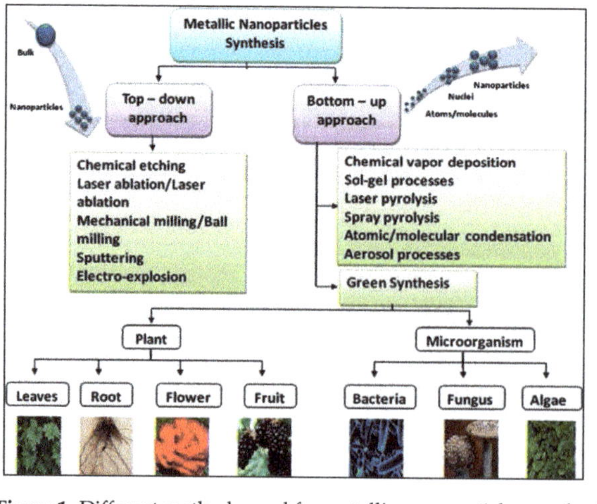

Figure 1. Different methods used for metallic nanoparticles synthesis [22].

Table 1. Advantages and Drawbacks of Physical and Chemical Methods.

Methods	Advantages	Disadvantages	Refs.
Physical Methods			
Ball milling	Low cost	Agglomeration of nanoparticles with low production	[68–72]
Evaporation	Large production of nanoparticles	Costly; Time-consuming; high operating temperature	[70]
Arc discharge	Found high purity of nanoparticles; large production	Large size distribution of nanoparticles	[70–73]
Laser ablation	Found small size of nanoparticles; narrow size distribution	Very expensive, large amount of energy used; Less production of nanoparticles	[70]
Spray pyrolysis	Found high purity of nanoparticles	Require very high operating temperature	[71,72]
Chemical methods			
Sol-gel process	High-purity nanoparticles synthesized	Limited industrial applications; costly precursors; time-consuming process; difficult to handle	[77]
Chemical vapor deposition	Single deposition step is required for nanoparticles synthesis	High process cost; difficult to handle; Low yield	[78]
Reverse micelle	Simple process and easy to handle	Nanoparticle's production is less with large size; toxic and hazardous chemicals are used as a stabilizing agent	[79]

Merits of Biological Methods for Sustainable Approach

Biological methods utilize natural resources, like plant extracts [88], bacteria [72,88], fungi, or algae [22,72,88], that act as reducing and stabilizing agents. They eliminate the use of hazardous chemicals, reducing environmental pollution [92–96]. Their non-toxicity makes them highly biocompatible, minimizing side effects and making them ideal for therapeutic applications like antibacterial and anticancer treatments [95–98]. In terms of cost-effectiveness, since biological approaches need softer reaction conditions, including ambient temperature and standard pressure, it therefore reduces energy consumption, thus making it cost-effective and more environmental friendly (i.e., carbon footprint). In fact, plant-based synthesis is particularly cost-effective, as easily accessible raw materials such as leaves, stems, and fruits [99–102]. Biological synthesis is also highly scalable and simple, requiring minimal equipment compared to physical and chemical methods. Moreover, the nanoparticles produced through this approach exhibit enhanced stability, as natural capping agents like proteins, carbohydrates, and polyphenols prevent aggregation and ensure prolonged efficacy [102–110]. Lastly, these methods support sustainability by reducing reliance on toxic chemical precursors, instead utilizing plant materials and agricultural waste, thereby lowering environmental risks and promoting a circular economy [111–122].

3. Methodology of Plant-Based Synthesis of Nanoparticles and Their Mechanisms

Many plant parts (such as fruits, leaves, stem, roots, and flowers) are rich in bioactive compounds like polyphenols, flavonoids, and alkaloids, which can act as both reducing and stabilizing agents [115–124].

Preparation of plant extract: The plant material is washed, dried, and chopped. The chopped material is boiled in distilled water to extract the bioactive compounds [125–128]. Addition of Metal Salts: A solution of metal salts (e.g., silver nitrate for Ag-NPs, gold chloride for Au-NPs, and zinc nitrate for ZnO-NPs) is prepared, preferably in distilled water. The biological extract is incorporated into the metal salt solution. The bioactive chemicals in the extract reduce metal ions to metal atoms, hence starting the formation of nanoparticles, starting by nucleation [129–132].

Reduction Process: The bioactive molecules in the plant extract act as reducing agents. They reduce metal ions (e.g., Ag^{+}, Au^{3+}, Zn^{2+}) to form metal nanoparticles (e.g., Ag^{0}, Au^{0}, Zn). The reduction process typically results in a color change in the solution, indicating nanoparticle formation [133–137].

Stabilization and Size Control: The bioactive compounds in the extract do not only reduce the metal ions but also stabilize the NPs by preventing agglomeration. Stabilizing agents such as proteins, polysaccharides, and other compounds are responsible for controlling the size and preventing the particles from clustering together [138–141].

Purification: Following nanoparticle synthesis, the solution is often processed using centrifugation, filtering, or dialysis to eliminate surplus biological material, unreacted metal ions, and other contaminants. The pure nanoparticles may then be dried and characterized for further investigations, as seen in (Figure 2a,b) [142–147].

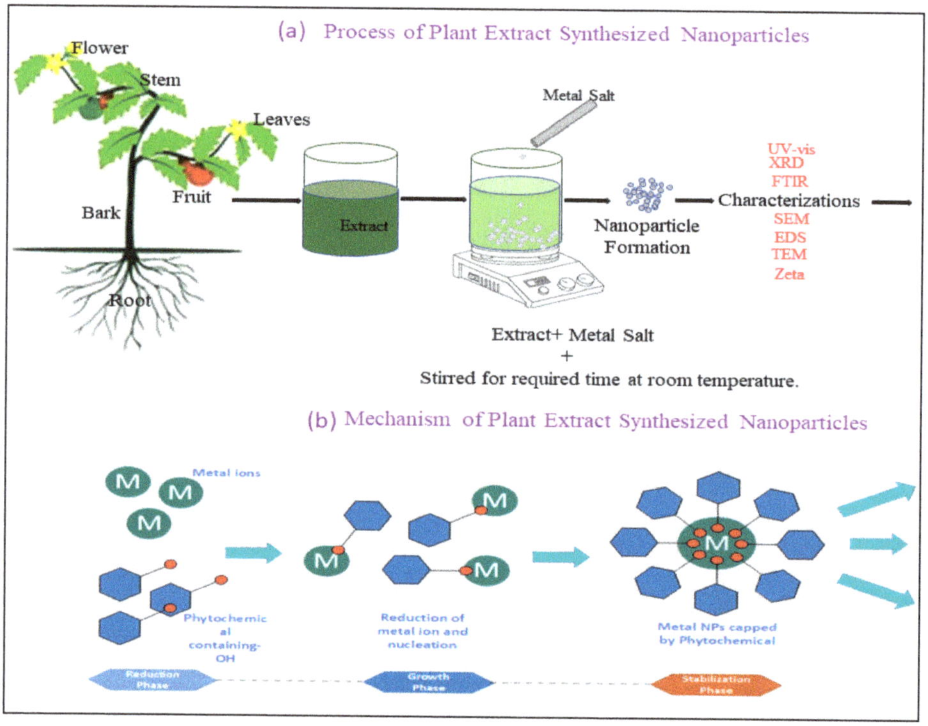

Figure 2. Process and mechanism of plant extract-synthesized nanoparticles.

3.1. Characterization Methods to Control the Stability and the Properties

The NPs are characterized to determine their size, shape, surface charge, and chemical composition. Some common characterization methods include UV-Vis Spectroscopy to observe the characteristic surface plasmon resonance (SPR) peak, indicating NPs formation;

transmission Electron Microscopy (TEM) to examine the size, shape, and morphology of the NPs; Scanning Electron Microscopy (SEM) to analyze surface structure; X-ray Diffraction (XRD) to determine the crystalline structure and phase, and to estimate the size using Scherrer equation; Fourier Transform Infrared Spectroscopy (FTIR) to identify the functional groups responsible for the reduction and stabilization.

UV-Vis spectrophotometry shows that particle size influences absorption, with smaller nanoparticles causing a blue shift (shorter wavelengths) and larger ones causing a red shift (longer wavelengths). For example, Au-NPs of around 10 nm exhibit absorption peaks near 520 nm, whereas particles larger than 80 nm shift the peak towards 600 nm due to increased scattering and plasmon resonance effects. Shape also influences the absorption, as spherical nanoparticles exhibit a single absorption peak, while anisotropic shapes like nanorods show multiple peaks at different wavelengths [5,7,21,49,65,106,147]. Aggregation of nanoparticles leads to plasmonic coupling, resulting in a red shift. Additionally, the refractive index of the surrounding medium affects the peak position, with a higher refractive index causing a red shift and a lower one a blue shift. Ag-NPs, Au-NPs, and ZnO-NPs exhibit distinct absorption peaks between 400 and 430 nm, 520 and 550 nm, and 375 and 400 nm, as seen in Figure 3a–c and Table 2. These UV-vis results serve as robust evidence for confirming the formation of green-synthesized Ag, Au, and ZnO NPs.

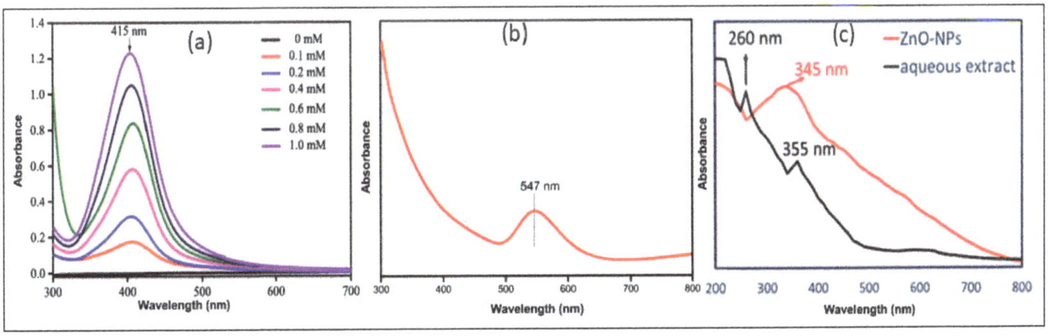

Figure 3. UV-vis analysis of plant extract-synthesized Ag-NPs (**a**) [21]; Au-NPs (**b**) [20]; and ZnO-NPs (**c**) [130].

Table 2. Summarized the UV-Vis absorption reported work on Ag-NPs, Au-NPs, and ZnO-NPs from different parts of the plant for biomedical applications.

Plant Species	Absorption (nm)	Refs.
	Ag Nanoparticles	
Euphorbia serpens Kunth	420	[2]
Areca catechu nut	415	[21]
Carica papaya	419	[92]
Ferula asafoetida	410–420	[101]
Ipomoea carnea Jacq.	390–410	[108]
	Au Nanoparticles	
Crataegus oxyacantha	520–530	[1]
Moringa oleifera	535	[6]
Flaxseed	540	[7]
Curcuma pseudomontana	542	[9]
Spondias dulcis	536	[49]

Table 2. *Cont.*

Plant Species	Absorption (nm)	Refs.
	ZnO Nanoparticles	
Phlomis Leaf	360	[29]
Cymbopogon citratus	370	[54]
Mentha piperata	385	[56]
Rivina humilis	370	[131]

FTIR spectroscopy analysis of Ag-NPs, Au-NPs, and ZnO-NPs synthesized from plant extracts reveals the presence of key functional groups. In the case of Ag-NPs synthesized from banana pulp [55], *Cassia tora* seed [95], *Jasminum nudiflorum* flower [109], Sesbania grandiflora leaves [124], *Vaccinium arctostaphylos* fruit [129], and peel, extracts show peaks at 820 C-H vibrations of cellulose, 502 C-H bending vibration, Ag-O bonds, 1635 to 1645 cm^{-1} ketone group due to -C=O stretching alkynes due to C-C stretching, and alcohol or phenol groups due to -OH stretching in the bands at 2134 cm^{-1} and 3317 cm^{-1}, respectively, as shown in (Figure 4a). Similarly, Au-NPs, commonly reveal peaks appear at 3200–3400 cm^{-1} (O-H stretch) hydroxyl groups, 1600 cm^{-1} (C=O stretch) carbonyl groups, 1400–1500 cm^{-1} (C-H bending) aromatic or aliphatic groups, 1000–1200 cm^{-1} (C-O stretch) alcohols, phenols, ethers, and 600–800 cm^{-1} (C-H bending) aromatic rings, corresponding to different plant extracts such as *Curcuma pseudomontana* flower [9], *Glaucium flavum* leaves [15], *Citrus* peel [32], and *Delphinium chitralense* tuber [82], as shown in (Figure 4b). Moreover, ZnO-NPs synthesized from *Lantana camara* flower [24], *Lepidium sativum* seed [31], *Ipomoea sagittifolia* Burm.f leaves [53], *Olive* fruit [86], and *Punica granatum* peel [130] typically appear at 3400–3500 cm^{-1} (O-H stretch), 1600 cm^{-1} (O-H bending), 500–700 cm^{-1} (Zn-O stretch), and 1100–1300 cm^{-1} (C-O stretch), indicating the presence of hydroxyl groups, water, and capping agents involved in stabilization, as shown in (Figure 4c). The findings underscore the importance of FTIR to identify the essential function of plant-derived bioactive compounds, including flavonoids, phenols, and proteins on the surface of green-synthesized nanoparticles, hence enabling their many biological and environmental uses [17,37,71].

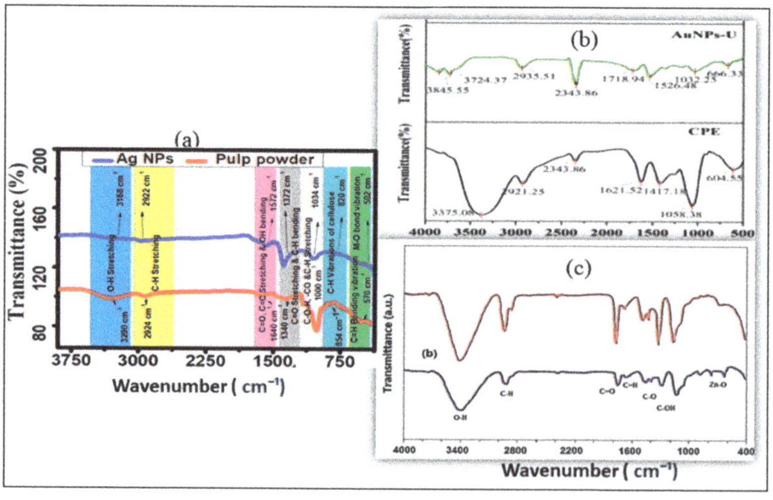

Figure 4. FTIR analysis of plant extract-synthesized Ag-NPs (**a**) [55]; Au-NPs (**b**) [32]; and ZnO-NPs (**c**) [130].

XRD study is a key routine method to characterize the crystalline structure and phase purity of Ag-NPs, Au-NPs, and ZnO-NPs produced by plant extracts and the estimation of the crystallite size, as seen in (Figure 5a–c). Distinct diffraction peaks for Ag-NPs were observed at 37–38° (111), 44–47° (200), 63–64.5° (220), and 77–79° (311), confirming the face-centered cubic structure, as evidenced in various studies involving extracts from Abelmoschus esculentus flower [27], *Withania coagulans* seed [35], banana pulp [54], *Carica papaya* leaf [92], and *Rosa canina* fruit [122]. Similarly, Au-NPs display characteristic peaks at 2θ values corresponding to 37–39° (111), 44–45° (200), 64–67° (220), and 77–78° (311), confirming the face-centered cubic crystalline structure in different plant extract-mediated Au-NPs, such as *Capsicum chinense* Leaves, stem and root [2]; *Flaxseed* seed [7]; *Jatropha integerrima* Jacq. flower [20]; and *Citrus* peel [32]. In addition, for XRD, peaks for ZnO-NPs typically appear at 28–32° (100), 34–35° (002), 36–39° (101), 44–48° (102), 56–57° (110), 62–63° (103), 66–67° (112), 72–73° (004), and 77–78° (220), confirming the hexagonal wurtzite structure with various reported studies, such as *Phlomis* leaf [29], *Andrographis alata* whole plant [13], Plantain peel [34], *Caesalpinia crista* seed [44], and *Capparis spinosa* fruit [52].

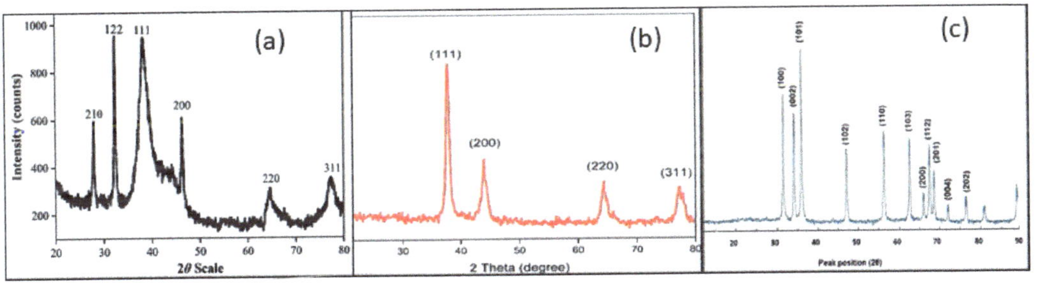

Figure 5. XRD analysis of plant extract-synthesized Ag-NPs (**a**) [21]; Au-NPs (**b**) [20]; and ZnO-NPs (**c**) [29].

The XRD findings conclusively demonstrate the successful production of crystalline nanoparticles with unique and stable stuctures, hence improving their efficacy in biological, catalytic, and environmental applications [8,36,93,135,137].

3.2. Morphology and Size Distribution

SEM + EDS (morphology and elemental distribution), TEM and DLS (morphology and size), and zetapotential (stability) can provide valuable insights information of plant extract-synthesized Ag-NPs, Au-NPs, and ZnO-NPs, as shown in Figure 6I(a–f). In the case of Ag-NPs, they show predominantly spherical, regular, circular shapes, and sizes with high stability, such as those synthesized from *Eucalyptus globulus* and *Salvia officinalis* (Spherical) [4], *Aervalanata* (Spherical) [17], *Areca catechu* nut (Regular spherical) [21], *Allium cepa* (spherical) [30], *Ferula asafoetida* (Circular) [101], and *Ocimumcanum* (Rod and Spherical) [115], demonstrating well-dispersed and stable nanoparticles. For Au-NPs, they mainly show spherical, hexagonal, and circular particles like in the case of studies on *Crataegus oxyacantha* (Spherical) [1], flaxseed (Spherical and Triangular) [7], *Jatropha integerrima* Jacq. (Spherical) [20], *Licorice* (Circular) [25], and saffron (Spherical and Oval) [51]. Similarly, ZnO-NPs synthesized from *Andrographis alata* (Spherical, Oval, and Hexagonal) [13], *Berberis aristate* (Needle) [23], *Cymbopogon citratus* (Spherical) [54], *Phlomis* leaf (Hexagonal) [85], *Punica granatum* peel (Spherical and Hexagonal) [90], and *Capsicum chinense* (Spherical) [94] extracts revealed a variety of shapes, including spherical, irregular, and hexagonal, with a strong influence of plant compounds in controlling nanoparticle morphology during the growth. These SEM and TEM observations underscore the effectiveness of plant-mediated green synthesis methods in controlling the size, shape, and uniformity of nanoparticles, which are crucial for their diverse applications in antimicrobial, antioxidant, catalytic, and biomedical fields. According to reported studies, particle size can vary depending on the characterization technique used, such as XRD, DLS, or TEM.

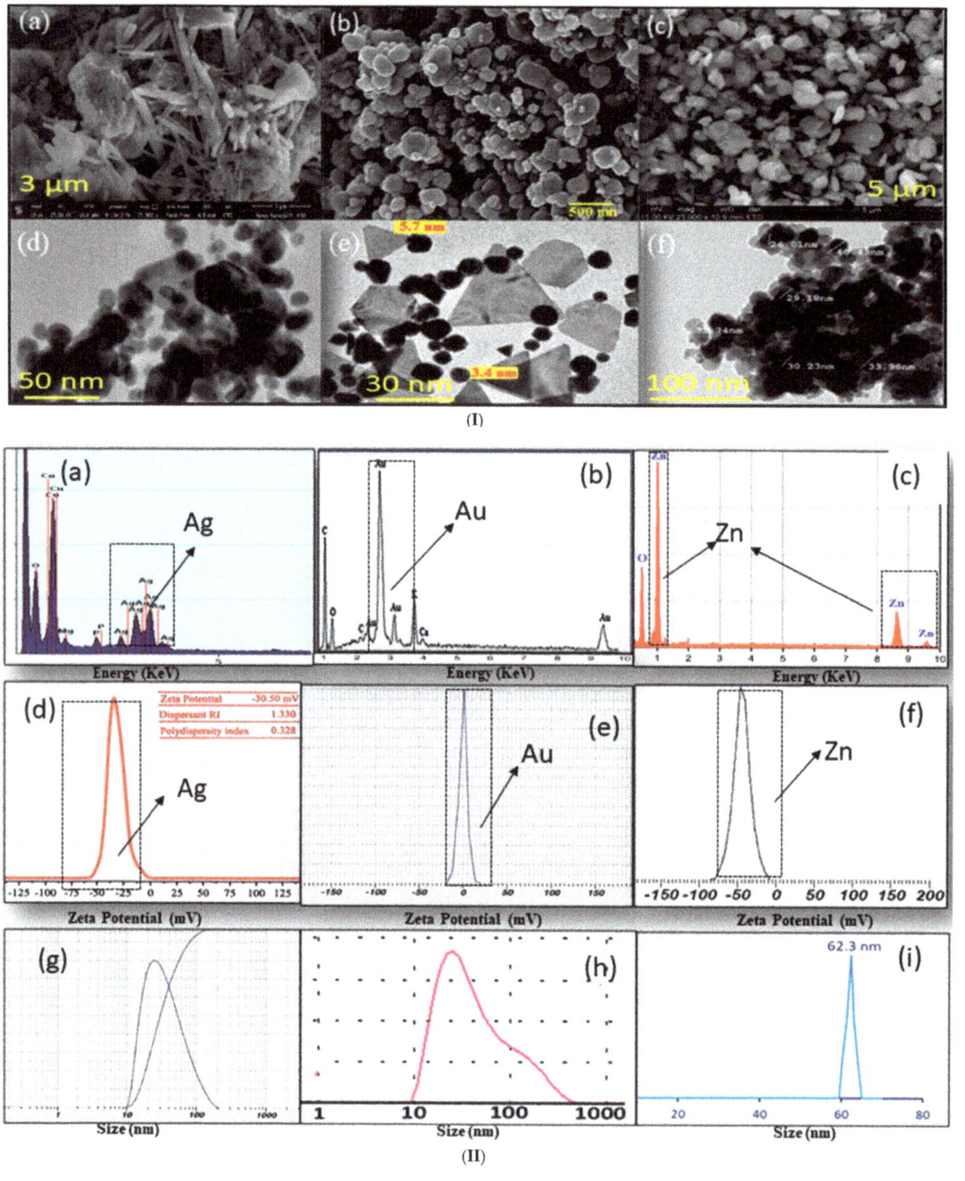

Figure 6. (**I**) SEM and TEM studies of plant extract-synthesized Ag-NPs ((**a**) [115], (**d**) [4]); Au-NPs ((**b**) [25], (**e**) [7]); and ZnO-NPs ((**c**) [94], (**f**) [90]). (**II**) EDS, zeta Potential, and DLS analysis of plant extract-synthesized Ag-NPs ((**a**) [4], (**d**) [21], (**g**) [100]); Au-NPs ((**b**) [7], (**e**) [20], (**h**) [6]); and ZnO-NPs ((**c**) [54], (**f**) [29], (**i**) [130]).

EDS is used to analyze the elemental composition of nanoparticles. Ag-NPs strong peak at approximately at 3.0 keV (Ag L_α); Au-NPs 2.2 keV (Au M_α), with additional peaks at 9.7 keV (Au L_α) and 11.4 keV (Au L_β); and ZnO-NPs peak at 1.0 keV (Zn K_α) are visible, as shown in Figure 6II(a–c) [4,7,54]. Zeta potential is used to evaluate the stability and dispersion of nanoparticles. However, Ag-NPs and Au-NPs exhibit values between −20 and −40 mV under pH 7–9, and ZnO-NPs range from −15 to −30 mV under pH 9–10. These values indicate stable, well-dispersed nanoparticles and good colloidal stability, as shown in Figure 6II(d–f) [4,7,20,21,29,54]. Moreover, DLS was used to determine the hydrodynamic size distribution of Ag-NPs, Au-NPs, and ZnO-NPs synthesized using a plant extract. This technique provides valuable insight into the dispersion quality and stability of the nanoparticles in solution, essential for biological and catalytic applications, as shown in Figure 6II(g–i) [6,100,130]. However, the main limitation using DLS is in the case of agitated nanoparticles or very small nanoparticles (<50 nm), which is usually the case for nanoparticles synthesized by plant extract methods.

4. Literature Study

During the last 10 years, many studies reported the synthesis of Ag-NPs, Au-NPs, and ZnO-NPs using plant extract-based synthesis. Ag-NPs and ZnO-NPs are known for their strong antibacterial properties, mainly attributed to their capacity to produce reactive oxygen species (ROS) under light exposure that interfere with microbial cell membranes and DNA. Au-NPs nanoparticles are more known for their compatibility with biological systems and their use in drug delivery and diagnostic applications. These nanoparticles demonstrate significant antioxidant activity, effectively scavenging free radicals and lowering oxidative stress, which positions them as beneficial agents in the fight against degenerative diseases. Studies consistently emphasize the environmentally friendly, cost-effective, and versatile nature of green-synthesized NPs, showcasing their immense potential for applications in biomedicine, environmental remediation, and other fields, as illustrated in Tables 3–5.

Table 3. Summarized the reported work on Ag-NPs from different parts of the plant for biomedical applications.

Plant Species	Part of Plant	Size/Shape	Application	Ref.
Euphorbia serpens	Leaves	~30–80 nm (Spherical)	Antimicrobial and Antibiofilm Activities	[2]
Lawsonia inermis	Leaves	~40 nm (Spherical)	Antimicrobial Activity	[3]
Eucalyptus globulus and *Salvia officinalis*	Leaves	~17.5 ± 5.89 nm and 34.3 ± 7.76 nm (Spherical)	Antioxidant and Antimicrobial Activities	[4]
Aerva lanata	Flower	~7 ± 3 nm (Spherical)	Antibacterial and Antioxidant Activities	[17]
Areca catechu nut	Fruit	~15–20 nm (Regular spherical)	Antioxidant and Antibacterial Activities	[21]
Abelmoschus esculentus	Flower	~5.52 to 31.96 nm (Spherical)	Antimicrobial Activity	[27]
Acacia cyanophylla	Leaves, Flowers and stems	~88.11 nm (Spherical)	Antibacterial Activity	[28]

Table 3. *Cont.*

Plant Species	Part of Plant	Size/Shape	Application	Ref.
Allium cepa L.	Bulb	~19.47 ± 1.12 nm (spherical)	Antioxidant, Antipathogenic, Anticholinesterase	[30]
Eupatorium adenophorum	Leaves	~117.75 nm (Spherical)	Antioxidant and Antibacterial Activities	[33]
Withania coagulans	seeds	~25 nm (spherical)	Antioxidant Activity	[35]
Camellia sinensis	Leaves	~8–26 nm (Spherical)	Antibacterial Activity	[37]
Ocimum basilicum L.	Leaves and stem	~35 nm (Oval)	Anticancer Activity	[38]
Moringa oleifera	Leaves	~25.235 ± 0.694 nm (spherical)	Antibacterial Activity	[40]
Annona muricata	Leaves	~19.63 ± 3.7 nm and 16.56 ± 4.1 nm	Antitumor Activity	[43]
Banana	Pulp	~42.97 nm (Spherical)	Antibacterial and Bioelectricity generation Activities	[55]
Cupressus macrocarpa	Leaves	~13.5–25.8 nm (Spherical)	Antibacterial Activity	[71]
Aloe vera	Leaves	~30–80 nm (spherical)	Antibacterial Activity	[83]
Origanum majorana L.	Leaves	~72.01 nm (Spherical)	Antioxidant Activity	[84]
Carica papaya	Leaves	10–25 nm (Round)	Anticancer Activity	[92]
Cassia tora	Seed	~50–60 nm (spherical)	Antibacterial Activity	[95]
Eugenia roxburghii	Leaves	~19–39 nm (Spherical)	Antimicrobial Activity	[96]
Euphorbia hirta	Leaves	~90–120 nm (spherical)	Antimicrobial Activity	[99]
Flaxseed	Seed	~46.98 ± 12.45 nm (Spherical)	Antibacterial and Antioxidant Activity	[100]
Ferula asafoetida	Leaves	~10 ± 2.77 nm (Circular)	Antibacterial and Cytotoxicity Activities	[101]
Ocimum tenuiflorum	Leaves	~10–65 nm (Round)	Helicoverpa Armigera Activity	[102]
Green Tea	……	~50 nm (spherical)	Antimicrobial Activity	[103]
Grape seed	Seed	~10 to 30 nm (spherical)	Antimicrobial Activity	[105]
Heteropyxis natalensis	Leaves	~5–60 nm (spherical)	Antibacterial Activity	[106]
Hibiscus rosasinensis	Leaves, Flower, and Bark	~200 nm to 1 μm (Spherical)	Antimicrobial Activity	[107]
Ipomoea carnea Jacq.	Leaves	~11.21–46.90 nm (Spherical)	Antimicrobial Activity	[108]

Table 3. *Cont.*

Plant Species	Part of Plant	Size/Shape	Application	Ref.
Jasminum nudiflorum	Flower	~13 nm (Spherical)	Antioxidant and Antifungal Activities	[109]
Morinda lucida	Leaves	~11 nm (spherical)	Antioxidant and Antimicrobial Activities	[110]
Mussaenda frondosa	Leaves	~30–60 nm (spherical)	Antioxidant Activity	[111]
Mikania cordata	Leaves	~46–50 nm (spherical)	Antioxidant, Antimicrobial, Cytotoxic Activities	[113]
Curcuma longa	Leaves	~5–25 nm (Spherical)	Antibacterial Activity	[114]
Ocimum canum	Leaves	~15.76 nm (Rod and Spherical)	Antibacterial Activity	[115]
Ocimum americanum	Leaves	~48.25 nm (spherical)	Antibacterial, Antioxidant, and Anticancer Activities	[116]
Otostegia persica	Leaves	~36.5 ± 2.0 nm (Spherical)	Antibacterial, Antifungal, and Anti-inflammatory Activities	[117]
Pisonia alba L.	Leaves	Not mentioned (Spherical)	Antioxidant Activity	[118]
Retama monosperma	Root	~9.87–21.16 nm (Spherical)	Antimicrobial Activity	[121]
Rosa canina	Fruit	~13–31 nm (Spherical)	Antioxidant, Antimicrobial, and DNA cleavage	[122]
Spilanthes acmella	Flower	~10–35 nm (Spherical and oval)	Antioxidant Activity	[123]
Sesbania grandiflora	Leaves	~10–25 nm (Granular-like)	Antibacterial and Antioxidant Activities	[124]
Sambucus ebulus	Leaves	~18.6 nm (Spherical)	Antioxidant and Antibacterial Activities	[125]
Thuja orientalis	Leaves	~85.77 nm (spherical)	Antimicrobial Activity	[126]
Terminalia chebula	Fruit	~10–30 nm (Spherical)	Antioxidant, Protein Leakage Analysis Antibacterial, Zebrafish Embryonic Toxicology	[127]
Tropaeolum majus	Leaves	~25 nm (Crystalline)	Antimicrobial Activity	[128]
Aloe vera	Stems	Cubical, Spherical, and Triangles	Antibacterial Activity	[137]

Table 4. Summarized the reported work on Au-NPs from different parts of the plant for biomedical applications.

Plant Species	Part of Plant	Size/Shape	Applications	Ref.
Crataegus oxyacantha	Twig	~85 nm (Spherical)	Urease inhibitory Activities	[1]
Moringa oleifera	Leaves	~14–30 nm	Antibacterial, Antioxidant, and Cytotoxicity Activities	[6]
Flaxseed	Seed	~3.4–6.9 nm (Spherical and Triangular) 10–30 Spherical	Anticancer Activity	[7]
Oak gum	Fruit	~10–15 nm (Spherical)	Antioxidant and Anti-colon cancer Activities	[8]
Curcuma pseudomontana	Flower	~20 nm (Spherical)	Antimicrobial, antioxidant, anti-inflammatory activities	[9]
Glaucium flavum	Leaves	~32 nm (Hexagonal, Triangular, and Spherical)	Anticancer Activity	[15]
Jatropha integerrima Jacq.	Flower	~38.8 nm (Spherical)	Antibacterial activity	[20]
Licorice	Root	~ 2.647 nm to 16.25 nm (Circular)	Antimicrobial and Anticancer Activities	[25]
Kaempferia parviflo	Rhizomes	~ 44 ± 3 nm (Spherical)	Antimicrobial and Antioxidant Activities	[26]
Citrus peel	Fruit	~13.65–16.80 nm (Spherical)	Anti-inflammatory activity	[32]
Cynodondactylon	Grass	~21.33 nm (Spherical and Irregular)	Cytotoxicity and Antibacterial activities	[39]
Spondias dulcis	Fruit	~36.75 ± 11.36 nm (Spherical)	Cytotoxic activity in human breast cancer cells	[49]
Saffron	Flower	~25 nm (Spherical and Oval) 15–50 Spherical	Antioxidant and Cytotoxicity Activities	[51]
Delphinium chitralense	Tuber	~100–300 nm (Cubic)	Enzyme inhibitory Activity	[82]
Combretum erythrophyllum	Leaves	~13.20 nm (Spherical)	Antibacterial, Cell viability Activities	[85]

Table 5. Summarized the reported work on ZnO-NPs from different parts of the plant for biomedical applications.

Plant Species	Part of Plant	Size/Shape	Applications	Ref.
Borreria hispida	Leaves	~21.87 nm (Hexagonal)	Antioxidant Activity	[10]
Ailanthus altissima	Leaves	~13.27 nm (Spherical)	Antibacterial and Antioxidant Activities	[11]
Passiflora subpeltata	Leaves	~40–50 nm (Irregular)	Antibacterial Activity	[12]
Andrographis alata	Whole plant	~45 ± 4.23 nm (Spherical, Oval, and Hexagonal)	Antibacterial, Antioxidant, Antidiabetic, and Anti-Alzheimer Activities	[13]
Mucuna pruriens	Peel	~21.60 to 47.16 nm (Spherical)	Anticancer, antioxidant activity	[14]
Bergenia ciliata	Rhizome	(Flower-like bundles)	Antibacterial, anticancer potential	[16]
Berberis aristata	Leaves	~5–25 nm (Needle)	Antibacterial activities and Antioxidant	[23]
Lantana camara	Flower	~25 nm (Spherical)	anti-inflammatory	[24]
Phlomis leaf	Leaves	~79 nm (Hexagonal)	Cytotoxicity and Antibacterial Activities	[29]
Lepidium sativum	Seeds	~36.96–44.50 (Spherical)	Anticancer activity	[31]
Plantain peel	Peel	~20 nm (Spherical)	Antibacterial Activity	[34]
Camellia sinensis	Leaves	~6 to 112 nm (Spherical)	Antioxidant, Antibacterial, and Anticancer Activity	[42]
Caesalpinia crista	Seed	~34.67 nm (Irregular)	Antibacterial, Antioxidant, and Anticancer Activities	[44]
Aquilegia pubiflora	Leaves	~34.23 nm (Spherical or Elliptical)	Antiproliferative Activity	[45]
Amygdalus scoparia	Bark	~15–40 nm (Spherical)	Antimicrobial, Anticancer, and Antidiabetic	[46]
Sauropus androgynus	Leaves	~12 to 23 nm (Spherical)	Antineoplastic Activity	[50]
Capparis spinosa	Fruit	~37.49 nm (Spherical)	Antioxidant Activity	[52]
Ipomoea Sagittifolia Burm.f	Leaves	~51.2 ± 8.5 nm (Hexagonal)	Antibacterial, Antioxidant, and Anticancer Activities	[53]
Cymbopogon citratus	Aerial Part	~21 nm (Spherical)	Antimicrobial and Anticancer Activity	[54]
Mentha piperata	Leaves	15 to 27 nm (Globular and Oblong)	UTI-resistant pathogens	[56]

Table 5. *Cont.*

Plant Species	Part of Plant	Size/Shape	Applications	Ref.
Mangifera indica	Seed	~40–70 nm (Cylindrical)	Antibacterial and Antioxidant Activities	[65]
Olive fruit	Fruit	~56.8 ± 0.6 (Spheroidal)	Antioxidant Activity	[86]
Punica granatum	Peel	~20–40 nm (Spherical and Hexagonal)	Antimicrobial Activity	[90]
Myrica esculenta	Fruit	~115 ± 3.21 nm (Pellets-like)	Antioxidant, antimicrobial, photocatalytic	[93]
Capsicum chinense	Fruit	~12.7 nm (Spherical)	Cytotoxicity and Antioxidant Activities	[94]
Myristica fragrans	Fruit	~43.3 to 83.1 nm (Semi spherical)	Antioxidant, antimicrobial, photocatalytic	[97]
Boerhaavia erecta	Leaves	~20.55 nm (Spherical)	Antioxidant and Antiglycation Activities	[104]
Brassica oleracea	Leaves	~52 nm (Flower-like)	Antimicrobial and Larvicidal Activities	[112]
Punica granatum	Peel	~10.45 nm (Spherical)	Antimicrobial Activity	[130]
Rivina humilis	Leaves	~14.4 nm (Irregular and Circular)	Anticancer, Cytotoxicity, and Antiproliferative Activities	[131]
Sphagneticola trilobata	Leaves	~29.83 (Spherical)	Colon cancer and Antioxidant Activities	[132]
Thryallis glauca	Leaves	~50 nm (Hexagonal wurtzite)	Antioxidant and Antibacterial	[133]
Thuja officinalis	Leaves	4–5 nm	Antimicrobial	[134]
Ocimum tenuiflorum	Leaves	~30 to 40 nm Flakes	Photocatalytic-Antimicrobial Activity	[136]
Ricinus Communis L.	Fruit	Hexagonal irregular	Antibacterial Activity	[138]

4.1. Pioneering Studies on Ag-NPs Derived from Plant Extracts

As shown in Table 3, various plant part extracts have been used to synthesize Ag-NPs, which have emerged as a pivotal innovation in nanotechnology, particularly due to their invaluable biomedical applications. *Euphorbia serpens* plant (leaves) extract-mediated Ag-NPs demonstrated strong antibacterial activity, particularly against *E. coli* (20 ± 06 mm) and *S. typhi* (18 ± 0.5 mm), with moderate antifungal effects against *C. albicans* and *A. alternata*. They also showed significant cytotoxicity against Artemia salina with LD_{50} values of 5.37 and 5.82 at 1–2 μg/mL concentrations [2]. Additionally, Aervalanata (flower) extract biosynthesized Ag-NPs exhibited antibacterial activity against both Gram-positive and Gram-negative bacteria, and demonstrated antioxidant potential at a concentration of 100 μg/mL in the DPPH radical scavenging assay [17]. Similarly, Ag-NPs Salvia officinalis plant extract exhibited antimicrobial properties, both Gram-positive and Gram-negative bacteria (four strains), and the antimicrobial properties were assessed using multiple methodologies [143]. Notably, *Allium cepa* L. (bulb) extract-mediated Ag-NPs exhibited strong antibacterial activity against *P. aeruginosa*, *B. subtilis*, and *S. aureus* strains, outperforming standard antibiotics. Their antioxidant properties were demonstrated through various assays, with IC_{50} values of 116.9 μg/mL in the β-carotene linoleic acid assay, and also showed inhibitory effects on acetylcholinesterase and butyrylcholinesterase enzymes [30]. Ag-NPs produced using banana (pulp) extra exhibited strong antibacterial activity, especially against *E. coli*, and enhanced voltage regulation in bio-electrochemical cells (BEC) t. Ag-NPs investigated through Flaxseed extract demonstrated significant antimicrobial activity against *A. baumannii*, with a minimum inhibitory concentration of 2 μg/mL, showing no toxicity at lower concentrations and enhanced effects compared to chemically synthesized Ag-NPs, though cytotoxicity increased at higher concentrations [100]. The antioxidant assessment revealed superior antioxidative activity (84.48%) in green-synthesized *Morinda lucida* (leaves) nanoparticles compared to chemically synthesized ones (75.87%). Additionally, the nanoparticles exhibited dose-dependent anti-inflammatory effects, with IC_{50} values of 30.19 μg/mL for plant extract-mediated Ag-NPs and 34.14 μg/mL for chemically synthesized Ag-NPs. The lower IC50 value of plant extract-mediated NPs (30.19 μg/mL) indicates stronger and more effective activity, likely due to the presence of phytochemicals in the green synthesis [110]. *Otostegia persica* phyto-fabricated Ag-NPs exhibited superior antioxidant activity and significant antibacterial effects against both Gram-positive and Gram-negative bacteria. They also demonstrated dose-dependent anti-inflammatory and antifungal properties, with MIC values of 18.75, 37.5, and 75 μg/mL against *Candida* species, making them a promising agent for medical treatments [117]. Furthermore, *Terminalia chebula* (fruit) extract-synthesized Ag-NPs exhibited strong antibacterial and antioxidant activities, with no significant toxic effects on zebrafish embryos, highlighting their potential as therapeutic agents [127]. However, Table 2 summarizes their shape, size, and potential for different biomedical applications.

4.2. Studies on Au-NPs Synthesized from Plant Extract

These studies highlight the potential and application of Au-NPs synthesized via green methods, demonstrating their broad biological, catalytic, and medicinal applications. *Crataegus oxyacantha* (twig) extract-mediated Au-NPs exhibited strong urease inhibitory activity (99.25% inhibition, $IC_{50} = 1.38 \pm 0.3$), surpassing the standard thiourea, highlighting their potential for biomedical applications in homeopathic and pharmaceutical industries [1]. Au-NPs synthesized from *M. oleifera* (leaves) extract exhibited low cytotoxicity and promoted neuronal cell regeneration in animal model studies [6]. *Curcuma pseudomontana* (flower) extract-mediated Au-NPs exhibited strong antibacterial activity against *P. aeruginosa*, *S. aureus*, *B. subtilis*, and *E. coli*, with a maximum inhibition efficiency of 94% in the human red blood cell method. They also demonstrated significant antioxidant and radical scavenging activities, highlighting their potential for biomedical applications [9]. Au-NPs exhibited strong antimicrobial activity against *E. coli*, *P. aeruginosa*, *S. aureus*, and *B. subtilis*, along with effective antioxidant activity (IC_{50} = 94.5 mg/mL) synthesized from *Kaempferia parviflo* (rhizomes) [26]. *Cynodondactylon* (grass) extract-loaded Au-NPs exhibited cytotoxicity against the MCF-7 cell line, with an IC_{50} of 31.34 μg/mL, and induced ROS generation, DNA fragmentation, and mitochondrial membrane changes. They also demonstrated significant antibacterial activity against clinically isolated pathogens like *Enterobacter cloacae*, *Staphylococcus haemolyticus*, and *Staphylococcus petrasiisubsp* [39]. Au-NPs exhibited potent dose-dependent anticholinesterase activity, with IC_{50} values of 15 ± 0.42 μg/mL for AChE and 12 ± 0.42 μg/mL for BChE synthesized from *Delphinium chitralense* (tuber) [82]. However, Table 5 summarizes their shape, size, and potential biomedical applications.

4.3. Exploratory Studies on ZnO-NPs Produced from Plant Extracts

ZnO-NPs synthesized using Ailanthus altissima (leaves) extract-mediated ZnO-NPs showed dose-dependent antibacterial and antioxidant activity at 20 mg/mL ZnO-NPs; the maximum bactericidal potential of ZnO-NPs was reported against *Staphylococcus aureus* (201.2 mm). These ZnO-NPs have also an IC_{50} value of 78.23 µg/mL, indicating that they are an effective antioxidant [11]. *Mucuna pruriens* (peel) extract-mediated ZnO-NPs exhibited selective cytotoxicity against HeLa and HEK 293 cancer cell lines in the MTT assay, with no toxicity to normal cells. This study presents a simple, non-toxic biosynthesis method for ZnO-NPs with promising anticancer and antioxidant applications [14]. ZnO-NPs synthesized from *C. spinosa* L. (fruit) extract showed strong antioxidant activity, low cytotoxicity on L929 fibroblast cells, and good RBC biocompatibility. With no hemolytic effects at 7.5–120 µg/mL, they offer a safe alternative for pharmaceutical and biomedical applications [52]. *Myrica* esculenta (Fruit) extract-fabricated ZnO-NPs exhibited superior antimicrobial activity (MIC: *B. subtilis* 0.031 mg/mL, *S. aureus* 0.062 mg/mL, *P. aeruginosa* 0.125 mg/mL). They also showed strong concentration-dependent antioxidant activity (IC_{50}-DPPH: 182.63 µg/mL; FRAP: 129.44 µM $FeSO_4$ equivalents) [93]. Biosynthesized ZnO-NPs exhibited cytotoxic activity against HEK 293 and HeLa cell lines and showed superior antioxidant potential in the DPPH assay compared to conventional ZnO-NPs. These findings highlight *C. chinense* (fruit) extract as an effective agent for ZnO-NP biosynthesis, with anticancer and antioxidant properties [94]. Boerhaavia *erecta* (leaves) extract phyto-fabricated ZnO-NPs inhibited advanced glycation end products (AGEs) formation by blocking Amadori products, trapping dicarbonyl intermediates, and breaking glycated protein cross-links. Additionally, they protected RBCs from MGO-induced damage, highlighting their potential in diabetes-related complications [104].

4.4. Biomedical Applications

4.4.1. Antimicrobial and Antioxidant Activities

The increasing resistance of bacteria such as *Bacillus subtilis, Staphylococcus aureus, Vibrio cholerae, Escherichia coli, Pseudomonas aeruginosa, Aeromonas hydrophila,* and *Proteus mirabilis* [133] to several commonly used antibiotics has become a critical worldwide medical concern that requires immediate solutions. Ag-NPs, Au-NPs, and ZnO-NPs have emerged as promising alternatives in the field of biomedical applications. Research showed that plant extracts-mediated NPs can effectively destroy harmful bacteria and inhibit their growth. Tables 3–5 summarize published studies that report the green synthesis of nanoparticles using plant extract and which parts of the plants that were used. In these studies, almost all the plant extracts-mediated NPs exhibited significant antimicrobial effects against the tested microorganisms. Mechanism: As shown in Figure 7a, antibacterial mechanisms of plant extracts-mediated NPs operate via multiple pathways. These include the generation of ROS, which disrupts bacterial membranes, damages proteins, and induces oxidative stress in bacterial DNA. This mechanism is known to be the main mechanism involved in ZnO-NPs. In the case of Ag-NPs, this is the release of Ag^+ ions that bind to thiol groups in bacterial enzymes, inhibiting essential metabolic functions. The release of Zn^{2+} ions by ZnO-NPs is lower, but Zn^{2+} ions disrupt bacterial ion homeostasis and catalyze additional ROS production under light exposure. The NPs attach to bacterial membranes via electrostatic interactions, leading to membrane disruption, leakage of cytoplasmic contents, and ultimately resulting in cell death. These NPs inhibit protein synthesis through binding to ribosomal subunits and impede biofilm formation by disrupting bacterial adhesion and the extracellular matrix. Furthermore, phytochemicals that originate from plant extracts work in conjunction with nanoparticles, improving their stability, bioavailability, and overall effectiveness against microbes [51,130,135]. However, some studies showed that the plant extract method can also shield the production of ROS if green-synthesized nanoparticles are synthesized using beta-diketonate precursors like zinc acetate [134], and nitrate should be preferred for the synthesis using plant extract.

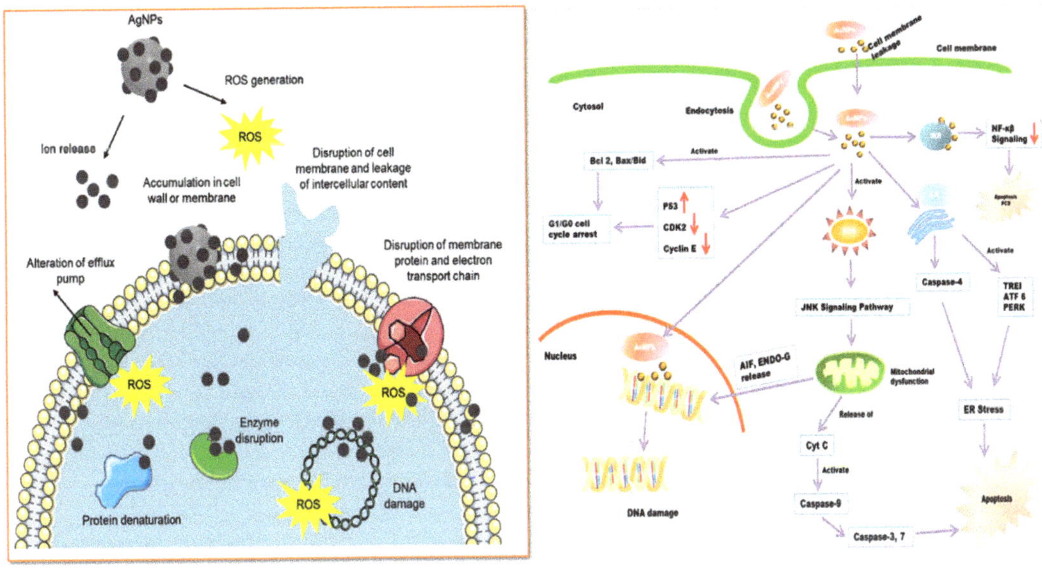

Figure 7. Reported antibacterial (**a**) and anticancer (**b**) mechanisms of plant extract-mediated NPs [19,135].

4.4.2. Anticancer Activities

Ag-NPs, Au-NPs, and ZnO-NPs have shown potential application in cancer therapy owing to their unique properties and mechanisms of action.

Mechanism: As shown in Figure 7a, Ag-NPs, Au-NPs, and ZnO-NPs demonstrate notable anticancer mechanisms due to their high surface-area-to-volume ratios, nanoscale properties, and functionalization with plant-derived bioactive compounds. Anticancer mechanism of green-synthesized Ag-NPs, Au-NPs, and ZnO-NPs are diverse and have been extensively studied. Three primary mechanisms are often highlighted: (1) interference with cell membranes, (2) ROS-induced apoptosis, and (3) disruption of the chemistry of proteins and DNA. Plant extracts-mediated NPs interact with cancer cells in several ways, and their surface properties play a critical role in cellular internalization. The positive charges on green-synthesized NPs interact with the negatively charged phosphate groups on the lipid bilayer of cell membranes, facilitating their uptake into cells through endocytosis. Once inside, NPs accumulate in the mitochondria, where they cause structural damage, impair the electron transport chain, and activate NADPH oxidases, leading to mitochondrial dysfunction and ROS generation. ROS production is a major factor behind the anticancer effects of NPs. These elevated ROS levels induce oxidative stress, resulting in DNA fragmentation, mitochondrial dysfunction, and eventual cell death through apoptosis, autophagy, or necroptosis. The oxidative stress caused by NPs depletes cellular antioxidants like glutathione and disrupts the mitochondrial membrane potential, triggering the release of apoptotic molecules such as cytochrome-c, apoptosis-inducing factor, and endonuclease G into the cytosol, leading to cell death.

In addition to their ROS-induced apoptosis, NPs can interfere with the metabolic activity of cancer cells. They have been shown to inhibit DNA biosynthesis by directly binding to DNA and reducing the expression of DNA repair proteins, thus enhancing the anticancer effect. Beyond the ROS generation and mitochondrial dysfunction, NPs also lead to the activation of endoplasmic reticulum stress responses. The complex relationship of ROS-mediated damage, ion release, and biofilm disruption positions green-synthesized Ag, Au, and ZnO NPs as highly effective agents against oxidative stress and bacterial pathogens, highlighting their considerable promise in both biomedical and environmental fields [7,14,31,92,131].

5. Future Outlook of Hybrid Nanocomposites with Graphitic Carbon Nitride (GCN)

Graphitic Carbon Nitride is a metal-free semiconductor material composed of C and N compounds, known for its unique layered structure and promising candidacy for different applications in various fields such as biomedical, photo-catalysis, energy conversion, and environmental remediation. Its structure resembles graphite, with nitrogen atoms integrated into the carbon frame work, offering excellent stability. Moreover, to date, Graphitic carbon nitride traditionally has been synthesized by thermal polymerization of nitrogen-rich precursors like urea, thiourea, and melamine for huge environmental applications. However, GCN and their nanocomposites have gained attention for dye degradation, demonstrating enhanced photocatalytic activity for methylene blue. These studies used conventional chemical methods to synthesize both GCN and ZnO-NPs [63,64,144]. GCN-ZnO-NPs hybrid nanocomposites have also been studied for possible applications in drug delivery, making it a promising candidate for breast cancer treatment [146,148]. GCN has already been combined with Ag-NPs using the ex-situ method through the use of ultrasonic bath [149] and through the in-situ method [150]. The optical properties of the nanocomposite were enhanced due to the plasmonic effect of Ag-NPs. These hybrid nanocomposites were tested for the catalytic reduction of 4-nitrophenol and the degradation of 2,4-Dichlorophenoxyacetic acid. In both cases, the combination of GCN with Ag-NPs showed enhanced photocatalytic properties. GCN-Ag-NPs hybrid nanocomposites were also studied for the degradation of dyes and H_2 production. The study showed that GCN-Ag-NPs hybrid nanocomposite exhibits superior photocatalytic activity [151]. However, in all these studies, Ag-NPs and GCN were prepared using conventional chemical methods [143,151]. Au NPs were combined with GCN and carbon nanotubes (CNTs) to study their photocatalytic properties applied to the effective photodegradation of organic pollutants and possible application in photoelectrochemical water splitting [152]. The authors showed that GCN-Au-NPs hybrid nanocomposite exhibits photocatalytic activity increased by a factor of almost 40 in the case of Rhodamine B photodegradation under visible light exposure. GCN-Au-NPs hybrid nanocomposite has been also studied for the development of fluorescent sensor for Fe^{3+} and dichromate ions detection in aqueous medium [153].

All these studies highlight the high potential to combine metal (oxide) nanoparticles with GCN. However, to date, graphitic carbon nitride has not been synthesized using plant extracts. Current investigations indicate that this approach holds significant potential for the future, similarly to metallic nanoparticles. With ongoing advancements, the effective formulation of plant extract-mediated GCN-NPs may pave the way for innovative, sustainable approaches in biomedical applications (see Table 6), pollution management, energy generation, and environmental restoration [140–144].

Table 6. Comparison: Traditional Methods between Nanoparticle-Based Methods in Biological Applications.

Application	Traditional Methods	Nanoparticle-Based Method	Advantages of Nanoparticles	Refs.
Cancer Treatment	Chemotherapy; radiotherapy	Targeted drug delivery; photothermal therapy	Minimize damage to healthy cells	[16,49,92,132]
Antibacterial	Broad-spectrum antibiotics	Ag-, ZnO-, Cu-NPs etc.	Effective against resistant strains	[2,27,138,154]
Biosensors	Enzyme and antibody-based detection	Nanoparticles-based sensors	Faster detection	[5,13,55,60]
Wound healing	Traditional dressing; antiseptic solution	Nanoparticles-embedded dressing such as Ag-, ZnO-NPs	Antibacterial, faster healing, reduced infection rate	[24,53,138]
Drug delivery	Oral intravenous drug	Targeted delivery	Site specific action, reduce side effects	[57,58]

Our Current Research Approach and Innovations

Impressed by the significance of plant-based nanoparticles and their nanocomposites, we synthesized plant-based GCN using *Ocimum tenuiflorum*, as shown in Figure 8. The plant-based GCN was derived by using *Ocimum tenuiflorum* leaves that had been kept in a hot-air oven at a temperature of 70 °C for 36 h. The dry leaves were then ground into a powder. The powder was then added to a closed ceramic crucible and heated in a muffle furnace at 550 °C for 3 h at a heating rate of 2 °C/min. GCN-NPs pellets were crushed into a fine powder with a mortar and pestle. The synthesized GCN-NPs was confirmed by the XRD. GCN-NPs exhibit a strong XRD peak at 2θ of 27.7° that corresponds to the (002) plane that is common to all GCN phases, which therefore confirms the Graphitic stacking of C_3N_4 in GCN-NPs. The next step is to combine plant-based synthesized GCN with metal (oxide) nanoparticles and study these hybrid nanocomposites fully produced using environmental friendly method of syntheses [155].

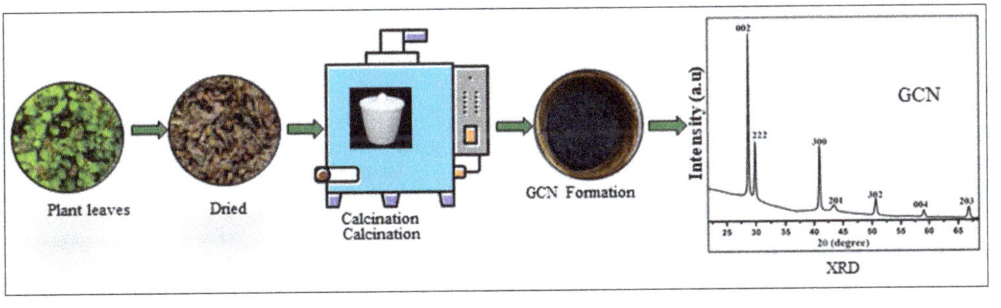

Figure 8. Overview of plant-based GCN nanoparticles formation.

6. Limitations of Green-Synthesized Nanoparticles

However, green synthesis is not without limitations. Its scalability and reproducibility can be affected by factors such as the regional and seasonal variability of plant materials, as well as differences in extract composition based on geographic origin. Also, we face the problem of the control of synthesis with a sharp size distribution, reproducibility due to possible variation of the plant extract composition depending on the origin. For this reason, the chemical method must still be considered for synthesis. These factors hinder, increase costs, and challenge the overall feasibility of industrial production [154,156–158]. Also, there is the presence of chlorine in plant extracts that can induce the formation of metal chlorine like AgCl as secondary phase during the synthesis process [98]. The nature of the plant not only influences the properties of the nanoparticles through functionalization by the bioactive compounds present in the extract but also affects their morphology and size during the synthesis process [156]. While many studies have shown that locally accessible plants like cotton leaves and Coffea arabica can serve as effective agents for nanoparticle synthesis, practical limitations exist. Many plants in green synthesis produce supplementary by-products that require additional treatment before they can be used, adding extra steps, time, and expense to the process. Additionally, a major scientific hurdle is the lack of a clear understanding of the chemical pathways involved in the synthesis process. While some studies have demonstrated the effects of plant extracts on nanoparticle formation, the exact mechanisms remain largely unknown. This limits the ability to control and optimize the synthesis, making it difficult to predict outcomes or scale the process reliably. Current research indicates large variations in particle diameter, which presents a major obstacle for industrial applications. Moreover, the characterization techniques used to evaluate the synthesized nanoparticles are sometimes insufficient to understand. Without standardized protocols and deeper insight into the reaction mechanisms, achieving consistent, high-quality products suitable for commercial use remains a challenge. Therefore, while green synthesis has tremendous potential, addressing these issues related to raw material sourcing, reaction conditions, process understanding, and product quality control is essential for its transition from laboratory research to viable industrial technology [57–62,154,156–158].

7. Conclusions

In this review, we systematically explore the synthesis approaches, significance, design, advancements, and biomedical applications of plant extract-mediated nanoparticles, with a specific focus on Ag-, Au-, and ZnO-NPs. The first section highlights the importance of nanomaterials in biomedical applications, and their various synthesis methods, followed by a discussion on the significance of plant extract-mediated nanomaterials and their properties. Therefore, different plant extracts were used in the study for the nanoparticle synthesis; however, no real studies were performed to identify a specific plant family that would be best to target a specific pathogen. So, it appears necessary to use machine learning software to identify such a pattern through the creation of a database. Such studies will shed a light on the best plant extract that must be used for a specific targeted application. The second section discusses the antibacterial, antioxidant, and anticancer properties of Ag-NPs, Au-NPs, and ZnO-NPs and highlights their potential biomedical applications. In the third section, we provide key insights and future perspectives on combining these nanoparticles with GCN to enhance their properties. This shows that the similar way of synthesis involving plant extracts should be applied to the synthesis of GCN to promote better sustainability in the production of these hybrid nanocomposites. Finally, we present our recent findings and research progress on plant-based synthesized GCN-NPs, highlighting their potential for future applications in biomedical applications and environmental remediation.

This review tries to highlight the recent progresses in the field and how it is important to now combine different nanomaterials to obtain a synergistic effect to obtain new or improved properties. Additionally, there is a clear need of studies using machine learning software to identify a pattern or the most promising plant extract that must be used for a targeted application. This identification can only be accomplished through the creation of a database using such software combined with artificial intelligence.

Author Contributions: Conceptualization, P.P., E.R. and P.R.; methodology, P.P. and E.R.; software, P.S., M.K. and G.H.; validation, P.P., E.R. and S.P.N.; formal analysis, S.P.N., M.K. and G.H.; investigation, P.P. and E.R.; resources, E.R., P.R., S.P.N., P.S., M.K. and G.H.; data curation, P.P. and E.R.; writing—original draft preparation, P.P.; writing—review and editing, E.R., P.R., S.P.N., P.R., M.K. and G.H.; visualization, P.P. and E.R.; supervision, E.R. and P.R.; project administration, E.R. and E.R.; funding acquisition, P.P. and E.R. All authors have read and agreed to the published version of the manuscript.

Funding: This work was supported by the following funding sources: the Estonian Research Agency (ETAg) Team grant PRG2115, Thematic research and development programs Tem-TA144 and Mobilitas + MOBJD1204. Priyanka Panchal and Erwan Rauwel received research support from the Estonian Research Council.

Institutional Review Board Statement: Not applicable.

Informed Consent Statement: Not applicable.

Data Availability Statement: The data presented in this study are included within the article.

Conflicts of Interest: The authors declare no competing interests.

References

1. Ali, S.; Bacha, M.; Shah, M.R.; Shah, W.; Kubra, K.; Khan, A.; Ahmad, M.; Latif, A.; Ali, M. Green synthesis of silver and gold nanoparticles using *Crataegus oxyacantha* extract and their urease inhibitory activities. *Biotechnol. Appl. Biochem.* **2021**, *68*, 992–1002. [CrossRef] [PubMed]
2. Ahmad, N.; Fozia Jabeen, M.; Haq, Z.U.; Ahmad, I.; Wahab, A.; Islam, Z.U.; Ullah, R.; Bari, A.; Abdel-Daim, M.M.; El-Demerdash, F.M.; et al. Green fabrication of silver nanoparticles using *Euphorbia serpens* Kunth aqueous extract, their characterization, and investigation of its in vitro antioxidative, antimicrobial, insecticidal, and cytotoxic activities. *BioMed Res. Int.* **2022**, *2022*, 5562849. [CrossRef]
3. Alhomaidi, E.; Jasim, S.A.; Amin, H.I.M.; Lima Nobre, M.A.; Khatami, M.; Jalil, A.T.; Hussain Dilfy, S. Biosynthesis of silver nanoparticles using *Lawsonia inermis* and their biomedical application. *IET Nanobiotechnol.* **2022**, *16*, 284–294. [CrossRef]
4. Balčiūnaitienė, A.; Liaudanskas, M.; Puzerytė, V.; Viškelis, J.; Janulis, V.; Viškelis, P.; Griškonis, E.; Jankauskaitė, V. *Eucalyptus globulus* and *Salvia officinalis* extracts mediated green synthesis of silver nanoparticles and their application as an antioxidant and antimicrobial agent. *Plants* **2022**, *11*, 1085. [CrossRef] [PubMed]
5. Anastas, P.; Eghbali, N. Green chemistry: Principles and practice. *Chem. Soc. Rev.* **2010**, *39*, 301–312. [CrossRef] [PubMed]
6. Boruah, J.S.; Devi, C.; Hazarika, U.; Reddy, P.V.B.; Chowdhury, D.; Barthakur, M.; Kalita, P. Green synthesis of gold nanoparticles using an antiepileptic plant extract: In vitro biological and photo-catalytic activities. *RSC Adv.* **2021**, *11*, 28029–28041. [CrossRef]
7. Al-Radadi, N.S. Green biosynthesis of flaxseed gold nanoparticles (Au-NPs) as a potent anti-cancer agent against breast cancer cells. *J. Saudi Chem. Soc.* **2021**, *25*, 101243. [CrossRef]
8. Lu, L.; Zhao, Q.; Wang, Z.; Ju, F. Oak gum mediated sustainable synthesis of gold nanoparticles (Au NPs): Evaluation of its antioxidant and anti-colon cancer effects. *J. Exp. Nanosci.* **2022**, *17*, 377–388. [CrossRef]
9. Muniyappan, N.; Pandeeswaran, M.; Amalraj, A. Green synthesis of gold nanoparticles using *Curcuma pseudomontana* isolated curcumin: Its characterization, antimicrobial, antioxidant and anti-inflammatory activities. *Environ. Chem. Ecotoxicol.* **2021**, *3*, 117–124. [CrossRef]
10. Ramesh, A.M.; Purushotham, D.; Kodandaram, A.; Shilpa, N.; Singh, S.B.; Aiyaz, M.; Gowtham, H.G.; Rahdar, A.; Kaviyarasu, K.; Murali, M. Visible light driven photocatalytic and competent antioxidant properties of phyto-fabricated zinc oxide nanoparticles (ZnO-NPs) from *Borreria hispida*. *J. Mol. Struct.* **2023**, *1293*, 136152. [CrossRef]
11. Awan, S.S.; Khan, R.T.; Mehmood, A.; Hafeez, M.; Abass, S.R.; Nazir, M.; Raffi, M. Ailanthus altissima leaf extract mediated green production of zinc oxide (ZnO) nanoparticles for antibacterial and antioxidant activity. *Saudi J. Biol. Sci.* **2023**, *30*, 103487. [CrossRef]
12. Chunchegowda, U.A.; Shivaram, A.B.; Mahadevamurthy, M.; Ramachndrappa, L.T.; Lalitha, S.G.; Krishnappa, H.K.N.; Anandan, S.; Sudarshana, B.S.; Chanappa, E.G.; Ramachandrappa, N.S. Biosynthesis of zinc oxide nanoparticles using leaf extract of *Passiflora subpeltata*: Characterization and antibacterial activity against *Escherichia coli* isolated from poultry faeces. *J. Clust. Sci.* **2021**, *32*, 1663–1672. [CrossRef]
13. Dappula, S.S.; Kandrakonda, Y.R.; Shaik, J.B.; Mothukuru, S.L.; Lebaka, V.R.; Mannarapu, M.; Amooru, G.D. Biosynthesis of zinc oxide nanoparticles using aqueous extract of *Andrographis alata*: Characterization, optimization and assessment of their antibacterial, antioxidant, antidiabetic and anti-Alzheimer's properties. *J. Mol. Struct.* **2023**, *1273*, 134264. [CrossRef]

14. Gamedze, N.P.; Mthiyane, D.M.N.; Mavengahama, S.; Singh, M.; Onwudiwe, D.C. Biosynthesis of ZnO nanoparticles using the aqueous extract of *Mucuna pruriens* (utilis): Structural characterization, and the anticancer and antioxidant activities. *Chem. Afr.* **2024**, *7*, 219–228. [CrossRef]
15. Dehghani, F.; Mosleh-Shirazi, S.; Shafee, M.; Kasaee, S.R.; Amani, A.M. Antiviral and antioxidant properties of green synthesized gold nanoparticles using *Glaucium favum* leaf extract. *Appl. Nanosci.* **2023**, *13*, 4395–4405. [CrossRef] [PubMed]
16. Dulta, K.; Ağçeli, G.K.; Chauhan, P.; Jasrotia, R.; Chauhan, P.K. A novel approach of synthesis zinc oxide nanoparticles by *Bergenia ciliata* rhizome extract: Antibacterial and anticancer potential. *J. Inorg. Organomet. Polym. Mater.* **2021**, *31*, 180–190. [CrossRef]
17. Palithya, S.; Gaddam, S.A.; Kotakadi, V.S.; Penchalaneni, J.; Golla, N.; Naidu Krishna, S.B.; Naidu, C.V. Green synthesis of silver nanoparticles using flower extracts of *Aerva lanata* and their biomedical applications. *Part. Sci. Technol.* **2022**, *40*, 84–96. [CrossRef]
18. Darwesh, M.Y.; Ibrahim, S.S.; Mohammed, M.A. Plant extract mediated green synthesis of zinc oxide nanoparticles and their biomedical applications. *Results Chem.* **2024**, *7*, 101368. [CrossRef]
19. Bharadwaj, K.K.; Rabha, B.; Pati, S.; Sarkar, T.; Choudhury, B.K.; Barman, A.; Bhattacharjya, D.; Srivastava, A.; Baishya, D.; Edinur, H.A.; et al. Green Synthesis of Gold Nanoparticles Using Plant Extracts as Beneficial Prospect for Cancer Theranostics. *Molecules* **2021**, *26*, 6389. [CrossRef]
20. Suriyakala, G.; Sathiyaraj, S.; Babujanarthanam, R.; Alarjani, K.M.; Hussein, D.S.; Rasheed, R.A.; Kanimozhi, K. Green synthesis of gold nanoparticles using *Jatropha integerrima* Jacq. flower extract and their antibacterial activity. *J. King Saud Univ.-Sci.* **2022**, *34*, 101830. [CrossRef]
21. Hu, X.; Wu, L.; Du, M.; Wang, L. Eco-friendly synthesis of size-controlled silver nanoparticles by using *Areca catechu* nut aqueous extract and investigation of their potent antioxidant and antibacterial activities. *Arab. J. Chem.* **2022**, *15*, 103763. [CrossRef]
22. Singh, J.; Dutta, T.; Kim, K.H.; Rawat, M.; Samddar, P.; Kumar, P. 'Green' synthesis of metals and their oxide nanoparticles: Applications for environmental remediation. *J. Nanobiotechnol.* **2018**, *16*, 84. [CrossRef]
23. Chandra, H.; Patel, D.; Kumar, P.; Jangwan, J.S.; Yadav, S. Phyto-mediated synthesis of zinc oxide nanoparticles of Berberis aristata: Characterization, antioxidant activity, and antibacterial activity with special reference to urinary tract pathogens. *Mater. Sci. Eng. C* **2019**, *102*, 212–220. [CrossRef] [PubMed]
24. Surendra, B.S.; Mallikarjunaswamy, C.; Pramila, S.; Rekha, N.D. Bio-mediated synthesis of ZnO nanoparticles using *Lantana Camara* flower extract: Its characterizations, photocatalytic, electrochemical and anti-inflammatory applications. *J. Clust. Sci.* **2021**, *32*, 647–663. [CrossRef]
25. Al-Radadi, N.S. Facile one-step green synthesis of gold nanoparticles (AuNp) using licorice root extract: Antimicrobial and anticancer study against HepG2 cell line. *Arab. J. Chem.* **2021**, *14*, 102956. [CrossRef]
26. Varghese, B.A.; Nair, R.V.R.; Jude, S.; Varma, K.; Amalraj, A.; Kuttappan, S. Green synthesis of gold nanoparticles using *Kaempferia parviflora* rhizome extract and their characterization and application as an antimicrobial, antioxidant, and catalytic degradation agent. *J. Taiwan Inst. Chem. Eng.* **2021**, *126*, 166–172. [CrossRef]
27. Devanesan, S.; Al Salhi, M.S. Green synthesis of silver nanoparticles using the flower extract of *Abelmoschus esculentus* for cytotoxicity and antimicrobial studies. *Int. J. Nanomed.* **2021**, *16*, 3343–3356. [CrossRef] [PubMed]
28. Jalab, J.; Abdelwahed, W.; Kitaz, A.; Al-Kayali, R. Green synthesis of silver nanoparticles using aqueous extract of *Acacia cyanophylla* and its antibacterial activity. *Heliyon* **2021**, *7*, e08033. [CrossRef]
29. Alyamani, A.A.; Albukhaty, S.; Aloufi, S.; AlMalki, F.A.; Al-Karagoly, H.; Sulaiman, G.M. Green fabrication of zinc oxide nanoparticles using *Phlomis* leaf extract: Characterization and in vitro evaluation of cytotoxicity and antibacterial properties. *Molecules* **2021**, *26*, 6140. [CrossRef]
30. Baran, M.F.; Keskin, C.; Baran, A.; Hatipoğlu, A.; Yildiztekin, M.; Küçükaydin, S.; Kurt, K.; Hoşgören, H.; Sarker, M.M.R.; Sufianov, A.; et al. Green synthesis of silver nanoparticles from *Allium cepa* L. peel extract, their antioxidant, antipathogenic, and anticholinesterase activity. *Molecules* **2023**, *28*, 2310. [CrossRef]
31. Efati, Z.; Shahangian, S.S.; Darroudi, M.; Amiri, H.; Hashemy, S.I.; Aghamaali, M.R. Green chemistry synthesized zinc oxide nanoparticles in *Lepidium sativum* L. seed extract and evaluation of their anticancer activity in human colorectal cancer cells. *Ceram. Int.* **2023**, *49*, 32568–32576. [CrossRef]
32. Gao, L.; Mei, S.; Ma, H.; Chen, X. Ultrasound-assisted green synthesis of gold nanoparticles using citrus peel extract and their enhanced anti-inflammatory activity. *Ultrason. Sonochem.* **2022**, *83*, 105940. [CrossRef] [PubMed]
33. Dua, T.K.; Giri, S.; Nandi, G.; Sahu, R.; Shaw, T.K.; Paul, P. Green synthesis of silver nanoparticles using *Eupatorium adenophorum* leaf extract: Characterizations, antioxidant, antibacterial and photocatalytic activities. *Chem. Pap.* **2023**, *77*, 2947–2956. [CrossRef] [PubMed]
34. Imade, E.E.; Ajiboye, T.O.; Fadiji, A.E.; Onwudiwe, D.C.; Babalola, O.O. Green synthesis of zinc oxide nanoparticles using plantain peel extracts and the evaluation of their antibacterial activity. *Sci. Afr.* **2022**, *16*, e01152. [CrossRef]
35. Hasan, M.; Zafar, A.; Imran, M.; Iqbal, K.J.; Tariq, T.; Iqbal, J.; Shaheen, A.; Hussain, R.; Anjum, S.I.; Shu, X. Crest to trough cellular drifting of green-synthesized zinc oxide and silver nanoparticles. *ACS Omega* **2022**, *7*, 34770–34778. [CrossRef]

36. Mishra, S.; Das, A.; Mohanty, S. Green synthesis of silver nanoparticles using *Phyllanthus niruri* and their biomedical applications. *ACS Omega* **2023**, *8*, 1528–1537. [CrossRef]
37. Nazir, R.; Malik, M. Green synthesis of silver nanoparticles using *Camellia sinensis* leaf extract and their antibacterial properties. *Environ. Nanotechnol. Monit. Manag.* **2022**, *17*, 100641. [CrossRef]
38. Qamar, S.U.R.; Virijević, K.; Arsenijević, D.; Avdović, E.; Živanović, M.; Filipović, N.; Ćirić, A.; Petrović, I. Silver nanoparticles from *Ocimum basilicum* L. tea: A green route with potent anticancer efficacy. *Colloid Interface Sci. Commun.* **2024**, *59*, 100771. [CrossRef]
39. Vinayagam, R.; Santhoshkumar, M.; Lee, K.E.; David, E.; Kang, S.G. Bioengineered gold nanoparticles using *Cynodond actylon* extract and its cytotoxicity and antibacterial activities. *Bioprocess Biosyst. Eng.* **2021**, *44*, 1253–1262. [CrossRef]
40. Younas, M.; Rasool, M.H.; Khurshid, M.; Khan, A.; Nawaz, M.Z.; Ahmad, I.; Lakhan, M.N. *Moringa oleifera* leaf extract mediated green synthesis of silver nanoparticles and their antibacterial effect against selected gram-negative strains. *Biochem. Syst. Ecol.* **2023**, *107*, 104605. [CrossRef]
41. Patil, T.; Gambhir, R.; Vibhute, A.; Tiwari, A.P. Gold nanoparticles: Synthesis methods, functionalization, and biological applications. *J. Clust. Sci.* **2023**, *34*, 705–725. [CrossRef]
42. Al-Ghamdi, S.A.; Albukhaty, S.; Aloufi, S.; AlMalki, F.A.; Al-Karagoly, H.; Sulaiman, G.M. Green synthesis and characterization of zinc oxide nanoparticles using *Camellia sinensis* tea leaf extract and their antioxidant, anti-bactericidal and anticancer efficacy. *Res. Chem. Intermed.* **2022**, *48*, 4769–4783. [CrossRef]
43. González-Pedroza, M.G.; Argueta-Figueroa, L.; García-Contreras, R.; Jiménez-Martínez, Y.; Martínez-Martínez, E.; Navarro-Marchal, S.A.; Marchal, J.A.; Morales-Luckie, R.A.; Boulaiz, H. Silver nanoparticles from *Annona muricata* peel and leaf extracts as a potential potent, biocompatible, and low-cost antitumor tool. *Nanomaterials* **2021**, *11*, 1273. [CrossRef] [PubMed]
44. Donga, S.; Chanda, S. Caesalpinia crista seeds mediated green synthesis of zinc oxide nanoparticles for antibacterial, antioxidant, and anticancer activities. *BioNanoScience* **2022**, *12*, 451–462. [CrossRef]
45. Jan, H.; Shah, M.; Andleeb, A.; Faisal, S.; Khattak, A.; Rizwan, M.; Drouet, S.; Hano, C.; Abbasi, B.H. Plant-based synthesis of zinc oxide nanoparticles (ZnO NPs) using aqueous leaf extract of *Aquilegia pubiflora*: Their antiproliferative activity against HepG2 cells inducing reactive oxygen species and other in vitro properties. *Oxid. Med. Cell. Longev.* **2021**, *2021*, 4786227. [CrossRef]
46. Jobie, F.N.; Ranjbar, M.; Moghaddam, A.H.; Kiani, M. Green synthesis of zinc oxide nanoparticles using *Amygdalus scoparia* Spach stem bark extract and their applications as an alternative antimicrobial, anticancer, and anti-diabetic agent. *Adv. Powder Technol.* **2021**, *32*, 2043–2052. [CrossRef]
47. Nour, M.; Hamdy, O.; Faid, A.H.; Eltayeb, E.A.; Zaky, A.A. Utilization of gold nanoparticles for the detection of squamous cell carcinoma of the tongue based on laser-induced fluorescence and diffuse reflectance characteristics: An in vitro study. *Laser Med. Sci.* **2022**, *37*, 3551–3560. [CrossRef]
48. Sargazi, S.; Laraib, U.; Er, S.; Rahdar, A.; Hassanisaadi, M.; Zafar, M.N.; Diez-Pascual, A.M.; Bilal, M. Application of green gold nanoparticles in cancer therapy and diagnosis. *Nanomaterials* **2022**, *12*, 1102. [CrossRef]
49. Pechyen, C.; Ponsanti, K.; Tangnorawich, B.; Ngernyuang, N. Biogenic synthesis of gold nanoparticles mediated by *Spondias dulcis* (Anacardiaceae) peel extract and its cytotoxic activity in human breast cancer cell. *Toxicol. Rep.* **2022**, *9*, 1092–1098. [CrossRef]
50. D'Souza, J.N.; Nagaraja, G.K.; Prabhu, A.; Navada, K.M.; Kouser, S.; Manasa, D.J. *Sauropus androgynus* (L.) leaf phytochemical activated biocompatible zinc oxide nanoparticles: An antineoplastic agent against human triple negative breast cancer and a potent nanocatalyst for dye degradation. *Appl. Surf. Sci.* **2021**, *552*, 149429. [CrossRef]
51. Alhumaydhi, F.A.; Khan, I.; Rauf, A.; Qureshi, M.N.; Aljohani, A.S.; Khan, S.A.; Khalil, A.A.; El-Esawi, M.A.; Muhammad, N. Synthesis, characterization, biological activities, and catalytic applications of alcoholic extract of saffron (*Crocus sativus*) flower stigma-based gold nanoparticles. *Green Process. Synth.* **2021**, *10*, 230–245. [CrossRef]
52. Neamah, S.A.; Albukhaty, S.; Falih, I.Q.; Dewir, Y.H.; Mahood, H.B. Biosynthesis of Zinc Oxide Nanoparticles Using *Capparis spinosa* L. Fruit Extract: Characterization, Biocompatibility, and Antioxidant Activity. *Appl. Sci.* **2023**, *13*, 6604. [CrossRef]
53. Muchinthala, P.K.; Losetty, V. Role of phytochemical mediated zinc oxide nanoparticles on biomedical and industrial wastewater treatment: A green approach by experimental and molecular docking exploration. *J. Environ. Chem. Eng.* **2025**, *13*, 115333. [CrossRef]
54. Abdelbaky, A.S.; Mohamed, A.M.H.A.; Sharaky, M.; Mohamed, N.A.; Diab, Y.M. Green approach for the synthesis of ZnO nanoparticles using *Cymbopogon citratus* aqueous leaf extract: Characterization and evaluation of their biological activities. *Chem. Biol. Technol. Agric.* **2023**, *10*, 63. [CrossRef]
55. Ohiduzzaman, M.; Khan, M.N.I.; Khan, K.A.; Paul, B. Biosynthesis of silver nanoparticles by banana pulp extract: Characterizations, antibacterial activity, and bioelectricity generation. *Heliyon* **2024**, *10*, e25520. [CrossRef]
56. Ahmad, N.; Ali, S.; Abbas, M.; Fazal, H.; Saqib, S.; Ali, A.; Ullah, Z.; Zaman, S.; Sawati, L.; Zada, A.; et al. Antimicrobial efficacy of *Mentha piperata*-derived biogenic zinc oxide nanoparticles against UTI-resistant pathogens. *Sci. Rep.* **2023**, *13*, 14972. [CrossRef]
57. Karnwal, A.; Jassim, A.Y.; Mohammed, A.A.; Sharma, V.; Al-Tawaha, A.R.M.S.; Sivanesan, I. Nanotechnology for healthcare: Plant-derived nanoparticles in disease treatment and regenerative medicine. *Pharmaceuticals* **2024**, *17*, 1711. [CrossRef]

58. Kaushal, P.; Maity, D.; Awasthi, R. Nano-green: Harnessing the potential of plant extracts for sustainable antimicrobial metallic nanoparticles. *J. Drug Deliv. Sci. Technol.* **2024**, *94*, 105488. [CrossRef]
59. Singh, H.; Desimone, M.F.; Pandya, S.; Jasani, S.; George, N.; Adnan, M.; Aldarhami, A.; Bazaid, A.S.; Alderhami, S.A. Revisiting the green synthesis of nanoparticles: Uncovering influences of plant extracts as reducing agents for enhanced synthesis efficiency and its biomedical applications. *Int. J. Nanomed.* **2023**, *18*, 4727–4750. [CrossRef]
60. Afonso, I.S.; Cardoso, B.; Nobrega, G.; Minas, G.; Ribeiro, J.E.; Lima, R.A. Green synthesis of nanoparticles from olive oil waste for environmental and health applications: A review. *J. Environ. Chem. Eng.* **2024**, *12*, 114022. [CrossRef]
61. Osman, A.I.; Zhang, Y.; Farghali, M.; Rashwan, A.K.; Eltaweil, A.S.; El-Monaem, E.M.A.; Mohamed, I.M.A.; Badr, M.M.; Ihara, I.; Rooney, D.W.; et al. Synthesis of green nanoparticles for energy, biomedical, environmental, agricultural, and food applications: A review. *Environ. Chem. Lett.* **2024**, *22*, 841–887. [CrossRef]
62. Kuznetsova, V.; Coogan, Á.; Botov, D.; Gromova, Y.; Ushakova, E.V.; Gun'ko, Y.K. Expanding the horizons of machine learning in nanomaterials to chiral nanostructures. *Adv. Mater.* **2024**, *36*, 2308912. [CrossRef] [PubMed]
63. Dahiya, S.; Sharma, R.; Gautam, P.; Panchal, P.; Chaudhary, S.; Sharma, A.; Almáši, M.; Nehra, S.P. Eco-friendly phytofabrication of *Ficus Benjamina*, L. based ZnO-doped g-C_3N_4 nanocomposites for remarkable photocatalysis and antibacterial applications. *Chemosphere* **2023**, *339*, 139707. [CrossRef]
64. Paul, D.R.; Gautam, S.; Panchal, P.; Nehra, S.P.; Choudhary, P.; Sharma, A. ZnO-modified g-C_3N_4: A potential photocatalyst for environmental application. *ACS Omega* **2020**, *5*, 3828–3838. [CrossRef]
65. Rajeshkumar, S.; Parameswari, R.P.; Sandhiya, D.; Al-Ghanim, K.A.; Nicoletti, M.; Govindarajan, M. Green Synthesis, Characterization and Bioactivity of Mangifera indica Seed-Wrapped Zinc Oxide Nanoparticles. *Molecules* **2023**, *28*, 2818. [CrossRef] [PubMed]
66. Paul, D.R.; Sharma, R.; Singh, S.; Singh, P.; Panchal, P.; Sharma, A.; Devi, P.; Nehra, S.P. Mg/Li Co-doped g-C_3N_4: An excellent photocatalyst for waste water remediation and hydrogen production applications towards sustainable development. *Int. J. Hydrogen Energy* **2023**, *48*, 37746–37761. [CrossRef]
67. Paul, D.R.; Sharma, R.; Rao, V.S.; Panchal, P.; Gautam, S.; Sharma, A.; Nehra, S.P. Mg/Li@ GCN as highly active visible light responding 2D photocatalyst for waste water remediation application. *Environ. Sci. Pollut. Res.* **2023**, *30*, 98540–98547. [CrossRef]
68. Ullah, M.; Ali, M.E.; Hamid, S.B.A. Surfactant-assisted ball milling: A novel route to novel materials with controlled nanostructure—A review. *Rev. Adv. Mater. Sci.* **2014**, *37*, 1–14.
69. Yeganeh-Faal, A.; Bordbar, M.; Negahdar, N.; Nasrollahzadeh, M. Green synthesis of the Ag/ZnO nanocomposite using *Valeriana officinalis* L. root extract: Application as a reusable catalyst for the reduction of organic dyes in a very short time. *IET Nanobiotechnol.* **2017**, *11*, 669–676. [CrossRef]
70. Kaabipour, S.; Hemmati, S. A review on the green and sustainable synthesis of silver nanoparticles and one-dimensional silver nanostructures. *Beilstein J. Nanotechnol.* **2021**, *12*, 102–136. [CrossRef]
71. Enan, E.T.; Ashour, A.A.; Basha, S.; Felemban, N.H.; Gad El-Rab, S.M.F. Antimicrobial activity of biosynthesized silver nanoparticles, amoxicillin, and glass-ionomer cement against *Streptococcus mutans* and *Staphylococcus aureus*. *Nanotechnology* **2021**, *32*, 215101. [CrossRef] [PubMed]
72. Natsuki, J. A review of silver nanoparticles: Synthesis methods, properties, and applications. *Int. J. Mater. Sci. Appl.* **2015**, *4*, 325–332. [CrossRef]
73. Ashkarran, A.A. A novel method for synthesis of colloidal silver nanoparticles by arc discharge in liquid. *Curr. Appl. Phys.* **2010**, *10*, 1442–1447. [CrossRef]
74. Uttayarat, P.; Eamsiri, J.; Tangthong, T.; Suwanmala, P. Radiolytic synthesis of colloidal silver nanoparticles for antibacterial wound dressings. *Adv. Mater. Sci. Eng.* **2015**, *2015*, 376082. [CrossRef]
75. Keskar, M.; Sabatini, C.; Cheng, C.; Swihart, M.T. Synthesis and characterization of silver nanoparticle-loaded amorphous calcium phosphate microspheres for dental applications. *Nanoscale Adv.* **2019**, *1*, 627–635. [CrossRef] [PubMed]
76. Rai, M.; Ingle, A.P.; Birla, S.; Yadav, A.; Dos Santos, C.A. Strategic role of selected noble metal nanoparticles in medicine. *Crit. Rev. Microbiol.* **2016**, *42*, 696–719. [CrossRef]
77. Milea, C.A.; Bogatu, C.; Duta, A. The influence of parameters in silica sol-gel process. *Eng. Sci.* **2011**, *4*, 59–66.
78. Piszczek, P.; Radtke, A. Silver nanoparticles fabricated using chemical vapor deposition and atomic layer deposition techniques: Properties, applications, and perspectives: Review. In *Noble Precious Metals: Properties, Nanoscale Effects, and Applications*; InTech: London, UK, 2018. [CrossRef]
79. Stark, W.J.; Stoessel, P.R.; Wohlleben, W.; Hafner, A. Industrial applications of nanoparticles. *Chem. Soc. Rev.* **2015**, *44*, 5793–5805. [CrossRef]
80. Rafique, M.; Sadaf, I.; Rafique, M.S.; Tahir, M.B. A review on green synthesis of silver nanoparticles and their applications. *Artif. Cells Nanomed. Biotechnol.* **2017**, *45*, 1272–1291. [CrossRef]

81. Shahzadi, S.; Fatima, S.; Qurat ul Ain Shafiq, Z.; Janjua, M.R.S.A. A review on green synthesis of silver nanoparticles (SNPs) using plant extracts: A multifaceted approach in photocatalysis, environmental remediation, and biomedicine. *RSC Adv.* **2025**, *15*, 3858–3903. [CrossRef]
82. Ahmad, S.; Ahmad, H.; Khan, I.; Alghamdi, S.; Almehmadi, M.; Ali, M.; Ullah, A.; Hussain, H.; Khan, N.; Ali, F. Green synthesis of gold nanoparticles using *Delphinium chitralense* tuber extracts, their characterization, and enzyme inhibitory potential. *Braz. J. Biol.* **2022**, *82*, e257622. [CrossRef] [PubMed]
83. Arshad, H.; Saleem, M.; Pasha, U.; Sadaf, S. Synthesis of Aloe vera-conjugated silver nanoparticles for use against multidrug-resistant microorganisms. *Electron. J. Biotechnol.* **2022**, *55*, 55–64. [CrossRef]
84. Erenler, R.; Dag, B. Biosynthesis of silver nanoparticles using *Origanum majorana* L. and evaluation of their antioxidant activity. *Inorg. Nano-Met. Chem.* **2022**, *52*, 485–492. [CrossRef]
85. Fanoro, O.T.; Parani, S.; Maluleke, R.; Lebepe, T.C.; Varghese, J.R.; Mavumengwana, V.; Oluwafemi, O.S. Facile green, room-temperature synthesis of gold nanoparticles using *Combretum erythrophyllum* leaf extract: Antibacterial and cell viability studies against normal and cancerous cells. *Antibiotics* **2021**, *10*, 893. [CrossRef] [PubMed]
86. Ghaffar, S.; Abbas, A.; Naeem-ul-Hassan, M.; Assad, N.; Sher, M.; Ullah, S.; Alhazmi, H.A.; Najmi, A.; Zoghebi, K.; Al Bratty, M.; et al. Improved photocatalytic and antioxidant activity of olive fruit extract-mediated ZnO nanoparticles. *Antioxidants* **2023**, *12*, 1201. [CrossRef]
87. Singh, D.; Lodhi, R.; Nayak, S. A review on green synthesis and physiological properties of silver nanoparticles. *Int. J. Res. Sci. Innov.* **2025**, *12*, 1470–1479. [CrossRef]
88. Küünal, S.; Rauwel, P.; Rauwel, E. Plant extract mediated synthesis of nanoparticles. In *Emerging Applications of Nanoparticles and Architecture Nanostructures: Current Prospects and Future Trends*; Elsevier: Amsterdam, The Netherlands, 2018; pp. 411–446. [CrossRef]
89. Jadoun, S.; Arif, R.; Jangid, N.K.; Meena, R.K. Green synthesis of nanoparticles using plant extracts: A review. *Environ. Chem. Lett.* **2021**, *19*, 355–374. [CrossRef]
90. Shaban, A.S.; Owda, M.E.; Basuoni, M.M.; Mousa, M.A.; Radwan, A.A.; Saleh, A.K. *Punica granatum* peel extract mediated green synthesis of zinc oxide nanoparticles: Structure and evaluation of their biological applications. *Biomass Convers. Biorefin.* **2024**, *14*, 12265–12281. [CrossRef]
91. Ehsan, M.; Waheed, A.; Ullah, A.; Kazmi, A.; Ali, A.; Raja, N.I.; Mashwani, Z.U.R.; Sultana, T.; Mustafa, N.; Ikram, M.; et al. Plant-based bimetallic silver-zinc oxide nanoparticles: A comprehensive perspective of synthesis, biomedical applications, and future trends. *BioMed Res. Int.* **2022**, *2022*, 1215183. [CrossRef]
92. Singh, S.P.; Mishra, A.; Shyanti, R.K.; Singh, R.P.; Acharya, A. Silver nanoparticles synthesized using *Carica papaya* leaf extract (AgNPs-PLE) causes cell cycle arrest and apoptosis in human prostate (DU145) cancer cells. *Biol. Trace Elem. Res.* **2021**, *199*, 1316–1331. [CrossRef]
93. Lal, S.; Verma, R.; Chauhan, A.; Dhatwalia, J.; Guleria, I.; Ghotekar, S.; Thakur, S.; Mansi, K.; Kumar, R.; Kumari, A.; et al. Antioxidant, antimicrobial, and photocatalytic activity of green synthesized ZnO-NPs from *Myrica esculenta* fruits extract. *Inorg. Chem. Commun.* **2022**, *141*, 109518. [CrossRef]
94. Mthana, M.S.; Mthiyane, D.M.N.; Onwudiwe, D.C.; Singh, M. Biosynthesis of ZnO nanoparticles using *Capsicum chinense* fruit extract and their in vitro cytotoxicity and antioxidant assay. *Appl. Sci.* **2022**, *12*, 4451. [CrossRef]
95. Nawabjohn, M.S.; Sivaprakasam, P.; Anandasadagopan, S.K.; Begum, A.A.; Pandurangan, A.K. Green synthesis and characterisation of silver nanoparticles using *Cassia tora* seed extract and investigation of antibacterial potential. *Appl. Biochem. Biotechnol.* **2022**, *194*, 464–478. [CrossRef] [PubMed]
96. Giri, A.K.; Jena, B.; Biswal, B.; Pradhan, A.K.; Arakha, M.; Acharya, S.; Acharya, L. Green synthesis and characterization of silver nanoparticles using *Eugenia roxburghii* DC. extract and activity against biofilm-producing bacteria. *Sci. Rep.* **2022**, *12*, 8383. [CrossRef]
97. Faisal, S.; Jan, H.; Shah, S.A.; Shah, S.; Khan, A.; Akbar, M.T.; Rizwan, M.; Jan, F.; Wajidullah; Akhtar, N.; et al. Green synthesis of zinc oxide (ZnO) nanoparticles using aqueous fruit extracts of *Myristica fragrans*: Their characterizations and biological and environmental applications. *ACS Omega* **2021**, *6*, 9709–9722. [CrossRef]
98. Küünal, S.; Visnapuu, M.; Volubujeva, O.; Soares Rosario, M.; Rauwel, P.; Rauwel, E. Optimisation of plant mediated synthesis of silver nanoparticles by common weed *Plantago major* and their antimicrobial properties. *IOP Conf. Ser. Mater. Sci. Eng.* **2019**, *613*, 012003. [CrossRef]
99. Edith, E.I.; Uzor, P.F. Green synthesis of silver nanoparticles using *Euphorbia hirta* leaf extract and the determination of their antimicrobial activity. *Adv. Biosci. Bioeng.* **2021**, *9*, 124–129. [CrossRef]
100. Alzubaidi, A.K.; Al-Kaabi, W.J.; Ali, A.A.; Albukhaty, S.; Al-Karagoly, H.; Sulaiman, G.M.; Asiri, M.; Khane, Y. Green Synthesis and Characterization of Silver Nanoparticles Using Flaxseed Extract and Evaluation of Their Antibacterial and Antioxidant Activities. *Appl. Sci.* **2023**, *13*, 2182. [CrossRef]

101. Abootalebi, S.N.; Mousavi, S.M.; Hashemi, S.A.; Shorafa, E.; Omidifar, N.; Gholami, A. Antibacterial effects of green-synthesized silver nanoparticles using *Ferula asafoetida* against *Acinetobacter baumannii* isolated from the hospital environment and assessment of their cytotoxicity on human cell lines. *J. Nanomater.* **2021**, *2021*, 6676555. [CrossRef]
102. Safi Ur Rehman Qamar, S.; Tanwir, S.; Khan, W.A.; Altaf, J.; Ahmad, J.N. Biosynthesis of silver nanoparticles using *Ocimum tenuiflorum* extract and its efficacy assessment against *Helicoverpa armigera*. *Int. J. Pest Manag.* **2024**, *70*, 375–383. [CrossRef]
103. Hassan Afandy, H.; Sabir, D.K.; Aziz, S.B. Antibacterial Activity of the Green Synthesized Plasmonic Silver Nanoparticles with Crystalline Structure against Gram-Positive and Gram-Negative Bacteria. *Nanomaterials* **2023**, *13*, 1327. [CrossRef] [PubMed]
104. Murali, M.; Thampy, A.; Anandan, S.; Aiyaz, M.; Shilpa, N.; Singh, S.B.; Gowtham, H.G.; Ramesh, A.M.; Rahdar, A.; Kyzas, G.Z. Competent antioxidant and antiglycation properties of zinc oxide nanoparticles (ZnO NPs) phyto-fabricated from aqueous leaf extract of *Boerhaavia erecta* L. *Environ. Sci. Pollut. Res.* **2023**, *30*, 56731–56742. [CrossRef]
105. Güngörmüş, M. Synthesis and characterization of silver nanoparticles using *Myrtus communis* (myrtle) and grape seed extract. *Gazi J. Eng. Sci.* **2021**, *7*, 320–329. [CrossRef]
106. Meera, S.D.; Naidoo, Y.; Dewir, Y.H.; Lin, J.; Rihan, H.Z. Green synthesis of silver nanoparticles from Heteropyxisnatalensis leaf extract and their potential antibacterial efficacy. *ScienceAsia* **2022**, *48*, 196–201. [CrossRef]
107. Periasamy, S.; Jegadeesan, U.; Sundaramoorthi, K.; Rajeswari, T.; Tokala, V.N.B.; Bhattacharya, S.; Muthusamy, S.; Sankoh, M.; Nellore, M.K. Comparative analysis of synthesis and characterization of silver nanoparticles extracted using leaf, flower, and bark of *Hibiscus rosasinensis* and examine its antimicrobicidal activity. *J. Nanomater.* **2022**, *2022*, 8123854. [CrossRef]
108. Singh, R.; Sagar, N.A.; Kumar, N. Bio-inspired green fabrication of silver nanoparticles (AgNPs) using aqueous leaf extract of *Ipomoea carnea* Jacq. to tackle multiple drug resistance MTCC bacterial strains. *Eur. J. Med. Chem. Rep.* **2022**, *6*, 100066. [CrossRef]
109. Yang, Q.; Guo, J.; Long, X.; Pan, C.; Liu, G.; Peng, J. Green Synthesis of Silver Nanoparticles Using *Jasminum nudiflorum* Flower Extract and Their Antifungal and Antioxidant Activity. *Nanomaterials* **2023**, *13*, 2558. [CrossRef]
110. Labulo, A.H.; David, O.A.; Terna, A.D. Green synthesis and characterization of silver nanoparticles using *Morinda lucida* leaf extract and evaluation of its antioxidant and antimicrobial activity. *Chem. Pap.* **2022**, *76*, 7313–7325. [CrossRef]
111. Sreelekha, E.; George, B.; Shyam, A.; Sajina, N.; Mathew, B. A comparative study on the synthesis, characterization, and antioxidant activity of green and chemically synthesized silver nanoparticles. *BioNanoScience* **2021**, *11*, 489–496. [CrossRef]
112. Manojkumar, U.; Kaliannan, D.; Srinivasan, V.; Balasubramanian, B.; Kamyab, H.; Mussa, Z.H.; Palaniyappan, J.; Mesbah, M.; Chelliapan, S.; Palaninaicker, S. Green synthesis of zinc oxide nanoparticles using *Brassica oleracea* var. botrytis leaf extract: Photocatalytic, antimicrobial and larvicidal activity. *Chemosphere* **2023**, *323*, 138263. [CrossRef]
113. Khatun, M.; Khatun, Z.; Karim, M.R.; Habib, M.R.; Rahman, M.H.; Aziz, M.A. Green synthesis of silver nanoparticles using extracts of *Mikania cordata* leaves and evaluation of their antioxidant, antimicrobial, and cytotoxic properties. *Food Chem. Adv.* **2023**, *3*, 100386. [CrossRef]
114. Fatimah, I.; Hidayat, H.; Purwiandono, G.; Khoirunisa, K.; Zahra, H.A.; Audita, R.; Sagadevan, S. Green Synthesis of Antibacterial Nanocomposite of Silver Nanoparticle-Doped Hydroxyapatite Utilizing *Curcuma longa* Leaf Extract and Land Snail (*Achatina fulica*) Shell Waste. *J. Funct. Biomater.* **2022**, *13*, 84. [CrossRef] [PubMed]
115. Tailor, G.; Yadav, B.L.; Chaudhary, J.; Joshi, M.; Suvalka, C. Green synthesis of silver nanoparticles using *Ocimum canum* and their antibacterial activity. *Biochem. Biophys. Rep.* **2020**, *24*, 100848. [CrossRef]
116. Manikandan, D.B.; Sridhar, A.; Sekar, R.K.; Perumalsamy, B.; Veeran, S.; Arumugam, M.; Karuppaiah, P.; Ramasamy, T. Green fabrication, characterization of silver nanoparticles using aqueous leaf extract of *Ocimum americanum* (Hoary Basil) and investigation of its in vitro antibacterial, antioxidant, anticancer, and photocatalytic reduction. *J. Environ. Chem. Eng.* **2021**, *9*, 104845. [CrossRef]
117. Sharifi-Rad, M.; Pohl, P.; Epifano, F. Phytofabrication of Silver Nanoparticles (AgNPs) with Pharmaceutical Capabilities Using *Otostegia persica* (Burm.) Boiss. Leaf Extract. *Nanomaterials* **2021**, *11*, 1045. [CrossRef]
118. Piramila, H.B.; Kavitha, R.; Oviya, B. Green synthesis of silver nanoparticles using *Pisonia alba* L. and its antioxidant activity. *Res. J. Agric. Sci.* **2024**, *15*, 1033–1036.
119. Saxena, R.; Kotnala, S.; Bhatt, S.C.; Uniyal, M.; Rawat, B.S.; Negi, P.; Riyal, M.K. A review on green synthesis of nanoparticles toward sustainable environment. *Sustain. Chem. Clim. Action* **2025**, *6*, 100071. [CrossRef]
120. Nandhini, J.; Karthikeyan, E.; Rajeshkumar, S. Green synthesis of zinc oxide nanoparticles: Eco-friendly advancements for biomedical marvels. *Resour. Chem. Mater.* **2024**, *3*, 294–316. [CrossRef]
121. Alyami, M.H.; Fakhry, A.M.; El Halfawy, N.M.; Toto, S.M.; Sedky, N.K.; Yassin, H.A.; Fahmy, S.A.; Mokhtar, F.A. *Retama monosperma* chemical profile, green synthesis of silver nanoparticles, and antimicrobial potential: A study supported by network pharmacology and molecular docking. *RSC Adv.* **2023**, *13*, 26213. [CrossRef]
122. Gulbagca, F.; Ozdemir, S.; Gulcan, M.; Sen, F. Synthesis and characterization of *Rosa canina*-mediated biogenic silver nanoparticles for antioxidant, antibacterial, antifungal, and DNA cleavage activities. *Heliyon* **2019**, *5*, e02980. [CrossRef]

123. Lalsangpuii, F.; Rokhum, S.L.; Nghakliana, F.; Fakawmi, L.; Ruatpuia, J.V.L.; Laltlanmawii, E.; Lalfakzuala, R.; Siama, Z. Green synthesis of silver nanoparticles using *Spilanthes acmella* leaf extract and its antioxidant-mediated ameliorative activity against doxorubicin-induced toxicity in Dalton's lymphoma ascites (DLA)-bearing mice. *ACS Omega* **2022**, *7*, 44346–44359. [CrossRef] [PubMed]
124. Rengarajan, S.; Thangavel, N.; Sivalingam, A.M.; Lakshmanan, G.; Selvakumari, J.; Pandian, A. Green synthesis and characterization of silver nanoparticles with different solvent extracts of *Sesbania grandiflora* (L.) Poiret and assessment of their antibacterial and antioxidant potentials. *Biomass Convers. Biorefin.* **2023**. [CrossRef]
125. Karan, T.; Gonulalan, Z.; Erenler, R.; Kolemen, U.; Eminagaoglu, O. Green synthesis of silver nanoparticles using *Sambucus ebulus* leaves extract: Characterization, quantitative analysis of bioactive molecules, antioxidant and antibacterial activities. *J. Mol. Struct.* **2024**, *1296 Pt 1*, 136836. [CrossRef]
126. Pérez-Marroquín, X.A.; Aguirre-Cruz, G.; Campos-Lozada, G.; Callejas-Quijada, G.; León-López, A.; Campos-Montiel, R.G.; García-Hernández, L.; Méndez-Albores, A.; Vázquez-Durán, A.; Aguirre-Álvarez, G. Green Synthesis of Silver Nanoparticles for Preparation of Gelatin Films with Antimicrobial Activity. *Polymers* **2022**, *14*, 3453. [CrossRef]
127. Tharani, M.; Rajeshkumar, S.; Al-Ghanim, K.A.; Nicoletti, M.; Sachivkina, N.; Govindarajan, M. *Terminalia chebula*-Assisted Silver Nanoparticles: Biological Potential, Synthesis, Characterization, and Ecotoxicity. *Biomedicines* **2023**, *11*, 1472. [CrossRef]
128. Bawazeer, S.; Rauf, A.; Shah, S.U.A.; Shawky, A.M.; Al-Awthan, Y.S.; Bahattab, O.S.; Uddin, G.; Sabir, J.; El-Esawi, M.A. Green synthesis of silver nanoparticles using *Tropaeolum majus*: Phytochemical screening and antibacterial studies. *Green Process. Synth.* **2021**, *10*, 85–94. [CrossRef]
129. Hamad, I.; Aleidi, S.M.; Alshaer, W.; Twal, S.; Al Olabi, M.; Bustanji, Y. Advancements and global perspectives in the green synthesis of silver nanoparticles: A two-decade analysis. *Pharmacia* **2025**, *72*, 1–13. [CrossRef]
130. Fouda, A.; Saied, E.; Eid, A.M.; Kouadri, F.; Alemam, A.M.; Hamza, M.F.; Alharbi, M.; Elkelish, A.; Hassan, S.E.-D. Green Synthesis of Zinc Oxide Nanoparticles Using an Aqueous Extract of *Punica granatum* for Antimicrobial and Catalytic Activity. *J. Funct. Biomater.* **2023**, *14*, 205. [CrossRef]
131. Annapoorani, A.; Koodalingam, A.; Beulaja, M.; Saiprasad, G.; Chitra, P.; Stephen, A.; Palanisamy, S.; Prabhu, N.M.P.; You, S.G.; Janarthanan, S.; et al. Eco-friendly synthesis of zinc oxide nanoparticles using *Rivina humilis* leaf extract and their biomedical applications. *Process Biochem.* **2022**, *112*, 192–202. [CrossRef]
132. Murali, M.; Manjula, S.; Shilpa, N.; Ravishankar, D.K.; Shivakumara, C.S.; Thampy, A.; Ayeshamariam, A.; Pandey, S.; Anandan, S.; Amruthesh, K.N.; et al. Facile synthesis of ZnO-NPs from yellow creeping daisy (*Sphagneticola trilobata* L.) attenuates cell proliferation by inducing cellular level apoptosis against colon cancer. *J. King Saud Univ.-Sci.* **2022**, *34*, 102084. [CrossRef]
133. Dey, A.; Somaiah, S. Green synthesis and characterization of zinc oxide nanoparticles using leaf extract of *Thryallis glauca* (Cav.) Kuntze and their role as antioxidant and antibacterial. *Microsc. Res. Tech.* **2022**, *85*, 2835–2847. [CrossRef] [PubMed]
134. Nagpal, K.; Rosario, M.S.; Praakle, K.; Rauwel, E.; Rauwel, P. Investigation of the optical and antibacterial properties of biosynthesized ZnO nanoparticles using Thuja tincture. *J. Phys. Conf. Ser.* **2022**, *2315*, 012016. [CrossRef]
135. Fahim, M.; Shahzaib, A.; Nishat, N.; Jahan, A.; Bhat, T.A.; Inam, A. Green synthesis of silver nanoparticles: A comprehensive review of methods, influencing factors, and applications. *JCIS Open* **2024**, *16*, 100125. [CrossRef]
136. Panchal, P.; Paul, D.R.; Sharma, A.; Choudhary, P.; Meena, P.; Nehra, S.P. Biogenic mediated Ag/ZnO nanocomposites for photocatalytic and antibacterial activities towards disinfection of water. *J. Colloid Interface Sci.* **2020**, *563*, 370–380. [CrossRef]
137. Panchal, P.; Paul, D.R.; Gautam, S.; Meena, P.; Nehra, S.P.; Maken, S.; Sharma, A. Photocatalytic and antibacterial activities of green synthesized Ag doped MgO nanocomposites towards environmental sustainability. *Chemosphere* **2022**, *297*, 134182. [CrossRef]
138. Panchal, P.; Paul, D.R.; Sharma, A.; Hooda, D.; Yadav, R.; Meena, P.; Nehra, S.P. Phytoextract mediated ZnO/MgO nanocomposites for photocatalytic and antibacterial activities. *J. Photochem. Photobiol. A Chem.* **2019**, *385*, 112049. [CrossRef]
139. Panchal, P.; Meena, P.; Nehra, S.P. A rapid green synthesis of Ag/AgCl-NC photocatalyst for environmental applications. *Environ. Sci. Pollut. Res. Int.* **2021**, *28*, 3972–3982. [CrossRef]
140. Panchal, P.; Sharma, R.; Sudarshan Reddy, A.; Nehra, K.; Sharma, A.; Nehra, S.P. Eco-friendly synthesis of Ag-doped ZnO/MgO as a potential photocatalyst for antimicrobial and dye degradation applications. *Coord. Chem. Rev.* **2023**, *493*, 215283. [CrossRef]
141. Paul, D.R.; Sharma, R.; Sharma, A.; Panchal, P.; Singh, A.; Chaudhary, S.; Nehra, S.P. Structural properties of Mg − x wt% Co (x = 0, 5, 10 & 20) nanocomposites for hydrogen storage applications. *Mater. Today Proc.* **2021**, *42*, 1713–1717. [CrossRef]
142. Garg, A.; Almáši, M.; Bednarčík, J.; Sharma, R.; Rao, V.S.; Panchal, P.; Jain, A.; Sharma, A. Gd(III) metal-organic framework as an effective humidity sensor and its hydrogen adsorption properties. *Chemosphere* **2022**, *305*, 135467. [CrossRef]
143. Rauwel, E.; Arya, G.; Praakle, K.; Rauwel, P. Use of Aloe Vera Gel as Media to Assess Antimicrobial Activity and Development of Antimicrobial Nanocomposites. *Int. J. Mol. Sci.* **2024**, *25*, 5599. [CrossRef] [PubMed]
144. Khan, S.; Qi, K.; Khan, I.; Wang, A.; Liu, J.; Humayun, M.; Khan, A.; Bahadur, A.; Alanazi, A.F.; Bououdina, M. Eco-friendly graphitic carbon nitride nanomaterials for the development of innovative biomaterials: Preparation, properties, opportunities, current trends, and future outlook. *J. Saudi Chem. Soc.* **2023**, *27*, 101753. [CrossRef]

145. Panthi, G.; Park, M. Graphitic carbon nitride/zinc oxide-based Z-scheme and S-scheme heterojunction photocatalysts for the photodegradation of organic pollutants. *Int. J. Mol. Sci.* **2023**, *24*, 15021. [CrossRef] [PubMed]
146. Padmanabhan, V.P.; Sivashanmugam, P.; Mubashera, S.M.; Sagadevan, S.; Kulandaivelu, R. The development of ZnO nanoparticle-embedded graphitic-carbon nitride towards triple-negative breast cancer therapy. *RSC Adv.* **2023**, *13*, 24333–24342. [CrossRef] [PubMed]
147. Tsekov, R.; Georgiev, P.; Simeonova, S.; Balashev, K. Quantifying the Blue Shift in the Light Absorption of Small Gold Nanoparticles. *arXiv* **2017**, arXiv:1702.04513.
148. Anjum, S.; Hashim, M.; Malik, S.A.; Khan, M.; Lorenzo, J.M.; Abbasi, B.H.; Hano, C. Recent advances in zinc oxide nanoparticles (ZnO NPs) for cancer diagnosis, target drug delivery, and treatment. *Cancers* **2021**, *13*, 4570. [CrossRef]
149. Lan, P.T.; Hao, N.H.; Hieu, N.T.; Ha, N.T.T.; Brown, C.T.; Cam, L.M. Graphitic carbon nitride supported silver nanoparticles (Ag-NPs/g-C_3N_4): Synthesis and photocatalytic behaviour in the degradation of 2,4-dichlorophenoxyacetic acid. *RSC Adv.* **2024**, *14*, 19014–19028. [CrossRef]
150. Wang, X.; Tan, F.; Wang, W.; Qiao, X.; Qiu, X.; Chen, J. Anchoring of silver nanoparticles on graphitic carbon nitride sheets for the synergistic catalytic reduction of 4-nitrophenol. *Chemosphere* **2017**, *172*, 147–154. [CrossRef]
151. Nunna, G.P.; Rosaiah, P.; Sangaraju, S.; Ramalingam, G.; Jwuiyad, A.; Adem, S.; Ko, T.J. Mesostructured graphitic carbon nitride composites with silver nanoparticle decoration as the best visible-light-driven photocatalysts for dye degradation and H_2 production. *Colloids Surf. A Physicochem. Eng. Asp.* **2024**, *680*, 132615. [CrossRef]
152. Pawar, R.C.; Kang, S.; Ahn, S.H.; Lee, C.S. Gold nanoparticle modified graphitic carbon nitride/multi-walled carbon nanotube (g-C_3N_4/CNTs/Au) hybrid photocatalysts for effective water splitting and degradation. *RSC Adv.* **2015**, *5*, 24281–24292. [CrossRef]
153. Arushi Sharma, A.; Arora, A.; Mehta, N.; Kataria, R.; Mehta, S.K. Au nanoparticles decorated graphitic carbon nitride nanosheets as a sensitive and selective fluorescence probe for Fe^{3+} and dichromate ions in aqueous medium. *Chemosphere* **2024**, *363*, 142834. [CrossRef] [PubMed]
154. Sukweenadhi, J.; Setiawan, K.I.; Avanti, C.; Kartini, K.; Rupa, E.J.; Yang, D.-C. Scale-up of green synthesis and characterization of silver nanoparticles using ethanol extract of Plantago major L. leaf and its antibacterial potential. *S. Afr. J. Chem. Eng.* **2021**, *38*, 1–8. [CrossRef]
155. Panchal, P.; Rauwel, P.; Sharma, S.; Nehra, S.P.; Estephan, E.; Praakle, K.; Rauwel, E. *Ocimum tenuiflorum* leaf-mediated graphitic carbon nitride and ZnO/GCN nanohybrid: A sustainable approach for environmental applications. *Environ. Sci. Pollut. Res.* **2025**, *32*, 9945–9965. [CrossRef] [PubMed]
156. Rauwel, P.; Küünal, S.; Ferdov, S.; Rauwel, E. A review on the green synthesis of silver nanoparticles and their morphologies studied via TEM. *Adv. Mater. Sci. Eng.* **2015**, *2015*, 682749. [CrossRef]
157. Ying, S.; Guan, Z.; Ofoegbu, P.C.; Clubb, P.; Rico, C.; He, F.; Hong, J. Green synthesis of nanoparticles: Current developments and limitations. *Environ. Technol. Innov.* **2022**, *26*, 102336. [CrossRef]
158. Sastry, A.B.S.; Karthik Aamanchi, R.B.; Sree Rama Linga Prasad, C.; Murty, B.S. Large-scale green synthesis of Cu nanoparticles. *Environ. Chem. Lett.* **2013**, *11*, 183–187. [CrossRef]

pharmaceuticals

MDPI

Review

Current Overview of Metal Nanoparticles' Synthesis, Characterization, and Biomedical Applications, with a Focus on Silver and Gold Nanoparticles

Ana Flavia Burlec [1,†], Andreia Corciova [1,†], Monica Boev [2,*], Denisa Batir-Marin [2,*], Cornelia Mircea [1], Oana Cioanca [1], Gabriela Danila [2], Marius Danila [2], Anca Florentina Bucur [2] and Monica Hancianu [1]

1 Faculty of Pharmacy, "Grigore T. Popa" University of Medicine and Pharmacy Iasi, 700115 Iasi, Romania; ana-flavia.l.burlec@umfiasi.ro (A.F.B.); maria.corciova@umfiasi.ro (A.C.); cornelia.mircea@umfiasi.ro (C.M.); oana.cioanca@umfiasi.ro (O.C.); monica.hancianu@umfiasi.ro (M.H.)

2 Research Centre in the Medical-Pharmaceutical Field, Faculty of Medicine and Pharmacy, "Dunarea de Jos" University of Galati, 800008 Galati, Romania; gd243@student.ugal.ro (G.D.); dumitru.danila@ugal.ro (M.D.); fb173@student.ugal.ro (A.F.B.)

* Correspondence: monica.boev@ugal.ro (M.B.); denisa.batir@ugal.ro (D.B.-M.)

† These authors contributed equally to this work.

Abstract: Metal nanoparticles (NPs) have garnered considerable attention, due to their unique physicochemical properties, that render them promising candidates for various applications in medicine and industry. This article offers a comprehensive overview of the most recent advancements in the manufacturing, characterization, and biomedical utilization of metal NPs, with a primary focus on silver and gold NPs. Their potential as effective anticancer, anti-inflammatory, and antimicrobial agents, drug delivery systems, and imaging agents in the diagnosis and treatment of a variety of disorders is reviewed. Moreover, their translation to therapeutic settings, and the issue of their inclusion in clinical trials, are assessed in light of over 30 clinical investigations that concentrate on administering either silver or gold NPs in conditions ranging from nosocomial infections to different types of cancers. This paper aims not only to examine the biocompatibility of nanomaterials but also to emphasize potential challenges that may limit their safe integration into healthcare practices. More than 100 nanomedicines are currently on the market, which justifies ongoing study into the use of nanomaterials in medicine. Overall, the present review aims to highlight the potential of silver and gold NPs as innovative and effective therapeutics in the field of biomedicine, citing some of their most relevant current applications.

Keywords: inorganic nanoparticles; biological synthesis; anticancer activity; antimicrobial activity; clinical trials

1. Introduction

Over the past two decades, significant advancements in nanotechnology have been recorded, since researchers have discovered multiple ways to design and use nanostructured materials in a variety of fields (e.g., medicine, pharmacy, biology, the food and textiles industries, agriculture and electronics) [1].

The ability to control and direct the behavior of atoms and molecules is a key component of nanoscience. The etymology of the word "nano" has its origin in the Greek word "nanos", which means "dwarf"; "nanotechnology" thus refers to particles with sizes between 1 and 100 nm, equivalent to three linearly aligned atoms' lengths [2], dispersed in a solid, liquid or gaseous phase [3,4].

The unique characteristics of such particles, including their size, structure, and composition, make them useful as disinfectants, genetic vector and medication delivery systems, nano-magnets, or as water depollution vectors [5,6].

Nanoparticles (NPs) can vary largely in size. Generally, they can be *zero-dimensional*, meaning that their length, thickness and height are all determined by a single point, as in the case of nano dots; *one-dimensional*, meaning that they only present a single parameter, as in the case of nanotubes; *two-dimensional*, such as graphene; and *three-dimensional*, such as metal NPs (silver, gold, copper, etc.). NPs can also have a wide range of structures and shapes (e.g., spherical, cylindrical, tubular, conical, helical), while some can be flat or even irregular. NPs can be aerated or compact, amorphous or crystalline, with one or more solid crystals [7].

The uniqueness and specificity of such materials are usually granted by the synthesis method. Even a small change in the technological process can lead to major differences in their essential properties. After performing the synthesis through one of the two major routes, NPs should be characterized in terms of size, composition, electric surface charge and surface morphology. This characterization can be achieved using various techniques, that range from visual inspection to XRD (X-Ray Diffraction) analysis [8,9].

Following a general classification of nanomaterials, this review focuses on the synthesis, characterization, and biomedical applications, as well as on the cytotoxicity, of metal NPs, with a specific focus on silver and gold NPs. Such materials have the appealing capacity of improving the stability and solubility of encapsulated cargos, of facilitating transport across cell membranes, and of facilitating the delivery at specific levels [10].

Noble elements such as gold, silver, and palladium in the form of NPs represent some of the most promising trends in nanotechnology, especially for creating bioengineering materials that could be used as cutting-edge diagnostic tools and devices to treat severe diseases. For instance, silver and gold NPs generally hold considerable significance in the field of nanotechnology and materials science due to their unique properties and versatile applications. These noble metal NPs stand out for their exceptional optical, catalytic and antimicrobial properties, and effective targeted drug delivery in various types of cancers [11]. Their distinctive plasmonic properties enable them to manipulate light at the nanoscale, making them invaluable in areas such as photonics and sensor technology. Additionally, silver NPs are renowned for their potent antimicrobial properties, thus finding widespread use in medical applications and as disinfectants [12]. Gold NPs, on the other hand, exhibit remarkable catalytic activity and are employed in various chemical reactions and drug delivery systems [13]. Compared to other metal or metal oxide NPs, silver and gold NPs often excel in terms of stability and tunable surface chemistry. Furthermore, they are also well-known for their capacity to efficiently convert and manipulate energy and matter at nanoscale. These qualities make them indispensable tools for researchers seeking innovative solutions in fields ranging from medicine to environmental science [14,15]. Moreover, such NPs can be found in a variety of everyday items, such as personal care items, food-storage containers, cosmetics and bandages [8].

The significant interest in nanotechnology can also be seen in the number of results obtained when searching for keywords such as 'nanoparticles' or 'metal nanoparticles', with the latter search currently generating around 170,000 results in SpringerLink and 85,000 results in PubMed, respectively. References from major databases such as PubMed, Science Direct, SpringerNature, Scopus, SpringerLink, Google Scholar, and Web of Science were used to assemble this review. The novelty of the current paper lies in revealing the most recent advancements in the medical use of gold and silver NPs synthesized in various manners, including the most up-to-date incidents of such nanomaterials being included in in vitro, in vivo and clinical studies.

2. Classification of NPs

Nanomaterials can generally be classified depending on the employed synthesis technique and on the origin of the materials used in the process: organic, inorganic, or carbon-based [16]. The main categories of NPs with potential applications in biomedicine are briefly presented below, along with some promising nanomaterial combinations.

2.1. Organic Nanoparticles

Organic NPs are widely employed in the biomedical and pharmaceutical areas, as systems for the controlled release of medicinal substances, notably by injection at the level of specific organs [17]. This category mainly includes micelles, liposomes, dendrimers, polymeric and protein nanoparticles, and nanogels.

Liposomes and micelles are biodegradable NPs that share the characteristic of having a hollow core region, which is also known as a nanocapsule [17,18]. These specific features make them good candidates as drug matrices. Micelles are the most-used release systems for water-insoluble therapeutic agents. The hydrophilic exterior protects them from physiological processes, while the hydrophobic core can contain chemicals that are water insoluble, making them versatile delivery systems for active compounds [19].

Liposomes are synthesized vesicles, that show great flexibility and can be conjugated with different lipid substances, thus having a wide variety of structures, sizes, and compositions. The most important advantage of liposomes is their ability to fuse with cell membranes, followed by the release of the active substance [20,21]. Multilamellar liposomes contain several lipid layers separated by hydrophilic layers, which makes them suitable for the encapsulation of both lipophilic and hydrophilic substances (Figure 1) [22]. Nowadays, liposomes are widely investigated for a variety of therapeutic applications, such as cancer diagnosis and therapy, targeted drug delivery, and antimicrobial therapy [21].

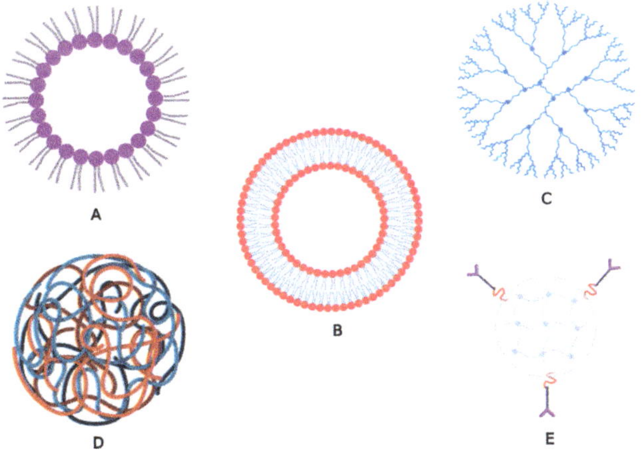

Figure 1. Most well-known organic NPs: (**A**) mycelium; (**B**) liposome; (**C**) dendrimer; (**D**) polymeric nanoparticle; (**E**) nanogel.

Dendrimers are extensively used for clinical applications due to their small size (1–5 nm) and structure, that allows them to be coupled with biocompatible compounds in order to decrease their cytotoxicity. They can act as vehicles for vaccines, genes, or drugs [23,24]. Despite their versatility, some have also shown a high risk of aggregation and toxicity, with only a few representatives being approved by the FDA (Food and Drug Administration) [10].

Polymeric NPs can be either synthetic or natural and can be classified as nanocapsules or nanospheres. These NPs can be synthesized so as to contain therapeutic agents in high concentrations that can be released in a targeted manner [25,26]. For instance, several studies focused on the production and practical applications of poly (methyl methacrylate) (PMMA) dispersions [27,28]. PMMA, a commonly used amorphous synthetic polymer, appears to be promising for biomedical applications due to its lack of toxicity, low cost, minimal inflammatory reactions on tissues and simple processability [29].

Furthermore, protein NPs can be formed through the self-assembly of protein polymers, which consist of isolated proteins derived from animals or plants, such as collagen, gelatin, albumin or elastin. Protein subunits are self-assembled into effective drug delivery vehicles with the help of polymer-based NPs via genetic engineering [30,31]. Abraxane® is one example of an FDA-approved protein nanoparticle drug that enables the delivery of paclitaxel by albumin in different types of breast cancer [32]. Ontak, an engineered protein combining IL-2 and the diphtheria toxin nanoparticle formulation used to treat persistent or recurrent cutaneous T-cell lymphoma, is another such example [33].

Nanogels are defined by the International Union of Pure and Applied Chemistry as "the particle of gel of any shape with an equivalent diameter of approximately 1 to 100 nm" [34]. When compared to other nanocarrier systems, nanogels offer several benefits, including a reduced rate of premature drug release, the capacity to encapsulate therapeutic compounds inside a single formulation, and simple delivery via the parenteral or mucosal routes. Some of their most researched uses include the administration of nucleic acids, cytokines and vaccines, including nasal vaccines, which appear to be the most promising [35].

2.2. *Inorganic Nanoparticles*

Inorganic nanoparticles generally include metal and metal oxide NPs, quantum dots, and silica-based NPs (Figure 2).

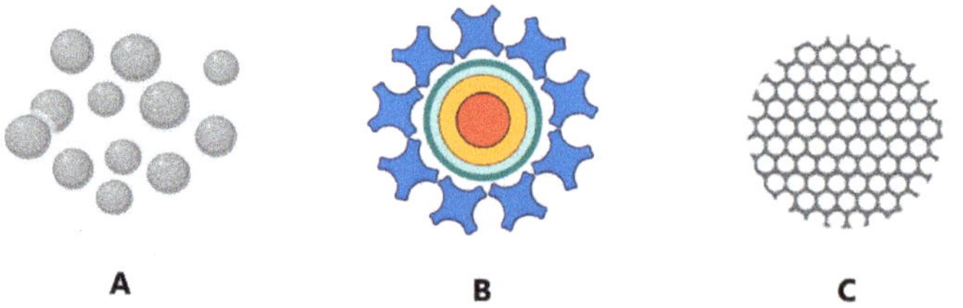

Figure 2. Most common inorganic nanoparticles: (**A**) metal NPs; (**B**) quantum dots, and (**C**) silica-based NPs.

Metal NPs can be synthesized using metals either by the top-down or bottom-up approaches [36]. Nearly all metals can be brought in the form of NPs [37]. The most used metals for the synthesis of NPs are gold (Au), silver (Ag), iron (Fe), aluminum (Al), cadmium (Cd), cobalt (Co), copper (Cu) and zinc (Zn) [16]. All NPs show distinctive general properties such as size, surface–volume ratio, pore size, electronic surface charge and surface density, crystalline or amorphous structure, spherical or cylindrical shape, and color, as well as reactivity and sensitivity to external factors (e.g., humidity, temperature, and light).

Metal oxide NPs are synthesized by modifying the properties of the corresponding metal. For instance, iron-based nanoparticles (FeNPs) spontaneously oxidize to iron oxide in the presence of oxygen, at room temperature, which generates an increase in reactivity compared to FeNPs. Most metal oxide NPs are manufactured using physical or chemical synthesis methods because they have the capacity to decrease in size when exposed to physical forces, such as those inside a rotating reactor, or due to their increased chemical reactivity [38]. Usually, aluminum oxide (Al_2O_3), iron oxide (Fe_2O_3), magnetite (Fe_3O_4), silicon dioxide (SiO_2) and titanium oxide (TiO_2) NPs are synthesized in this manner. These particles possess improved properties compared to those of the metals from which they originate [16,39].

Quantum dots are a certain type of nanoscale particle characterized by their small size and unique optical properties. They are made of semiconductor materials, typically cadmium selenide (CdSe), cadmium telluride (CdTe), or lead sulfide (PbS), and have diameters in the range of 2 to 10 nm. They are used for various applications, including displays, solar cells and bioimaging [40].

Silica-based NPs represent another type of inorganic nanomaterial composed of silicon and oxygen atoms in a crystalline configuration. Biocompatibility, chemical and thermal stability, as well as low toxicity make them promising materials for biomedical applications such as drug delivery, imaging, and biosensing [41].

Semimetal NPs also represent a compelling subgroup within the diverse domain of inorganic NPs. These nanomaterials possess intriguing electronic and optical properties, as well as potential therapeutic effects. Notable examples include selenium [42], antimony [43] and bismuth-based nanomaterials, that possess versatile biomedical applications [44].

Nonetheless, combining metal NPs with other nanostructures has shown remarkable potential for various specific medical applications. One notable example is the integration of gold NPs with silica nanoshells to create photothermal agents. These particles can selectively accumulate in tumor tissues, and when exposed to near-infrared light, they generate localized heat, effectively and specifically destroying cancer cells. This approach, known as photothermal therapy, holds promise in cancer treatment [45]. Gold NPs can also be encapsulated within liposomes, creating versatile drug delivery systems. These hybrid nanostructures enable the controlled release of therapeutic agents and can be guided to specific disease sites such as tumors, enhancing the effectiveness of chemotherapy while minimizing side effects [46,47]. Moreover, mesoporous silica-coated gold nanorods have been used as a theranostic nanoplatform for standard radiation diagnosis and therapy [48]. Similarly, incorporating silver NPs into hydrogel matrices can result in effective antimicrobial wound dressings. These wound dressings release silver ions over time, effectively preventing bacterial infections and promoting faster wound-healing [49].

2.3. Carbon-Based Nanoparticles

Carbon-based nanoparticles can be classified into fullerenes, graphene, carbon nanotubes, carbon nanofibers and carbon black (Figure 3). NPs that are made entirely of carbon are known as carbonaceous NPs [50].

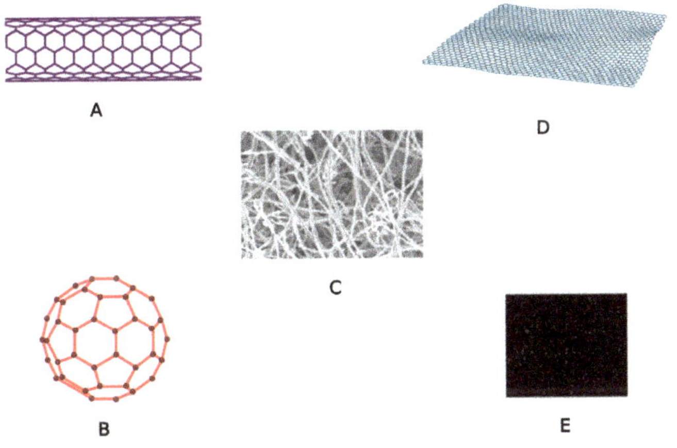

Figure 3. Carbon-based nanomaterials: (**A**) carbon nanotubes; (**B**) fullerenes; (**C**) carbon nanofibers; (**D**) graphene; (**E**) carbon black.

Fullerenes are spherical molecules composed of up to 1500 carbon atoms connected by sp^2 hybridization [51]. Studies have shown that fullerenes can be used to deliver antibiotics and antiviral agents, being also extensively employed in the development of high-performance MRI contrast agents, X-ray imaging contrast agents, photodynamic treatment, and gene delivery [52,53].

Graphene is an allotrope of carbon that consists of carbon atoms arranged in a hexagonal arrangement in a two-dimensional plane [54].

Carbon nanotubes are carbon-based tubular structures with a diameter of 1 nm and a length of 1–100 nm. Given that they can enter the cell by endocytosis or by direct insertion through the cell membrane, their size and shape make them desirable carriers for active substances [55]. In the case of carbon nanofibers, graphene sheets used to make nanotubes can be shaped like a cone or a cup, instead of a cylindrical shape [56].

Carbon black is an amorphous carbon material, generally spherical in shape, with a diameter ranging between 20 to 70 nm. The interaction between particles is so intense that they aggregate, thus forming agglomerates of around 500 nm [57].

3. Synthesis of Metal Nanoparticles

Metal NPs are generally obtained with the use of nanotechnology, by reducing the respective metal to its nuclear size. Their synthesis can be performed using several physical, chemical, and biological techniques.

Physical approaches prevent NPs from becoming contaminated with solvents but need a significant amount of energy for condensation and evaporation. Implicitly, the exceptionally high modulation of temperature and pressure also raises the cost of synthesis.

In chemical engineering, reducing and protective agents are used for the synthesis of NPs and agglomeration is prevented, thus generating NPs with high purity and stability. However, the contamination of synthetic NPs can be caused by a significant number of strong chemicals.

The interest in the biological synthesis of NPs is based on the fact that it offers an eco-friendly and effective alternative to the chemical and physical methods. The biological technique employed in green synthesis varies depending on the type of reducing agent that is used, such as microorganisms (bacteria and fungi) and plants (Figure 4).

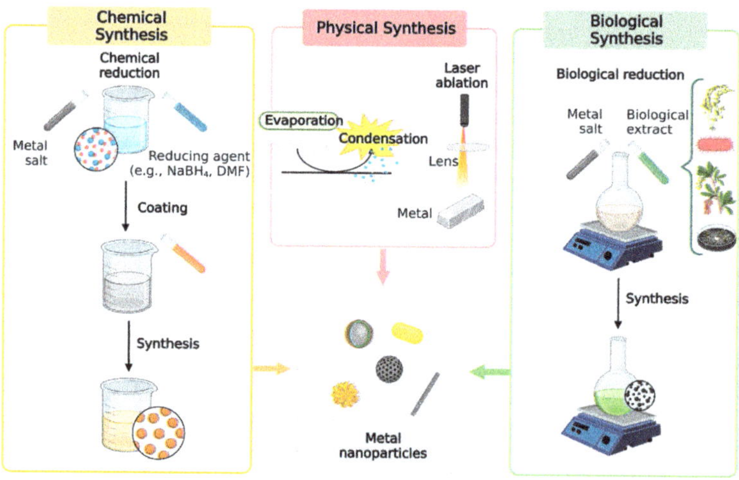

Figure 4. Common synthesis techniques employed for metal NPs.

One of the most cost-effective and environmentally friendly methods of synthesizing metal NPs is through the use of plants, which act as biofactories for NP- production. The potential scale-up of metal NPs obtained with the use of plants is immense due to their abundance, easy availability, and ability to be grown under diverse environmental conditions. Additionally, plant-based synthesis eliminates the need for expensive and hazardous chemicals typically used in conventional methods, making it a safer and more sustainable option. However, there are potential drawbacks associated with the use of plants for NP synthesis, such as variability in the size and shape of the produced NPs due to differences in plant species, plant-growth conditions, and harvesting times [58]. Nonetheless, considering the quantitative aspect, recent studies have reported yields of up to several grams of NPs per kilogram of plant material, making plant-based NP synthesis a viable option for industrial-scale production [59]. Consequently, synthesis using plant extracts is an interesting alternative for the large-scale manufacturing of NPs and has a great deal of potential for medical applications [60].

Considering the applied methods, the synthesis usually follows one of two well-known approaches: bottom-up or top-down.

The *bottom-up approach* is a constructive process that gradually leads to the synthesis of nanomaterials, starting from atoms to clusters, then to NPs. This method generally includes processes such as sol-gel, spinning and biosynthesis [61]. Other examples include sedimentation, reduction, green synthesis, centrifugation, biochemical synthesis, atomic layer deposition and molecular self-assembly [62].

On the other hand, the *top-down approach* seeks to reduce unprocessed, raw material to nanometric dimensions by destroying the initial structures using various physical forces to synthesize NPs [63]. Mechanical milling, nanolithography, laser ablation, sputtering, and thermal decomposition are some of the most well-known techniques that correspond to this approach [63].

Silver, gold, iron, cobalt, nickel, and some of their corresponding oxides (e.g., magnetite, cobalt ferrite) make up the majority of metal NPs employed in medical applications. They can be synthesized and modified with functional chemicals, which contain versatile groups that allow them to be combined with a variety of molecules, including therapeutic agents and biomolecules such as peptides, proteins, and DNA (deoxyribonucleic acid) [64]. For example, AuNPs have unique photothermal properties due to the free electrons on the surface that continually oscillate at a frequency dependent on their size and shape. Iron oxide NPs are among the majority of FDA-approved inorganic nanomedicines. Magnetic iron oxide NPs, composed of magnetite (Fe_3O_4) or maghemite (Fe_2O_3), possess superparamagnetic properties at certain sizes and have demonstrated efficacy as contrast agents, and drug delivery systems, for DNA extraction and thermal-based therapeutics [65].

3.1. Synthesis of Silver Nanoparticles (AgNPs)

Due to their exceptional physical and chemical properties, AgNPs are frequently used in numerous fields, such as medicine, agriculture, and different industries [66]. Such properties include optical, electrical, thermal, and high electrical conductivity, as well as biological properties [67–69]. Therefore, in order to meet the increasing demand for AgNPs, numerous synthesis techniques have been developed.

Both the *top-down* and *bottom-up* approaches demonstrate a high yield of synthesized AgNPs without the use of toxic chemicals that endanger human health and the environment. In the absence of capping agents, however, aggregation is frequently a challenging task. In addition, both methods consume more energy, require a longer synthesis time, and call for complex equipment, all of which increase operating expenses [70].

Typically, the evaporation–condensation method employs a gas-phase route that enables a tube furnace to produce nanospheres at atmospheric pressure. This technique has been used to synthesize nanospheres made from a variety of substances, including Au, Ag, and PbS [71]. At the center of the tube furnace, there is a vessel containing a source of base metal, which is evaporated into the carrier gas to enable the final synthesis. By modifying the design of the reaction facilities, the size, shape, and yield of NPs can be controlled. The tube furnace occupies a large area, has a high energy output, increases the surrounding temperature of the metal source, and requires more time to maintain thermal stability. In order to overcome these drawbacks, Jung et al. demonstrated that a ceramic heater can be effectively implemented in the synthesis of high concentration AgNPs [72].

Laser ablation is another technique applied in physical synthesis. Laser ablation of a bulk metal source in a liquid medium can be used to synthesize AgNPs. After irradiation with a pulsed laser, the medium contains only NPs of the source base metal, with no additional ions or compounds [73]. Therefore, in contrast to chemical synthesis, the synthesis of NPs by laser ablation is pure and contaminant-free, as this technique uses mild surfactants in the solvent and no other chemical reagents [74].

The advantages of physical methods include speed, the use of radiation as a reducing agent, and the lack of hazardous chemicals, whereas the disadvantages include low yield and high energy consumption, solvent contamination, and lack of uniform distribution [75–77].

Chemical synthesis methods have been commonly applied for metal NPs synthesis as a colloidal dispersion in an aqueous solution or organic solvent by reducing their metal salts [78]. For the synthesis of metal nanospheres, such as gold, silver, iron, zinc oxide, copper, palladium, platinum, various metal salts are used [79]. Reducing and capping agents can be easily modified in order to achieve the desired particles' characteristics (e.g., size distribution, shape, dispersion rate) [80].

AgNPs are chemically synthesized mainly by Brust–Schiffrin synthesis (BSS) or the Turkevich method [74,81,82]. The chemical reduction of silver metal salts can be achieved using various reducing agents such as glucose, hydrazine, ascorbate, dimethylformamide, hydrogen, dextrose (Turkevich method) or sodium borohydride (BSS method) [83,84].

The chemical synthesis process typically uses three main components: metal precursors, reducing agents and stabilizing or coating agents. Essentially, the reduction of silver salts involves a nucleation in two steps and a subsequent growth. The main advantage of chemical methods is their high yield, in contrast to the low yield of physical methods. Nonetheless, these methods are extremely expensive, and the materials used for AgNPs synthesis, such as citrate, borohydride, thio-glycerol and 2-mercaptoethanol are toxic and unsafe [85]. Apart from these disadvantages, the obtained particles lack the desired purity, as numerous chemicals have been detected on their surfaces. Additionally, it is quite difficult to prepare AgNPs with a well-defined size, as this would require an additional step to prevent particle aggregation [86].

In order to overcome the limitations of chemical methods, viable biological alternatives have emerged. Recently, it has been demonstrated that biologically mediated synthesis is a simple, cost-effective, reliable, and environmentally friendly method. Moreover, much attention has been paid to the high-yield production of size-defined AgNPs using various biological systems such as bacteria, fungi, plant extracts or even small biomolecules (e.g., vitamins, amino acids) as an alternative to chemical methods, not only for AgNPs, but also for the synthesis of several other types of NPs [87–90]. The biosynthesized NPs' yield is a pivotal metric, often ranging from 60% to 90% or higher. Achieving a high yield is essential for the economic viability and scalability of nanoparticle production. Therefore, researchers diligently optimize parameters such as temperature, pH, and precursor concentrations to enhance production [59]. Selecting appropriate plant extracts or sustainable materials is equally crucial. A high yield in green synthesis not only ensures efficient production but also supports the sustainability and applicability of these nanomaterials across various domains, ranging from medicine to catalysis and environmental remediation [91].

Biological synthesis depends mostly upon three elements: the solvent, the reducing agent, and the non-toxic material. The main advantage of biological methods is the presence of amino acids, proteins, and secondary metabolites in the synthesis process, which eliminates the need for an additional step so as to prevent particle aggregation. Furthermore, such techniques appear to provide a controlled particle size and shape, which is crucial for a variety of biomedical applications [87]. Given that nowadays the focus is set on biosynthesized nanomaterials, some examples of biological reducing agents used for obtaining silver and gold NPs, respectively, and the corresponding optimal conditions, can be found in Table 1.

Nevertheless, the presence of some biomaterials in the extract could raise issues during the production of metal NPs, including delayed reaction kinetics that could further result in low stability. These concerns are usually counteracted by plant polyphenols [92]. Therefore, in order to overcome such potential limitations, recent studies have also focused on fabricating particles with the help of isolated biocompounds, such as curcumin, kaempferol, apigenin and quercetin [93,94].

Table 1. Examples of biological agents and conditions used for AgNPs/AuNPs synthesis.

NPs Type	Species	Biological material, conditions	Salt concentration ($AgNO_3$)	Mixing conditions	Reference
			Plants		
AgNPs	*Adenia trilobata*	leaves, 5 g % (70 °C for 25 min)	1–5 mM	60 °C, 30 min and kept at room temperature for 1 h	[95]
	Buchanania lanzan	leaves, 5 g % (boiled for 5 min)	0.5 mM	70 °C, 1 h	[96]
	Cannabis sativa	leaves, 1 g % (60 °C)	3 mM	37 °C, 24 h	[97]
	Embelia ribes	fruits, extraction of embelin	1 mM	30 min in the dark and incubation for 24 h	[98]
	Lythrum salicaria	aerial part, 10 g % (40 °C)	3 mM	60 °C, pH = 8, 180 min	[99]
	Moringa oleifera	leaves, 10 g % (100 °C for 20–30 min)	1 mM	60–80 °C, 60 min	[100]
	Picea abies, *Pinus nigra*	bark, 10 g % (60 °C)	1 mM	room temperature, 60 min	[101]
	Sanvitalia procumbens	plant, 10 g % (70 °C)	0.01 M	70 °C, pH = 8, 8 h	[102]
	Tagetes erecta	flowers, 10 g %	3 mM	40 °C, pH = 8, 75 min	[103]
	Theobroma cacao	cocoa pods and leaves, 15 g % (boiled for 10 min)	1 mM	room temperature, 4–24 h	[104]
			Microorganisms		
	Bacillus cereus	37 °C, 24 h, 120 rpm	1 mM	48.5 °C, pH = 9, 69 h	[105]
	Bacillus subtilis, *B. cereus*, *B. megaterium*, *B. pumilus*, *B. circulans*	isolates grown aerobically, 37 °C, 24 h	1 mM	30 °C, 24 h	[106]
	Klebsiella pneumonia	37 °C, 24 h	4 mM	37 °C, pH = 10, 24 h	[107]
	Lactobacillus bulgaricus	isolated colonies from raw milk, 37 °C	1 mM	room temperature	[108]
	Aspergillus terreus	strain isolated from soil, 25 °C, 5 days, 200 rpm	1 mM	25 °C, 200 rpm	[109]
	Penicillium verrucosum	strain isolated from vegetable-cultivated greenhouse soil, 25 °C, 7 days, 150 rpm	10 mM	25 °C, in the dark	[110]
			Algae		
	Chlorella vulgaris	algal suspension	10 mM	25 °C, 48 h	[111]
	Dunaliella salina	culture grown for 2 weeks, 12 h in light and 12 h in dark	1 mM	in the dark, 48 h	[112]
	Ulva rigida (green alga) *Cystoseira myrica* (brown alga) *Gracilaria foliifera* (red alga)	ethanolic and aqueous extracts obtained from dried algal powder, room temperature, 24 h	1 mM	70 °C, 15 min	[113]
	Species	**Biological material, conditions**	**Salt concentration ($HAuCl_4$)**	**Mixing conditions**	**Reference**
			Plants		
AuNPs	*Allium ampeloprasum*	leaves, 10 g % (boiled)	1 mM	room temperature, 30 min	[114]
	Annona muricata	leaves, ~ 13 g % (20 g–150 mL) (boiled 20 min)	1 mM	room temperature, 22 h	[115]
	Citrus maxima	peel, 1 g % (boiled for 10 min)	100 mM	room temperature, 1 h	[116]
	Dillenia indica	leaves, 1 g % (100 °C for 30 min)	1 mM	dark, 30 min	[117]
	Dracocephalum kotschyi	leaves, ~ 7 g % (5 g–75 mL) boiled	1 mM	10 min	[118]
	Lonicera japonica	flowers, 10 g % (boiled 20 min)	1 mM	70 °C, 30 min	[119]
	Mentha and *Pelargonium*	plant (boiled 20 min)	-	boiling	[120]
	Pergularia daemia	leaves, 10 g % (boiling at 60 °C for 30 min)	1 mM	15 min	[121]
	Platycodon grandiflorum	leaves, 10 g % (boiled for 20 min)	1 mM	50 °C, 15 min	[122]
	Tecoma capensis	leaves, 5 g % (heated till boiling ~20 min)	0.011 M	heating, 10 min	[123]
			Microorganisms		
	Bacillus cereus *Fusarium oxysporum*	bacterial culture incubated at 37 °C, 150 rpm, 24 h fungal culture incubated at 30 °C, 150 rpm, 72 h	1 mM	37 °C, 200 rpm, 24 h	[124]
	Pseudomonas	culture, 24 h	1 mM	80–85 °C, 30 min	[125]
	Candida albicans	cytosolic extract	1 mM	24 h	[126]

Table 1. *Cont.*

NPs Type			Plants		
			Algae		
	Chlorella sorokiniana	aqueous extract, 80 °C, 20 min	1 mM	80 °C, pH = 8, 60 min	[127]
	Galaxaura elongata	powder form and ethanolic extract	1 mM	room temperature, 120 rpm, 12 h	[128]
	Oscillatoria sp. *Spirulina platensis*	microalgal cultures	5 mM	room temperature, 24 h	[129]

Current research highlights the benefits that the use of plant material (plants, microorganisms, algae) confers in obtaining NPs. However, the limitations they may have must also be taken into account. Table 2 summarizes these advantages and disadvantages.

Table 2. Advantages and limitations of using biological agents for NPs synthesis.

Biological Material	Advantages	Ref.	Limitations	Ref.
Plants	• natural source of antioxidant compounds	[95]	• the amount of material depends on seasonality of plants • the quality of NPs depends on the growth, harvesting conditions and geographical distribution of plants	[130]
	• easy to grow and safe to handle	[100]		
	• biocompatibility	[96]		
	• wide range of applications in different industries and medicine	[98]	• specific process conditions such as temperature, pressure, reaction time, chemical reagents additions • variation in NPs quality due to irregular shape and size	[59]
	• small particle size and zeta potential indicate good stability • enhanced biodistribution, efficacy and safety for NPs compared to raw extracts	[130]		
	• biodegradable post-materials • non-toxic products, safer solvents • and reaction conditions	[97]		
	• waste prevention	[100]	• possible neurotoxicity, tissue toxicity • only a few are approved by EMA and FDA, thus needing further analyses.	[131]
	• good antifungal activity and promising antibacterial activity on both Gram-positive and Gram-negative strains	[99]		
	• promising results in cancer therapy.	[100]		
Microorganisms	• very good yield due to faster processes • environmental safety	[132]	• tedious purification steps • poor/lack of understanding of the competitive antimicrobial mechanisms • challenges in translating the production of NPs to industrial scale • potentially diminished yield caused by the growth rate of bacteria, enzymatic activity, biochemical pathways and purification processes.	[133]
	• cost effectiveness • the preferred material consists of bacteria/fungi due to their capacity for producing higher concentrations of reductase enzyme	[134]		
	• ease of cultivation and adjustment of size, shape and morphology of NPs • effective penetration through matrix biofilm produced by pathogenic agents, NPs play the role of a "Trojan horse" for bacterial cells.	[134]		
Algae	• renewable and sustainable natural resources • natural products and biodegradable post-materials	[135]	• limited applicability, low biocompatibility and, in some cases, elevated toxicity • no clear reaction mechanism during NPs formation	[136]
	• biocompatibility • cost-effectiveness • versatility	[137]		
	• scalability • reduced negative environmental impact.	[138]	• irregular shapes and sizes depending on solvents, vehicle, pH, exposure time and age of algal extract.	[59]

3.2. Synthesis of Gold Nanoparticles (AuNPs)

The synthesis of AuNPs generally adheres to the two above-mentioned methods.

Physical methods, including pyrolysis, nanolithography, thermolysis, and radiation-induced methods involve the controlled processes of cutting, milling, and modeling the materials into the desired order and shape. The high cost of the process, which requires a massive amount of energy to maintain the high pressure and high temperature conditions, is, however, a limitation of these methods [139–141].

In the second approach, AuNPs are created through the self-assembly of individual species (atoms, molecules, or clusters) using chemical or biological processes. This method is less expensive and provides greater control over the final development of a product with a more uniform size, shape, and chemical composition [142].

The chemical synthesis of AuNPs involves the reduction of a gold salt and subsequent stabilization of NPs. Turkevich and Brust–Schiffrin are the two conventional chemical techniques for obtaining AuNPs. Turkevich described a simple method for producing spherical AuNPs by reducing hydrogentetrachloroaurate(III) ($HAuCl_4$) with trisodium citrate, which acts as both a reducing agent and a stabilizer [143]. The discovery of the Brust–Schiffrin method allowed a simple approach to synthesize stable AuNPs of controlled size. Using tetraoctylammonium bromide as a phase transfer agent, $AuCl_4^-$ was transferred to toluene from an aqueous phase and reduced, in the presence of dodecanethiol, with $NaBH_4$. The addition of the reducing agent caused the organic phase to change color from orange to dark brown, which unequivocally demonstrated the formation of AuNPs [82].

Although chemical methods are the most common way to synthesize metal nanomaterials, the use of expensive and toxic reagents as reducing and stabilizing agents restricts their applications. Additionally, such NPs may also have negative effects when used for biomedical applications [144,145]. Consequently, there is a growing need to develop environmentally friendly and cost-effective nanoparticle synthesis techniques that do not employ toxic substances. Therefore, in recent years, biological synthesis has gained significant attention as a green and environmentally friendly approach, that uses microorganisms, enzymes, and plants to produce NPs [146,147].

4. Characterization of Metal NPs

The appropriate characterization of metal NPs is essential for obtaining in-depth information that will influence their application in various fields, as well as for assessing their toxicity [148]. Figure 5 depicts the variety of techniques used to characterize metal NPs, such as UV-Vis spectroscopy, SEM (Scanning Electron Microscopy), TEM (Transmission Electron Microscopy), FTIR (Fourier-Transform Infrared Spectroscopy) and EDX (Energy Dispersive X-Ray) analysis, techniques that can reveal their composition, morphology, and size [149].

4.1. Visual Inspection and UV-Vis Spectroscopy

For the initial observation of NPs formation, the color change of the extract-metal salt mixture can be observed. The color varies from the initial color of the extract to orange-brown (AgNPs obtained using *Eucalyptus camaldulensis*) or dark brown (AgNPs obtained using *Prunus persica*) in the case of AgNPs, whereas for AuNPs, the color indicating the obtaining of NPs can be pink (AuNPs from Tamarindus indica) or ruby-red (AuNPs from *Mentha piperita, Melissa officinalis* and *Salvia officinalis*) [150–153].

Using UV-is spectroscopy, the formation of AgNPs (by reducing Ag^+ to Ag^0) and AuNPs (by reducing Au^{3+} to Au^0) with the help of biomolecules present in plant extracts is demonstrated by the appearance of SPR (Surface Plasmon Resonance) that is highlighted in the UV-Vis spectra by a maximum absorbance around specific wavelengths. In general, the maximum absorption of AgNPs occurs between 400 and 500 nm, while that of AuNPs occurs between 500 and 600 nm. For instance, in the case of AgNPs synthesized from *Gleichenia pectinata* (obtained using a 10 g % methanolic extract, 5 mM silver nitrate and 1:5 extract–silver nitrate volume ratio), the SPR band appeared at 460 nm [154]. However,

AgNPs obtained from *Euterpe oleracea*, *Oenocarpus bacaba* and *Mauritia flexuosa* (obtained using a 100 mg/mL aqueous extract, 1 mmol/L silver nitrate and incubation under low light conditions at 75 °C for 90 min), the SPR band could be found in the 420–430 nm range [155].

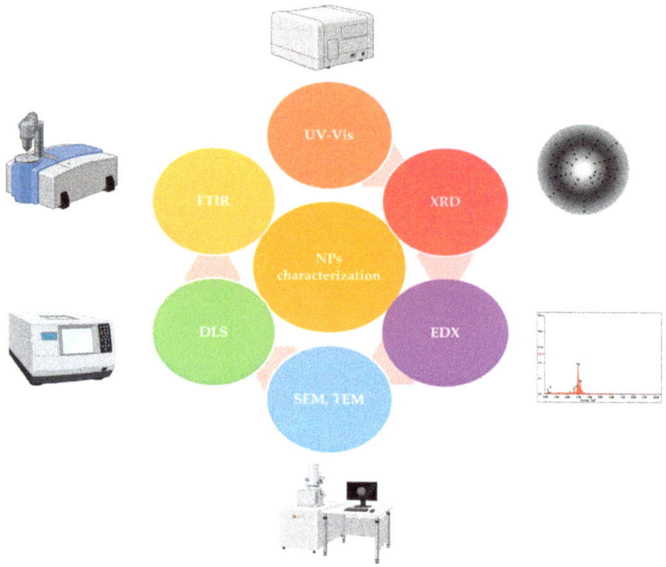

Figure 5. The most used characterization techniques: UV-Vis—UV-Vis Spectroscopy; XRD—X-Ray Diffraction; EDX—Energy Dispersive X-Ray Analysis; SEM—Scanning Electron Microscopy, TEM—Transmission Electron Microscopy; DLS—Dynamic Light Scattering; FTIR—Fourier-Transform Infrared Spectroscopy.

For AuNPs obtained from *Dodonaea viscose* leaves (in the presence of 1000 ppm tetrachloroaurate), the SPR band appears at 600 nm [156], and around 530–540 nm for those obtained from green tea and *Juniperus communis* extract [157]. Another example shows the characteristic peak for AuNPs derived from *Platycodon grandiflorum* extract (with 1 mM $HAuCl_4$ solution) at 545 nm [122].

In addition, the nature of the colloidal dispersion and the shape of AgNPs can be deduced using the data obtained during UV-Vis analysis. For example, according to Mie's theory, the presence of a single resonance band corresponds to spherical particles [158]. Moreover, according to Miskovská et al., a peak absorption around 420 nm indicates monodisperse and smaller AgNPs, whereas a peak at 440 nm indicates polydisperse AgNPs [159]. Other studies also confirm the relationship between NPs size and shape and the increase or decrease of wavelength, through the observation of different red or blue shifts [160,161]. Additionally, for AuNPs, smaller particles are typically associated with shorter wavelengths [162].

4.2. FTIR Spectroscopy

FTIR analysis can be used to highlight functional groups of different active substances or of plant secondary metabolites that are responsible for the reduction, capping, and stabilization of both AgNPs and AuNPs. Absorption bands around 3300–3400 cm^{-1}, 2940 cm^{-1}, 1600–1700 cm^{-1}, 1350–1450 cm^{-1} and 1000–1100 cm^{-1} correspond to O-H stretching and C-H stretching of aromatic compounds, C=O stretching of carbonyl compounds, to C-C, C-O-C, C-N, C-O bonds, N-H stretch vibration of amides, and to C=C groups. Such bonds can be easily found in phytochemicals such as polyphenols, proteins, alcohols, terpenoids, enzymes and alkaloids [154,155].

When comparing the FTIR spectra of nanoparticles to those of extracts, shifted or diminished peaks are revealed, which can be explained by the phytochemicals' participation in the synthesis and capping processes [163]. Moreover, the literature describes a new band around 640–670 cm^{-1} for AuNPs, that can be attributed to metal formation [153,156].

4.3. Dynamic Light Scattering (DLS) and Zeta Potential Analysis

The surface charge of particles determines the interactions with the targeted site. In order to measure the electric charge of the surface and the electrostatic stability of the dispersion in the liquid phase, the zeta potentiometer is used [164].

Due to the ionizable or adsorbed groups on the surface, the zeta potential indicates the surface charge of particles. As a result, it is possible to study the aggregation tendency and understand the stability of NPs in suspension. A zeta potential value greater than +30 mV or less than −30 mV determines the production of a very stable nanoparticle dispersion [165]. A zeta potential closer to 0 mV suggests a larger probability of aggregation [155]. Generally, the zeta potential has a negative value, which implies electrostatic repulsion between particles, thus avoiding agglomeration and flocculation and ensuring stability.

The PdI (polydispersity index) represents the width of particle size distribution. The PdI range is between zero and one, with zero suggesting a monodisperse distribution and uniform size and one indicating a polydisperse distribution with various nanoparticle sizes [166].

DLS is used to determine the hydrodynamic diameter of NPs in a medium (by measuring dynamic fluctuations of light-scattering intensity due to the Brownian motion of particles). For instance, there are differences between NPs as they can be coated with plant extract biomolecules that have varying molecular weights [167].

4.4. SEM, TEM, EDX

The elemental composition can be determined using EDX analysis. Strong absorption peaks at 3 keV and 2 keV indicate the binding energies of silver and gold, respectively. In the case of biosynthesized NPs, additional elements that may be present, such as carbon, oxygen, aluminum, and potassium, are a result of plant phytochemicals that act as capping elements on the surface of AgNPs [154,163,168].

The selection of NPs for various applications is influenced significantly by the variety of forms and surface conformations that they exhibit. The surface morphology can be crystalline or amorphous, smooth, or rough, and geometries range from spherical, flat, cylindrical, tubular, or conical to irregular. Electron microscopy techniques, such as SEM or TEM, are used to analyze the surface morphology [169]. For example, AgNPs obtained from *M. oleifera* leaves extract (1 mM $AgNO_3$, 60–80 °C, 1 h) showed spherical nanoparticles, with a particle size in the 10–25 nm range [100]. On the other hand, for AuNPs synthesized from *Citrullus colocynthis* seed extract (1 mM chloroauric acid, extract–salt 1:2 volume ratio, 70 °C, 20–30 min), the obtained particles presented spherical shape and a particle size in the 7–33 nm range [170].

Moreover, the size, morphology and shape of NPs can be determined by TEM analysis. In the case of AgNPs synthesized from *Euterpe oleracea*, *Oenocarpus bacaba* and *Mauritia flexuosa* leaves extract, TEM analysis revealed polydisperse AgNPs, mostly spherical in shape, without the formation of large agglomerates [155]. Another given example is that of AgNPs obtained from parts of *Crataegus ambigua*, with TEM analysis demonstrating the obtaining of particles with an average size of 32 nm, that are spherical in shape [171].

In the case of AuNPs, the shape may be more variable. For example, AuNPs obtained from green tea have an hexagonal shape, those from *Juniperus communis* extract are circular and hexagonal, while those from green coconut water are circular, triangular, hexagonal and rod-shaped [157]. AuNPs from *Salvia officinalis* extract (obtained using 200 mg/l $HAuCl_4$) were spherical, triangular, star-shaped, and rod-shaped, whereas AuNPs from *Melissa officinalis* extract were triangular, star- and rod-shaped [153]. Botteon et al. correlated different geometries of AuNPs obtained from Brazilian red propolis with their size and concluded that particles with triangle, pentagon, hexagon and rod shapes have large sizes, while those with spherical shapes are usually smaller [172]. This can be explained by the amount of phytochemicals that coat and stabilize the surface of nanoparticles, which determines their anisotropic nature [173].

SEM and TEM can reveal the coating of biosynthesized NPs, which is generally due to plant phytochemicals that cover the particles' surface [99,155,174]. Verma and Mehata demonstrated that NPs size is influenced by synthesis conditions. Thus, for AgNPs synthesized from *Azadirachta indica*, an increase in the extract concentration and pH results in an increase in particle size, whereas an increase in temperature results in a decrease in size [175]. Additionally, Bhaskaran et al. demonstrated that the gold salt concentration influences the size and form of AuNPs generated from *Medicago sativa* cell suspension cultures. If, at a concentration of 10 ppm gold salt, the majority of the formed NPs are spherical, the percentage of AuNPs with other shapes, such as hexagons, pentagons, and triangles, will grow as the salt concentration rises. Increasing the concentration of the gold salt will also simultaneously increases the size of particles [176].

4.5. XRD Analysis

The XRD technique analyzes the nature of nanoparticles. According to Bragg's model of diffraction given by the Joint Committee on Powder Diffraction Standards—International Center for Diffraction, sharp and clear peaks (2Θ) of particles are located around 38, 44, 64° which can be assigned to (101), (111), (200) and (220) planes that indicate a high degree of crystallinity of AgNPs (JCPDS No 65-2871) [154,177].

For AuNPs, the pattern shows peaks around 38, 44, 64 and 77°, with corresponding lattice plane at (111), (200), (220) and (311) of face-centered cubic crystal structure (JCPDS No. 01-089-3697) [153].

The Scherrer equation, originally formulated by Paul Scherrer in 1918, has remained a cornerstone in X-ray crystallography and materials science. When applied to the XRD analysis of NPs, it provides critical insights into their structural characteristics, size and crystalline quality. This equation is particularly relevant in the study of metal NPs because their properties are often closely linked to their size and structure [178].

One important aspect to note is that the broadening of XRD peaks is primarily influenced by the finite size of the crystalline domains in the sample. Larger nanocrystals will exhibit narrower peaks, while smaller ones will result in broader peaks. This relationship between crystallite size and peak-broadening represents the very basis of the Scherrer equation. Moreover, the equation can be used to estimate the overall size distribution within a sample by analyzing multiple diffraction peaks. Implicitly, the Scherrer equation provides an average size, and in actual samples, size distributions are often present due to variations in the synthesis process [179].

4.6. Other Characterization Techniques

Other techniques such as inductively coupled plasma-mass spectrometry and capillary electrophoresis can be used for detection or for the study of other NPs characteristics, such as mass distribution, electrical conductivity or resistance [180,181]. The applied techniques are different depending on the complexity of the matrix, and the concentration of the analyte, as well as on the physicochemical characteristics. In certain cases, sample dilution may be required additionally so as to accurately detect the concentration of the analyte. The characteristics that can be established using some of the previously mentioned techniques are summarized below.

Particle size represents one of the fundamental characteristics of NPs and particles can be classified as nano or micro. Usually, the particle size and distribution can be determined by electron microscopy. The images obtained by SEM or TEM are employed in order to determine the size of particles and clusters, while laser diffraction techniques are used to measure the raw material in the solid phase [182].

The surface area is another essential characteristic of nanomaterials. In describing the properties and performances of such particles, the area-to-volume ratio is of critical importance. Typically, the surface area is determined using Brunauer–Emmett–Teller (BET) analysis. In a liquid phase, conventional titration can be used to analyze the surface of NPs, but the process is laborious and time-consuming; therefore, NMR is usually preferred. The differential mobility analyzer is used to measure the NPs' gas-phase surface area [183].

The purity and effectiveness of NPs are determined by their chemical or elemental composition. Large concentrations of secondary or undesirable components might impair performance or potentially cause adverse effects, and chemicals could contaminate the final product. Besides FTIR spectroscopy, X-ray photoelectron spectroscopy can also be used to analyze the composition [184].

In other cases, the particles initially undergo a process of chemical digestion before being submitted to wet chemical examination in specific techniques such as chromatography, atomic emission spectroscopy, and mass spectroscopy. The particles detected in the gas phase are collected using filtration or electrostatic collection, and then they are subjected to spectrometric or wet chromatographic examination [17].

The experimental science of crystallography examines the configuration of atoms and molecules in crystalline materials. To emphasize the structural arrangement, X-ray, electron, or neutron diffraction methods are typically used in the crystallography of NPs [185].

Testing the stability of NPs is also crucial to ensure their performance and safety in various applications. Nanoparticle stability can be assessed through a combination of physical, chemical, and biological methods, ranging from visual inspection to biological compatibility assessment. However, one of the most common methods is represented by UV-Vis spectroscopy that can monitor changes in the absorption spectrum of the NPs [186]. Aggregation or precipitation often results in a shift or broadening of the absorption peaks [187].

5. Biomedical Applications of Metal NPs

Metal NPs have gained significant attention due to their unique physical, chemical, and optical properties. These nanoscale metal structures have demonstrated a wide range of applications in various medical areas. Figure 6 schematically shows the most promising biomedical uses of metal NPs and the mechanisms involved.

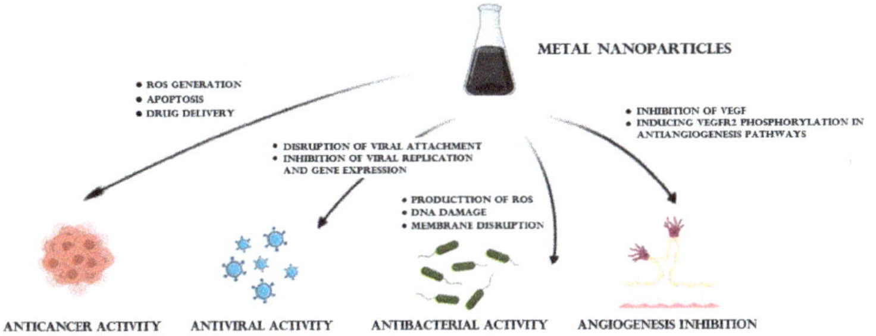

Figure 6. Biological activities of metal NPs and their potential mechanisms (VEGF—vascular endothelial growth factor; VEGFR—vascular endothelial growth factor receptor).

One of the major applications is in targeted drug delivery, where metal particles can be functionalized with specific ligands or antibodies to selectively target and deliver therapeutic agents to cells or tissues. Additionally, metal NPs have shown promise in medical imaging as contrast agents, enhancing the visualization of tissues and enabling early disease detection. They also find utility in biosensing and diagnostic applications, where their optical properties facilitate sensitive and rapid detection of biomarkers. Furthermore, metal

NPs have been explored for their antimicrobial properties, with potential applications in combating bacterial infections and developing novel antimicrobial agents.

5.1. Diagnostics

Some nanoparticles are generally rendered as stable and biocompatible, but they also have special qualities, including magnetic properties. As a result, an externally generated magnetic field can be used to direct magnetic NPs to a specific position inside the body. Regarding their application in medicine, a crucial parameter is their magnetic susceptibility, which is defined by the relationship between the applied field and the induced magnetization. For instance, super-paramagnetic iron oxide nanoparticles (SPIONs) are frequently used in clinics as contrast agents for magnetic resonance imaging (MRI) [188].

Additionally, it has been observed that, when administered intravenously, iron oxide NPs with diameters ranging from 50 to 150 nm are taken up by macrophages and accumulate in the liver, spleen, lymph nodes and bone marrow. This could prove useful for generating contrast in several diagnostic techniques [189]. In this regard, Ferrucci, Marchal and Deng used this feature to visualize the primary lesions of the liver in cases of hepatocellular carcinoma and liver metastases [190–192]. In contrast to the healthy liver tissue, tumor lesions are characterized by a lower uptake of SPIONs in the macrophages compared to the healthy liver parenchyma. Santoro et al. also used SPIONs to distinguish malignant from benign neoplastic liver lesions [193].

Metal NPs with diameters between 20–30 nm or less, called ultra-small superparamagnetic iron oxides (USPIO), have a long circulation time and have been used in lymphography primarily because, after intravenous administration, they extravasate into tissues, where they are picked up by the lymphatic circulation and reach the macrophages in the lymph nodes. Lymph node staging, which is essential for both classifying cancer and determining the best treatment plan, can be carried out using MRI [194]. Nonetheless, USPIO compounds could also serve for imaging parotid and other salivary gland tumors [195].

The ability of metal NPs to accumulate in macrophages can also be used to gather data on the development and stage of cardiovascular diseases such as atherosclerosis. Macrophages can be loaded with iron oxide NPs and viewed using MRI in order to identify inflammatory lesions [196,197]. In rabbits with susceptible or ruptured atherosclerotic plaques, Kaneko et al. noticed a rise in the absorption of USPIO/SPION [198]. Similar characteristics were also observed in humans [199]. Moreover, in the diagnosis and staging of multiple sclerosis, MRI has been used to assess the integrity of the blood-brain barrier and to localize inflammatory lesions in the brain. Tourdias et al. demonstrated that, by combining gadolinium with iron NPs, a much greater precision was achieved in the detection of such lesions [200].

The development of new AuNP-based contrast agents for MRI is rapidly evolving, and in addition to gadolinium chelates, several new and highly effective contrast agents have been reported [201]. For instance, gadolinium chelates now used for clinical diagnosis can be transported via AuNPs to increase the sensitivity of MRI [202].

Given the variety of processes used to control their size, shape, and distribution, AuNPs represent promising substrates for building prototypes for diagnosis and treatment. In addition, the organic coating that stabilizes the metal surface can be easily modified so as to control the interaction of NPs with biological systems [203].

Taking into account their optical and magnetic properties, AuNPs, as well as AgNPs, have also been intensively used in the diagnosis and bioimaging of cancer cells. Several studies focused on the potential use of the magnetic properties of AgNPs in detecting cancer cells [204]. Moreover, other researchers have developed a gold nanoparticle-based sensor capable of detecting lung cancer by analyzing the patient's breath [205]. A more recent clinical study on AuNPs showed that such materials can be used for the screening of gastrointestinal tract tumors [206].

Other potential uses of silver and gold NPs focus on the detection of certain viruses. One example is represented by the detection of human immunodeficiency virus-1 (HIV-1) protease using electrochemical impedance spectroscopy [207]. AuNPs conjugated with hepatitis B virus (HBV) DNA can be used to directly detect the presence of HBV DNA using fluorescence-based methods, which are simple and low-cost [208]. Moreover, Yen et al. have successfully investigated AgNPs conjugated with antibodies for the detection of yellow fever and Dengue virus proteins and the Zaire ebolavirus glycoprotein [209].

Using a colorimetric method, Balakumar et al. also used AgNPs for the recognition and detection of lead, which could represent a possible application in the diagnosis of saturnism [210]. The use of AuNPs for the detection of *Mycobacterium tuberculosis*, the etiological agent that causes tuberculosis, also employing colorimetry, was another successful application of such particles. This type of approach also aided defining diabetes as a complex disease [211].

Furthermore, AuNPs were used to determine the concentration of amyloid-derivative-B-ligand, a potential marker for Alzheimer's disease (AD), in the cerebrospinal fluid of affected individuals [212]. AuNPs optical biosensors were also used to monitor the interaction between the antigen (amyloid β-derived diffusible ligands) and its specific antibodies, with the aim of characterizing AD [213]. Interestingly, a recent study showed that AuNPs might also play an important role not only in the diagnosis of AD, but also in its treatment, by preventing neuroinflammation and impaired cognition on an okadaic acid-induced model in rats [214]. Calorimetric biosensors based on gold NPs also have rather promising detection limits for proteins and nucleic acids [215]. Nonetheless, several new biotechnological methods based on the expression of specific peptides on the surface of phages have been developed to improve the detection limit [216,217].

5.2. Anticancer Activity

NPs are used as vectors for diagnostic, hydrophobic medicine, therapeutics and especially in the delivery of antineoplastic agents to cancerous tissues, where NPs can penetrate and deliver the drug to a specific targeted site [218]. NPs are also attracting significant interest as carriers for the applications in drug and gene delivery, or as biosensors, where a direct contact with blood occurs [219]. Jagtap et al. investigated the in vitro anticancer potential of embelin fabricated AgNPs, the results confirming a promising anticancer effect specifically against A549 lung cancer cells [98].

Metal NPs are more cytotoxic to cancerous cell lines than to normal cells. The cytotoxicity of metal NPs has been explained by a number of different mechanisms, including the production of reactive oxygen species (ROS), permeabilization of the mitochondrial outer membrane, and specific DNA cleavage, all of which result in the death of the cancer cell via apoptosis, autophagy, and necrosis [220].

Many metal NPs are used in medicine, particularly in the treatment and diagnosis of cancer, considering optical and localized surface plasmon resonance, as well as their rather low cytotoxicity. Due to their characteristics, the photothermal conversion and heating of the targeted tumor tissue happen when the right wavelength of light is applied as an external stimulus, thus killing cancer cells. Iron oxide NPs can be used in this anticancer procedure as well, when exposed to an alternating external magnetic field that causes particle movement and local heating. This hyperthermic effect that has been employed in tumor therapy can cause tissue damage in the proximity of NPs [221].

The use of magnetic drug targeting to improve SPION accumulation in pathological regions has been investigated. This implies the existence of a strong external magnetic field gradient that guides SPIONs to a specific target region, such as a tumor, where an alternating magnetic field can produce hyperthermia or inducible drug release [222].

Amatya et al. studied the photothermal activity of iron oxide NPs modified with polyethylene glycol (PEG) and starch in order to prevent aggregation and improve biocompatibility. The modified particles were injected intravenously into U87 MG xenograft tumor-bearing mice (human cerebral glioblastoma) and were used as agents to treat cancer at a wavelength of 885 nm (NIR domain). The tumor site was irradiated with a laser, and the modified nanoparticles demonstrated the most prominent reduction in tumor size compared to control groups [223]. Moreover, Ge and collaborators proposed the encapsulation and transport of medicinal substances in iron oxide NPs to a magnetically targeted site. Using complementary photothermal therapy in conjunction with laser irradiation could lead to both a direct and indirect (via immune activation) means of significantly reducing the tumor's volume [224].

Metal NPs also hold immense promise in the fields of radiotherapy and electric field-induced hyperthermia. In radiotherapy, NPs can be precisely delivered to tumor sites, enhancing the therapeutic effect while minimizing damage to healthy tissue [225]. By loading particles with radiation-absorbing materials, such as gold or hafnium oxide, they become radiation amplifiers, intensifying their tumor-killing capacity [226]. Additionally, when exposed to external electromagnetic fields, the particles generate localized heat, ideal for electric field-induced hyperthermia. This controlled heating can selectively damage cancer cells, improving treatment outcomes. The integration of NPs into these therapies represents a futuristic approach to cancer treatment, offering a potent combination of precision and potency in the fight against this disease [227].

Moreover, metal NPs have been used as intracellular carriers of nano-drugs into specific cells using various methods such as mechanical bombardment or electroporation [228]. Electroporation is one of the most successfully applied methods, given its improved transfer efficiency. For instance, the combination of functionalized gold NPs and electroporation can lead to the improvement of radiotherapy efficacy in cancers associated with human papillomavirus [229].

AuNPs are used for drug delivery, where light irradiation can trigger the drug's release at the targeted site [230]. As previously mentioned, different ligands, such as peptides, proteins, or DNA, can optimize the surface of such particles. Moreover, in vivo studies of AuNPs demonstrated that they readily diffuse into tumor cells and therefore are highly effective in tumor targeting [231]. One proposed treatment involves targeted chemotherapy that delivers a tumor-killing agent, TNFα, to malignant tumors. CYT-6901 represents a first in nanomedicine, consisting of AuNPs coated with TNFα factor. Goel R et al. performed an in vivo study of this drug in human prostate tumor-bearing mice and the results indicated the specific accumulation of TNFα in tumor cells within 4 h after its injection [232].

Coelho et al. investigated the conjugation of bortezomib with pegylated gold NPs (PEGAuNPs), in vitro, on pancreatic and lung cancer cells and reported that this conjugation increased the inhibitory effect of the drug on cancer cells and decreased its toxicity in normal cells [233].

Another chemotherapeutic drug, afatinib, which was approved for the treatment of lung cancer, was conjugated in vitro with AuNPs to improve its efficacy and biocompatibility in cancer cells [233]. This conjugation significantly inhibited the proliferation of lung cancer cells. Additionally, the alveolar epithelial type I cells maintained their viability and released less pro-inflammatory cytokines when compared to the unconjugated drug [234].

Furthermore, Ramalingam et al. investigated the conjugation of doxorubicin on the surface of AuNPs with polyvinylpyrrolidone for the treatment of lung cancer. They reported a targeted release of the chemotherapy drug, with a strong inhibition of lung cancer cell-growth compared to the unconjugated doxorubicin [235].

Several studies also focused on obtaining NPs by the use of natural compounds that might prove beneficial in the treatment of different types of cancer. For instance, Balakrishnan et al. developed AuNPs conjugated with quercetin, and reported their role in inhibiting the migration, invasion, angiogenesis and metastasis of breast cancer cells [236]. Additionally, another study conducted by Chen et al. demonstrated the inhibition of breast cancer MDA-MB-231 cells by AuNPs capped with gallic acid [237].

The analysis of cancer cell morphology indicated that AgNPs can cause cell death. Due to their properties, magnetic AgNPs of various sizes were successfully used to target breast cancer cells (SKBR3) and floating leukemia cells (SP2/O), as produced by Jun et al. [238].

On the other hand, AgNPs coated with folic acid and conjugated with doxorubicin were produced by Wang et al., with cell death occurring after 8 h [239]. In one other study, Fang et al. obtained polyconjugated AgNPs with tertiary amines, doxorubicin, guanidines and imidazoles, that were responsible for increasing their intracellular uptake and cytotoxicity in lymphoma cells [240].

AshaRani et al. investigated the cellular and molecular mechanisms induced by AgNPs using IMR-90 normal human lung tissue and the U251 glioblastoma cell line. It was discovered that the synthesized NPs were able to adsorb on their surface proteins from the cytosol, proteins that can influence the function of intracellular factors and can regulate gene expression and pro-inflammatory cytokines [241]. One study even showed that AgNPs could alter the regulation of more than 1000 genes [242]. Recently, it has also been demonstrated that the autophagy of cancer cells is another mechanism induced by such materials [243].

Nonetheless, AgNPs synthesized from plant extracts showed a more pronounced cytotoxic effect on lung carcinoma cells (A549) than on healthy lung cells, indicating that the AgNPs could target a cell-specific toxicity, which could be indicated by a lower pH level in cancer cells [244]. Moreover, AgNPs obtained using *E. sylvaticum* proved cytotoxic efficacy against osteosarcoma cell line MG-63 [90].

5.3. Angiogenesis Inhibition

The process of new blood vessel formation is also a remarkable opportunity for the use of gold nanoparticles in cancer therapy [245]. By targeting NPs, a variety of therapeutic agents, such as antibodies and small compounds, could be employed as therapeutic solutions [246].

Mukherjee et al. found that AuNPs inhibit angiogenesis by inhibiting the protein phosphorylation in a dose-dependent manner, with almost complete inhibition observed at concentrations in the 335–670 nM range. It has been suggested that the inhibition mechanism involves the direct binding of AuNPs to heparin-binding growth factors [247].

A more recent study on AuNPs biosynthesized with the use of apigenin has also emphasized their potential in angiogenesis suppression, as well as in mediating anti-proliferation of cholangiocarcinoma, thus pointing to their dual activity [248]. Moreover, another novel research study that focused on investigating the effects of sorafenib derivatives-capped AuNPs showed in vitro suppression of angiogenesis, as well as of tumor migration and mesenchymal–epithelial transition [249].

When discussing AgNPs, a study performed by Kalishwaralal et al. demonstrated the inhibition of angiogenesis by nanoparticles biologically synthesized using *Bacillus licheniformis* on bovine retinal endothelial cells (BRECs) [250]. A further investigation demonstrated that AgNPs inhibited VEGF-induced cell proliferation, migration, and tube formation in BRECs, that were used as an in vitro model system [251].

Another study focused on observing the effects posed by eco-friendly synthesized AgNPs with the use of an extract obtained from a brown alga, *Dictyota ciliolate*. Besides enhanced cytotoxicity on a lung adenocarcinoma cell line, the NPs also exhibited antiangiogenic effects, considering the inhibition of tertiary blood vessel development [252].

Moreover, AgNPs synthesized from *Achillea biebersteinii* flower extract at a concentration of 200 g/mL showed a 50% reduction in the newly formed vessels on a rat aortic ring model [253]. One other example is that of AgNPs produced using a *Saliva officinalis* extract that exhibited dose-dependent antiangiogenic properties on a chicken embryo [254]. The inhibitory effect of such nanomaterials on the induced angiogenic activity in human breast cancer cells has also been shown [255].

5.4. Antimicrobial Properties

Metal NPs appear to be promising antibacterial agents given their high surface-to-volume ratios, as well as their diverse mechanisms for bacterial inhibition and bactericidal action, which decreases bacterial resistance. Silver, copper, gold, titanium, magnesium, and zinc NPs demonstrated the most encouraging results among the metal NPs that were studied [256,257].

Nonetheless, more complex nanoparticles, such as Ag/Cu_2MoO_4 NPs, were also shown to possess an inhibitory activity against *Pseudomonas aeruginosa* and *Streptococcus pneumoniae*, as well as the capacity for glucose detection [258], while $LaNiO_3/SrCeO_3$ and NiV_2O_6/CeO_2 nanocomposites showed promising activity against *Klebsiella pneumoniae* and *Bacillus cereus*, and *Streptococcus pneumoniae*, *Enterococcus faecalis*, *Pseudomonas aeruginosa*, and *Legionella pneumophila*, respectively, as well as photocatalytic activity [259,260]. Moreover, iron- and nickel-introduced bimetallic metal-organic frameworks were also tested on *Mycobacterium tuberculosis* and *Helicobacter pylori*, presenting promising activity [261]. Silver indium sulfide/nickel molybdenum sulfide nanostructures also exhibited antibacterial activity on *Escherichia coli* and *Staphylococcus aureus* under light irradiation, as well as enhanced peroxidase-like activity for the detection of acid uric [262]. Furthermore, $NiCo_2O_4$-Bi_2O_3-Ag_2ZrO_3 nanocomposites also revealed excellent bactericidal activity against *P. aeruginosa* and *S. pneumoniae*, which proves the multifaceted actions such nanomaterials possess [263].

Metal NPs have an advantage over bulk metals due to their small size and high surface area, which allows for greater interaction with microorganisms. Several mechanisms are responsible for their antimicrobial activity, including oxidative stress, membrane damage, and DNA damage [264].

Oxidative stress is one of the most common mechanisms explaining antimicrobial activity. The ROS produced by these nanoparticles, such as superoxide anions and hydroxyl radicals, can damage the bacterial cell membrane and proteins. For instance, AgNPs have been shown to produce ROS that result in lipid peroxidation of bacterial membranes and inhibition of bacterial growth [265]. Similarly, it has been demonstrated that copper NPs generate ROS that damage the cell membrane and induce oxidative stress, leading to bacterial cell death [266].

Membrane damage is another mechanism of antimicrobial activity. Metal NPs can interact with bacterial membranes and compromise their integrity, resulting in intracellular component leakage and ultimately cell death. It has been demonstrated, for instance, that AuNPs penetrate the bacterial membranes and change their permeability, eventually resulting in cell lysis and bacterial death [267]. It has also been shown that AgNPs, as well as zinc oxide NPs, adhere to and penetrate bacterial membranes, altering their properties [268,269].

Lastly, metal NPs can cause DNA damage, which results in cell death or mutations. These particles can enter bacterial cells and interact with DNA, causing strand breaks and other forms of damage. It has been established, for instance, that AgNPs induce DNA damage, resulting in decreased bacterial viability [270].

Sondi and Salopek-Sondi revealed the excellent antimicrobial action of AgNPs against *Escherichia coli* cultures treated with AgNPs that accumulated in the cell wall and caused the formation of gaps in bacterial walls, thus resulting in cell death [271,272].

Various kinds of AgNPs with an average size of around 15 nm were synthesized, and all of them exhibited promising antimicrobial and bactericidal activity against both Gram-positive and Gram-negative bacteria, including multi-resistant strains such as methicillin-resistant *Staphylococcus aureus* [273,274]. One other study investigated the effectiveness of AgNPs against *E. coli* and *S. aureus* and yeast. The obtained results indicated complete inhibition of *E. coli* and yeast at a low concentration of AgNPs, while only a minor effect was noted on *S. aureus* [275].

Since antibiotic resistance has become a major threat to public health, there is a need for novel and innovative methods to combat bacterial infections. In the field of antimicrobial therapy, the association of NPs with antibiotics has become a significant area of research [276]. Such an association has several advantages over using antibiotics alone.

First of all, it can increase the antimicrobial activity of antibiotics by enhancing their effectiveness against bacterial pathogens. It has been demonstrated that combining AgNPs with antibiotics such as amoxicillin and tetracycline enhances their antibacterial activity against *E. coli*, *S. aureus* and *Salmonella typhimurium* resistant strains [277,278]. Similarly, it has been demonstrated that the association of AuNPs with antibiotics such as streptomycin or azithromycin increases their antibacterial activity against *Pseudomonas aeruginosa* [279,280].

The combination of NPs and antibiotics can also boost the drug's bioavailability and decrease its toxicity. Antibiotics can be carried by NPs, allowing for a targeted delivery to the site of infection. Consequently, the risk of toxicity and side effects is reduced [281].

Thirdly, this association can also inhibit the emergence of antibiotic resistance. NPs can function as synergistic agents, enhancing the antibacterial activity of antibiotics and decreasing the likelihood of antibiotic resistance. It has been shown that the combination of AgNPs with antibiotics inhibits the development of resistance in *E. coli* and *S. aureus* [282].

Likewise, AgNPs biologically synthesized using *Klebsiella pneumoniae* culture supernatant increased the efficacy of various antibiotics, such as amoxicillin, clindamycin, erythromycin, penicillin G, and vancomycin against *S. aureus* and *E. coli* [283].

Regarding the antifungal properties of nanomaterials, fungus-mediated synthesized AgNPs showed enhanced activity against *Phoma glomerata, P. herbarum, Fusarium semitectum, Trichoderma* sp. and *Candida albicans,* especially when associated with fluconazole [284]. Moreover, an antimicrobial gel formulation containing AgNPs also displayed good antifungal activity against strains of *Aspergillus niger* and *Candida albicans* [285].

One other study showed that AgNPs synthesized using ribose and stabilized by sodium dodecyl sulfate highlighted a good antifungal activity against *Candida albicans* and *C. tropicalis,* comparable to that of conventional antifungals, such as amphotericin B [286].

Other examples of biosynthesized AgNPs with promising antifungal activity, especially on *Candida albicans,* are those obtained using plants such as *Brassica oleracea, Camellia sinensis, Glycine max, Mentha piperita,* and *Tagetes erecta* [103,287–289].

Nevertheless, AuNPs can also present antifungal activity. For example, nanoparticles synthesized using the solvothermal method show size-dependent antifungal activity against *Candida* isolates, with smaller sized particles (7 nm) exhibiting greater fungicidal activity compared to larger ones (15 nm) [290]. Moreover, AuNPs biosynthesized using olive leaf extract and *Allium sativum* also presented good antifungal activity on *Candida* species, which was mainly due to ROS formation [291,292].

5.5. Antiviral Action

Due to their unique physicochemical properties, such as small size, high surface area, and catalytic activity, metal nanoparticles have emerged as a promising group of antiviral agents. These characteristics allow them to interact with viral components (e.g., proteins and membranes) and to disrupt their structure or function, thereby preventing viral replication and transmissibility [293].

The antiviral activity of AgNPs against a wide variety of viruses, such as herpes simplex virus (HSV), HIV-1, and influenza virus, has been extensively studied [294,295]. By interfering with viral envelope proteins or RNA, AgNPs have been shown to prevent viral entry, replication, and release. Moreover, by inducing production of cytokines and chemokines that activate antiviral defenses, AgNPs can also boost the immune response [296]. Due to the varying methods of synthesis and sizes of these NPs, it is extremely difficult to determine their precise mechanism of action. The research conducted by Gaikwad et al. confirmed that the smaller their size, the greater their inhibition efficacy [297]. Although the antiviral mechanism remains somewhat unclear, it appears that AgNPs interact with viral surface proteins to prevent their penetration or to destroy them, thereby affecting the structure and integrity of the virion. For instance, a study conducted in vitro on AgNPs revealed that they prevented the SARS-CoV-2 virus from entering cells and thereby prevented infection [298].

The anti-HIV activity was demonstrated by Lara et al. at an early stage of viral replication. Polyvinyl pyrrolidone (PVP)-coated AgNPs inhibited both HIV-1 cell-free and cell-associated infection and showed that, as virucidal agents, they prevent the infection regardless of viral tropism or resistance profile, bind to envelope glycoprotein gp120 and therefore prevent CD4-dependent virion binding and infectivity. Therefore, AgNPs proved to be effective inhibitors of HIV, as well as of HBV [299–301].

A study involving the intranasal administration of AgNPs to mice revealed that the tested sample had a significant effect on survival, as evidenced by the decrease in lung viral titer levels, minor pathological lesions in the lung tissue, and a remarkable survival advantage after H3N2 influenza virus infection [302]. The size and zeta potential of the biologically synthesized AgNPs were responsible for inhibiting the viability of HSV types 1 and 2 and human parainfluenza virus type 3 [297].

The treatment of Vero cells with non-cytotoxic concentrations of AgNPs also significantly inhibited the Peste des petits ruminants virus replication. The mechanism on viral replication is due to the interaction of AgNPs with the surface and core of the virion [303].

AuNPs have also demonstrated antiviral activity against multiple viruses, such as hepatitis B virus, human papillomavirus, human rhinovirus and even SARS-CoV-2 [304,305]. One recent study also demonstrated that AuNPs inhibited the influenza virus' pre- and post-entry life-cycle phases, implicitly providing promising antiviral activity [306]. Nevertheless, AuNPs' mechanism of action is not fully understood, but it is believed to involve disruption of viral attachment and entry into host cells, as well as the inhibition of viral replication and gene expression [307]. Moreover, such particles could also prove to be useful for virus detection applications [308].

Even though metal NPs hold great promise as antiviral agents, there are still obstacles to overcome, such as optimizing their size, shape, and surface for maximum efficacy and minimizing their potential toxicity to host cells. In spite of this, their unique properties and broad-spectrum activity make them a promising option for the development of novel antiviral therapies [309].

The antiviral action of other different types of NPs was reported by Murugan et al. against the Dengue virus, and by Gutierrez et al. against rotavirus [310,311]. Using in vitro methods, Kumar et al. investigated the antiviral action of synthesized FeNPs on the H1N1 influenza A virus, and obtained promising results [312]. Another molecular study concluded that iron oxide NPs effectively interacted with the viral proteins of SARS-CoV-2 and the hepatitis C virus and prevented the penetration and binding in the host cells. This finding was proposed to the FDA for inclusion in clinical trials [313].

5.6. Anti-Inflammatory Activity

Inflammation is an immunological response sustained by the production of pro-inflammatory cytokines, the activation of the immune system, and the release of prostaglandins and chemotactic substances, including complement cells, interleukin-1 (IL-1), TNF-α and TGF-β [314,315].

Metal NPs have emerged as a promising material class due to their potential anti-inflammatory properties. According to numerous studies, such nanomaterials can suppress the production of pro-inflammatory cytokines, chemokines, and ROS to reduce inflammation. These characteristics make them promising therapeutic candidates for the treatment of a variety of inflammatory diseases [316–319].

The ability of AgNPs to inhibit the production of pro-inflammatory cytokines, such as TNF-α and IL-1 in macrophages, has been attributed to their potent anti-inflammatory activity. In addition, AgNPs reduce the expression of the inducible nitric oxide synthase (iNOS) enzyme, which plays a crucial role in the inflammatory response [320].

AuNPs inhibit the production of pro-inflammatory cytokines and chemokines, such as interleukin-6 (IL-6), interleukin-8 (IL-8), and monocyte chemoattractant protein-1 (MCP-1) in human endothelial cells, according to several studies [321–324].

Moreover, copper NPs have also exhibited anti-inflammatory activity by inhibiting the production of pro-inflammatory cytokines such as IL-1, TNF-α, and IL-6 in lipopolysaccharide (LPS)-stimulated macrophages, whereas zinc oxide nanoparticles (ZnO NPs) have been found to inhibit the activation of nuclear factor-kappa B (NF-kB) [325,326].

The anti-inflammatory activity of AgNPs produced using ginger oil was tested via an inhibition of albumin denaturation assay. The synthesized particles exhibited very good anti-inflammatory activity, with 90% of inhibition at a dose of 60 μL [327]. In another study, AgNPs and AuNPs obtained using *Prunus serrulata* fruit extract demonstrated in vitro anti-inflammatory properties, with a significant suppression of lipopolysaccharide-induced activation of the NF-κB signaling pathway via p38 MAPK on a RAW 264.7 murine macrophage cell line [328].

This anti-inflammatory activity was also highlighted in in vivo experiments. For instance, one study reported anti-inflammatory activity in rats treated intracolonically or orally with nanocrystalline silver (NPI 32101), showing substantially reduced colonic inflammation and rapid dose-dependent healing [329].

AgNPs that were delivered topically on wound mice models, and showed significant antimicrobial properties, also displayed a decrease in wound inflammation. Therefore, using both in vitro and in vivo models, Wong et al. discovered that AgNPs were able to reduce inflammatory markers, indicating that such particles could suppress inflammatory effects in the early stages of wound-healing [330].

AgNPs treatment significantly increased apoptosis in inflammatory cells and decreased the level of proinflammatory cytokines in a porcine model of contact dermatitis. Biosynthesized AgNPs can also inhibit UV-B-induced cytokine production in cells and reduce edema and cytokine levels in different pig tissues [331,332].

5.7. AgNPs and AuNPs in Clinical Trials

Due to their unique features, silver and gold NPs have become potential candidates for a variety of biomedical applications. Recent clinical trials have established the viability of these nanomaterials as antibacterial agents, drug delivery vehicles, and imaging agents for the diagnosis and treatment of a variety of disorders [333,334]. For example, AgNPs have demonstrated significant promise in wound-healing, whereas AuNPs have been exploited in cancer diagnosis and treatment (Table 3) [335,336].

Table 3. Lists of clinical trials of silver and gold NPs.

NPs Type	Study Title	Material	Targeted Condition	Clinical trials.gov Identifier
AgNPs	Topical Silver Nanoparticles for Microbial Activity	Topical cream containing silver NPs	• Fungal foot infection • Bacterial infection	NCT03752424
	Topical Application of Silver Nanoparticles and Oral Pathogens in Ill Patients	Innocuous gel compound with 12 ppm of AgNPs	• Dental diseases	NCT02761525
	Silver Nanoparticles in Multidrug Resistant Bacteria	AgNPs	• Multidrug resistant bacteria	NCT04431440
	Nanosilver Fluoride to Prevent Dental Biofilms Growth	Nanosilver fluoride	• Dental caries	NCT01950546
	Remineralization of Dentine Caries Using Two Remineralizing Agents Which Are Nanosilver Fluoride and Casein phosphopeptides amorphous Calcium Phosphate	Nanosilver fluoride vs. sodium fluoride with casein phosphopeptides amorphous calcium phosphate on carious lesion	• Dental caries	NCT04930458
	Radiographic Assessment of Glass Ionomer Restorations with and Without Prior Application of Nano Silver Fluoride in Occlusal Carious Molars Treated with Partial Caries Removal Technique	Nano silver fluoride solution	• Partial dentin caries removal	NCT03193606
	Clinical Evaluation of Silver Nanoparticles in Comparison to Silver Diamine Fluoride in Management of Deep Carious Lesions	AgNPs vs. silver diamine fluoride	• Caries, dental therapy	NCT05231330
	Addition of Silver Nanoparticles to an Orthodontic Primer in Preventing Enamel Demineralization Adjacent Brackets	AgNPs incorporated into the primer orthodontic Transbond XT	• Tooth demineralization	NCT02400957
	Evaluation of Antimicrobial Efficacy and Adaptability of Bioceramic Sealer Containing Nanoparticles	Bioceramic sealers with classic mix vs. AgNPs vs. chitosan	• Endodontic disease	NCT04481945
	Silver Nanoparticle Investigation for Treating Chronic Sinusitis	Topical colloidal silver	• Chronic rhinosinusitis	NCT03243201

Table 3. *Cont.*

NPs Type	Study Title	Material	Targeted Condition	Clinical trials.gov Identifier
	The Antibacterial Effect of Nanosilver Fluoride on Primary Teeth	Topic solution of silver fluoride NPs	• Microbial colonization for preventing dental caries in children	NCT05221749
	Colloidal Silver, Treatment of COVID-19	Inhalation colloidal silver solution	• SARS (severe acute respiratory syndrome)	NCT04978025
	Evaluation of Silver Nanoparticles for the Prevention of COVID-19	AgNPs solution with 1 wt% concentration (0.6 mg/mL metallic silver) for mouthwash and nose rinse	• Coronavirus disease • (COVID-19)	NCT04894409
	SARS-CoV-2-Spike-Ferritin-Nanoparticle (SpFN) Vaccine with ALFQ Adjuvant for Prevention of COVID-19 in Healthy Adults	SpFN COVID-19 vaccine with Army Liposomal Formulation QS21 (ALFQ) adjuvant	• SARS-CoV-2 infection	NCT04784767
	Fluor Varnish with Silver Nanoparticles for Dental Remineralization in Patients with Trisomy 21	Fluor dental varnish plus 25% 50 nm AgNPs vs. fluor dental varnish	• Dental remineralization, Down syndrome	NCT01975545
	Comparison of Central Venous Catheters with Silver Nanoparticles Versus Conventional Catheters	Central venous catheter impregnated with AgNPs (AgTive®)	• Central venous catheter related infections	NCT00337714
	Evaluation of Diabetic Foot Wound Healing Using Hydrogel/Nano Silver-based Dressing vs. Traditional Dressing	Hydrogel/nano silver-based dressing	• Diabetic foot wound	NCT04834245
	Efficacy of Silver Nanoparticle Gel Versus a Common Antibacterial Hand Gel	Nano-silver gel exposed vs. alcohol-based hand gel	• Antibacterial activity in normal patients' skin	NCT00659204
	Post-operative Pain Reduction	Conventional calcium hydroxide paste vs. combined calcium hydroxide with AgNPs	• Effects of the elements	NCT04338633
	A Phase I/II Double-blind Safety and Efficacy Evaluation of Nowarta110 in Patients with Plantar Warts	Nowarta 110, a colloidal silver fig extract for topical administration	• Plantar warts	NCT02338336
	Research on the Key Technology of Burn Wound Treatment	Nano-silver ion gel and dressings for wounds	• Burn wounds	NCT03279549
	Altrazeal Range of Motion Study Comparing with Typical Carboxymethyl	Hydrogel nanoparticle wound dressing	• Burn wounds	NCT01062191
	Assessment of Postoperative Pain After Using Various Intracanal Medication in Patients with Necrotic Pulp	AgNPs in gel form vs. calcium hydroxide intracanal medication	• Postoperative pain	NCT03692286
	Preadmission Skin Wipe Use for Surgical Site Infection Prophylaxis in Adult Orthopaedic Surgery Patients	Bath wipes impregnated with multiple ingredients, among which colloid silver	• Surgical site infection	NCT03401749
	Thyme and Carvacroll Nanoparticle Effect on Fungi	Thyme and carvacrol AgNPs	• Aspergillosis	NCT04431804
	Clinical and Radiographic Evaluation of the Synergistic Effect of Nano Silver Particles and Calcium Hydroxide Versus Triple Antibiotic Paste as Antibacterial Agents for Lesion Sterilization and Tissue Repair (LSTR) in Necrotic Primary Molars	Nano-silver particles and calcium hydroxide vs. triple antibiotic paste	• Root canal infection	NCT05681221
	Effect of Metallic Nanoparticles on Nosocomial Bacteria	AgNPs and CuNPs	• Nosocomial infections	NCT04775238
	Evaluation of the Antimicrobial Fiber Reinforced Composite Resin Space Maintainer Modified With Silver Nano Particles	Fiber reinforced composite resin space maintainer modified with silver nano particles	• Dental caries in children	NCT05902975
	Gold Factor on Knee Joint Health and Function	Dietary supplement: Gold factor (AuNPs)	• Knee arthritis • Knee osteoarthritis • Knee pain chronic • Knee discomfort • Rheumatoid arthritis • Knee pain swelling	NCT05347602
	Diagnostic and Prognostic Accuracy of Gold Nanoparticles in Salivary Gland Tumors	CD24-Gold nanocomposites	• Carcinoma ex pleomorphic adenoma of salivary glands • Pleomorphic adenoma of salivary glands	NCT04907422
	Plasmonic Photothermal and Stem Cell Therapy of Atherosclerosis Versus Stenting	Stenting and micro-infusion of NP	• Coronary artery disease • Atherosclerosis	NCT01436123

Table 3. *Cont.*

NPs Type	Study Title	Material	Targeted Condition	Clinical trials.gov Identifier
AuNPs	Enhanced Epidermal Antigen Specific Immunotherapy Trial -1	C19-A3 GNP drug	• Type 1 diabetes	NCT02837094
	NU-0129 in Treating Patients with Recurrent Glioblastoma or Gliosarcoma Undergoing Surgery	NU-0129 drug	• Gliosarcoma • Recurrent glioblastoma	NCT03020017
	Plasmonic Nanophotothermal Therapy of Atherosclerosis	Transplantation of NPs	• Stable angina • Heart failure • Atherosclerosis • Multivessel coronary artery disease	NCT01270139
	Effect of Nano Care Gold on Marginal Integrity of Resin Composite	Nano Care Gold (silver and gold NPs) suspended in 70% isopropyl alcohol, for cavity pre-treatment.	• Caries class ii	NCT03669224
	A Phase-I Study of a Nanoparticle-based Peptide Vaccine Against Dengue Virus	Nanoparticle-based peptide vaccine	• Dengue fever	NCT04935801
	A Phase-I Study of a Nanoparticle-based Peptide Vaccine Against SARS-CoV-2	Nanoparticle-based peptide vaccine	• Coronavirus • SARS-CoV-2 infection • COVID-19	NCT05113862
	Diagnosis of Gastric Lesions from Exhaled Breath and Saliva	Chemical nanosensors based on organically functionalized AuNPs and carbon nanotubes	• Stomach diseases	NCT01420588
	Exploratory Study Using Nanotechnology to Detect Biomarkers of Parkinson's Disease from Exhaled Breath	Combinations of nanomaterial-based sensors	• Parkinson's disease • Parkinsonism	NCT01246336

However, other types of metal particles, such as iron oxide NPs, have attracted significant attention in recent years due to their unique magnetic properties, biocompatibility, and versatile applications in clinical settings. Their properties have been extensively studied in preclinical and clinical trials for various biomedical applications, including magnetic resonance imaging (MRI), drug delivery, and hyperthermia therapy [337].

Overall, metal NPs hold enormous potential for the creation of new and successful treatments, but their clinical applications must be examined and monitored with considerable care.

6. Cytotoxicity of Metal NPs

The adverse effects of nanomaterials on human organs have gradually been noticed, and may have an impact on their use in biomedical applications [338,339]. A metal-based NP can easily cross cell membranes due to its size comparable to that of protein molecules. As a result, they can interact with proteins or subcellular organelle structures, leading to neurotoxicity, immunotoxicity, and genotoxicity. Implicitly, the biosafety of such materials and technologies has received significant attention considering their widespread use [339].

AgNPs can be ingested, injected intravenously, or absorbed through the skin. The absorbed AgNPs are distributed in numerous organs and systems, including the respiratory tract, digestive and urinary systems, and nervous system, as well as the immune and reproductive systems, but primarily in the spleen, liver, kidney, and lungs, with a lesser deposition in the teeth and bones [338,340]. The non-specific distribution of AgNPs can result in skin, eye, respiratory, and reproductive system toxicity and even neurotoxicity, thereby restricting their application.

The potential cytotoxicity of AgNPs depends on properties such as size, shape, and concentration, and routes of administration [341]. For instance, Wang et al. employed TEM and the integration of synchrotron radiation beam transmission X-ray microscopy (SR-TXM) with 3D tomographic imaging to collect data on the uptake, buildup, cellular degradation, chemical transformation, and removal of AgNPs at a cellular level. The experiment demonstrated that the cytotoxicity was caused by the chemical transformation of AgNPs to Ag+, Ag-O- and Ag-S- species, which could induce biological modifications [342].

Similarly, the cytotoxicity of AuNPs has been examined and reviewed by several researchers [343–345]. Different routes of administration can have varying effects on the biodistribution of drug carriers [346]. The in vivo distribution after administration largely depends on the NPs' size, surface charge and surface hydrophobicity. The influence of these factors on NPs uptake by the mononuclear phagocyte system has also been described [347].

The size-dependent permeability of AuNPs through the skin and intestine has been demonstrated [348]. AuNPs of different sizes (15, 50, 100, and 200 nm) were administered intravenously to mice, and their biodistribution was studied. AuNPs of all sizes primarily accumulated in the liver, lungs, and spleen, whereas the accumulation in other tissues was dependent on the size of AuNPs. Large amounts of 15 nm AuNPs were detected in all tissues, including blood, liver, lungs, spleen, and kidney, and it was discovered that they could cross the blood-brain barrier, similar to the 50 nm AuNPs. On the other hand, only a small amount of the 200 nm AuNPs could be found in blood, brain, stomach, and pancreas [349].

Moreover, the shape also poses an important influence of the observed toxicity. For instance, AuNPs stars were the most cytotoxic type of investigated particle, whereas AuNPs spheres seem to be safest, but had a reduced anticancer potential [350].

At the same time, iron oxide NPs are associated with a low toxicity in the human body [351–354]. For example, an in vitro study comparing several metal oxide nanoparticles showed that iron oxide ones are not cytotoxic at any concentration smaller than 100 mg/mL [355]. Nevertheless, the absence of cytotoxicity does not guarantee that they do not pose a risk to human health in specific applications, as several recent studies have reported potentially negative cellular effects, such as DNA damage, and mitochondrial membrane dysfunction, as well as changes in gene expression [356,357].

Among the several processes that might be linked to the toxicity of NPs in the body, it appears that magnetic iron oxide NPs' toxicity is caused by excessive ROS production [358–361]. High levels of ROS can damage the cells through lipid peroxidation, mitochondrial damage, DNA disruption, modulation of gene transcription and protein oxidation, which can trigger mechanisms that result in decreased physiological functions and cell apoptosis [362]. Other toxic effects include cell-cycle changes and the induction of apoptosis, and can occur in various cell types after iron oxide NPs treatment [363,364].

The cytotoxicity of NPs on normal cells has significant implications for their use in various applications, such as medicine and consumer goods. The aspects described in this section emphasize the need for careful consideration of the potential health risks of NPs and the creation of safer alternatives. For example, since surface coating is one of the most popular modification strategies to reduce the potential toxicity of metal NPs, coating AgNPs with biocompatible materials, such as PEG, could help reduce their cytotoxicity on normal cells [365].

Furthermore, the non-biodegradability of metal NPs raises significant environmental concerns due to their persistence and potential long-term impact on ecosystems. To address this issue, researchers are actively exploring various strategies. One approach involves the development of biodegradable coatings or encapsulation techniques, thus enabling controlled release and gradual degradation over time [366].

Consequently, complex further studies in the area of metal NPs are highly recommended. The first step in ensuring safe clinical integration is a thorough investigation of biocompatibility and toxicity profiles. A second way to improve stability and targeting-accuracy is by developing synthesis methods and surface modifications. Moreover, a deeper comprehension of interactions between biological systems at a molecular level is required [367].

7. Key Challenges

Metal NPs represent a fascinating class of nanomaterials with diverse and promising applications across numerous fields. However, as previously mentioned, their unique properties also bring forth several key challenges that must be overcome to fully exploit their potential. One of the foremost challenges is the precise control of their size and shape during synthesis. Achieving uniformity in size and shape is critical for tailoring their properties and functionalities. This often demands complex synthesis techniques and a deep understanding of the underlying chemistry and physics involved. Moreover, scaling up the production of well-defined metal NPs while maintaining this control can be challenging and costly [368].

Another critical issue is the tendency of metal NPs to agglomerate or aggregate, leading to reduced stability and diminished performance. The forces driving agglomeration, such as van der Waals forces and electrostatic interactions, must be mitigated through strategies like surface functionalization or the use of stabilizing agents [369].

In biomedical applications, the toxicity of certain metal NPs, particularly heavy metals like cadmium or mercury, poses a significant challenge. Understanding the potential health risks and developing safe handling procedures is essential for their use in drug delivery, imaging, and diagnostics.

Furthermore, the long-term stability and behavior of such materials under different environmental conditions remain complex and not fully elucidated. Variations in temperature, pH, and exposure to other chemicals can alter their properties, potentially rendering them less effective or even hazardous. Consequently, robust characterization and predictive models are needed to ensure consistent performance.

Lastly, the scalability of production and the environmental impact of their waste disposal are pressing concerns. Large-scale production methods need to be developed that are cost-effective and environmentally sustainable. Additionally, strategies for recycling or disposing of used NPs in an eco-friendly manner are crucial to minimize their ecological footprint [370].

Addressing these challenges demands an interdisciplinary collaboration among material scientists, biologists, clinicians, engineers, and regulatory bodies. As research continues to advance, the potential benefits of metal nanoparticles in catalysis, electronics, and various medical applications make overcoming these hurdles all the more imperative. Ultimately, solving these challenges will not only unlock the full potential of metal NPs, but also ensure their safe and responsible utilization in a wide range of innovative technologies [367].

8. Conclusions

Within the last two decades, nanotechnology has continuously and rapidly evolved, since the emergence of multiple methods of obtaining and using raw materials for nanostructures in a variety of fields such as medicine, pharmacy, agriculture, electronics, and many others.

Due to their unique physicochemical and biological features, silver and gold NPs have been the focus of substantial study in the field of nanomedicine. Their exceptional biocompatibility, stability, and comparatively low toxicity make them attractive candidates for a variety of biological applications. In addition, AgNPs have demonstrated outstanding antibacterial activity against a broad spectrum of pathogens, making them particularly interesting for application in medical device coatings, wound-healing, and water-purification. In contrast, AuNPs have remarkable optical and electrical capabilities, rendering them valuable for imaging, sensing, and drug delivery applications. Several clinical trials have also included such representatives to better understand their implications in various conditions.

Nevertheless, the toxicity, long-term stability, and non-biodegradability of metal NPs represent key issues to weigh when considering their use. Resolving these challenges necessitates interdisciplinary cooperation and robust regulatory oversight to harness the full potential of metal NPs in biomedical applications, while ensuring safety and sustainability. Overall, this review emphasizes the importance and prospects of metal NPs in the biomedical field, considering some of the most recent studies. However, further research is encouraged in order to fully understand their safety and efficacy in clinical settings and to optimize their physicochemical properties for specific applications.

Author Contributions: Conceptualization, A.F.B. (Ana Flavia Burlec), M.B. and A.C.; methodology, A.F.B. (Ana Flavia Burlec), C.M. and O.C.; software, D.B.-M.; validation, M.B., A.C. and M.H.; resources, D.B.-M. and M.H.; writing—original draft preparation, M.B. and A.C.; writing—review and editing, A.F.B. (Ana Flavia Burlec); visualization, A.F.B. (Ana Flavia Burlec), G.D., M.D. and A.F.B. (Anca Florentina Bucur); supervision, A.F.B. (Ana Flavia Burlec) and A.C. All authors have read and agreed to the published version of the manuscript.

Funding: This research received no external funding.

Institutional Review Board Statement: Not applicable.

Informed Consent Statement: Not applicable.

Data Availability Statement: Not applicable.

Acknowledgments: The figures of this article were created using BioRender.com.

Conflicts of Interest: The authors declare no conflict of interest.

References

1. Rana, A.; Yadav, K.; Jagadevan, S. A Comprehensive Review on Green Synthesis of Nature-Inspired Metal Nanoparticles: Mechanism, Application and Toxicity. *J. Clean. Prod.* **2020**, *272*, 122880. [CrossRef]
2. Thakkar, K.N.; Mhatre, S.S.; Parikh, R.Y. Biological synthesis of metallic nanoparticles. *Nanomed. Nanotechnol. Biol. Med.* **2010**, *6*, 257–262. [CrossRef]
3. Ghosh, S.K.; Pal, T. Interparticle Coupling Effect on the Surface Plasmon Resonance of Gold Nanoparticles: From Theory to Applications. *Chem. Rev.* **2007**, *107*, 4797–4862. [CrossRef]
4. Buzea, C.; Pacheco, I.I.; Robbie, K. Nanomaterials and Nanoparticles: Sources and Toxicity. *Biointerphases* **2007**, *2*, MR17–MR71. [CrossRef]
5. Jacob, S.J.P.; Prasad, V.L.S.; Sivasankar, S.; Muralidharan, P. Biosynthesis of Silver Nanoparticles Using Dried Fruit Extract of *Ficus Carica*—Screening for Its Anticancer Activity and Toxicity in Animal Models. *Food Chem. Toxicol.* **2017**, *109*, 951–956. [CrossRef]
6. Khan, I.; Saeed, K.; Khan, I. Nanoparticles: Properties, Applications and Toxicities. *Arab. J. Chem.* **2019**, *12*, 908–931. [CrossRef]
7. Machado, S.; Pacheco, J.G.; Nouws, H.P.A.; Albergaria, J.T.; Delerue-Matos, C. Characterization of Green Zero-Valent Iron Nanoparticles Produced with Tree Leaf Extracts. *Sci. Total Environ.* **2015**, *533*, 76–81. [CrossRef]
8. Talarska, P.; Boruczkowski, M.; Żurawski, J. Current Knowledge of Silver and Gold Nanoparticles in Laboratory Research—Application, Toxicity, Cellular Uptake. *Nanomaterials* **2021**, *11*, 2454. [CrossRef]
9. Modena, M.M.; Rühle, B.; Burg, T.P.; Wuttke, S. Nanoparticle Characterization: What to Measure? *Adv. Mater.* **2019**, *31*, 1901556. [CrossRef]
10. Mitchell, M.J.; Billingsley, M.M.; Haley, R.M.; Wechsler, M.E.; Peppas, N.A.; Langer, R. Engineering Precision Nanoparticles for Drug Delivery. *Nat. Rev. Drug Discov.* **2021**, *20*, 101–124. [CrossRef]
11. Yaqoob, S.B.; Adnan, R.; Rameez Khan, R.M.; Rashid, M. Gold, Silver, and Palladium Nanoparticles: A Chemical Tool for Biomedical Applications. *Front. Chem.* **2020**, *8*, 376. [CrossRef]
12. Loiseau, A.; Asila, V.; Boitel-Aullen, G.; Lam, M.; Salmain, M.; Boujday, S. Silver-Based Plasmonic Nanoparticles for and Their Use in Biosensing. *Biosensors* **2019**, *9*, 78. [CrossRef]
13. Khan, A.; Rashid, R.; Murtaza, G.; Zahra, A. Gold Nanoparticles: Synthesis and Applications in Drug Delivery. *Trop. J. Pharm. Res.* **2014**, *13*, 1169. [CrossRef]
14. Păduraru, D.N.; Ion, D.; Niculescu, A.-G.; Mușat, F.; Andronic, O.; Grumezescu, A.M.; Bolocan, A. Recent Developments in Metallic Nanomaterials for Cancer Therapy, Diagnosing and Imaging Applications. *Pharmaceutics* **2022**, *14*, 435. [CrossRef]
15. Heinemann, M.G.; Rosa, C.H.; Rosa, G.R.; Dias, D. Biogenic Synthesis of Gold and Silver Nanoparticles Used in Environmental Applications: A Review. *Trends Environ. Anal. Chem.* **2021**, *30*, e00129. [CrossRef]
16. Ealia, S.A.M.; Saravanakumar, M.P. A Review on the Classification, Characterisation, Synthesis of Nanoparticles and Their Application. *IOP Conf. Ser. Mater. Sci. Eng.* **2017**, *263*, 032019. [CrossRef]
17. Khan, Y.; Sadia, H.; Shah, S.Z.A.; Khan, M.N.; Shah, A.A.; Ullah, N.; Ullah, M.F.; Bibi, H.; Bafakeeh, O.T.; Khedher, N.B.; et al. Classification, Synthetic, and Characterization Approaches to Nanoparticles, and Their Applications in Various Fields of Nanotechnology: A Review. *Catalysts* **2022**, *12*, 1386. [CrossRef]
18. Esakkimuthu, T.; Sivakumar, D.; Akila, S. Application of Nanoparticles in Wastewater Treatment. *Pollut. Res.* **2014**, *33*, 567–571.
19. Ahmad, Z.; Shah, A.; Siddiq, M.; Kraatz, H.-B. Polymeric Micelles as Drug Delivery Vehicles. *RSC Adv.* **2014**, *4*, 17028–17038. [CrossRef]
20. Buse, J.; El-Aneed, A. Properties, Engineering and Applications of Lipid-Based Nanoparticle Drug-Delivery Systems: Current Research and Advances. *Nanomedicine* **2010**, *5*, 1237–1260. [CrossRef]
21. Daraee, H.; Etemadi, A.; Kouhi, M.; Alimirzalu, S.; Akbarzadeh, A. Application of liposomes in medicine and drug delivery. *Artif. Cells Nanomed. Biotechnol.* **2016**, *44*, 381–391. [CrossRef] [PubMed]

22. Oberholzer, T.; Luisi, P.L. The Use of Liposomes for Constructing Cell Models. *J. Biol. Phys.* **2002**, *28*, 733–744. [CrossRef] [PubMed]
23. Hsu, H.; Bugno, J.; Lee, S.; Hong, S. Dendrimer-based nanocarriers: A versatile platform for drug delivery. *WIREs Nanomed. Nanobiotechnology* **2017**, *9*, e1409. [CrossRef] [PubMed]
24. Palmerston Mendes, L.; Pan, J.; Torchilin, V.P. Dendrimers as Nanocarriers for Nucleic Acid and Drug Delivery in Cancer Therapy. *Molecules* **2017**, *22*, 1401. [CrossRef]
25. Spirescu, V.A.; Chircov, C.; Grumezescu, A.M.; Andronescu, E. Polymeric Nanoparticles for Antimicrobial Therapies: An up-to-date Overview. *Polymers* **2021**, *13*, 724. [CrossRef] [PubMed]
26. Vardaxi, A.; Kafetzi, M.; Pispas, S. Polymeric Nanostructures Containing Proteins and Peptides for Pharmaceutical Applications. *Polymers* **2022**, *14*, 777. [CrossRef]
27. Nemtsev, I.V.; Shabanova, O.V.; Shestakov, N.P.; Cherepakhin, A.V.; Zyryanov, V.Y. Morphology Stability of Polymethylmethacrylate Nanospheres Formed in Water–Acetone Dispersion Medium. *Appl. Phys. A* **2019**, *125*, 738. [CrossRef]
28. Shabanova, O.V.; Korshunov, M.A.; Nemtsev, I.V.; Shabanov, A.V. Features of Self-Assembly of Opal-like Structures Based on Poly(Methyl Methacrylate) Submicron Dispersions. *Nanotechnologies Russ.* **2016**, *11*, 633–639. [CrossRef]
29. Batista, F.A.; Fontele, S.B.C.; Santos, L.K.B.; Filgueiras, L.A.; Nascimento, S.Q.; Sousa, J.M.d.C.e.; Gonçalves, J.C.R.; Mendes, A.N. Synthesis, characterization of α-terpineol-loaded PMMA nanoparticles as proposed of therapy for melanoma. *Mater. Today Commun.* **2020**, *22*, 100762. [CrossRef]
30. Aljabali, A.A.; Rezigue, M.; Alsharedeh, R.H.; Obeid, M.A.; Mishra, V.; Serrano-Aroca, Á.; El-Tanani, M.; Tambuwala, M.M. Protein-Based Nanomaterials: A New Tool for Targeted Drug Delivery. *Ther. Deliv.* **2022**, *13*, 321–338. [CrossRef]
31. Kianfar, E. Protein Nanoparticles in Drug Delivery: Animal Protein, Plant Proteins and Protein Cages, Albumin Nanoparticles. *J. Nanobiotechnol.* **2021**, *19*, 159. [CrossRef] [PubMed]
32. Estrada, L.P.H.; Champion, J.A. Protein nanoparticles for therapeutic protein delivery. *Biomater. Sci.* **2015**, *3*, 787–799. [CrossRef] [PubMed]
33. Hong, S.; Choi, D.W.; Kim, H.N.; Park, C.G.; Lee, W.; Park, H.H. Protein-Based Nanoparticles as Drug Delivery Systems. *Pharmaceutics* **2020**, *12*, 604. [CrossRef] [PubMed]
34. Alemán, J.V.; Chadwick, A.V.; He, J.; Hess, M.; Horie, K.; Jones, R.G.; Kratochvíl, P.; Meisel, I.; Mita, I.; Moad, G.; et al. Definitions of Terms Relating to the Structure and Processing of Sols, Gels, Networks, and Inorganic-Organic Hybrid Materials (IUPAC Recommendations 2007). *Pure Appl. Chem.* **2007**, *79*, 1801–1829. [CrossRef]
35. Neamtu, I.; Rusu, A.G.; Diaconu, A.; Nita, L.E.; Chiriac, A.P. Basic Concepts and Recent Advances in Nanogels as Carriers for Medical Applications. *Drug Deliv.* **2017**, *24*, 539–557. [CrossRef]
36. Khan, M.U.; Ullah, H.; Honey, S.; Talib, Z.; Abbas, M.; Umar, A.; Ahmad, T.; Sohail, J.; Sohail, A.; Makgopa, K.; et al. Metal Nanoparticles: Synthesis Approach, Types and Applications—A Mini Review. *Nano-Horizons* **2023**, *2*, 1–21. [CrossRef]
37. Salavati-Niasari, M.; Davar, F.; Mir, N. Synthesis and Characterization of Metallic Copper Nanoparticles via Thermal Decomposition. *Polyhedron* **2008**, *27*, 3514–3518. [CrossRef]
38. Tai, C.Y.; Tai, C.-T.; Chang, M.-H.; Liu, H.-S. Synthesis of Magnesium Hydroxide and Oxide Nanoparticles Using a Spinning Disk Reactor. *Ind. Eng. Chem. Res.* **2007**, *46*, 5536–5541. [CrossRef]
39. Vargas-Ortiz, J.R.; Gonzalez, C.; Esquivel, K. Magnetic Iron Nanoparticles: Synthesis, Surface Enhancements, and Biological Challenges. *Processes* **2022**, *10*, 2282. [CrossRef]
40. Le, N.; Zhang, M.; Kim, K. Quantum Dots and Their Interaction with Biological Systems. *Int. J. Mol. Sci.* **2022**, *23*, 10763. [CrossRef]
41. Bruckmann, F.d.S.; Nunes, F.B.; Salles, T.d.R.; Franco, C.; Cadoná, F.C.; Rhoden, C.R.B. Biological Applications of Silica-Based Nanoparticles. *Magnetochemistry* **2022**, *8*, 131. [CrossRef]
42. Shahmoradi, S.; Shariati, A.; Amini, S.M.; Zargar, N.; Yadegari, Z.; Darban-Sarokhalil, D. The Application of Selenium Nanoparticles for Enhancing the Efficacy of Photodynamic Inactivation of Planktonic Communities and the Biofilm of Streptococcus Mutans. *BMC Res. Notes* **2022**, *15*, 84. [CrossRef] [PubMed]
43. Chen, Y.; Yu, Z.; Zheng, K.; Ren, Y.; Wang, M.; Wu, Q.; Zhou, F.; Liu, C.; Liu, L.; Song, J.; et al. Degradable Mesoporous Semimetal Antimony Nanospheres for Near-Infrared II Multimodal Theranostics. *Nat. Commun.* **2022**, *13*, 539. [CrossRef] [PubMed]
44. Shahbazi, M.-A.; Faghfouri, L.; Ferreira, M.P.A.; Figueiredo, P.; Maleki, H.; Sefat, F.; Hirvonen, J.; Santos, H.A. The Versatile Biomedical Applications of Bismuth-Based Nanoparticles and Composites: Therapeutic, Diagnostic, Biosensing, and Regenerative Properties. *Chem. Soc. Rev.* **2020**, *49*, 1253–1321. [CrossRef]
45. Vines, J.B.; Yoon, J.-H.; Ryu, N.-E.; Lim, D.-J.; Park, H. Gold Nanoparticles for Photothermal Cancer Therapy. *Front. Chem.* **2019**, *7*, 167. [CrossRef]
46. Koga, K.; Tagami, T.; Ozeki, T. Gold Nanoparticle-Coated Thermosensitive Liposomes for the Triggered Release of Doxorubicin, and Photothermal Therapy Using a near-Infrared Laser. *Colloids Surfaces A Physicochem. Eng. Asp.* **2021**, *626*, 127038. [CrossRef]
47. Amini, S.M.; Rezayat, S.M.; Dinarvand, R.; Kharrazi, S.; Jaafari, M.R. Gold Cluster Encapsulated Liposomes: Theranostic Agent with Stimulus Triggered Release Capability. *Med. Oncol.* **2023**, *40*, 126. [CrossRef]
48. Koosha, F.; Farsangi, Z.J.; Samadian, H.; Amini, S.M. Mesoporous Silica Coated Gold Nanorods: A Multifunctional Theranostic Platform for Radiotherapy and X-Ray Imaging. *J. Porous Mater.* **2021**, *28*, 1961–1968. [CrossRef]

49. Pangli, H.; Vatanpour, S.; Hortamani, S.; Jalili, R.; Ghahary, A. Incorporation of Silver Nanoparticles in Hydrogel Matrices for Controlling Wound Infection. *J. Burn Care Res.* **2021**, *42*, 785–793. [CrossRef]
50. Bhaviripudi, S.; Mile, E.; Steiner, S.A.; Zare, A.T.; Dresselhaus, M.S.; Belcher, A.M.; Kong, J. CVD Synthesis of Single-Walled Carbon Nanotubes from Gold Nanoparticle Catalysts. *J. Am. Chem. Soc.* **2007**, *129*, 1516–1517. [CrossRef]
51. Nasrollahzadeh, M.; Issaabadi, Z.; Sajjadi, M.; Sajadi, S.M.; Atarod, M. Types of Nanostructures. In *Interface Science and Technology*; Elsevier: Amsterdam, The Netherlands, 2019; pp. 29–80.
52. Lu, Z.; Dai, T.; Huang, L.; Kurup, D.B.; Tegos, G.P.; Jahnke, A.; Wharton, T.; Hamblin, M.R. Photodynamic Therapy with a Cationic Functionalized Fullerene Rescues Mice from Fatal Wound Infections. *Nanomedicine* **2010**, *5*, 1525–1533. [CrossRef] [PubMed]
53. Gaur, M.; Misra, C.; Yadav, A.B.; Swaroop, S.; Maolmhuaidh, F.Ó.; Bechelany, M.; Barhoum, A. Biomedical Applications of Carbon Nanomaterials: Fullerenes, Quantum Dots, Nanotubes, Nanofibers, and Graphene. *Materials* **2021**, *14*, 5978. [CrossRef] [PubMed]
54. Geim, A.K. Graphene: Status and Prospects. *Science* **2009**, *324*, 1530–1534. [CrossRef]
55. Behzadi, S.; Serpooshan, V.; Tao, W.; Hamaly, M.A.; Alkawareek, M.Y.; Dreaden, E.C.; Brown, D.; Alkilany, A.M.; Farokhzad, O.C.; Mahmoudi, M. Cellular Uptake of Nanoparticles: Journey inside the Cell. *Chem. Soc. Rev.* **2017**, *46*, 4218–4244. [CrossRef] [PubMed]
56. Yadav, D.; Amini, F.; Ehrmann, A. Recent Advances in Carbon Nanofibers and Their Applications—A Review. *Eur. Polym. J.* **2020**, *138*, 109963. [CrossRef]
57. Fan, Y.; Fowler, G.D.; Zhao, M. The Past, Present and Future of Carbon Black as a Rubber Reinforcing Filler—A Review. *J. Clean. Prod.* **2020**, *247*, 119115. [CrossRef]
58. Silva, L.P.; Reis, I.G.; Bonatto, C.C. Green Synthesis of Metal Nanoparticles by Plants: Current Trends and Challenges. In *Green Processes for Nanotechnology*; Springer International Publishing: Cham, Switzerland, 2015; pp. 259–275.
59. Ying, S.; Guan, Z.; Ofoegbu, P.C.; Clubb, P.; Rico, C.; He, F.; Hong, J. Green Synthesis of Nanoparticles: Current Developments and Limitations. *Environ. Technol. Innov.* **2022**, *26*, 102336. [CrossRef]
60. Ramanathan, S.; Gopinath, S.C.B.; Arshad, M.K.M.; Poopalan, P.; Perumal, V. Nanoparticle Synthetic Methods: Strength and Limitations. In *Nanoparticles in Analytical and Medical Devices*; Elsevier: Amsterdam, The Netherlands, 2021; pp. 31–43. [CrossRef]
61. Krishnia, L.; Thakur, P.; Thakur, A. Synthesis of Nanoparticles by Physical Route. In *Synthesis and Applications of Nanoparticles*; Springer Nature: Singapore, 2022; pp. 45–59.
62. Biswas, A.; Bayer, I.S.; Biris, A.S.; Wang, T.; Dervishi, E.; Faupel, F. Advances in Top–down and Bottom–up Surface Nanofabrication: Techniques, Applications & Future Prospects. *Adv. Colloid Interface Sci.* **2012**, *170*, 2–27. [CrossRef]
63. Abid, N.; Khan, A.M.; Shujait, S.; Chaudhary, K.; Ikram, M.; Imran, M.; Haider, J.; Khan, M.; Khan, Q.; Maqbool, M. Synthesis of Nanomaterials Using Various Top-down and Bottom-up Approaches, Influencing Factors, Advantages, and Disadvantages: A Review. *Adv. Colloid Interface Sci.* **2022**, *300*, 102597. [CrossRef]
64. Ali, A.; Shah, T.; Ullah, R.; Zhou, P.; Guo, M.; Ovais, M.; Tan, Z.; Rui, Y. Review on Recent Progress in Magnetic Nanoparticles: Synthesis, Characterization, and Diverse Applications. *Front. Chem.* **2021**, *9*, 629054. [CrossRef]
65. Stolyar, S.V.; Komogortsev, S.V.; Gorbenko, A.S.; Knyazev, Y.V.; Yaroslavtsev, R.N.; Olkhovskiy, I.A.; Neznakhin, D.S.; Tyumentseva, A.V.; Bayukov, O.A.; Iskhakov, R.S. Maghemite Nanoparticles for DNA Extraction: Performance and Blocking Temperature. *J. Supercond. Nov. Magn.* **2022**, *35*, 1929–1936. [CrossRef]
66. Ivănescu, B.; Burlec, A.F.; Crivoi, F.; Roșu, C.; Corciovă, A. Secondary Metabolites from *Artemisia* Genus as Biopesticides and Innovative Nano-Based Application Strategies. *Molecules* **2021**, *26*, 3061. [CrossRef] [PubMed]
67. Gurunathan, S.; Park, J.H.; Han, J.W.; Kim, J.-H. Comparative Assessment of the Apoptotic Potential of Silver Nanoparticles Synthesized by *Bacillus tequilensis* and *Calocybe indica* in MDA-MB-231 Human Breast Cancer Cells: Targeting P53 for Anticancer Therapy. *Int. J. Nanomed.* **2015**, *10*, 4203–4223. [CrossRef] [PubMed]
68. Li, W.-R.; Xie, X.-B.; Shi, Q.-S.; Zeng, H.-Y.; OU-Yang, Y.-S.; Chen, Y.-B. Antibacterial Activity and Mechanism of Silver Nanoparticles on *Escherichia Coli*. *Appl. Microbiol. Biotechnol.* **2010**, *85*, 1115–1122. [CrossRef] [PubMed]
69. Mukherjee, P.; Ahmad, A.; Mandal, D.; Senapati, S.; Sainkar, S.R.; Khan, M.I.; Parishcha, R.; Ajaykumar, P.V.; Alam, M.; Kumar, R.; et al. Fungus-Mediated Synthesis of Silver Nanoparticles and Their Immobilization in the Mycelial Matrix: A Novel Biological Approach to Nanoparticle Synthesis. *Nano Lett.* **2001**, *1*, 515–519. [CrossRef]
70. Lee, S.; Jun, B.-H. Silver Nanoparticles: Synthesis and Application for Nanomedicine. *Int. J. Mol. Sci.* **2019**, *20*, 865. [CrossRef]
71. Kruis, F.E.; Fissan, H.; Rellinghaus, B. Sintering and Evaporation Characteristics of Gas-Phase Synthesis of Size-Selected PbS Nanoparticles. *Mater. Sci. Eng. B* **2000**, *69–70*, 329–334. [CrossRef]
72. Jung, J.H.; Oh, C.H.; Noh, H.S.; Ji, J.H.; Kim, S.S. Metal nanoparticle generation using a small ceramic heater with a local heating area. *J. Aerosol Sci.* **2006**, *37*, 1662–1670. [CrossRef]
73. Chen, Y.-H.; Yeh, C.-S. Laser Ablation Method: Use of Surfactants to Form the Dispersed Ag Nanoparticles. *Colloids Surfaces A Physicochem. Eng. Asp.* **2002**, *197*, 133–139. [CrossRef]
74. Chugh, H.; Sood, D.; Chandra, I.; Tomar, V.; Dhawan, G.; Chandra, R. Role of Gold and Silver Nanoparticles in Cancer Nano-Medicine. *Artif. Cells Nanomed. Biotechnol.* **2018**, *46*, 1210–1220. [CrossRef]
75. Elsupikhe, R.F.; Shameli, K.; Ahmad, M.B.; Ibrahim, N.A.; Zainudin, N. Green Sonochemical Synthesis of Silver Nanoparticles at Varying Concentrations of κ-Carrageenan. *Nanoscale Res. Lett.* **2015**, *10*, 302. [CrossRef] [PubMed]
76. Shameli, K.; Ahmad, M.B.; Yunis, W.Z.; Ibrahim, N.A.; Darroudi, M.; Gharayebi, Y.; Sedaghat, S. Synthesis of Silver/Montmorillonite Nanocomposites Using γ-Irradiation. *Int. J. Nanomed.* **2010**, *5*, 1067–1077. [CrossRef]

77. El-Nour, K.M.M.A.; Eftaiha, A.; Al-Warthan, A.; Ammar, R.A.A. Synthesis and Applications of Silver Nanoparticles. *Arab. J. Chem.* **2010**, *3*, 135–140. [CrossRef]
78. Iravani, S.; Korbekandi, H.; Mirmohammadi, S.V.; Zolfaghari, B. Synthesis of Silver Nanoparticles: Chemical, Physical and Biological Methods. *Res. Pharm. Sci.* **2014**, *9*, 385–406.
79. Kinnear, C.; Moore, T.L.; Rodriguez-Lorenzo, L.; Rothen-Rutishauser, B.; Petri-Fink, A. Form Follows Function: Nanoparticle Shape and Its Implications for Nanomedicine. *Chem. Rev.* **2017**, *117*, 11476–11521. [CrossRef] [PubMed]
80. Pillai, Z.S.; Kamat, P.V. What Factors Control the Size and Shape of Silver Nanoparticles in the Citrate Ion Reduction Method? *J. Phys. Chem. B* **2004**, *108*, 945–951. [CrossRef]
81. Turkevich, J. Colloidal Gold. Part I. *Gold Bull.* **1985**, *18*, 86–91. [CrossRef]
82. Brust, M.; Walker, M.; Bethell, D.; Schiffrin, D.J.; Whyman, R. Synthesis of Thiol-Derivatised Gold Nanoparticles in a Two-Phase Liquid–Liquid System. *J. Chem. Soc. Chem. Commun.* **1994**, *7*, 801–802. [CrossRef]
83. Evanoff, D.D.; Chumanov, G. Synthesis and Optical Properties of Silver Nanoparticles and Arrays. *ChemPhysChem* **2005**, *6*, 1221–1231. [CrossRef]
84. Goulet, P.J.G.; Lennox, R.B. New Insights into Brust−Schiffrin Metal Nanoparticle Synthesis. *J. Am. Chem. Soc.* **2010**, *132*, 9582–9584. [CrossRef]
85. Mallick, K.; Witcomb, M.J.; Scurrell, M.S. Polymer Stabilized Silver Nanoparticles: A Photochemical Synthesis Route. *J. Mater. Sci.* **2004**, *39*, 4459–4463. [CrossRef]
86. Malik, M.A.; O'Brien, P.; Revaprasadu, N. A Simple Route to the Synthesis of Core/Shell Nanoparticles of Chalcogenides. *Chem. Mater.* **2002**, *14*, 2004–2010. [CrossRef]
87. Gurunathan, S.; Han, J.W.; Kwon, D.-N.; Kim, J.-H. Enhanced Antibacterial and Anti-Biofilm Activities of Silver Nanoparticles against Gram-Negative and Gram-Positive Bacteria. *Nanoscale Res. Lett.* **2014**, *9*, 373. [CrossRef] [PubMed]
88. Gurunathan, S.; Han, J.W.; Kim, J.-H. Green chemistry approach for the synthesis of biocompatible graphene. *Int. J. Nanomed.* **2013**, *8*, 2719–2732. [CrossRef]
89. Gurunathan, S.; Han, J.W.; Park, J.-H.; Kim, E.S.; Choi, Y.-J.; Kwon, D.-N.; Kim, J.-H. Reduced graphene oxide–silver nanoparticle nanocomposite: A potential anticancer nanotherapy. *Int. J. Nanomed.* **2015**, *10*, 6257–6276. [CrossRef]
90. Batir-Marin, D.; Mircea, C.; Boev, M.; Burlec, A.F.; Corciova, A.; Fifere, A.; Iacobescu, A.; Cioanca, O.; Verestiuc, L.; Hancianu, M. In Vitro Antioxidant, Antitumor and Photocatalytic Activities of Silver Nanoparticles Synthesized Using *Equisetum* Species: A Green Approach. *Molecules* **2021**, *26*, 7325. [CrossRef] [PubMed]
91. Noah, N.M.; Ndangili, P.M. Green Synthesis of Nanomaterials from Sustainable Materials for Biosensors and Drug Delivery. *Sensors Int.* **2022**, *3*, 100166. [CrossRef]
92. Amini, S.M.; Pour, M.S.S.; Vahidi, R.; Kouhbananinejad, S.M.; Bardsiri, M.S.; Farsinejad, A.; Mirzaei-Parsa, M.J. Green Synthesis of Stable Silver Nanoparticles Using Teucrium Polium Extract: In-Vitro Anticancer Activity on NALM-6. *Nanomed. Res. J.* **2021**, *6*, 170–178. [CrossRef]
93. Sysak, S.; Czarczynska-Goslinska, B.; Szyk, P.; Koczorowski, T.; Mlynarczyk, D.T.; Szczolko, W.; Lesyk, R.; Goslinski, T. Metal Nanoparticle-Flavonoid Connections: Synthesis, Physicochemical and Biological Properties, as Well as Potential Applications in Medicine. *Nanomaterials* **2023**, *13*, 1531. [CrossRef]
94. Amini, S.M.; Mohammadi, E.; Askarian-Amiri, S.; Azizi, Y.; Shakeri-Zadeh, A.; Neshastehriz, A. Investigating the in Vitro Photothermal Effect of Green Synthesized Apigenin-coated Gold Nanoparticle on Colorectal Carcinoma. *IET Nanobiotechnology* **2021**, *15*, 329–337. [CrossRef]
95. Islam, M.J.; Kamaruzzaman; Hossain, M.F.; Awal, M.A.; Rahman, M.R.; Alam, M.M. Green Synthesis of Silver Nanoparticles (AgNPs) Incorporated with *Adenia Trilobata* Leaf Extracts and Its Anti-Bacterial Application. *AIP Adv.* **2022**, *12*, 115116. [CrossRef]
96. Purohit, A.; Sharma, R.; Ramakrishnan, R.S.; Sharma, S.; Kumar, A.; Jain, D.; Kushwaha, H.S.; Maharjan, E. Biogenic Synthesis of Silver Nanoparticles (AgNPs) Using Aqueous Leaf Extract of *Buchanania lanzan* Spreng and Evaluation of Their Antifungal Activity against Phytopathogenic Fungi. *Bioinorg. Chem. Appl.* **2022**, *2022*, 6825150. [CrossRef] [PubMed]
97. Chouhan, S.; Guleria, S. Green Synthesis of AgNPs Using *Cannabis sativa* Leaf Extract: Characterization, Antibacterial, Anti-Yeast and α-Amylase Inhibitory Activity. *Mater. Sci. Energy Technol.* **2020**, *3*, 536–544. [CrossRef]
98. Jagtap, R.R.; Garud, A.; Puranik, S.S.; Rudrapal, M.; Ansari, M.A.; Alomary, M.N.; Alshamrani, M.; Salawi, A.; Almoshari, Y.; Khan, J.; et al. Biofabrication of Silver Nanoparticles (AgNPs) Using Embelin for Effective Therapeutic Management of Lung Cancer. *Front. Nutr.* **2022**, *9*, 960674. [CrossRef] [PubMed]
99. Corciovă, A.; Mircea, C.; Burlec, A.F.; Fifere, A.; Moleavin, I.T.; Sarghi, A.; Tuchiluș, C.; Ivănescu, B.; Macovei, I. Green Synthesis and Characterization of Silver Nanoparticles Using a *Lythrum salicaria* Extract and in Vitro Exploration of Their Biological Activities. *Life* **2022**, *12*, 1643. [CrossRef] [PubMed]
100. Asif, M.; Yasmin, R.; Asif, R.; Ambreen, A.; Mustafa, M.; Umbreen, S. Green Synthesis of Silver Nanoparticles (AgNPs), Structural Characterization, and Their Antibacterial Potential. *Dose-Response* **2022**, *20*, 155932582210887. [CrossRef]
101. Macovei, I.; Luca, S.V.; Skalicka-Woźniak, K.; Sacarescu, L.; Pascariu, P.; Ghilan, A.; Doroftei, F.; Ursu, E.-L.; Rimbu, C.M.; Horhogea, C.E.; et al. Phyto-Functionalized Silver Nanoparticles Derived from Conifer Bark Extracts and Evaluation of Their Antimicrobial and Cytogenotoxic Effects. *Molecules* **2022**, *27*, 217. [CrossRef]

102. Aslam, M.; Fozia, F.; Gul, A.; Ahmad, I.; Ullah, R.; Bari, A.; Mothana, R.A.; Hussain, H. Phyto-Extract-Mediated Synthesis of Silver Nanoparticles Using Aqueous Extract of *Sanvitalia procumbens*, and Characterization, Optimization and Photocatalytic Degradation of Azo Dyes Orange g and Direct Blue-15. *Molecules* **2021**, *26*, 6144. [CrossRef]
103. Burlec, A.F.; Hăncianu, M.; Macovei, I.; Mircea, C.; Fifere, A.; Turin-Moleavin, I.-A.; Tuchiluș, C.; Robu, S.; Corciovă, A. Eco-Friendly Synthesis and Comparative in Vitro Biological Evaluation of Silver Nanoparticles Using *Tagetes erecta* Flower Extracts. *Appl. Sci.* **2022**, *12*, 887. [CrossRef]
104. Efavi, J.K.; Nyankson, E.; Kyeremeh, K.; Manu, G.P.; Asare, K.; Yeboah, N. Monodispersed AgNPs Synthesized from the Nanofactories of *Theobroma Cacao* (Cocoa) Leaves and Pod Husk and Their Antimicrobial Activity. *Int. J. Biomater.* **2022**, *2022*, 4106558. [CrossRef]
105. Ibrahim, S.; Ahmad, Z.; Manzoor, M.Z.; Mujahid, M.; Faheem, Z.; Adnan, A. Optimization for Biogenic Microbial Synthesis of Silver Nanoparticles through Response Surface Methodology, Characterization, Their Antimicrobial, Antioxidant, and Catalytic Potential. *Sci. Rep.* **2021**, *11*, 770. [CrossRef] [PubMed]
106. Alsamhary, K.I. Eco-Friendly Synthesis of Silver Nanoparticles by *Bacillus subtilis* and Their Antibacterial Activity. *Saudi J. Biol. Sci.* **2020**, *27*, 2185–2191. [CrossRef]
107. Saleh, M.N.; Alwan, S.K. Bio-Synthesis of Silver Nanoparticles from Bacteria Klebsiella Pneumonia: Their Characterization and Antibacterial Studies. *J. Phys. Conf. Ser.* **2020**, *1664*, 012115. [CrossRef]
108. Naseer, Q.A.; Xue, X.; Wang, X.; Dang, S.; Din, S.U.; Kalsoom; Jamil, J. Synthesis of Silver Nanoparticles Using *Lactobacillus bulgaricus* and Assessment of Their Antibacterial Potential. *Braz. J. Biol.* **2022**, *82*, e232434. [CrossRef] [PubMed]
109. Lotfy, W.A.; Alkersh, B.M.; Sabry, S.A.; Ghozlan, H.A. Biosynthesis of Silver Nanoparticles by Aspergillus Terreus: Characterization, Optimization, and Biological Activities. *Front. Bioeng. Biotechnol.* **2021**, *9*, 633468. [CrossRef]
110. Yassin, M.A.; Elgorban, A.M.; El-Samawaty, A.E.-R.M.A.; Almunqedhi, B.M.A. Biosynthesis of Silver Nanoparticles Using *Penicillium verrucosum* and Analysis of Their Antifungal Activity. *Saudi J. Biol. Sci.* **2021**, *28*, 2123–2127. [CrossRef]
111. Soleimani, M.; Habibi-Pirkoohi, M. Biosynthesis of Silver Nanoparticles Using *Chlorella vulgaris* and Evaluation of the Antibacterial Efficacy against *Staphylococcus aureus*. *Avicenna J. Med. Biotechnol.* **2017**, *9*, 120–125.
112. Shantkriti, S.; Pradeep, M.; Unish, K.; Viji Das, M.; Nidhin, S.; Gugan, K.; Murugan, A. Bioynthesis of Silver Nanoparticles Using *Dunaliella Salina* and Its Antibacterial Applications. *Appl. Surf. Sci. Adv.* **2023**, *13*, 100377. [CrossRef]
113. Algotiml, R.; Gab-Alla, A.; Seoudi, R.; Abulreesh, H.H.; El-Readi, M.Z.; Elbanna, K. Anticancer and Antimicrobial Activity of Biosynthesized Red Sea Marine Algal Silver Nanoparticles. *Sci. Rep.* **2022**, *12*, 2421. [CrossRef]
114. Hatipoğlu, A. Rapid Green Synthesis of Gold Nanoparticles: Synthesis, Characterization, and Antimicrobial Activities. *Prog. Nutr.* **2021**, *23*, e2021242. [CrossRef]
115. Folorunso, A.; Akintelu, S.; Oyebamiji, A.K.; Ajayi, S.; Abiola, B.; Abdusalam, I.; Morakinyo, A. Biosynthesis, Characterization and Antimicrobial Activity of Gold Nanoparticles from Leaf Extracts of *Annona muricata*. *J. Nanostruct. Chem.* **2019**, *9*, 111–117. [CrossRef]
116. Yuan, C.; Huo, C.; Gui, B.; Cao, W. Green Synthesis of Gold Nanoparticles Using *Citrus Maxima* Peel Extract and Their Catalytic/Antibacterial Activities. *IET Nanobiotechnol.* **2017**, *11*, 523–530. [CrossRef] [PubMed]
117. Huang, Q.; Luo, A.; Jiang, L.; Zhou, Y.; Yang, Y.; Liu, Q.; Zhang, C. Disinfection Efficacy of Green Synthesized Gold Nanoparticles for Medical Disinfection Applications. *Afr. Health Sci.* **2019**, *19*, 1441–1448. [CrossRef] [PubMed]
118. Dorosti, N.; Jamshidi, F. Plant-Mediated Gold Nanoparticles by *Dracocephalum kotschyi* as Anticholinesterase Agent: Synthesis, Characterization, and Evaluation of Anticancer and Antibacterial Activity. *J. Appl. Biomed.* **2016**, *14*, 235–245. [CrossRef]
119. Nagajyothi, P.C.; Lee, S.-E.; An, M.; Lee, K.-D. Green Synthesis of Silver and Gold Nanoparticles Using *Lonicera Japonica* Flower Extract. *Bull. Korean Chem. Soc.* **2012**, *33*, 2609–2612. [CrossRef]
120. Jafarizad, A.; Safaee, K.; Gharibian, S.; Omidi, Y.; Ekinci, D. Biosynthesis and In-Vitro Study of Gold Nanoparticles Using *Mentha* and *Pelargonium* Extracts. *Procedia Mater. Sci.* **2015**, *11*, 224–230. [CrossRef]
121. Rajendran, A. Antibacterial Properties and Mechanism of Gold Nanoparticles Obtained from *Pergularia daemia* Leaf Extract. *J. Nanomed. Res.* **2017**, *6*, 00146. [CrossRef]
122. Anbu, P.; Gopinath, S.C.; Jayanthi, S. Synthesis of Gold Nanoparticles Using *Platycodon Grandiflorum* Extract and Its Antipathogenic Activity under Optimal Conditions. *Nanomater. Nanotechnol.* **2020**, *10*, 184798042096169. [CrossRef]
123. Hosny, M.; Fawzy, M.; El-Badry, Y.A.; Hussein, E.E.; Eltaweil, A.S. Plant-Assisted Synthesis of Gold Nanoparticles for Photocatalytic, Anticancer, and Antioxidant Applications. *J. Saudi Chem. Soc.* **2022**, *26*, 101419. [CrossRef]
124. Pourali, P.; Badiee, S.H.; Manafi, S.; Noorani, T.; Rezaei, A.; Yahyaei, B. Biosynthesis of Gold Nanoparticles by Two Bacterial and Fungal Strains, *Bacillus cereus* and *Fusarium oxysporum*, and Assessment and Comparison of Their Nanotoxicity in Vitro by Direct and Indirect Assays. *Electron. J. Biotechnol.* **2017**, *29*, 86–93. [CrossRef]
125. Nadhim, K.; Mahmood, N.N.; Mustafa, A. Synthesized Gold Nanoparticles Using *Pseudomonas* Supernatant and Study the Physical Characterization–Antproliferative Activity of Breast Cancer Cells (MCF-7). *Medico-Legal Updat.* **2021**, *21*, 1281–1286. [CrossRef]
126. Owais, M.; Chauhan, A.; Tufail; Sherwani; Sajid, M.; Suri, C.R.; Owais, M. Fungus-Mediated Biological Synthesis of Gold Nanoparticles: Potential in Detection of Liver Cancer. *Int. J. Nanomed.* **2011**, *6*, 2305–2319. [CrossRef] [PubMed]
127. Gürsoy, N.; Yilmaz Öztürk, B.; Dağ, İ. Synthesis of Intracellular and Extracellular Gold Nanoparticles with a Green Machine and Its Antifungal Activity. *Turk. J. Biol.* **2021**, *45*, 196–213. [CrossRef]

128. Abdel-Raouf, N.; Al-Enazi, N.M.; Ibraheem, I.B.M. Green Biosynthesis of Gold Nanoparticles Using *Galaxaura Elongata* and Characterization of Their Antibacterial Activity. *Arab. J. Chem.* **2017**, *10*, S3029–S3039. [CrossRef]
129. El-Sheekh, M.M.; Shabaan, M.T.; Hassan, L.; Morsi, H.H. Antiviral Activity of Algae Biosynthesized Silver and Gold Nanoparticles against Herps Simplex (HSV-1) Virus in Vitro Using Cell-Line Culture Technique. *Int. J. Environ. Health Res.* **2022**, *32*, 616–627. [CrossRef]
130. Pal, K.; Chakroborty, S.; Nath, N. Limitations of Nanomaterials Insights in Green Chemistry Sustainable Route: Review on Novel Applications. *Green Process. Synth.* **2022**, *11*, 951–964. [CrossRef]
131. Khan, M.; Khan, M.S.A.; Borah, K.K.; Goswami, Y.; Hakeem, K.R.; Chakrabartty, I. The Potential Exposure and Hazards of Metal-Based Nanoparticles on Plants and Environment, with Special Emphasis on ZnO NPs, TiO_2 NPs, and AgNPs: A Review. *Environ. Adv.* **2021**, *6*, 100128. [CrossRef]
132. Qamar, S.U.R.; Ahmad, J.N. Nanoparticles: Mechanism of Biosynthesis Using Plant Extracts, Bacteria, Fungi, and Their Applications. *J. Mol. Liq.* **2021**, *334*, 116040. [CrossRef]
133. Lahiri, D.; Nag, M.; Sheikh, H.I.; Sarkar, T.; Edinur, H.A.; Pati, S.; Ray, R.R. Microbiologically-Synthesized Nanoparticles and Their Role in Silencing the Biofilm Signaling Cascade. *Front. Microbiol.* **2021**, *12*, 636588. [CrossRef]
134. Iravani, S. Bacteria in Nanoparticle Synthesis: Current Status and Future Prospects. *Int. Sch. Res. Not.* **2014**, *2014*, 359316. [CrossRef]
135. Khan, F.; Shahid, A.; Zhu, H.; Wang, N.; Javed, M.R.; Ahmad, N.; Xu, J.; Alam, M.A.; Mehmood, M.A. Prospects of Algae-Based Green Synthesis of Nanoparticles for Environmental Applications. *Chemosphere* **2022**, *293*, 133571. [CrossRef] [PubMed]
136. Michael, A.; Singh, A.; Roy, A.; Islam, M.R. Fungal- and Algal-Derived Synthesis of Various Nanoparticles and Their Applications. *Bioinorg. Chem. Appl.* **2022**, *2022*, 3142674. [CrossRef] [PubMed]
137. Nahvi, I.; Belkahla, S.; Asiri, S.M.; Rehman, S. Overview and Prospectus of Algal Biogenesis of Nanoparticles. In *Microbial Nanotechnology: Green Synthesis and Applications*; Springer: Singapore, 2021; pp. 121–134.
138. Velgosová, O.; Mražíková, A. Limitations and Possibilities of Green Synthesis and Long-Term Stability of Colloidal Ag Nanoparticles. *AIP Conf. Proc.* **2017**, *1918*, 020004.
139. Košević, M.G.; Zarić, M.M.; Stopić, S.R.; Stevanović, J.S.; Weirich, T.E.; Friedrich, B.G.; Panić, V.V. Structural and Electrochemical Properties of Nesting and Core/Shell Pt/TiO2 Spherical Particles Synthesized by Ultrasonic Spray Pyrolysis. *Metals* **2019**, *10*, 11. [CrossRef]
140. Yu, X.; Pham, J.T.; Subramani, C.; Creran, B.; Yeh, Y.-C.; Du, K.; Patra, D.; Miranda, O.R.; Crosby, A.J.; Rotello, V.M. Direct Patterning of Engineered Ionic Gold Nanoparticles via Nanoimprint Lithography. *Adv. Mater.* **2012**, *24*, 6330–6334. [CrossRef] [PubMed]
141. Davies, G.-L.; O'Brien, J.; Gun'ko, Y.K. Rare Earth Doped Silica Nanoparticles via Thermolysis of a Single Source Metallasilsesquioxane Precursor. *Sci. Rep.* **2017**, *7*, 45862. [CrossRef]
142. Ielo, I.; Rando, G.; Giacobello, F.; Sfameni, S.; Castellano, A.; Galletta, M.; Drommi, D.; Rosace, G.; Plutino, M.R. Synthesis, Chemical–Physical Characterization, and Biomedical Applications of Functional Gold Nanoparticles: A Review. *Molecules* **2021**, *26*, 5823. [CrossRef]
143. Turkevich, J.; Stevenson, P.C.; Hillier, J. A Study of the Nucleation and Growth Processes in the Synthesis of Colloidal Gold. *Discuss. Faraday Soc.* **1951**, *11*, 55–75. [CrossRef]
144. Noruzi, M.; Zare, D.; Khoshnevisan, K.; Davoodi, D. Rapid Green Synthesis of Gold Nanoparticles Using *Rosa Hybrida* Petal Extract at Room Temperature. *Spectrochim. Acta Part A Mol. Biomol. Spectrosc.* **2011**, *79*, 1461–1465. [CrossRef]
145. Shankar, S.S.; Rai, A.; Ahmad, A.; Sastry, M. Rapid Synthesis of Au, Ag, and Bimetallic Au Core–Ag Shell Nanoparticles Using Neem (*Azadirachta indica*) Leaf Broth. *J. Colloid Interface Sci.* **2004**, *275*, 496–502. [CrossRef]
146. Mohanpuria, P.; Rana, N.K.; Yadav, S.K. Biosynthesis of Nanoparticles: Technological Concepts and Future Applications. *J. Nanoparticle Res.* **2008**, *10*, 507–517. [CrossRef]
147. Singh, M.; Kalaivani, R.; Manikandan, S.; Sangeetha, N.; Kumaraguru, A.K. Facile Green Synthesis of Variable Metallic Gold Nanoparticle Using *Padina gymnospora*, a Brown Marine Macroalga. *Appl. Nanosci.* **2013**, *3*, 145–151. [CrossRef]
148. Saleh, T.A. Nanomaterials: Classification, Properties, and Environmental Toxicities. *Environ. Technol. Innov.* **2020**, *20*, 101067. [CrossRef]
149. Shnoudeh, A.J.; Hamad, I.; Abdo, R.W.; Qadumii, L.; Jaber, A.Y.; Surchi, H.S.; Alkelany, S.Z. Synthesis, Characterization, and Applications of Metal Nanoparticles. In *Biomaterials and Bionanotechnology*; Elsevier: Amsterdam, The Netherlands, 2019; pp. 527–612.
150. Liaqat, N.; Jahan, N.; Rahman, K.U.; Anwar, T.; Qureshi, H. Green synthesized silver nanoparticles: Optimization, characterization, antimicrobial activity, and cytotoxicity study by hemolysis assay. *Front. Chem.* **2022**, *10*, 952006. [CrossRef] [PubMed]
151. Kumar, R.; Ghoshal, G.; Jain, A.; Goyal, M. Rapid Green Synthesis of Silver Nanoparticles (AgNPs) Using (*Prunus persica*) Plants Extract: Exploring Its Antimicrobial and Catalytic Activities. *J. Nanomed. Nanotechnol.* **2017**, *8*, 1000452. [CrossRef]
152. Correa, S.N.; Naranjo, A.M.; Herrera, A.P. Biosynthesis and Characterization of Gold Nanoparticles Using Extracts of *Tamarindus indica* L Leaves. *J. Phys. Conf. Ser.* **2016**, *687*, 012082. [CrossRef]
153. Dzimitrowicz, A.; Jamróz, P.; DiCenzo, G.C.; Sergiel, I.; Kozlecki, T.; Pohl, P. Preparation and Characterization of Gold Nanoparticles Prepared with Aqueous Extracts of Lamiaceae Plants and the Effect of Follow-up Treatment with Atmospheric Pressure Glow Microdischarge. *Arab. J. Chem.* **2019**, *12*, 4118–4130. [CrossRef]

154. Femi-Adepoju, A.G.; Dada, A.O.; Otun, K.O.; Adepoju, A.O.; Fatoba, O.P. Green Synthesis of Silver Nanoparticles Using Terrestrial Fern (*Gleichenia pectinata* (Willd.) C. Presl.): Characterization and Antimicrobial Studies. *Heliyon* **2019**, *5*, e01543. [CrossRef]
155. Lima, A.K.O.; Vasconcelos, A.A.; Kobayashi, R.K.T.; Nakazato, G.; Braga, H.d.C.; Taube, P.S. Green Synthesis: Characterization and Biological Activity of Silver Nanoparticles Using Aqueous Extracts of Plants from the Arecaceae Family. *Acta Sci. Technol.* **2021**, *43*, e52011. [CrossRef]
156. TYAl-Abdullah, Z.; Al-Shawi, A.A.A.; Aboud, M.N.; Abdulaziz, B.A.A.; Al-Furaiji, H.Q.M.; Luaibi, I.N. Synthesis and Analytical Characterization of Gold Nanoparticles Using Microwave-Assisted Extraction System and Study Their Application in Degradation. *J. Nanostruct.* **2020**, *10*, 682–690. [CrossRef]
157. Geraldes, A.N.; da Silva, A.A.; Leal, J.; Estrada-Villegas, G.M.; Lincopan, N.; Katti, K.V.; Lugão, A.B. Green Nanotechnology from Plant Extracts: Synthesis and Characterization of Gold Nanoparticles. *Adv. Nanoparticles* **2016**, *05*, 176–185. [CrossRef]
158. Reda, M.; Ashames, A.; Edis, Z.; Bloukh, S.; Bhandare, R.; Abu Sara, H. Green Synthesis of Potent Antimicrobial Silver Nanoparticles Using Different Plant Extracts and Their Mixtures. *Processes* **2019**, *7*, 510. [CrossRef]
159. Miškovská, A.; Rabochová, M.; Michailidu, J.; Masák, J.; Čejková, A.; Lorinčík, J.; Maťátková, O. Antibiofilm Activity of Silver Nanoparticles Biosynthesized Using Viticultural Waste. *PLoS ONE* **2022**, *17*, e0272844. [CrossRef]
160. Lee, K.-C.; Lin, S.-J.; Lin, C.-H.; Tsai, C.-S.; Lu, Y.-J. Size Effect of Ag Nanoparticles on Surface Plasmon Resonance. *Surf. Coatings Technol.* **2008**, *202*, 5339–5342. [CrossRef]
161. Sau, T.K.; Rogach, A.L.; Jäckel, F.; Klar, T.A.; Feldmann, J. Properties and Applications of Colloidal Nonspherical Noble Metal Nanoparticles. *Adv. Mater.* **2010**, *22*, 1805–1825. [CrossRef] [PubMed]
162. Saputra, I.; Suhartati, S.; Yulizar, Y.; Sudirman, S. Synthesis and Characterization of Gold Nanoparticles (AuNPs) by Utilizing Bioactive Compound of *Imperata cylndrica* L. *Indones. J. Appl. Chem.* **2020**, *22*, 1–7.
163. Jamil, S.; Dastagir, G.; Foudah, A.I.; Alqarni, M.H.; Yusufoglu, H.S.; Alkreathy, H.M.; Ertürk, Ö.; Shah, M.A.R.; Khan, R.A. *Carduus edelbergii* Rech. f. Mediated Fabrication of Gold Nanoparticles; Characterization and Evaluation of Antimicrobial, Antioxidant and Antidiabetic Potency of the Synthesized AuNPs. *Molecules* **2022**, *27*, 6669. [CrossRef]
164. Akkus, Z.B.; Imran Nazir, A.J.; Tribus, M.; Bernkop-Schnürch, A. Zeta Potential Changing Polyphosphate Nanoparticles: A Promising Approach to Overcome the Mucus and Epithelial Barrier. *Mol. Pharm.* **2019**, *16*, 2817–2825. [CrossRef]
165. Erdogan, O.; Abbak, M.; Demirbolat, G.M.; Birtekocak, F.; Aksel, M.; Pasa, S.; Cevik, O. Green Synthesis of Silver Nanoparticles via *Cynara scolymus* Leaf Extracts: The Characterization, Anticancer Potential with Photodynamic Therapy in MCF7 Cells. *PLoS ONE* **2019**, *14*, e0216496. [CrossRef]
166. Elamawi, R.M.; Al-Harbi, R.E.; Hendi, A.A. Biosynthesis and Characterization of Silver Nanoparticles Using *Trichoderma longibrachiatum* and Their Effect on Phytopathogenic Fungi. *Egypt. J. Biol. Pest Control* **2018**, *28*, 28. [CrossRef]
167. Mukherjee, S.; Sushma, V.; Patra, S.; Barui, A.K.; Bhadra, M.P.; Sreedhar, B.; Patra, C.R. Green Chemistry Approach for the Synthesis and Stabilization of Biocompatible Gold Nanoparticles and Their Potential Applications in Cancer Therapy. *Nanotechnology* **2012**, *23*, 455103. [CrossRef] [PubMed]
168. Riaz, M.; Ismail, M.; Ahmad, B.; Zahid, N.; Jabbour, G.; Khan, M.S.; Mutreja, V.; Sareen, S.; Rafiq, A.; Faheem, M.; et al. Characterizations and Analysis of the Antioxidant, Antimicrobial, and Dye Reduction Ability of Green Synthesized Silver Nanoparticles. *Green Process. Synth.* **2020**, *9*, 693–705. [CrossRef]
169. Hodoroaba, V.; Rades, S.; Unger, W. Inspection of Morphology and Elemental Imaging of Single Nanoparticles by High-resolution SEM/EDX in Transmission Mode. *Surf. Interface Anal.* **2014**, *46*, 945–948. [CrossRef]
170. Mubeen, B.; Rasool, M.G.; Ullah, I.; Rasool, R.; Imam, S.S.; Alshehri, S.; Ghoneim, M.M.; Alzarea, S.I.; Nadeem, M.S.; Kazmi, I. Phytochemicals Mediated Synthesis of AuNPs from *Citrullus colocynthis* and Their Characterization. *Molecules* **2022**, *27*, 1300. [CrossRef]
171. Ojemaye, M.O.; Okoh, S.O.; Okoh, A.I. Silver Nanoparticles (AgNPs) Facilitated by Plant Parts of *Crataegus ambigua* Becker AK Extracts and Their Antibacterial, Antioxidant and Antimalarial Activities. *Green Chem. Lett. Rev.* **2021**, *14*, 51–61. [CrossRef]
172. Botteon, C.E.A.; Silva, L.B.; Ccana-Ccapatinta, G.V.; Silva, T.S.; Ambrosio, S.R.; Veneziani, R.C.S.; Bastos, J.K.; Marcato, P.D. Biosynthesis and Characterization of Gold Nanoparticles Using Brazilian Red Propolis and Evaluation of Its Antimicrobial and Anticancer Activities. *Sci. Rep.* **2021**, *11*, 1974. [CrossRef]
173. Philip, D. Rapid Green Synthesis of Spherical Gold Nanoparticles Using *Mangifera indica* Leaf. *Spectrochim. Acta Part A Mol. Biomol. Spectrosc.* **2010**, *77*, 807–810. [CrossRef]
174. Corciovă, A.; Mircea, C.; Burlec, A.-F.; Cioancă, O.; Tuchiluş, C.; Fifere, A.; Lungoci, A.-L.; Marangoci, N.; Hăncianu, M. Antioxidant, Antimicrobial Activities and Photocatalytic Degradation Efficacy of Silver Nanoparticles Obtained by Bee Propolis Extract Assisted Biosynthesis. *Farmacia* **2019**, *67*, 482–489. [CrossRef]
175. Verma, A.; Mehata, M.S. Controllable Synthesis of Silver Nanoparticles Using Neem Leaves and Their Antimicrobial Activity. *J. Radiat. Res. Appl. Sci.* **2016**, *9*, 109–115. [CrossRef]
176. Bhaskaran, S.; Sharma, N.; Tiwari, P.; Singh, S.R.; Sahi, S.V. Fabrication of Innocuous Gold Nanoparticles Using Plant Cells in Culture. *Sci. Rep.* **2019**, *9*, 12040. [CrossRef]
177. Siakavella, I.K.; Lamari, F.; Papoulis, D.; Orkoula, M.; Gkolfi, P.; Lykouras, M.; Avgoustakis, K.; Hatziantoniou, S. Effect of Plant Extracts on the Characteristics of Silver Nanoparticles for Topical Application. *Pharmaceutics* **2020**, *12*, 1244. [CrossRef] [PubMed]

178. Meva, F.E.; Ntoumba, A.A.; Kedi, P.B.E.; Tchoumbi, E.; Schmitz, A.; Schmolke, L.; Klopotowski, M.; Moll, B.; Kökcam-Demir, Ü.; Mpondo, E.A.M.; et al. Silver and palladium nanoparticles produced using a plant extract as reducing agent, stabilized with an ionic liquid: Sizing by X-ray powder diffraction and dynamic light scattering. *J. Mater. Res. Technol.* **2019**, *8*, 1991–2000. [CrossRef]
179. Monshi, A.; Foroughi, M.R.; Monshi, M.R. Modified Scherrer Equation to Estimate More Accurately Nano-Crystallite Size Using XRD. *World J. Nano Sci. Eng.* **2012**, *2*, 154–160. [CrossRef]
180. Sajid, M.; Płotka-Wasylka, J. Nanoparticles: Synthesis, Characteristics, and Applications in Analytical and Other Sciences. *Microchem. J.* **2020**, *154*, 104623. [CrossRef]
181. Mansor, M.; Drabesch, S.; Bayer, T.; Van Le, A.; Chauhan, A.; Schmidtmann, J.; Peiffer, S.; Kappler, A. Application of Single-Particle ICP-MS to Determine the Mass Distribution and Number Concentrations of Environmental Nanoparticles and Colloids. *Environ. Sci. Technol. Lett.* **2021**, *8*, 589–595. [CrossRef]
182. Teulon, J.-M.; Godon, C.; Chantalat, L.; Moriscot, C.; Cambedouzou, J.; Odorico, M.; Ravaux, J.; Podor, R.; Gerdil, A.; Habert, A.; et al. On the Operational Aspects of Measuring Nanoparticle Sizes. *Nanomaterials* **2018**, *9*, 18. [CrossRef]
183. Sharma, J.N.; Pattadar, D.K.; Mainali, B.P.; Zamborini, F.P. Size Determination of Metal Nanoparticles Based on Electrochemically Measured Surface-Area-to-Volume Ratios. *Anal. Chem.* **2018**, *90*, 9308–9314. [CrossRef]
184. Salem, S.S.; Fouda, A. Green Synthesis of Metallic Nanoparticles and Their Prospective Biotechnological Applications: An Overview. *Biol. Trace Elem. Res.* **2021**, *199*, 344–370. [CrossRef]
185. Li, Y.; Jin, R. Seeing Ligands on Nanoclusters and in Their Assemblies by X-ray Crystallography: Atomically Precise Nanochemistry and Beyond. *J. Am. Chem. Soc.* **2020**, *142*, 13627–13644. [CrossRef]
186. Badirzadeh, A.; Alipour, M.; Najm, M.; Vosoogh, A.; Vosoogh, M.; Samadian, H.; Hashemi, A.S.; Farsangi, Z.J.; Amini, S.M. Potential Therapeutic Effects of Curcumin Coated Silver Nanoparticle in the Treatment of Cutaneous Leishmaniasis Due to Leishmania Major In-Vitro and in a Murine Model. *J. Drug Deliv. Sci. Technol.* **2022**, *74*, 103576. [CrossRef]
187. Blaškovičová, J.; Vyskočil, V.; Augustín, M.; Purdešová, A. Ethanol and NaCl-Induced Gold Nanoparticle Aggregation Toxicity toward DNA Investigated with a DNA/GCE Biosensor. *Sensors* **2023**, *23*, 3425. [CrossRef]
188. Cuenca, A.G.; Jiang, H.; Hochwald, S.N.; Delano, M.; Cance, W.G.; Grobmyer, S.R. Emerging Implications of Nanotechnology on Cancer Diagnostics and Therapeutics. *Cancer* **2006**, *107*, 459–466. [CrossRef] [PubMed]
189. Kiessling, F.; Mertens, M.E.; Grimm, J.; Lammers, T. Nanoparticles for Imaging: Top or Flop? *Radiology* **2014**, *273*, 10–28. [CrossRef] [PubMed]
190. Ferrucci, J.T.; Stark, D.D. Iron Oxide-Enhanced MR Imaging of the Liver and Spleen: Review of the First 5 Years. *Am. J. Roentgenol.* **1990**, *155*, 943–950. [CrossRef] [PubMed]
191. Marchal, G.; Van Hecke, P.; Demaerel, P.; Decrop, E.; Kennis, C.; Baert, A.; van der Schueren, E. Detection of Liver Metastases with Superparamagnetic Iron Oxide in 15 Patients: Results of MR Imaging at 1.5 T. *Am. J. Roentgenol.* **1989**, *152*, 771–775. [CrossRef] [PubMed]
192. Deng, H.; Shang, W.; Wang, K.; Guo, K.; Liu, Y.; Tian, J.; Fang, C. Targeted-Detection and Sequential-Treatment of Small Hepatocellular Carcinoma in the Complex Liver Environment by GPC-3-Targeted Nanoparticles. *J. Nanobiotechnology* **2022**, *20*, 156. [CrossRef] [PubMed]
193. Santoro, L.; Grazioli, L.; Filippone, A.; Grassedonio, E.; Belli, G.; Colagrande, S. Resovist Enhanced MR Imaging of the Liver: Does Quantitative Assessment Help in Focal Lesion Classification and Characterization? *J. Magn. Reson. Imaging* **2009**, *30*, 1012–1020. [CrossRef]
194. Neuwelt, E.A.; Hamilton, B.E.; Varallyay, C.G.; Rooney, W.R.; Edelman, R.D.; Jacobs, P.M.; Watnick, S.G. Ultrasmall Superparamagnetic Iron Oxides (USPIOs): A Future Alternative Magnetic Resonance (MR) Contrast Agent for Patients at Risk for Nephrogenic Systemic Fibrosis (NSF)? *Kidney Int.* **2009**, *75*, 465–474. [CrossRef]
195. Johnson, J.M.; Mohamed, A.S.R.; Ding, Y.; Wang, J.; Lai, S.Y.; Fuller, C.D.; Shah, R.; Butler, R.T.; Weber, R.S. Ultra-small Superparamagnetic Iron Oxide (USPIO) Magnetic Resonance Imaging in Benign Mixed Tumor of the Parotid Gland. *Clin. Case Rep.* **2021**, *9*, 123–127. [CrossRef]
196. Weissleder, R.; Nahrendorf, M.; Pittet, M.J. Imaging Macrophages with Nanoparticles. *Nat. Mater.* **2014**, *13*, 125–138. [CrossRef]
197. Lobatto, M.E.; Fuster, V.; Fayad, Z.A.; Mulder, W.J.M. Perspectives and Opportunities for Nanomedicine in the Management of Atherosclerosis. *Nat. Rev. Drug Discov.* **2011**, *10*, 835–852. [CrossRef] [PubMed]
198. Kaneko, C.; Nitta, N.; Tsuchiya, K.; Watanabe, S.; Nitta-Seko, A.; Ohta, S.; Otani, H.; Sonoda, A.; Murata, K.; Shiomi, M. MRI Study of Atherosclerotic Plaque Progression Using Ultrasmall Superparamagnetic Iron Oxide in Watanabe Heritable Hyperlipidemic Rabbits. *Br. J. Radiol.* **2015**, *88*, 20150167. [CrossRef] [PubMed]
199. Kooi, M.E.; Cappendijk, V.C.; Cleutjens, K.B.J.M.; Kessels, A.G.H.; Kitslaar, P.J.E.H.M.; Borgers, M.; Frederik, P.M.; Daemen, M.J.A.P.; van Engelshoven, J.M.A. Accumulation of Ultrasmall Superparamagnetic Particles of Iron Oxide in Human Atherosclerotic Plaques Can Be Detected by in Vivo Magnetic Resonance Imaging. *Circulation* **2003**, *107*, 2453–2458. [CrossRef] [PubMed]
200. Tourdias, T.; Roggerone, S.; Filippi, M.; Kanagaki, M.; Rovaris, M.; Miller, D.H.; Petry, K.G.; Brochet, B.; Pruvo, J.-P.; Radüe, E.-W.; et al. Assessment of Disease Activity in Multiple Sclerosis Phenotypes with Combined Gadolinium- and Superparamagnetic Iron Oxide–Enhanced MR Imaging. *Radiology* **2012**, *264*, 225–233. [CrossRef]
201. Fatima, A.; Ahmad, M.W.; Al Saidi, A.K.A.; Choudhury, A.; Chang, Y.; Lee, G.H. Recent Advances in Gadolinium Based Contrast Agents for Bioimaging Applications. *Nanomaterials* **2021**, *11*, 2449. [CrossRef]

202. Boisselier, E.; Astruc, D. Gold Nanoparticles in Nanomedicine: Preparations, Imaging, Diagnostics, Therapies and Toxicity. *Chem. Soc. Rev.* **2009**, *38*, 1759–1782. [CrossRef]
203. Şologan, M.; Padelli, F.; Giachetti, I.; Aquino, D.; Boccalon, M.; Adami, G.; Pengo, P.; Pasquato, L. Functionalized Gold Nanoparticles as Contrast Agents for Proton and Dual Proton/Fluorine MRI. *Nanomaterials* **2019**, *9*, 879. [CrossRef]
204. Eskandarinezhad, S.; Wani, I.A.; Nourollahileilan, M.; Khosla, A.; Ahmad, T. Review—Metal and Metal Oxide Nanoparticles/Nanocomposites as Electrochemical Biosensors for Cancer Detection. *J. Electrochem. Soc.* **2022**, *169*, 047504. [CrossRef]
205. Babu, A.; Templeton, A.K.; Munshi, A.; Ramesh, R. Nanoparticle-Based Drug Delivery for Therapy of Lung Cancer: Progress and Challenges. *J. Nanomater.* **2013**, *2013*, 863951. [CrossRef]
206. Nejati, K.; Rastegar, M.; Fathi, F.; Dadashpour, M.; Arabzadeh, A. Nanoparticle-Based Drug Delivery Systems to Overcome Gastric Cancer Drug Resistance. *J. Drug Deliv. Sci. Technol.* **2022**, *70*, 103231. [CrossRef]
207. Bowman, M.-C.; Ballard, T.E.; Ackerson, C.J.; Feldheim, D.L.; Margolis, D.M.; Melander, C. Inhibition of HIV Fusion with Multivalent Gold Nanoparticles. *J. Am. Chem. Soc.* **2008**, *130*, 6896–6897. [CrossRef] [PubMed]
208. Xi, D.; Luo, X.; Ning, Q.; Lu, Q.; Yao, K.; Liu, Z. The Detection of HBV DNA with Gold Nanoparticle Gene Probes. *J. Nanjing Med. Univ.* **2007**, *21*, 207–212. [CrossRef]
209. Yen, C.-W.; de Puig, H.; Tam, J.O.; Gómez-Márquez, J.; Bosch, I.; Hamad-Schifferli, K.; Gehrke, L. Multicolored Silver Nanoparticles for Multiplexed Disease Diagnostics: Distinguishing Dengue, Yellow Fever, and Ebola Viruses. *Lab Chip* **2015**, *15*, 1638–1641. [CrossRef] [PubMed]
210. Balakumar, V.; Prakash, P.; Muthupandi, K.; Rajan, A. Nanosilver for Selective and Sensitive Sensing of Saturnism. *Sensors Actuators B Chem.* **2017**, *241*, 814–820. [CrossRef]
211. Baptista, P.V.; Koziol-Montewka, M.; Paluch-Oles, J.; Doria, G.; Franco, R. Gold-Nanoparticle-Probe–Based Assay for Rapid and Direct Detection of *Mycobacterium tuberculosis* DNA in Clinical Samples. *Clin. Chem.* **2006**, *52*, 1433–1434. [CrossRef]
212. Georganopoulou, D.G.; Chang, L.; Nam, J.-M.; Thaxton, C.S.; Mufson, E.J.; Klein, W.L.; Mirkin, C.A. Nanoparticle-Based Detection in Cerebral Spinal Fluid of a Soluble Pathogenic Biomarker for Alzheimer's Disease. *Proc. Natl. Acad. Sci. USA* **2005**, *102*, 2273–2276. [CrossRef]
213. Haes, A.J.; Chang, L.; Klein, W.L.; Van Duyne, R.P. Detection of a Biomarker for Alzheimer's Disease from Synthetic and Clinical Samples Using a Nanoscale Optical Biosensor. *J. Am. Chem. Soc.* **2005**, *127*, 2264–2271. [CrossRef]
214. dos Santos Tramontin, N.; da Silva, S.; Arruda, R.; Ugioni, K.S.; Canteiro, P.B.; de Bem Silveira, G.; Mendes, C.; Silveira, P.C.L.; Muller, A.P. Gold Nanoparticles Treatment Reverses Brain Damage in Alzheimer's Disease Model. *Mol. Neurobiol.* **2020**, *57*, 926–936. [CrossRef]
215. Esmaeili-bandboni, A.; Amini, S.M.; Faridi-majidi, R.; Bagheri, J.; Mohammadnejad, J.; Sadroddiny, E. Cross-linking Gold Nanoparticles Aggregation Method Based on Localised Surface Plasmon Resonance for Quantitative Detection of MiR-155. *IET Nanobiotechnol.* **2018**, *12*, 453–458. [CrossRef]
216. Fortunati, S.; Pedrini, F.; Del Grosso, E.; Baranda Pellejero, L.; Bertucci, A. Design of Specific Nucleic Acid-Based Biosensors for Protein Binding Activity. *Anal. Sens.* **2022**, *2*, e202200037. [CrossRef]
217. Fatemi, F.; Amini, S.M.; Kharrazi, S.; Rasaee, M.J.; Mazlomi, M.A.; Asadi-Ghalehni, M.; Rajabibazl, M.; Sadroddiny, E. Construction of Genetically Engineered M13K07 Helper Phage for Simultaneous Phage Display of Gold Binding Peptide 1 and Nuclear Matrix Protein 22 ScFv Antibody. *Colloids Surf. B Biointerfaces* **2017**, *159*, 770–780. [CrossRef] [PubMed]
218. Andleeb, A.; Andleeb, A.; Asghar, S.; Zaman, G.; Tariq, M.; Mehmood, A.; Nadeem, M.; Hano, C.; Lorenzo, J.M.; Abbasi, B.H. A Systematic Review of Biosynthesized Metallic Nanoparticles as a Promising Anti-Cancer-Strategy. *Cancers* **2021**, *13*, 2818. [CrossRef] [PubMed]
219. Li, X.; Wang, L.; Fan, Y.; Feng, Q.; Cui, F. Biocompatibility and Toxicity of Nanoparticles and Nanotubes. *J. Nanomater.* **2012**, *2012*, 548389. [CrossRef]
220. Singh, A. Comparative Therapeutic Effects of Plant-Extract Synthesized and Traditionally Synthesized Gold Nanoparticles on Alcohol-Induced Inflammatory Activity in SH-SY5Y Cells In Vitro. *Biomedicines* **2017**, *5*, 70. [CrossRef] [PubMed]
221. Thiesen, B.; Jordan, A. Clinical Applications of Magnetic Nanoparticles for Hyperthermia. *Int. J. Hyperth.* **2008**, *24*, 467–474. [CrossRef] [PubMed]
222. Tassa, C.; Shaw, S.Y.; Weissleder, R. Dextran-Coated Iron Oxide Nanoparticles: A Versatile Platform for Targeted Molecular Imaging, Molecular Diagnostics, and Therapy. *Acc. Chem. Res.* **2011**, *44*, 842–852. [CrossRef]
223. Amatya, R.; Hwang, S.; Park, T.; Min, K.A.; Shin, M.C. In Vitro and in Vivo Evaluation of PEGylated Starch-Coated Iron Oxide Nanoparticles for Enhanced Photothermal Cancer Therapy. *Pharmaceutics* **2021**, *13*, 871. [CrossRef]
224. Ge, R.; Liu, C.; Zhang, X.; Wang, W.; Li, B.; Liu, J.; Liu, Y.; Sun, H.; Zhang, D.; Hou, Y.; et al. Photothermal-Activatable Fe3O4 Superparticle Nanodrug Carriers with PD-L1 Immune Checkpoint Blockade for Anti-Metastatic Cancer Immunotherapy. *ACS Appl. Mater. Interfaces* **2018**, *10*, 20342–20355. [CrossRef]
225. Mohamadkazem, M.; Neshastehriz, A.; Amini, S.M.; Moshiri, A.; Janzadeh, A. Radiosensitising Effect of Iron Oxide-gold Nanocomplex for Electron Beam Therapy of Melanoma in Vivo by Magnetic Targeting. *IET Nanobiotechnol.* **2023**, *17*, 212–223. [CrossRef]
226. Zhou, R.; Zhao, D.; Beeraka, N.M.; Wang, X.; Lu, P.; Song, R.; Chen, K.; Liu, J. Novel Implications of Nanoparticle-Enhanced Radiotherapy and Brachytherapy: Z-Effect and Tumor Hypoxia. *Metabolites* **2022**, *12*, 943. [CrossRef]

227. Rezaeian, A.; Amini, S.M.; Najafabadi, M.R.H.; Farsangi, Z.J.; Samadian, H. Plasmonic Hyperthermia or Radiofrequency Electric Field Hyperthermia of Cancerous Cells through Green-Synthesized Curcumin-Coated Gold Nanoparticles. *Lasers Med. Sci.* **2022**, *37*, 1333–1341. [CrossRef] [PubMed]
228. Kim, K.; Lee, W.G. Electroporation for Nanomedicine: A Review. *J. Mater. Chem. B* **2017**, *5*, 2726–2738. [CrossRef] [PubMed]
229. Ahmadi Kamalabadi, M.; Neshastehriz, A.; Ghaznavi, H.; Amini, S.M. Folate Functionalized Gold-Coated Magnetic Nanoparticles Effect in Combined Electroporation and Radiation Treatment of HPV-Positive Oropharyngeal Cancer. *Med. Oncol.* **2022**, *39*, 196. [CrossRef]
230. Tian, L.; Lu, L.; Qiao, Y.; Ravi, S.; Salatan, F.; Melancon, M. Stimuli-Responsive Gold Nanoparticles for Cancer Diagnosis and Therapy. *J. Funct. Biomater.* **2016**, *7*, 19. [CrossRef]
231. Puvanakrishnan, P. In Vivo Tumor Targeting of Gold Nanoparticles: Effect of Particle Type and Dosing Strategy. *Int. J. Nanomed.* **2012**, *7*, 1251–1258. [CrossRef] [PubMed]
232. Goel, R.; Shah, N.; Visaria, R.; Paciotti, G.F.; Bischof, J.C. Biodistribution of TNF-α-Coated Gold Nanoparticles in an in Vivo Model System. *Nanomedicine* **2009**, *4*, 401–410. [CrossRef]
233. Coelho, S.C.; Almeida, G.M.; Pereira, M.C.; Santos-Silva, F.; Coelho, M.A. Functionalized Gold Nanoparticles Improve Afatinib Delivery into Cancer Cells. *Expert Opin. Drug Deliv.* **2016**, *13*, 133–141. [CrossRef]
234. Cryer, A.M.; Chan, C.; Eftychidou, A.; Maksoudian, C.; Mahesh, M.; Tetley, T.D.; Spivey, A.C.; Thorley, A.J. Tyrosine Kinase Inhibitor Gold Nanoconjugates for the Treatment of Non-Small Cell Lung Cancer. *ACS Appl. Mater. Interfaces* **2019**, *11*, 16336–16346. [CrossRef]
235. Ramalingam, V.; Varunkumar, K.; Ravikumar, V.; Rajaram, R. Target Delivery of Doxorubicin Tethered with PVP Stabilized Gold Nanoparticles for Effective Treatment of Lung Cancer. *Sci. Rep.* **2018**, *8*, 3815. [CrossRef]
236. Balakrishnan, S.; Bhat, F.A.; Raja Singh, P.; Mukherjee, S.; Elumalai, P.; Das, S.; Patra, C.R.; Arunakaran, J. Gold Nanoparticle-Conjugated Quercetin Inhibits Epithelial-Mesenchymal Transition, Angiogenesis and Invasiveness via EGFR/VEGFR-2-Mediated Pathway in Breast Cancer. *Cell Prolif.* **2016**, *49*, 678–697. [CrossRef]
237. Chen, Y.-J.; Lee, Y.-C.; Huang, C.-H.; Chang, L.-S. Gallic Acid-Capped Gold Nanoparticles Inhibit EGF-Induced MMP-9 Expression through Suppression of P300 Stabilization and NFκB/c-Jun Activation in Breast Cancer MDA-MB-231 Cells. *Toxicol. Appl. Pharmacol.* **2016**, *310*, 98–107. [CrossRef] [PubMed]
238. Jun, B.-H.; Noh, M.S.; Kim, J.; Kim, G.; Kang, H.; Kim, M.-S.; Seo, Y.-T.; Baek, J.; Kim, J.-H.; Park, J.; et al. Multifunctional Silver-Embedded Magnetic Nanoparticles as SERS Nanoprobes and Their Applications. *Small* **2010**, *6*, 119–125. [CrossRef] [PubMed]
239. Wang, Y.; Newell, B.B.; Irudayaraj, J. Folic Acid Protected Silver Nanocarriers for Targeted Drug Delivery. *J. Biomed. Nanotechnol.* **2012**, *8*, 751–759. [CrossRef] [PubMed]
240. Fang, C.; Kievit, F.M.; Cho, Y.-C.; Mok, H.; Press, O.W.; Zhang, M. Effect of Cationic Side-Chains on Intracellular Delivery and Cytotoxicity of PH Sensitive Polymer–Doxorubicin Nanocarriers. *Nanoscale* **2012**, *4*, 7012–7020. [CrossRef] [PubMed]
241. AshaRani, P.; Sethu, S.; Lim, H.; Balaji, G.; Valiyaveettil, S.; Hande, M.P. Differential Regulation of Intracellular Factors Mediating Cell Cycle, DNA Repair and Inflammation Following Exposure to Silver Nanoparticles in Human Cells. *Genome Integr.* **2012**, 3, 2. [CrossRef] [PubMed]
242. Foldbjerg, R.; Irving, E.S.; Hayashi, Y.; Sutherland, D.S.; Thorsen, K.; Autrup, H.; Beer, C. Global Gene Expression Profiling of Human Lung Epithelial Cells after Exposure to Nanosilver. *Toxicol. Sci.* **2012**, *130*, 145–157. [CrossRef] [PubMed]
243. Mundekkad, D.; Cho, W.C. Mitophagy Induced by Metal Nanoparticles for Cancer Treatment. *Pharmaceutics* **2022**, *14*, 2275. [CrossRef]
244. Gurunathan, S.; Jeong, J.-K.; Han, J.W.; Zhang, X.-F.; Park, J.H.; Kim, J.-H. Multidimensional Effects of Biologically Synthesized Silver Nanoparticles in *Helicobacter pylori*, *Helicobacter felis*, and Human Lung (L132) and Lung Carcinoma A549 Cells. *Nanoscale Res. Lett.* **2015**, *10*, 35. [CrossRef]
245. Yeh, Y.-C.; Creran, B.; Rotello, V.M. Gold Nanoparticles: Preparation, Properties, and Applications in Bionanotechnology. *Nanoscale* **2012**, *4*, 1871–1880. [CrossRef]
246. Hashemi Goradel, N.; Ghiyami-Hour, F.; Jahangiri, S.; Negahdari, B.; Sahebkar, A.; Masoudifar, A.; Mirzaei, H. Nanoparticles as New Tools for Inhibition of Cancer Angiogenesis. *J. Cell. Physiol.* **2018**, *233*, 2902–2910. [CrossRef]
247. Bhattacharya, R.; Mukherjee, P. Biological Properties of "Naked" Metal Nanoparticles. *Adv. Drug Deliv. Rev.* **2008**, *60*, 1289–1306. [CrossRef] [PubMed]
248. Ngernyuang, N.; Wongwattanakul, M.; Charusirisawad, W.; Shao, R.; Limpaiboon, T. Green Synthesized Apigenin Conjugated Gold Nanoparticles Inhibit Cholangiocarcinoma Cell Activity and Endothelial Cell Angiogenesis in Vitro. *Heliyon* **2022**, *8*, e12028. [CrossRef] [PubMed]
249. Huang, W.; Xing, Y.; Zhu, L.; Zhuo, J.; Cai, M. Sorafenib Derivatives-Functionalized Gold Nanoparticles Confer Protection against Tumor Angiogenesis and Proliferation via Suppression of EGFR and VEGFR-2. *Exp. Cell Res.* **2021**, *406*, 112633. [CrossRef]
250. Kalishwaralal, K.; Banumathi, E.; Pandian, S.R.K.; Deepak, V.; Muniyandi, J.; Eom, S.H.; Gurunathan, S. Silver Nanoparticles Inhibit VEGF Induced Cell Proliferation and Migration in Bovine Retinal Endothelial Cells. *Colloids Surf. B Biointerfaces* **2009**, *73*, 51–57. [CrossRef] [PubMed]
251. Gurunathan, S.; Lee, K.-J.; Kalishwaralal, K.; Sheikpranbabu, S.; Vaidyanathan, R.; Eom, S.H. Antiangiogenic Properties of Silver Nanoparticles. *Biomaterials* **2009**, *30*, 6341–6350. [CrossRef]

252. Pavan, S.R.; Venkatesan, J.; Prabhu, A. Anticancer Activity of Silver Nanoparticles from the Aqueous Extract of *Dictyota Ciliolata* on Non-Small Cell Lung Cancer Cells. *J. Drug Deliv. Sci. Technol.* **2022**, *74*, 103525. [CrossRef]
253. Baharara, J.; Namvar, F.; Ramezani, T.; Hosseini, N.; Mohamad, R. Green Synthesis of Silver Nanoparticles Using *Achillea Biebersteinii* Flower Extract and Its Anti-Angiogenic Properties in the Rat Aortic Ring Model. *Molecules* **2014**, *19*, 4624–4634. [CrossRef]
254. Baharara, J.; Namvar, F.; Mousavi, M.; Ramezani, T.; Mohamad, R. Anti-Angiogenesis Effect of Biogenic Silver Nanoparticles Synthesized Using *Saliva officinalis* on Chick Chorioalantoic Membrane (CAM). *Molecules* **2014**, *19*, 13498–13508. [CrossRef]
255. Baharara, J.; Namvar, F.; Ramezani, T.; Mousavi, M.; Mohamad, R. Silver Nanoparticles Biosynthesized Using *Achillea biebersteinii* Flower Extract: Apoptosis Induction in MCF-7 Cells via Caspase Activation and Regulation of Bax and Bcl-2 Gene Expression. *Molecules* **2015**, *20*, 2693–2706. [CrossRef]
256. Ozdal, M.; Gurkok, S. A Recent advances in nanoparticles as antibacterial agent. *ADMET DMPK* **2022**, *10*, 115–129. [CrossRef]
257. Lei, C.; Sun, N.; Wu, H.; Zhao, Y.; Yu, C.; Janani, B.J.; Fakhri, A. Bio-Photoelectrochemical Degradation, and Photocatalysis Process by the Fabrication of Copper Oxide/Zinc Cadmium Sulfide Heterojunction Nanocomposites: Mechanism, Microbial Community and Antifungal Analysis. *Chemosphere* **2022**, *308*, 136375. [CrossRef] [PubMed]
258. Liu, Z.; Hadi, M.A.; Aljuboory, D.S.; Ali, F.A.; Jawad, M.A.; AL-Alwany, A.; Hadrawi, S.K.; Mundher, T.; Riadi, Y.; Amer, R.F.; et al. High Efficiency of Ag0 Decorated Cu2MoO4 Nanoparticles for Heterogeneous Photocatalytic Activation, Bactericidal System, and Detection of Glucose from Blood Sample. *J. Photochem. Photobiol. B Biol.* **2022**, *236*, 112571. [CrossRef] [PubMed]
259. Chen, Y.; Jihad, A.; Hussam, F.; Al-Abdeen, S.H.Z.; Hussein, J.M.; Adhab, Z.H.; Alzahraa, Z.H.A.; Ahmad, I.; Fatolahi, L.; Janani, B.J. A Facile Preparation Method for Efficiency a Novel $LaNiO_3/SrCeO_3$ (p-n Type) Heterojunction Catalyst in Photocatalytic Activities, Bactericidal Assessment and Dopamine Detection. *Surf. Interfaces* **2023**, *38*, 102830. [CrossRef]
260. Mahdi, A.A.; Obeid, R.A.; Abdullah, K.; Mohammed, S.; Kadhim, A.J.; Ramadan, M.F.; Hussien, B.M.; Alkahtani, A.; Ali, F.A.; Alkhathami, A.G.; et al. A Facile Construction of NiV2O6/CeO2 Nano-Heterojunction for Photo-Operated Process in Water Remediation Reaction, Antibacterial Studies, and Detection of D-Amino Acid in Peroxidase System. *Surf. Interfaces* **2023**, *40*, 102970. [CrossRef]
261. Aldhalmi, A.K.; Alkhayyat, S.; Younis Albahadly, W.K.; Jawad, M.A.; Alsaraf, K.M.; Riyad Muedii, Z.A.-H.; Ali, F.A.; Ahmed, M.; Asiri, M.; Al-Fatolahi, L.; et al. A Novel Fabricate of Iron and Nickel-Introduced Bimetallic MOFs for Quickly Catalytic Degradation via the Peroxymonosulfate, Antibacterial Efficiency, and Cytotoxicity Assay. *Inorg. Chem. Commun.* **2023**, *153*, 110823. [CrossRef]
262. Lai, Y.; Fakhri, A.; Janani, B.J. Synergistic Activities of Silver Indium Sulfide/Nickel Molybdenum Sulfide Nanostructures Anchored on Clay Mineral for Light-Driven Bactericidal Performance, and Detection of Uric Acid from Gout Patient Serum. *J. Photochem. Photobiol. B Biol.* **2022**, *234*, 112526. [CrossRef]
263. Syed, A.; Elgorban, A.M.; Bahkali, A.H.; Eswaramoorthy, R.; Verma, M.; Varma, R.S.; Janani, B.J. Highly-Impressive Performances of Novel $NiCo_2O_4/Bi_2O_3/Ag_2ZrO_3$ Nanocomposites in Photocatalysis System, Removal Pathway, Toxicity Estimation, and Antibacterial Activities. *J. Taiwan Inst. Chem. Eng.* **2023**, *149*, 105004. [CrossRef]
264. Maťátková, O.; Michailidu, J.; Miškovská, A.; Kolouchová, I.; Masák, J.; Čejková, A. Antimicrobial Properties and Applications of Metal Nanoparticles Biosynthesized by Green Methods. *Biotechnol. Adv.* **2022**, *58*, 107905. [CrossRef]
265. Slavin, Y.N.; Asnis, J.; Häfeli, U.O.; Bach, H. Metal Nanoparticles: Understanding the Mechanisms behind Antibacterial Activity. *J. Nanobiotechnol.* **2017**, *15*, 65. [CrossRef]
266. Chatterjee, A.K.; Chakraborty, R.; Basu, T. Mechanism of Antibacterial Activity of Copper Nanoparticles. *Nanotechnology* **2014**, *25*, 135101. [CrossRef]
267. Tao, C. Antimicrobial Activity and Toxicity of Gold Nanoparticles: Research Progress, Challenges and Prospects. *Lett. Appl. Microbiol.* **2018**, *67*, 537–543. [CrossRef] [PubMed]
268. Durán, N.; Durán, M.; de Jesus, M.B.; Seabra, A.B.; Fávaro, W.J.; Nakazato, G. Silver Nanoparticles: A New View on Mechanistic Aspects on Antimicrobial Activity. *Nanomed. Nanotechnol. Biol. Med.* **2016**, *12*, 789–799. [CrossRef] [PubMed]
269. Sirelkhatim, A.; Mahmud, S.; Seeni, A.; Kaus, N.H.M.; Ann, L.C.; Bakhori, S.K.M.; Hasan, H.; Mohamad, D. Review on Zinc Oxide Nanoparticles: Antibacterial Activity and Toxicity Mechanism. *Nano-Micro Lett.* **2015**, *7*, 219–242. [CrossRef] [PubMed]
270. Ahmad, S.A.; Das, S.S.; Khatoon, A.; Ansari, M.T.; Afzal, M.; Hasnain, S.; Nayak, A.K. Bactericidal activity of silver nanoparticles: A mechanistic review. *Mater. Sci. Energy Technol.* **2020**, *3*, 756–769. [CrossRef]
271. Sondi, I.; Salopek-Sondi, B. Silver Nanoparticles as Antimicrobial Agent: A Case Study on E. Coli as a Model for Gram-Negative Bacteria. *J. Colloid Interface Sci.* **2004**, *275*, 177–182. [CrossRef]
272. Baker, C.; Pradhan, A.; Pakstis, L.; Pochan, D.; Shah, S.I. Synthesis and Antibacterial Properties of Silver Nanoparticles. *J. Nanosci. Nanotechnol.* **2005**, *5*, 244–249. [CrossRef]
273. Morones, J.R.; Elechiguerra, J.L.; Camacho, A.; Holt, K.; Kouri, J.B.; Ramírez, J.T.; Yacaman, M.J. The Bactericidal Effect of Silver Nanoparticles. *Nanotechnol.* **2005**, *16*, 2346–2353. [CrossRef]
274. Yuan, Y.-G.; Peng, Q.-L.; Gurunathan, S. Effects of Silver Nanoparticles on Multiple Drug-Resistant Strains of *Staphylococcus aureus* and *Pseudomonas aeruginosa* from Mastitis-Infected Goats: An Alternative Approach for Antimicrobial Therapy. *Int. J. Mol. Sci.* **2017**, *18*, 569. [CrossRef]
275. Kim, J.S.; Kuk, E.; Yu, K.N.; Kim, J.-H.; Park, S.J.; Lee, H.J.; Kim, S.H.; Park, Y.K.; Park, Y.H.; Hwang, C.-Y.; et al. Antimicrobial Effects of Silver Nanoparticles. *Nanomed. Nanotechnol. Biol. Med.* **2007**, *3*, 95–101. [CrossRef]

276. Allahverdiyev, A.M.; Kon, K.V.; Abamor, E.S.; Bagirova, M.; Rafailovich, M. Coping with Antibiotic Resistance: Combining Nanoparticles with Antibiotics and Other Antimicrobial Agents. *Expert Rev. Anti Infect. Ther.* **2011**, *9*, 1035–1052. [CrossRef]
277. Deng, H.; McShan, D.; Zhang, Y.; Sinha, S.S.; Arslan, Z.; Ray, P.C.; Yu, H. Mechanistic Study of the Synergistic Antibacterial Activity of Combined Silver Nanoparticles and Common Antibiotics. *Environ. Sci. Technol.* **2016**, *50*, 8840–8848. [CrossRef]
278. Ipe, D.S.; Kumar, P.T.S.; Love, R.M.; Hamlet, S.M. Silver Nanoparticles at Biocompatible Dosage Synergistically Increases Bacterial Susceptibility to Antibiotics. *Front. Microbiol.* **2020**, *11*, 1074. [CrossRef] [PubMed]
279. León-Buitimea, A.; Garza-Cárdenas, C.R.; Román-García, M.F.; Ramírez-Díaz, C.A.; Ulloa-Ramírez, M.; Morones-Ramírez, J.R. Nanomaterials-Based Combinatorial Therapy as a Strategy to Combat Antibiotic Resistance. *Antibiotics* **2022**, *11*, 794. [CrossRef] [PubMed]
280. Nirmala Grace, A.; Pandian, K. Antibacterial Efficacy of Aminoglycosidic Antibiotics Protected Gold Nanoparticles—A Brief Study. *Colloids Surfaces A Physicochem. Eng. Asp.* **2007**, *297*, 63–70. [CrossRef]
281. Ribeiro, A.I.; Dias, A.M.; Zille, A. Synergistic Effects between Metal Nanoparticles and Commercial Antimicrobial Agents: A Review. *ACS Appl. Nano Mater.* **2022**, *5*, 3030–3064. [CrossRef]
282. Franci, G.; Falanga, A.; Galdiero, S.; Palomba, L.; Rai, M.; Morelli, G.; Galdiero, M. Silver Nanoparticles as Potential Antibacterial Agents. *Molecules* **2015**, *20*, 8856–8874. [CrossRef]
283. Shahverdi, A.R.; Fakhimi, A.; Shahverdi, H.R.; Minaian, S. Synthesis and Effect of Silver Nanoparticles on the Antibacterial Activity of Different Antibiotics against *Staphylococcus aureus* and *Escherichia coli*. *Nanomed. Nanotechnol. Biol. Med.* **2007**, *3*, 168–171. [CrossRef]
284. Gajbhiye, M.; Kesharwani, J.; Ingle, A.; Gade, A.; Rai, M. Fungus-Mediated Synthesis of Silver Nanoparticles and Their Activity against Pathogenic Fungi in Combination with Fluconazole. *Nanomed. Nanotechnol. Biol. Med.* **2009**, *5*, 382–386. [CrossRef]
285. Jain, J.; Arora, S.; Rajwade, J.M.; Omray, P.; Khandelwal, S.; Paknikar, K.M. Silver Nanoparticles in Therapeutics: Development of an Antimicrobial Gel Formulation for Topical Use. *Mol. Pharm.* **2009**, *6*, 1388–1401. [CrossRef]
286. Mallmann, E.J.J.; Cunha, F.A.; Castro, B.N.M.F.; Maciel, A.M.; Menzes, E.A.; Fechine, P.B.A. Antifungal Activity of Silver Nanoparticles Obtained by Green Synthesis. *Rev. Inst. Med. Trop. Sao Paulo* **2015**, *57*, 165–167. [CrossRef]
287. Peralta, L.C.F.; Almeida, N.L.M.; Pontes, F.M.L.; Rinaldo, D.; Carneiro, C.A.; Neppelenbroek, K.H.; Lara, V.S.; Porto, V.C. Silver Nanoparticles in Denture Adhesive: An Antimicrobial Approach against *Candida albicans*. *J. Dent.* **2023**, *131*, 104445. [CrossRef] [PubMed]
288. Golli, R.; Thummaneni, C.; Pabbathi, D.D.; Srungarapu, T.; Jayasri, G.; Vangalapati, M. Silver Nanoparticles Synthesized by *Brassica oleracea* (Broccoli) Acting as Antifungal Agent against *Candida albicans*. *Mater. Today Proc.* **2023**, *80*, 1495–1500. [CrossRef]
289. Robles-Martínez, M.; Patiño-Herrera, R.; Pérez-Vázquez, F.J.; Montejano-Carrizales, J.M.; González, J.F.C.; Pérez, E. *Mentha Piperita* as a Natural Support for Silver Nanoparticles: A New Anti-*Candida albicans* Treatment. *Colloid Interface Sci. Commun.* **2020**, *35*, 100253. [CrossRef]
290. Ahmad, T.; Wani, I.A.; Lone, I.H.; Ganguly, A.; Manzoor, N.; Ahmad, A.; Ahmed, J.; Al-Shihri, A.S. Antifungal Activity of Gold Nanoparticles Prepared by Solvothermal Method. *Mater. Res. Bull.* **2013**, *48*, 12–20. [CrossRef]
291. Umamaheswari, K.; Abirami, M. Assessment of Antifungal Action Mechanism of Green Synthesized Gold Nanoparticles (AuNPs) Using *Allium sativum* on Candida Species. *Mater. Lett.* **2023**, *333*, 133616. [CrossRef]
292. Kareem, H.A.; Samaka, H.M.; Abdulridha, W.M. Evaluation of the gold nanoparticles prepared by green chemistry in the treatment of cutaneous candidiasis. *Curr. Med. Mycol.* **2021**, *7*, 1–5. [CrossRef]
293. Rai, M.; Deshmukh, S.D.; Ingle, A.P.; Gupta, I.R.; Galdiero, M.; Galdiero, S. Metal Nanoparticles: The Protective Nanoshield against Virus Infection. *Crit. Rev. Microbiol.* **2016**, *42*, 46–56. [CrossRef]
294. Ratan, Z.A.; Mashrur, F.R.; Chhoan, A.P.; Shahriar, S.M.; Haidere, M.F.; Runa, N.J.; Kim, S.; Kweon, D.-H.; Hosseinzadeh, H.; Cho, J.Y. Silver Nanoparticles as Potential Antiviral Agents. *Pharmaceutics* **2021**, *13*, 2034. [CrossRef]
295. Wei, X.; Zhang, G.; Ran, D.; Krishnan, N.; Fang, R.H.; Gao, W.; Spector, S.A.; Zhang, L. T-cell-mimicking Nanoparticles Can Neutralize HIV Infectivity. *Adv. Mater.* **2018**, *30*, 1802233. [CrossRef]
296. Orłowski, P.; Kowalczyk, A.; Tomaszewska, E.; Ranoszek-Soliwoda, K.; Węgrzyn, A.; Grzesiak, J.; Celichowski, G.; Grobelny, J.; Eriksson, K.; Krzyzowska, M. Antiviral Activity of Tannic Acid Modified Silver Nanoparticles: Potential to Activate Immune Response in Herpes Genitalis. *Viruses* **2018**, *10*, 524. [CrossRef]
297. Gaikwad, S.; Ingle, A.; Gade, A.; Rai, M.; Falanga, A.; Incoronato, N.; Russo, L.; Galdiero, S.; Galdiero, M. Antiviral Activity of Mycosynthesized Silver Nanoparticles against Herpes Simplex Virus and Human Parainfluenza Virus Type 3. *Int. J. Nanomed.* **2013**, *8*, 4303–4314. [CrossRef]
298. Jeremiah, S.S.; Miyakawa, K.; Morita, T.; Yamaoka, Y.; Ryo, A. Potent Antiviral Effect of Silver Nanoparticles on SARS-CoV-2. *Biochem. Biophys. Res. Commun.* **2020**, *533*, 195–200. [CrossRef] [PubMed]
299. Lara, H.H.; Ayala-Nuñez, N.V.; Ixtepan-Turrent, L.; Rodriguez-Padilla, C. Mode of Antiviral Action of Silver Nanoparticles against HIV-1. *J. Nanobiotechnol.* **2010**, *8*, 1. [CrossRef] [PubMed]
300. Inagaki, S.; Ghirlando, R.; Grisshammer, R. Biophysical Characterization of Membrane Proteins in Nanodiscs. *Methods* **2013**, *59*, 287–300. [CrossRef]
301. Dos Santos, C.A.; Seckler, M.M.; Ingle, A.P.; Gupta, I.; Galdiero, S.; Galdiero, M.; Gade, A.; Rai, M. Silver Nanoparticles: Therapeutical Uses, Toxicity, and Safety Issues. *J. Pharm. Sci.* **2014**, *103*, 1931–1944. [CrossRef] [PubMed]

302. Xiang, D.; Zheng, C.; Zheng, Y.; Li, X.; Yin, J.; O'Conner, M.; Marappan, M.; Miao, Y.; Xiang, B.; Duan, W.; et al. Inhibition of A/Human/Hubei/3/2005 (H3N2) Influenza Virus Infection by Silver Nanoparticles In Vitro and In Vivo. *Int. J. Nanomed.* **2013**, *8*, 4103–4114. [CrossRef] [PubMed]
303. Khandelwal, N.; Kaur, G.; Chaubey, K.K.; Singh, P.; Sharma, S.; Tiwari, A.; Singh, S.V.; Kumar, N. Silver Nanoparticles Impair Peste Des Petits Ruminants Virus Replication. *Virus Res.* **2014**, *190*, 1–7. [CrossRef]
304. Mehranfar, A.; Izadyar, M. Theoretical Design of Functionalized Gold Nanoparticles as Antiviral Agents against Severe Acute Respiratory Syndrome Coronavirus 2 (SARS-CoV-2). *J. Phys. Chem. Lett.* **2020**, *11*, 10284–10289. [CrossRef]
305. Chen, L.; Liang, J. An Overview of Functional Nanoparticles as Novel Emerging Antiviral Therapeutic Agents. *Mater. Sci. Eng. C* **2020**, *112*, 110924. [CrossRef]
306. Babaei, A.; Mousavi, S.M.; Ghasemi, M.; Pirbonyeh, N.; Soleimani, M.; Moattari, A. Gold Nanoparticles Show Potential in Vitro Antiviral and Anticancer Activity. *Life Sci.* **2021**, *284*, 119652. [CrossRef]
307. Mikhailova, E.O. Gold Nanoparticles: Biosynthesis and Potential of Biomedical Application. *J. Funct. Biomater.* **2021**, *12*, 70. [CrossRef] [PubMed]
308. Draz, M.S.; Shafiee, H. Applications of Gold Nanoparticles in Virus Detection. *Theranostics* **2018**, *8*, 1985–2017. [CrossRef] [PubMed]
309. Maduray, K.; Parboosing, R. Metal Nanoparticles: Apromising Treatment for Viral and Arboviral Infections. *Biol. Trace Elem. Res.* **2021**, *199*, 3159–3176. [CrossRef] [PubMed]
310. Murugan, K.; Wei, J.; Alsalhi, M.S.; Nicoletti, M.; Paulpandi, M.; Samidoss, C.M.; Dinesh, D.; Chandramohan, B.; Paneerselvam, C.; Subramaniam, J.; et al. Magnetic Nanoparticles Are Highly Toxic to Chloroquine-Resistant Plasmodium Falciparum, Dengue Virus (DEN-2), and Their Mosquito Vectors. *Parasitol. Res.* **2017**, *116*, 495–502. [CrossRef]
311. Gutierrez, L.; Li, X.; Wang, J.; Nangmenyi, G.; Economy, J.; Kuhlenschmidt, T.B.; Kuhlenschmidt, M.S.; Nguyen, T.H. Adsorption of Rotavirus and Bacteriophage MS2 Using Glass Fiber Coated with Hematite Nanoparticles. *Water Res.* **2009**, *43*, 5198–5208. [CrossRef]
312. Kumar, R.; Nayak, M.; Sahoo, G.C.; Pandey, K.; Sarkar, M.C.; Ansari, Y.; Das, V.N.R.; Topno, R.K.; Bhawna; Madhukar, M.; et al. Iron Oxide Nanoparticles Based Antiviral Activity of H1N1 Influenza A Virus. *J. Infect. Chemother.* **2019**, *25*, 325–329. [CrossRef]
313. Abo-zeid, Y.; Ismail, N.S.M.; McLean, G.R.; Hamdy, N.M. A Molecular Docking Study Repurposes FDA Approved Iron Oxide Nanoparticles to Treat and Control COVID-19 Infection. *Eur. J. Pharm. Sci.* **2020**, *153*, 105465. [CrossRef]
314. Wong, C.K.; Cheung, P.F.Y.; Ip, W.K.; Lam, C.W.K. Intracellular Signaling Mechanisms Regulating Toll-like Receptor–Mediated Activation of Eosinophils. *Am. J. Respir. Cell Mol. Biol.* **2007**, *37*, 85–96. [CrossRef]
315. Eming, S.A.; Krieg, T.; Davidson, J.M. Inflammation in Wound Repair: Molecular and Cellular Mechanisms. *J. Investig. Dermatol.* **2007**, *127*, 514–525. [CrossRef]
316. Aththanayaka, S.; Thiripuranathar, G.; Ekanayake, S. Green Synthesized Nanomaterials as Antioxidant and Antiinflammatory Substances. In *Synthesis of Bionanomaterials for Biomedical Applications*; Elsevier: Amsterdam, The Netherlands, 2023; pp. 299–317.
317. Lan Chi, N.T.; Narayanan, M.; Chinnathambi, A.; Govindasamy, C.; Subramani, B.; Brindhadevi, K.; Pimpimon, T.; Pikulkaew, S. Fabrication, Characterization, Anti-Inflammatory, and Anti-Diabetic Activity of Silver Nanoparticles Synthesized from Azadirachta Indica Kernel Aqueous Extract. *Environ. Res.* **2022**, *208*, 112684. [CrossRef]
318. Londhe, S.; Haque, S.; Patra, C.R. Silver and Gold Nanoparticles: Potential Cancer Theranostic Applications, Recent Development, Challenges, and Future Perspectives. In *Gold and Silver Nanoparticles*; Elsevier: Amsterdam, The Netherlands, 2023; pp. 247–290.
319. Pauna, A.-M.R. The Prolongation of the Antinociceptive Effects of Novel Nanosystems Entrapping Diclofenac Sodium in Somatic Pain in Rats. *Rev. Med. Chir.* **2022**, *126*, 557–564. [CrossRef]
320. Afshari, A.R.; Sanati, M.; Mollazadeh, H.; Kesharwani, P.; Johnston, T.P.; Sahebkar, A. Nanoparticle-Based Drug Delivery Systems in Cancer: A Focus on Inflammatory Pathways. *Semin. Cancer Biol.* **2022**, *86*, 860–872. [CrossRef] [PubMed]
321. de Araújo, R.F.; de Araújo, A.A.; Pessoa, J.B.; Freire Neto, F.P.; da Silva, G.R.; Leitão Oliveira, A.L.C.S.; de Carvalho, T.G.; Silva, H.F.O.; Eugênio, M.; Sant'Anna, C.; et al. Anti-Inflammatory, Analgesic and Anti-Tumor Properties of Gold Nanoparticles. *Pharmacol. Rep.* **2017**, *69*, 119–129. [CrossRef] [PubMed]
322. Somasuntharam, I.; Yehl, K.; Carroll, S.L.; Maxwell, J.T.; Martinez, M.D.; Che, P.-L.; Brown, M.E.; Salaita, K.; Davis, M.E. Knockdown of TNF-α by DNAzyme Gold Nanoparticles as an Anti-Inflammatory Therapy for Myocardial Infarction. *Biomaterials* **2016**, *83*, 12–22. [CrossRef] [PubMed]
323. Sulaiman, G.M.; Waheeb, H.M.; Jabir, M.S.; Khazaal, S.H.; Dewir, Y.H.; Naidoo, Y. Hesperidin Loaded on Gold Nanoparticles as a Drug Delivery System for a Successful Biocompatible, Anti-Cancer, Anti-Inflammatory and Phagocytosis Inducer Model. *Sci. Rep.* **2020**, *10*, 9362. [CrossRef]
324. Gao, W.; Wang, Y.; Xiong, Y.; Sun, L.; Wang, L.; Wang, K.; Lu, H.Y.; Bao, A.; Turvey, S.E.; Li, Q.; et al. Size-Dependent Anti-Inflammatory Activity of a Peptide-Gold Nanoparticle Hybrid in Vitro and in a Mouse Model of Acute Lung Injury. *Acta Biomater.* **2019**, *85*, 203–217. [CrossRef]
325. Thiruvengadam, M.; Chung, I.-M.; Gomathi, T.; Ansari, M.A.; Khanna, V.G.; Babu, V.; Rajakumar, G. Synthesis, characterization and pharmacological potential of green synthesized copper nanoparticles. *Bioprocess Biosyst. Eng.* **2019**, *42*, 1769–1777. [CrossRef]
326. Agarwal, H.; Shanmugam, V. A Review on Anti-Inflammatory Activity of Green Synthesized Zinc Oxide Nanoparticle: Mechanism-Based Approach. *Bioorg. Chem.* **2020**, *94*, 103423. [CrossRef]

327. Aafreen, M.M.; Anitha, R.; Preethi, R.C.; Rajeshkumar, S.; Lakshmi, T. Anti-Inflammatory Activity of Silver Nanoparticles Prepared from Ginger Oil—An Invitro Approach. *Indian J. Public Health Res. Dev.* **2019**, *10*, 145. [CrossRef]
328. Singh, P.; Ahn, S.; Kang, J.-P.; Veronika, S.; Huo, Y.; Singh, H.; Chokkaligam, M.; El-Agamy Farh, M.; Aceituno, V.C.; Kim, Y.J.; et al. In Vitro Anti-Inflammatory Activity of Spherical Silver Nanoparticles and Monodisperse Hexagonal Gold Nanoparticles by Fruit Extract of *Prunus serrulata*: A Green Synthetic Approach. *Artif. Cells Nanomed. Biotechnol.* **2017**, *46*, 2022–2032. [CrossRef]
329. Bhol, K.C.; Schechter, P.J. Effects of Nanocrystalline Silver (NPI 32101) in a Rat Model of Ulcerative Colitis. *Dig. Dis. Sci.* **2007**, *52*, 2732–2742. [CrossRef] [PubMed]
330. Tian, J.; Wong, K.K.Y.; Ho, C.-M.; Lok, C.-N.; Yu, W.-Y.; Che, C.-M.; Chiu, J.-F.; Tam, P.K.H. Topical Delivery of Silver Nanoparticles Promotes Wound Healing. *ChemMedChem* **2007**, *2*, 129–136. [CrossRef] [PubMed]
331. David, L.; Moldovan, B.; Vulcu, A.; Olenic, L.; Perde-Schrepler, M.; Fischer-Fodor, E.; Florea, A.; Crisan, M.; Chiorean, I.; Clichici, S.; et al. Green Synthesis, Characterization and Anti-Inflammatory Activity of Silver Nanoparticles Using European Black Elderberry Fruits Extract. *Colloids Surf. B Biointerfaces* **2014**, *122*, 767–777. [CrossRef] [PubMed]
332. Nadworny, P.L.; Wang, J.; Tredget, E.E.; Burrell, R.E. Anti-Inflammatory Activity of Nanocrystalline Silver in a Porcine Contact Dermatitis Model. *Nanomed. Nanotechnol. Biol. Med.* **2008**, *4*, 241–251. [CrossRef] [PubMed]
333. Singh, P.; Pandit, S.; Mokkapati, V.R.S.S.; Garg, A.; Ravikumar, V.; Mijakovic, I. Gold Nanoparticles in Diagnostics and Therapeutics for Human Cancer. *Int. J. Mol. Sci.* **2018**, *19*, 1979. [CrossRef]
334. Boroumand, Z.; Golmakani, N.; Boroumand, S. Clinical Trials on Silver Nanoparticles for Wound Healing. *Nanomed. J* **2018**, *5*, 186–191. [CrossRef]
335. Carrasco-Esteban, E.; Domínguez-Rullán, J.A.; Barrionuevo-Castillo, P.; Pelari-Mici, L.; Leaman, O.; Sastre-Gallego, S.; López-Campos, F. Current Role of Nanoparticles in the Treatment of Lung Cancer. *J. Clin. Transl. Res.* **2021**, *7*, 140–155.
336. National Library of Medicine; National Institutes of Health; U.S. Department of Health and Human Services. ClinicalTrials.gov. Available online: https://clinicaltrials.gov/ct2/results?cond=&term=gold+nanoparticles&cntry=&state=&city=&dist= (accessed on 20 June 2023).
337. Dadfar, S.M.; Roemhild, K.; Drude, N.I.; von Stillfried, S.; Knüchel, R.; Kiessling, F.; Lammers, T. Iron Oxide Nanoparticles: Diagnostic, Therapeutic and Theranostic Applications. *Adv. Drug Deliv. Rev.* **2019**, *138*, 302–325. [CrossRef]
338. Singh, S.P.; Bhargava, C.S.; Dubey, V.; Mishra, A.; Singh, Y. Silver Nanoparticles: Biomedical Applications, Toxicity, and Safety Issues. *Int. J. Res. Pharm. Pharm. Sci.* **2017**, *2*, 2455–2698.
339. Xiong, P.; Huang, X.; Ye, N.; Lu, Q.; Zhang, G.; Peng, S.; Wang, H.; Liu, Y. Cytotoxicity of Metal-based Nanoparticles: From Mechanisms and Methods of Evaluation to Pathological Manifestations. *Adv. Sci.* **2022**, *9*, 2106049. [CrossRef]
340. Lansdown, A.B.G. Silver in Health Care: Antimicrobial Effects and Safety in Use. In *Biofunctional Textiles and the Skin*; KARGER: Basel, Switzerland, 2006; pp. 17–34.
341. Xu, L.; Wang, Y.-Y.; Huang, J.; Chen, C.-Y.; Wang, Z.-X.; Xie, H. Silver Nanoparticles: Synthesis, Medical Applications and Biosafety. *Theranostics* **2020**, *10*, 8996–9031. [CrossRef]
342. Wang, L.; Zhang, T.; Li, P.; Huang, W.; Tang, J.; Wang, P.; Liu, J.; Yuan, Q.; Bai, R.; Li, B.; et al. Use of Synchrotron Radiation-Analytical Techniques to Reveal Chemical Origin of Silver-Nanoparticle Cytotoxicity. *ACS Nano* **2015**, *9*, 6532–6547. [CrossRef]
343. Kus-Liśkiewicz, M.; Fickers, P.; Ben Tahar, I. Biocompatibility and Cytotoxicity of Gold Nanoparticles: Recent Advances in Methodologies and Regulations. *Int. J. Mol. Sci.* **2021**, *22*, 10952. [CrossRef] [PubMed]
344. Lewinski, N.; Colvin, V.; Drezek, R. Cytotoxicity of Nanoparticles. *Small* **2008**, *4*, 26–49. [CrossRef]
345. Murphy, C.J.; Gole, A.M.; Stone, J.W.; Sisco, P.N.; Alkilany, A.M.; Goldsmith, E.C.; Baxter, S.C. Gold Nanoparticles in Biology: Beyond Toxicity to Cellular Imaging. *Acc. Chem. Res.* **2008**, *41*, 1721–1730. [CrossRef] [PubMed]
346. Hawley, A.E.; Davis, S.S.; Illum, L. Targeting of Colloids to Lymph Nodes: Influence of Lymphatic Physiology and Colloidal Characteristics. *Adv. Drug Deliv. Rev.* **1995**, *17*, 129–148. [CrossRef]
347. Sabourian, P.; Yazdani, G.; Ashraf, S.S.; Frounchi, M.; Mashayekhan, S.; Kiani, S.; Kakkar, A. Effect of Physico-Chemical Properties of Nanoparticles on Their Intracellular Uptake. *Int. J. Mol. Sci.* **2020**, *21*, 8019. [CrossRef] [PubMed]
348. Sonavane, G.; Tomoda, K.; Sano, A.; Ohshima, H.; Terada, H.; Makino, K. In Vitro Permeation of Gold Nanoparticles through Rat Skin and Rat Intestine: Effect of Particle Size. *Colloids Surf. B Biointerfaces* **2008**, *65*, 1–10. [CrossRef]
349. Sonavane, G.; Tomoda, K.; Makino, K. Biodistribution of Colloidal Gold Nanoparticles after Intravenous Administration: Effect of Particle Size. *Colloids Surf. B Biointerfaces* **2008**, *66*, 274–280. [CrossRef]
350. Steckiewicz, K.P.; Barcinska, E.; Malankowska, A.; Zauszkiewicz–Pawlak, A.; Nowaczyk, G.; Zaleska-Medynska, A.; Inkielewicz-Stepniak, I. Impact of Gold Nanoparticles Shape on Their Cytotoxicity against Human Osteoblast and Osteosarcoma in in Vitro Model. Evaluation of the Safety of Use and Anti-Cancer Potential. *J. Mater. Sci. Mater. Med.* **2019**, *30*, 22. [CrossRef]
351. Hussain, S.M.; Hess, K.L.; Gearhart, J.M.; Geiss, K.T.; Schlager, J.J. In Vitro Toxicity of Nanoparticles in BRL 3A Rat Liver Cells. *Toxicol. Vitr.* **2005**, *19*, 975–983. [CrossRef]
352. Ling, D.; Hyeon, T. Chemical Design of Biocompatible Iron Oxide Nanoparticles for Medical Applications. *Small* **2013**, *9*, 1450–1466. [CrossRef]
353. Jeng, H.A.; Swanson, J. Toxicity of Metal Oxide Nanoparticles in Mammalian Cells. *J. Environ. Sci. Health Part A* **2006**, *41*, 2699–2711. [CrossRef] [PubMed]
354. Karlsson, H.L.; Gustafsson, J.; Cronholm, P.; Möller, L. Size-Dependent Toxicity of Metal Oxide Particles—A Comparison between Nano- and Micrometer Size. *Toxicol. Lett.* **2009**, *188*, 112–118. [CrossRef] [PubMed]

355. Karlsson, H.L.; Cronholm, P.; Gustafsson, J.; Möller, L. Copper Oxide Nanoparticles Are Highly Toxic: A Comparison between Metal Oxide Nanoparticles and Carbon Nanotubes. *Chem. Res. Toxicol.* **2008**, *21*, 1726–1732. [CrossRef] [PubMed]
356. Singh, N.; Jenkins, G.J.S.; Asadi, R.; Doak, S.H. Potential Toxicity of Superparamagnetic Iron Oxide Nanoparticles (SPION). *Nano Rev.* **2010**, *1*, 5358. [CrossRef]
357. Valdiglesias, V.; Kiliç, G.; Costa, C.; Fernández-Bertólez, N.; Pásaro, E.; Teixeira, J.P.; Laffon, B. Effects of Iron Oxide Nanoparticles: Cytotoxicity, Genotoxicity, Developmental Toxicity, and Neurotoxicity. *Environ. Mol. Mutagen.* **2015**, *56*, 125–148. [CrossRef]
358. Nel, A.; Xia, T.; Mädler, L.; Li, N. Toxic Potential of Materials at the Nanolevel. *Science* **2006**, *311*, 622–627. [CrossRef]
359. Shubayev, V.I.; Pisanic, T.R.; Jin, S. Magnetic Nanoparticles for Theragnostics. *Adv. Drug Deliv. Rev.* **2009**, *61*, 467–477. [CrossRef]
360. Unfried, K.; Albrecht, C.; Klotz, L.-O.; Von Mikecz, A.; Grether-Beck, S.; Schins, R.P.F. Cellular Responses to Nanoparticles: Target Structures and Mechanisms. *Nanotoxicology* **2007**, *1*, 52–71. [CrossRef]
361. Malhotra, N.; Lee, J.-S.; Liman, R.A.D.; Ruallo, J.M.S.; Villaflores, O.B.; Ger, T.-R.; Hsiao, C.-D. Potential Toxicity of Iron Oxide Magnetic Nanoparticles: A Review. *Molecules* **2020**, *25*, 3159. [CrossRef] [PubMed]
362. Juan, C.A.; Pérez de la Lastra, J.M.; Plou, F.J.; Pérez-Lebeña, E. The Chemistry of Reactive Oxygen Species (ROS) Revisited: Outliningtheir Role in Biological Macromolecules (DNA, Lipids and Proteins) and Induced Pathologies. *Int. J. Mol. Sci.* **2021**, *22*, 4642. [CrossRef] [PubMed]
363. Wu, J.; Ding, T.; Sun, J. Neurotoxic Potential of Iron Oxide Nanoparticles in the Rat Brain Striatum and Hippocampus. *Neurotoxicology* **2013**, *34*, 243–253. [CrossRef]
364. Soenen, S.J.H.; Himmelreich, U.; Nuytten, N.; De Cuyper, M. Cytotoxic Effects of Iron Oxide Nanoparticles and Implications for Safety in Cell Labelling. *Biomaterials* **2011**, *32*, 195–205. [CrossRef] [PubMed]
365. Pinzaru, I.; Coricovac, D.; Dehelean, C.; Moacă, E.-A.; Mioc, M.; Baderca, F.; Sizemore, I.; Brittle, S.; Marti, D.; Calina, C.D.; et al. Stable PEG-Coated Silver Nanoparticles—A Comprehensive Toxicological Profile. *Food Chem. Toxicol.* **2018**, *111*, 546–556. [CrossRef] [PubMed]
366. Samuel, M.S.; Ravikumar, M.; John, A.J.; Selvarajan, E.; Patel, H.; Chander, P.S.; Soundarya, J.; Vuppala, S.; Balaji, R.; Chandrasekar, N. A Review on Green Synthesis of Nanoparticles and Their Diverse Biomedical and Environmental Applications. *Catalysts* **2022**, *12*, 459. [CrossRef]
367. Joseph, T.M.; Mahapatra, D.K.; Esmaeili, A.; Piszczyk, Ł.; Hasanin, M.S.; Kattali, M.; Haponiuk, J.; Thomas, S. Nanoparticles: Taking a Unique Position in Medicine. *Nanomaterials* **2023**, *13*, 574. [CrossRef]
368. Anjum, S.; Ishaque, S.; Fatima, H.; Farooq, W.; Hano, C.; Abbasi, B.H.; Anjum, I. Emerging Applications of Nanotechnology in Healthcare Systems: Grand Challenges and Perspectives. *Pharmaceuticals* **2021**, *14*, 707. [CrossRef]
369. Zhang, N.; Xiong, G.; Liu, Z. Toxicity of Metal-Based Nanoparticles: Challenges in the Nano Era. *Front. Bioeng. Biotechnol.* **2022**, *10*, 1001572. [CrossRef]
370. Murali, M.; Gowtham, H.G.; Shilpa, N.; Singh, S.B.; Aiyaz, M.; Sayyed, R.Z.; Shivamallu, C.; Achar, R.R.; Silina, E.; Stupin, V.; et al. Zinc Oxide Nanoparticles Prepared through Microbial Mediated Synthesis for Therapeutic Applications: A Possible Alternative for Plants. *Front. Microbiol.* **2023**, *14*, 1227951. [CrossRef]

pharmaceuticals

Article

In Vitro, Oral Acute, and Repeated 28-Day Oral Dose Toxicity of a Mixed-Valence Polyoxovanadate Cluster

Mariana de M. Barbosa [1], Lidiane M. A. de Lima [2], Widarlane A. da S. Alves [1], Eucilene K. B. de Lima [1], Luzia A. da Silva [1*], Thiago D. da Silva [1], Kahoana Postal [3], Mohammad Ramadan [4], Kateryna Kostenkova [4], Dayane A. Gomes [1], Giovana G. Nunes [3], Michelly C. Pereira [1], Wagner E. da Silva [2], Mônica F. Belian [2], Debbie C. Crans [4,5,*] and Eduardo C. Lira [1,*]

1 Centro de Biociências, Departamento de Fisiologia e Farmacologia, Universidade Federal de Pernambuco, Recife 50670-901, PE, Brazil; mariana.melob@ufpe.br (M.d.M.B.); widarlane.angela@ufpe.br (W.A.d.S.A.); kelitabl@gmail.com (E.K.B.d.L.); luzia.abilio@ufrpe.br (L.A.d.S.); thiago.douberin@ufpe.br (T.D.d.S.); dayanejooh@gmail.com (D.A.G.); michelly.pereira@ufpe.br (M.C.P.)

2 Departamento de Química, Universidade Federal Rural de Pernambuco, Recife 52171-900, PE, Brazil; lidianelimaa@gmail.com (L.M.A.d.L.); wesdqr@gmail.com (W.E.d.S.); mfbelian@gmail.com (M.F.B.)

3 Centro Politécnico, Departamento de Química, Universidade Federal do Paraná, Curitiba 81530-900, PR, Brazil; kahoanapostal@gmail.com (K.P.); nunesgg@ufpr.br (G.G.N.)

4 Department of Chemistry, Colorado State University, Fort Collins, CO 80523, USA; mramadan@rams.colostate.edu (M.R.); kostenk@rams.colostate.edu (K.K.)

5 Cell and Molecular Biology Program, Colorado State University, Fort Collins, CO 80523, USA

* Correspondence: debbie.crans@colostate.edu (D.C.C.); eduardo.carvalholira@ufpe.br (E.C.L.); Tel.: +1-(970)-491-7635 (D.C.C.); +55-2126-8000 (E.C.L.); Fax: +1-(970)-491-1801 (D.C.C.); +55-2126-8000 (E.C.L.)

Abstract: Polyoxovanadates (POV) are a subgroup of polyoxometalates (POM), which are nanosized clusters with reported biological activities. This manuscript describes the first toxicity evaluation of a mixed-valence polyoxovanadate, pentadecavanadate, $(Me_4N)_6[V_{15}O_{36}Cl]$, abbreviated as V_{15}. Cytotoxicity experiments using peripheral blood mononuclear cells (PBMC), larvae of *Artemia salina* Leach, and in vivo oral acute and repeated 28-day doses in mice was carried out. The LC_{50} values in PBMC cells and *A. salina* were 17.5 ± 5.8 µmol L^{-1}, and 17.9 µg L^{-1}, respectively, which indicates high cytotoxic activity. The toxicity in mice was not observed upon acute exposure in a single dose, however, the V_{15} repeated 28-day oral administration demonstrated high toxicity using 25 mg/kg, 50 mg/kg and, 300 mg/kg doses. The biochemical and hematological analyses during the 28-day administration of V_{15} showed significant alteration of the metabolic parameters related to the kidney and liver, suggesting moderate toxicity. The V_{15} toxicity was attributed to the oxidative stress and lipid peroxidation, once thiobarbituric acid (TBAR) levels significantly increased in both males and females treated with high doses of the POV and also in males treated with a lower dose of the POV. This is the first study reporting a treatment-related mortality in animals acutely administrated with a mixed-valence POV, contrasting with the well-known, less toxic decavanadate. These results document the toxicity of this mixed-valence POV, which may not be suitable for biomedical applications.

Keywords: polyoxovanadates; pentadecavanadate; mixed valence; toxicology; oral acute toxicity

1. Introduction

Polyoxovanadates (POVs) belong to a subclass of polyanionic molecules made up by group 5 and 6 metal-oxo clusters referred to as polyoxometalates or polyoxidometalates (POM) [1–4]. Several studies describing POVs have demonstrated anticancer, antimicrobial, and antiviral applications [5–10]. Moreover, POVs trigger many biochemical effects such as lipoperoxidation and oxidative stress, affecting metabolic pathways and cell cycle arrest. It interferes with ions transport systems and apoptosis, induces cell morphology changes, inhibits phosphorylases and redox enzymes, and activates or deactivates cell signaling [11–13]. Regarding its future clinical use, POV's toxic effects on various biological systems remain a topic in need of more experimental data. In this work, we followed the guidelines from the Organization of Economic Co-operation and Development (OECD) for measuring the acute oral toxicity and long-term toxicity of chemicals [14]. This study deviates from biological studies showing inhibition of diseased cells or simply measuring inhibition of growth or other biological activities [15]. There are currently few studies in the literature using the OECD guidelines for POVs or other vanadium compounds on healthy cells, which is the focus of this work [16–18].

Among POVs, decavanadate ($[H_nV_{10}O_{28}{}^{(6-n)-}]$, abbreviated V_{10}) is the most commonly investigated POV regarding its chemical, physical, and biological properties. It is found to be stable at acidic pH (pH 3–6) and has shown biological effects, such as inducing oxidative stress processes, stimulating enzymes, and interfering in lipid structures and cellular function [19–22]. It also has effects as an antidiabetic and anticancer agent [23]. Recently, the transition-metal-monosubstituted decavanadates V_9Mo and V_9Pt were found to be effective in inhibiting the growth of Mycobacterium smegmatis [24] and Chinese Hamster Ovary (CHO) cells [12]. In both studies, the V_{10} cluster was reported to be a more effective growth inhibitor; however, in bacteria, the cluster is bound more tightly and hence more cell associated than in the mammalian system, where V_9Mo and V_9Pt are readily washed off [12,24].

Mixed-valence polyoxovanadates (MVPs) have contributed to an increase in the number of potential therapeutic agents for metabolic diseases and cancer [25–28]. The MVP $[(Me_4N)_6[V_{15}O_{36}Cl]$ is a polyoxovanadate (abbreviated V_{15}, see Figure 1) that contains six Me_4N^+ counterions, and the vanadium-oxygen cage of V_{15} consists of one central chlorine atom, eight vanadium(IV) and seven vanadium(V) atoms, and 36 oxygen atoms (Figure 1). V_{15} was effective against the deleterious effect of the deoxyribonucleic acid (DNA) plasmid pUC19 alkylating agents diethyl sulphate and dimethyl sulphate [29]. A chemoprotective effect of 30–40% against diethyl sulphate was additionally shown when using a more complex model of *Escherichia coli* DH5α cell cultures [30]. V_{15} has also demonstrated inhibitory growth effects on microorganisms including *Mycobacterium smegmatis* [31]. Furthermore, the luteinizing hormone receptor (LHR), a G-protein-coupled receptor (GPCR), was used to evaluate V_{15} and other POVs in their interaction with the cell membrane lipid interface of CHO cells [9]. Cell responses for all MVPs were greater than those seen for cells treated with the human chorionic gonadotropin (hCG) hormone, which was used in reference [9]. Despite the promising biological effects described for V_{15}, to the best of our knowledge, no evidence of toxicity (in vivo assays) was reported [2,3,29–32].

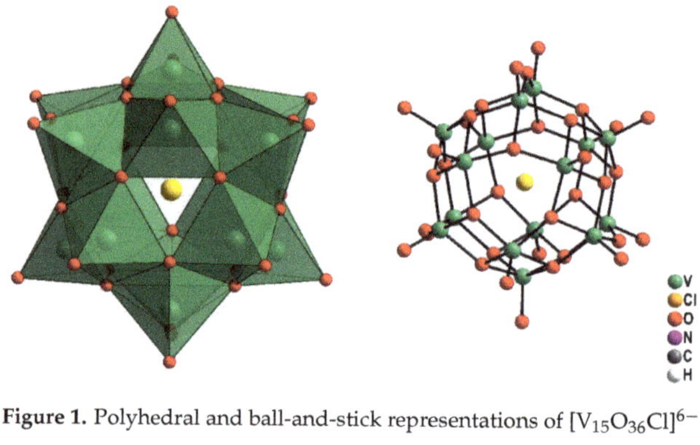

Figure 1. Polyhedral and ball-and-stick representations of $[V_{15}O_{36}Cl]^{6-}$ (V_{15}) anion.

Although several POV compounds have demonstrated favorable biological properties, many aspects of the toxicity remain to be further investigated [33]. The toxicological effects of vanadium compounds in biological systems are of great concern and have been well documented in in vitro assays [34–37]. However, less information on vanadium compounds and POV toxicity is available for in vivo systems, which is critical for any potential biomedical applications [26,27].

Some animal studies have demonstrated that toxicological effects of vanadium compounds are related to the oxidovanadate ion, complex nature, dose, and administration route [35,36,38]. Cellular assays suggest some levels of toxicity are associated with the vanadium compounds due to their commonly observed redox active properties and hence potential for causing oxidative tissue damage, which can be produced by both reactive oxygen species (ROS) and reactive nitrogen species (RNS) [39]. Excess ROS and RNS species result in extensive damage to tissues, nucleic acids, lipids, and proteins through the depletion of sulfhydryl groups, lipids, and reactive parts of biomolecules and their metabolites [39,40].

In this study, toxicological evaluations in both in vitro and in vivo models have been performed to determine any potential human health risk of this multivalent POV cluster, herein abbreviated V_{15}. This is the first report of oral single-dose toxicity of a mixed-valence POV in mice.

2. Results

2.1. Preparation and Characterization of the Pentadecavanadate V_{15}

The synthesis of $[(CH_3)_4N]_6[V^{IV}{}_8V^{V}{}_7O_{36}(Cl)]$ (V_{15}) was first described by Muller [41]. More recently, a methodology that uses cheaper and commercially available reactants was developed by Nunes and coworkers [29]. Aiming to produce enough V_{15} to perform in vivo studies, some modifications were made to the previously described procedure to increase scale. Special care must be taken with the control of temperature, because the fresnoite type oxide $(NH_4)_2VO_3$ is the only product obtained when the reaction takes place at 70 °C [29], and V_{15} can be produced as the major product under reflux.

The material was characterized in both solid and solution states, with the latter studies being performed in aqueous solution and under the conditions of biological studies. Solid-state characterization includes simulated and experimental powder X-ray diffraction studies. The pattern in powder X-ray diffraction experiments of V_{15} was compared to previously reported single-crystal X-ray diffraction structure, showing comparable results (Figure S1, Supplementary Materials). In addition, the FTIR (infrared) spectrum of V_{15} was recorded and showed characteristic bands of the polyoxoanion [42] (Figure S2, Supplementary Materials). The EPR (electronic paramagnetic resonance) spectrum of a pulverized sample of V_{15} was collected at 77 K and found to show the expected broad line centered at 339 mT (g = 1.963 and with Δ_{p-p} of 35 mT, assigned to a $\Delta MS = \pm 1$ transition (Figure S3, Supplementary Materials). This pattern confirms a partial reduction of the V(V) to V(IV) and is compatible with the intra- and intermolecular exchange interactions between vanadium(IV) centers present in the $[V^{IV}{}_8V^{V}{}_7O_{36}(Cl)]^{6-}$ polyoxoanion [29,43].

Solution studies included both ^{51}V NMR (nuclear magnetic resonance) and EPR studies, as well as UV-Vis spectroscopy. The multivalence of the V_{15} anion affects the ^{51}V NMR spectra, since the intact V_{15} is not observed because of the unpaired electrons of the V^{IV}-atoms in the cluster. The ^{51}V NMR and EPR spectra were recorded in aqueous solutions by dissolving the cluster in water and then adding phosphate-buffered saline (PBS pH 7.4) containing 1.0 mmol L^{-1} $CaCl_2$ and 0.50 mmol L^{-1} $MgCl_2$. The spectroscopic signature of V_{15} was monitored in a time-dependent study up to 34 h [19] and showed that some V_{15} was hydrolyzed in an aqueous solution and in media to give a mixture of the intact and oxidized V_{15}, other oxidovanadium(V) species ($H_2VO_4^-$, $H_2V_2O_7^{2-}$, $V_4O_{12}^{4-}$, $V_5O_{15}^{5-}$), and the oxidovanadium(IV) complex $[VO_2(OH_2)_4]^+$. Other speciation studies were carried out in Luria–Bertani (LB) broth in the presence of other additives at lower pH values (down to 6.0) [44].

The electronic absorption UV-Vis spectra of a 0.025 mmol L^{-1} aqueous solution of V_{15} showed a ligand-to-metal charge transfer (LMCT, p(O)→d(V)) band below 400 nm, and a broad band that extended from 600 nm to the near-infrared was attributed to an intervalence charge transfer transition (IVCT, $V^{IV} \rightarrow V^{V}$) in the POV anion. The spectrum did not change over a period of 24 h, which is consistent with V_{15} remaining stable in solution (Figure S4, Supplementary Materials).

2.2. *In Vitro Toxicity*

V_{15} with the formula $(Me_4N)_6[V_{15}O_{36}Cl]$ was screened for its cytotoxicity using peripheral blood mononuclear cells (PMBC) using a 3-(4,5-dimethyl-2-thiazolyl)-2,5-diphenyl-2H-tetrazolium bromide (MTT) cytotoxicity assay and a brine shrimp bioassay. The PMBC lethality assay measured in vitro toxicity of V_{15}, as shown in Figure 2A at 50 mol·L^{-1}. The inhibition stabilized, reaching 50%. In this assay, PBMCs showed decreasing viability at all concentrations tested (Figure 2A). Mean values of viability were 70.43 ± 3.4% at a concentration of 10 μM; 48.61 ± 3.3% at 25 μM; 41.79 ± 5.1% at 50 μM; and 28.06 ± 3.6% at 100 μM, with a dose-dependent reduction in viability. The IC_{50} value of V_{15} was 12.95 μM (10.99–14.91 μmol·L^{-1}, R^2 = 0.9440), indicating a significant inhibition of cell viability (Figure 2A).

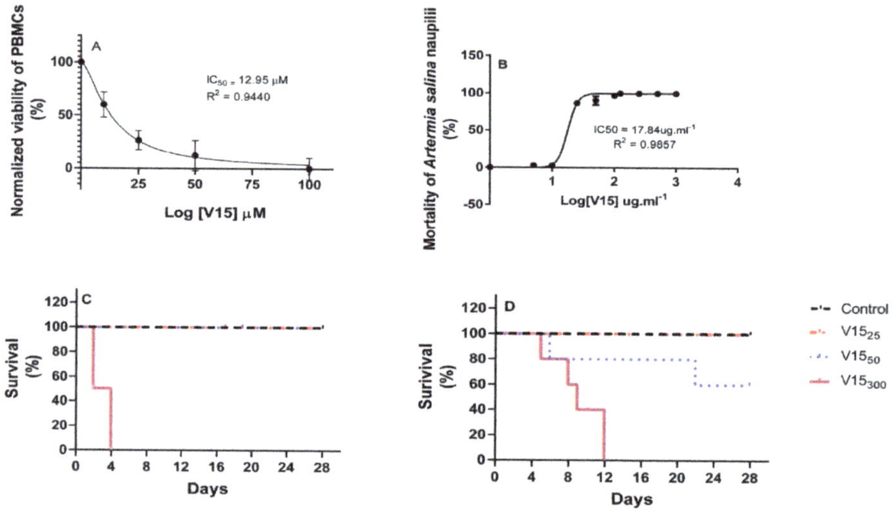

Figure 2. Toxicity effect of the V_{15} on PBMC viability (**A**), *A. saline* nauplii mortality (**B**), female (**C**) and (**D**) male Kaplan–Meier curve of Swiss mice treated with repeated 28-day oral dose. Mean value of triplicate experiments for each countertraction ± SEM (standard error of the mean).

The percentage of mortality in the *A. salina* nauplii response increased as the POV concentration increased (Figure 2B). The V_{15} showed a high cytotoxic activity against *A. salina*, with an IC_{50} value of 17.84 (15.75–20.21 μg·mL^{-1}, R^2 = 0.9857), causing 100% mortality in brine shrimp nauplii at a concentration of 125 μg mL^{-1}. See below for discussion of the results shown in Figure 2C,D.

2.3. *Acute Oral Toxicity of V_{15}*

In this study, the potential for mortality by acute treatment with a POV in animals was evaluated. The animals were examined for abnormal clinical signs and macroscopic findings, all of which were not observed. V_{15} induced a transitory behavioral alteration such as piloerection and lethargy immediately after dosing in the treated animals. The body weight gain, food, and organ mass were not adversely affected by a single oral V_{15} dose. As shown in Table 1, minor differences were observed, and the fluid intake increased by 20% ($p < 0.05$) in the V_{15}-treated mice at 300 and 2000 mg/kg treatment levels. The lung weight increased by 25% ($p < 0.05$) for the 300 and 2000 mg/kg levels in the V_{15}-treated groups, and liver weight increased by ~10% ($p < 0.05$) only at a higher treatment dose compared to the untreated control group (Table 2). The biochemical analysis of the biological assay is shown in Table 3. V_{15} treatment caused an increase in the aspartate aminotransferase (AST) by 50% ($p < 0.05$) in the 50 mg/kg group and at 2.0-fold ($p < 0.05$) in the 300 and 2000 mg/kg groups. The alanine aminotransferase (ALT) levels were increased 2.5-fold ($p < 0.05$) for the 300 and 2000 mg/kg groups compared to untreated mice. On the other hand, V_{15} reduced albumin content by ~25% ($p < 0.05$) in mice treated with higher doses of V_{15} compared with the control group, without any alteration in blood urea nitrogen (BUN) and total protein levels.

Table 1. Effects of the acute oral administration of V_{15} on total body weight (g), food intake (mL), and fluid intake (mL) after 14 days in different female mice groups.

Parameters	Control	V15–50	V15–300	V15–2000
Total body weight (g)	470.5 ± 11.7 ^{a}	484.3 ± 15.8 ^{a}	463.0 ± 14.8 ^{a}	503.0 ± 20.0 ^{a}
Total food intake (g)	96.5 ± 3.8 ^{a}	89.7 ± 4.1 ^{a}	102.7 ± 4.2 ^{a}	103.3 ± 1.9 ^{a}
Total fluid intake (mL)	194.0 ± 9.9 ^{a}	189.3 ± 3.7 ^{a}	234.0 ± 4.5 ^{b}	235.0 ± 9.5 ^{b}

The data are presented as the mean ± SEM (n = 3). The values with different superscript letters are statistically significant at $p < 0.05$ compared to control group, using one-way ANOVA.

Table 2. Effects of the acute oral administration of V_{15} on relative organ weight (mg/100 g of body weight) after 14 days in different female mice groups.

Organ	Control	$V_{15–50}$	$V_{15–300}$	$V_{15–2000}$
Heart	0.56 ± 0.13 ^{a}	0.69 ± 0.26 ^{a}	0.54 ± 0.06 ^{a}	0.57 ± 0.02 ^{a}
Lung	0.54 ± 0.01 ^{a}	0.58 ± 0.07 ^{a}	0.72 ± 0.05 ^{b}	0.68 ± 0.07 ^{b}
Liver	5.93 ± 0.19 ^{a}	5.15 ± 0.15 ^{a}	5.94 ± 0.18 ^{a}	6.70 ± 0.29 ^{b}
Kidney	1.46 ± 0.04 ^{a}	1.29 ± 0.03 ^{a}	1.38 ± 0.036 ^{a}	1.38 ± 0.04 ^{a}
Spleen	0.54 ± 0.02 ^{a}	0.46 ± 0.04 ^{a}	0.64 ± 0.03 ^{a}	0.66 ± 0.04 ^{a}
Tibial anterior	0.13 ± 0.01 ^{a}	0.15 ± 0.02 ^{a}	0.12 ± 0.01 ^{a}	0.13 ± 0.01 ^{a}
Soleus	0.024 ± 0.002 ^{a}	0.021 ± 0.001 ^{a}	0.016 ± 0.002 ^{a}	0.018 ± 0.001 ^{a}

The data are presented as the mean ± SEM (n = 3). The values with different superscript letters are statistically significant at $p < 0.05$ compared to control group, using one-way ANOVA.

Table 3. Effects of the acute oral administration of V_{15} on serum biochemical profile after 14 days in different female mice groups.

Parameters	Control	$V_{15–50}$	$V_{15–300}$	$V_{15–2000}$
AST (mmol/L)	58.5 ± 1.1 ^{a}	87.0 ± 3.1 ^{b}	117.9 ± 15.4 ^{c}	139.7 ± 19.4 ^{c}
ALT (mmol/L)	66.1 ± 5.3 ^{a}	59.9 ± 0.6 ^{a}	141.6 ± 8.9 ^{b}	165.5 ± 14.4 ^{b}
Total protein (mg/dL)	45.6 ± 4.2 ^{a}	41.0 ± 1.1 ^{a}	49.5 ± 2.5 ^{a}	48.0 ± 3.6 ^{a}
Albumin (µmol/L)	314.0 ± 29.0 ^{a}	357.4 ± 4.8 ^{a}	236.8 ± 11.3 ^{b}	218.7 ± 5.4 ^{c}
BUN (mg/dL)	11.2 ± 0.6 ^{a}	11.2 ± 0.6 ^{a}	11.2 ± 0.7 ^{a}	10.5 ± 0.4 ^{a}

The data are presented as the mean ± SEM (n = 5). Values with different superscript letters are statistically significant at $p < 0.05$ compared to control group, using one-way ANOVA. Abbreviations used are: ALT—alanine aminotransferase; AST—aspartate aminotransferase; BUN—blood urea nitrogen.

The V_{15} cluster has an acute oral lethal dose (LD_{50}) for female mice estimated above the 2000 mg/kg body weight based on an analysis of the dose–mortality relationship. Thus, this POV has a low order of acute toxicity by oral administration in mice, making it a category 5 of acute toxicity material [45].

2.4. *Repeated-Dose 28-Day Oral Toxicity Study in Rodents*

The effects of repeated doses were studied during the 28-day treatment. During the entire experimental period, mice were placed under observation, and signs of toxicity were recorded. No treatment-related changes in the external appearance or general condition were observed in either male or female mice treated with a 25 or a 50 mg/kg b.w. daily dose. The results are displayed in a Kaplan–Meier curve of both Swiss female (Figure 2C) and male mice (Figure 2D). All animals dosed at 300 mg/kg of V_{15} did not die until the 12th day of the treatment. Furthermore, 20% of female mice did not die at the 50 mg/kg level of V_{15} until the 22nd day (Figure 2C). At the high dose, the female mice only survived for 4 days. The survival for the female mice was 100% until the mice were treated with 100 μmol L^{-1}; the survival for the male mice was 100% until the mice were treated with 50 or 100 μmol L^{-1}, at which point the survival rate was 20 or 8 days. The most common symptoms were diarrhea, lethargy, and weight loss. Body weight and fluid intake of all mice treated with 25 and 50 mg/kg of V_{15} did not show any alteration compared to the untreated control group (Figure 3). However, male and female mice treated with 50 mg/kg had fluctuations in food intake during dosing (Figure 3), but the total food intake in this group was slightly less than that of the $V_{15\text{-}25}$ and control groups (Table 4).

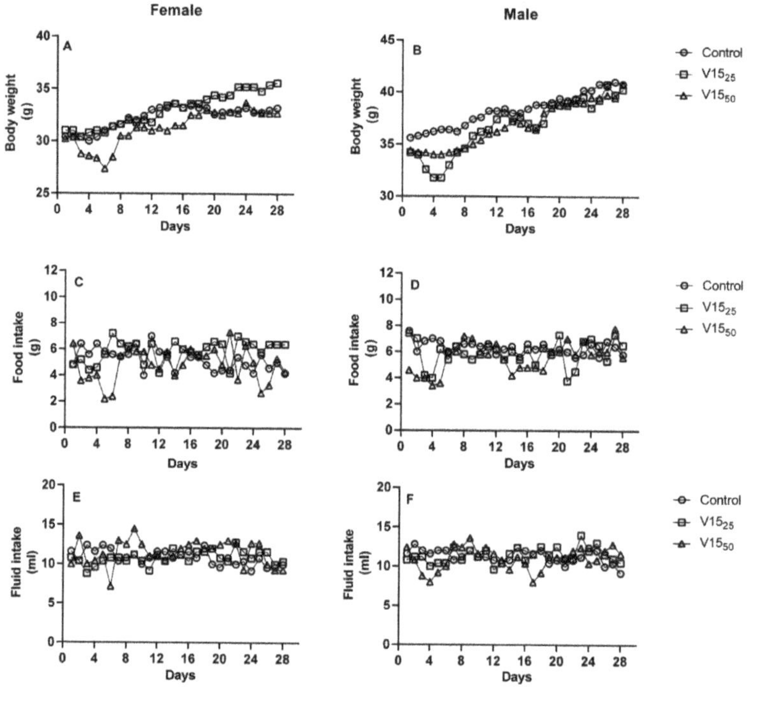

Figure 3. Effect of the 28-day repeated-dose treatment with V_{15} on body weight, food intake, and fluid intake in female (**A**,**C**,**E**) and male (**B**,**D**,**F**) mice. Mean values are presented as mean ± SEM. Mean value with different superscript letters is statistically different at $p < 0.05$ and was analyzed using a one-way ANOVA.

The necropsy did not reveal any gross pathological changes compared to the corresponding control groups. In addition, male mice did not show any effects on relative organ weight after V_{15} treatment, except for a slight reduction in teste (20%, $p < 0.05$) and kidney (15%, $p < 0.05$) mass at 50 mg/kg. In female mice, a reduction in tibial (~40%, $p < 0.05$) and reproductive organ mass (uterus and ovaries ~30%, $p < 0.05$) was observed at both doses evaluated (Table 5).

Table 4. Accumulative effects of the oral repeated administration of V_{15} on total body weight (g), food intake (g), and fluid intake (mL) in different male and female mice groups for 28-day toxicity evaluation.

Male			
Parameters	**Control**	V_{15-25}	V_{15-50}
Body weight (g)	1071.80 ± 21.8 [a]	1.08000 ± 12.1 [a]	1.07000 ± 29.2 [a]
Food intake (g)	177.0 ± 6.6 [a]	170.8 ± 2.8 [a]	155.6 ± 5.4 [b]
Fluid intake (mL)	316.0 ± 5.89 [a]	327.5 ± 14.2 [a]	308.40 ± 10.5 [a]
Female			
Parameters	**Control**	V_{15-25}	V_{15-50}
Body weight (g)	941.20 ± 28.7 [a]	928.80 ± 15.4 [a]	912.70 ± 15.5 [a]
Food intake (g)	146.8 ± 6.9 [a]	154.2 ± 5.7 [a]	136.3 ± 5.4 [b]
Fluid intake (mL)	303.6 ± 9.0 [a]	303.2 ± 18.1 [a]	322.7 ± 3.5 [a]

The data are presented as the mean ± SEM (n = 5). The values with different superscript letters are statistically significant at $p < 0.05$ compared to control group, using one-way ANOVA.

Table 5. Toxicity evaluation of the oral repeated administration of V_{15} on relative organ weight (mg/100 g of body weight) in different male and female mice groups after 28 days.

Male Mice			
Organ (mg/100 g of b.w.)	**Control**	V_{15-25}	V_{15-50}
Brain	1.13 ± 0.03 [a]	1.10 ± 0.06 [a]	1.05 ± 0.03 [a]
Heart	0.41 ± 0.01 [a]	0.51 ± 0.11 [a]	0.45 ± 0.01 [a]
Lung	0.62 ± 0.02 [a]	0.63 ± 0.05 [a]	0.52 ± 0.06 [a]
Liver	4.69 ± 0.24 [a]	5.41 ± 0.10 [a]	4.99 ± 0.66 [a]
Kidney	1.41 ± 0.05 [a]	1.44 ± 0.08 [a]	1.22 ± 0.08 [b]
Spleen	0.35 ± 0.04 [a]	0.42 ± 0.07 [a]	0.39 ± 0.02 [a]
Tibial anterior	0.15 ± 0.01 [a]	0.14 ± 0.01 [a]	0.13 ± 0.01 [a]
Gastrocnemius	0.36 ± 0.01 [a]	0.36 ± 0.02 [a]	0.34 ± 0.02 [a]
Teste	0.74 ± 0.08 [a]	0.64 ± 0.08 [b]	0.60 ± 0.01 [b]
Female mice			
Organ (mg/100 g of b.w.)	**Control**	V_{15-25}	V_{15-50}
Brain	1.24 ± 0.09 [a]	1.23 ± 0.04 [a]	1.25 ± 0.03 [a]
Heart	0.48 ± 0.04 [a]	0.41 ± 0.01 [a]	0.39 ± 0.02 [a]
Lung	0.67 ± 0.09 [a]	0.64 ± 0.05 [a]	0.70 ± 0.05 [a]
Liver	5.03 ± 0.22 [a]	5.30 ± 0.23 [a]	5.78 ± 0.05 [a]
Kidney	1.34 ± 0.06 [a]	1.16 ± 0.04 [a]	1.14 ± 0.05 [a]
Spleen	0.47 ± 0.04 [a]	0.48 ± 0.04 [a]	0.50 ± 0.05 [a]
Tibial anterior	0.24 ± 0.06 [a]	0.14 ± 0.01 [b]	0.13 ± 0.01 [b]
Gastrocnemius	0.25 ± 0.01 [a]	0.28 ± 0.01 [a]	0.31 ± 0.03 [a]
Ovaries and uterus	1.32 ± 0.07 [a]	0.94 ± 0.04 [b]	0.92 ± 0.07 [b]

The data are presented as the mean ± SEM (n = 5 for both male and female mice). Values with different superscript letters are statistically significant at $p < 0.05$ compared to control group, using one-way ANOVA.

Hematological values of mice treated orally with V_{15} for 28 days are shown in Table 6. In this study, a significant alteration was observed in some hematimetric indices in both male and female mice. V_{15} treatment reduced hemoglobin (~15%, $p < 0.05$), MCV (~20%, $p < 0.05$), and HCT in a dose-dependent manner (25 and 40%, respectively) and stimulated platelet content in a dose-dependent manner (~45 and ~80%, respectively) in treated male mice groups compared to untreated mice. In female mice treated with V_{15} daily for 28 days, a reduction in hemoglobin (~20%, $p < 0.05$), MCV (~15%, $p < 0.05$), and MCH (~15%, $p < 0.05$) was observed after the treatment with both doses, and a reduction in HCT (~70%, $p < 0.05$) was detected only at a higher treatment dose. V_{15} treatment also stimulated platelet content in a dose-dependent manner (~1.5 and ~2-fold, respectively). In addition, V_{15} reduced the WBC (~50%, $p < 0.05$) at a higher dose, and it increased monocytes as the dose increased compared to untreated female mice (Table 6). For other hematological parameters, no significant differences were detected between treated male mice and the control group.

Table 6. Evaluation of the effects of the repeated administration of V_{15} on hematological profile in different male and female mice groups after 28 days.

Male Mice			
Parameters	**Control**	V_{15-25}	V_{15-50}
RBC ($\times 10^3/\mu$)	6.80 ± 0.50 [a]	6.90 ± 0.50 [a]	5.50 ± 0.70 [a]
Hemoglobin (%)	15.20 ± 1.30 [a]	12.50 ± 0.50 [b]	13.30 ± 0.60 [b]
MCV (μm^3)	58.50 ± 2.70 [a]	46.70 ± 0.60 [b]	47.70 ± 0.40 [b]
MCH (pg)	20.60 ± 0.70 [a]	18.70 ± 0.70 [a]	17.40 ± 3.20 [b]
MCHC (g/dL)	35.50 ± 1.80 [a]	38.60 ± 0.10 [a]	37.40 ± 6.80 [a]
HCT (%)	43.40 ± 4.70 [a]	32.20 ± 1.90 [b]	26.40 ± 3.20 [c]
Platelet count ($\times 10^3/\mu$)	531.00 ± 86.94 [a]	778.00 ± 23.35 [b]	942.75 ± 17.33 [c]
WBC ($\times 10^3/\mu$)	14.00 ± 0.70 [a]	13.30 ± 0.10 [a]	14.50 ± 0.40 [a]
Monocytes (%)	6.90 ± 1.00 [a]	7.70 ± 0.30 [a]	7.60 ± 2.50 [a]
Lymphocyte (%)	89.50 ± 2.10 [a]	89.00 ± 0.30 [a]	87.40 ± 4.00 [a]
Female mice			
Parameters	**Control**	V_{15-25}	V_{15-50}
RBC ($\times 10^3/\mu$)	14.80 ± 0.50 [a]	13.90 ± 0.20 [a]	15.2 ± 0.20 [a]
Hemoglobin (%)	13.80 ± 0.60 [a]	11.20 ± 0.80 [b]	11.40 ± 0.70 [b]
MCV (μm^3)	56.80 ± 5.40 [a]	47.30 ± 0.80 [b]	42.40 ± 0.40 [c]
MCH (pg)	20.90 ± 1.10 [a]	15.10 ± 0.40 [b]	15.30 ± 7.00 [c]
MCHC (g/dL)	37.30 ± 1.60 [a]	35.30 ± 2.40 [a]	33.50 ± 17.50 [a]
HCT (%)	37.30 ± 2.30 [a]	31.80 ± 0.40 [a]	11.4 ± 5.80 [c]
Platelet count ($\times 10^3/\mu$)	438.82 ± 13.82 [a]	499.00 ± 14.47 [b]	911.66 ± 16.23 [b]
WBC ($\times 10^3/\mu$)	6.70 ± 0.60 [a]	7.40 ± 0.60 [a]	3.60 ± 0.90 [b]
Monocytes (%)	3.50 ±0.60 [a]	4.90 ± 0.60 [a]	6.10 ± 1.30 [b]
Lymphocyte (%)	88.30 ± 4.70 [a]	91.1 ± 0.60 [a]	90.40 ± 2.40 [a]

The data are presented as the mean ± SEM (n = 5). Values with different superscript letters are statistically significant at $p < 0.05$ compared to control group, using one-way ANOVA. Abbreviations used are: HCT—hematocrit; MCH—mean corpuscular hemoglobin; MCV—mean corpuscular volume; MCHC—mean corpuscular hemoglobin concentration; RBC—red blood cell; WBC—white blood cells.

The data shown in Table 7 indicate that V_{15} caused an increase in AST levels in male (~30%, $p < 0.05$) and female (95%, $p < 0.05$) mice. Creatinine was increased only in male mice treated with 50 mg/kg of V_{15} (2.5-fold $p < 0.05$), whereas uric acid levels increased in both male and female mice treated with 25 mg/kg of V_{15} (50%, $p < 0.05$) and male (70%, $p < 0.05$) and female (~80%, $p < 0.05$) mice groups treated with 50 mg/kg of V_{15} compared with the untreated control group. Considering the lipid profile, V_{15} reduced triglyceride (TG) levels in male mice at a higher V_{15} treatment dose (~20%, $p < 0.05$), and there was a reduction in total cholesterol (TC, ~30%, $p < 0.05$) and TG (~20%, $p < 0.05$) in treated female mice compared with the control group.

Table 7. Effects of the oral administration of V_{15} on serum biochemical profile in different male and female mice groups after 28 days in repeated-dose toxicity evaluation.

Male Mice			
Parameters	**Control**	V_{15-25}	V_{15-50}
ALT (mmol/L)	39.90 ± 1.80 [a]	103.70 ± 5.40 [b]	122.50 ± 8.30 [c]
AST (mmol/L)	74.80 ± 2.40 [a]	81.90 ± 2.90 [a]	145.40 ± 8.30 [b]
Total protein (g/L)	46.60 ± 5.60 [a]	46.00 ± 11.90 [a]	51.60 ± 2.10 [a]
BUN (mg/dL)	12.00 ± 0.20 [a]	13.40 ± 0.90 [a]	11.60 ± 0.01 [a]
Creatinine (μmol/L)	29.80 ± 2.40 [a]	23.86 ± 2.60 [a]	71.90 ± 20.50 [b]
Uric acid (μmol/L)	122.31 ± 15.30 [a]	185.30 ± 17.30 [b]	211.80 ± 32.70 [b]
Total cholesterol (mmol/L)	1.60 ± 0.18 [a]	1.58 ± 0.41 [a]	1.80 ± 0.19 [a]
Triglycerides (mmol/L)	1.62 ± 0.11 [a]	1.84 ± 0.13 [a]	1.30 ± 0.10 [b]

Table 7. *Cont.*

Female mice			
Parameters	**Control**	$V_{15\text{–}25}$	$V_{15\text{–}50}$
ALT (mmol/L)	35.40 ± 2.70 [a]	55.40 ± 3.10 [b]	76.30 ± 12.10 [c]
AST (mmol/L)	60.30 ± 3.10 [a]	73.20 ± 4.70 [a]	76.30 ± 12.10 [b]
Total protein (g/L)	41.80 ± 5.40 [a]	47.60 ± 6.00 [a]	50.70 ± 1.50 [a]
BUN (mg/dL)	11.49 ± 1.10 [a]	13.41 ± 0.28 [a]	11.74 ± 0.22 [a]
Creatinine (μmol/L)	37.53 ± 6.44 [a]	36.60 ± 3.60 [a]	34.50 ± 1.80 [a]
Uric acid (μmol/L)	114.92 ± 19.72 [a]	170.60 ± 42.0 [b]	204.10 ± 68.1 [b]
Total cholesterol (mmol/L)	2.03 ± 0.14 [a]	1.46 ± 0.17 [b]	1.83 ± 0.06 [a]
Triglycerides (mmol/L)	1.50 ± 0.14 [a]	1.21 ± 0.12 [b]	1.40 ± 0.07 [a]

The data are presented as the mean ± SEM (n = 5). Values with different superscript letters are statistically significant at $p < 0.05$ compared to control group, using one-way ANOVA. Abbreviations used are: ALT—alanine aminotransferase; AST—aspartate aminotransferase; BUN—blood urea nitrogen.

2.5. Oxidative Stress

The oxidative stress parameters evaluated were reduced glutathione (GSH), oxidized glutathione (pGSSG), malondialdehyde (MDA), superoxide dismutase (SOD), and catalase (CAT). The effect of the V_{15} compound on the generation of ROS species in the liver of male and female mice treated for 28 consecutive days is shown in Table 8. In male mice, the V_{15} treatment increased hepatic MDA content, and the SOD activity decreased (~30%, $p < 0.05$). On the other hand, in V_{15}-treated female mice, the GSSG levels increased, and the GSH/GSSG ratio decreased, which is consistent with oxidative damage. In addition, at elevated MDA content at a higher treatment dose (~30%, $p < 0.05$), the SOD and CAT activities in female mice were increased compared with the untreated group. The V_{15} treatment clearly impacted female and male mice differently, although evidence of oxidative stress in both male and female mice was apparent.

Table 8. Effects of the repeated oral administration of V_{15} on oxidative stress parameters in different male and female mice groups and after 28 days.

Male Mice			
Parameters	**Control**	$V_{15\text{–}25}$	$V_{15\text{–}50}$
GSH (nmol/mg ptn)	10.2 ± 0.3 [a]	10.4 ± 0.4 [a]	10.1 ± 0.6 [a]
GSSG (nmol/mg ptn)	7.4 ± 0.4 [a]	7.4 ± 0.4 [a]	7.3 ± 0.6 [a]
GSH/GSSG (nmol/mg ptn)	1.4 ± 0.1 [a]	1.4 ± 0.1 [a]	1.4 ± 0.05 [a]
MDA (U/mg ptn)	1.3 ± 0.1 [a]	5.3 ± 0.3 [b]	4.8 ± 0.5 [b]
SOD (U/mg ptn)	19.9 ± 3.0 [a]	14.0 ± 2.2 [b]	15.7 ± 1.3 [b]
CAT (U/mg ptn)	0.023 ± 0.005 [a]	0.027 ± 0.010 [a]	0.028 ± 0.010 [a]
Female mice			
Parameters	**Control**	$V_{15\text{–}25}$	$V_{15\text{–}50}$
GSH (nmol/mg ptn)	6.6 ± 0.2 [a]	7.2 ± 0.2 [a]	6.9 ± 0.2 [a]
GSSG (nmol/mg ptn)	2.4 ± 0.4 [a]	5.0 ± 0.3 [b]	4.2 ± 0.1 [b]
GSH/GSSG (nmol/mg ptn)	2.8 ± 0.5 [a]	1.5 ± 0.1 [b]	1.6 ± 0.1 [c]
MDA (U/mg ptn)	0.9 ± 0.1 [a]	1.0 ± 0.2 [a]	1.3 ± 0.1 [b]
SOD (U/mg ptn)	16.6 ± 3.6 [a]	33.7 ± 3.8 [b]	24.3 ± 1.9 [c]
CAT (U/mg ptn)	0.011 ± 0.003 [a]	0.029 ± 0.008 [b]	0.026 ± 0.004 [b]

The data are presented as the mean ± SEM (n = 5). The values with different superscript letters are statistically significant at $p < 0.05$ compared to control group, using one-way ANOVA. Abbreviations used are: GSH—glutathione reduced; GSSG—glutathione oxidized; GSH/GSSG—ratio between GSH and GSSG; MDA—malondialdehyde; SOD—superoxide dismutase; CAT—catalase.

Alterations in mice liver and kidneys after treatment were confirmed by hematoxylin and eosin (H&E) histology analysis in $V_{15\text{–}25}$ and $V_{15\text{–}50}$ groups, as well as by comparison to the untreated mice group. The histopathological findings shown in Figures 4 and 5 include vacuolization, mononuclear cell infiltrates, local macrovesicular steatosis, hypertrophy of hepatocyte nuclei, focal degeneration, hepatocyte degeneration, structural architecture damage, and hepatic remodeling in male and female mice treated with both doses. A microscopic evaluation of kidney tissue sections revealed focal tubular damage, swelling in the renal glomerulus, thickening of basement membranes, cell peeling, vacuolization, and cytoplasmic debris in male and female mice treated with both doses. Control animals of both sexes exhibited normal hepatic and kidney architecture.

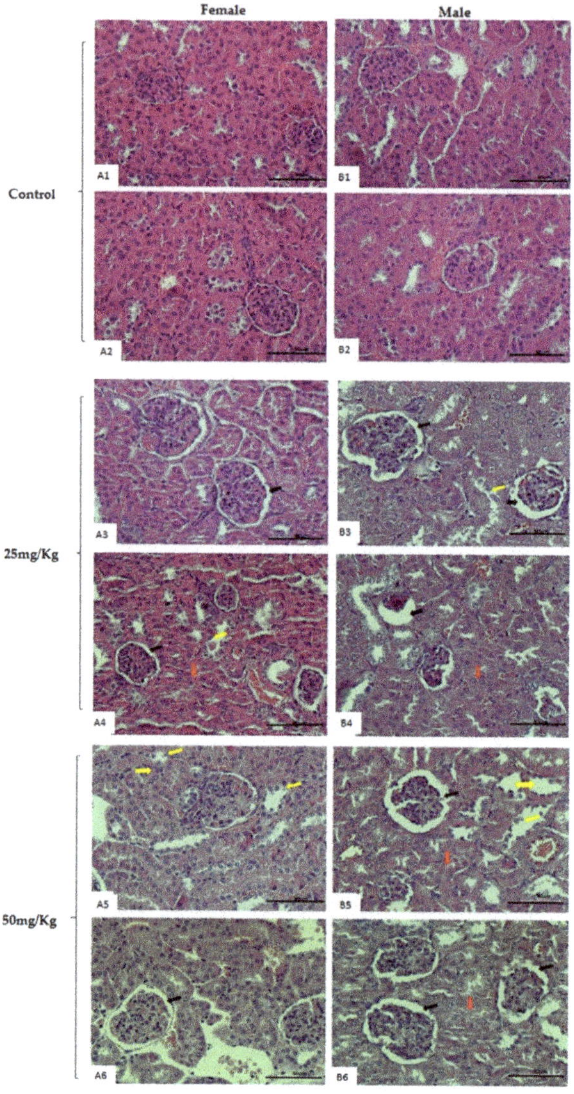

Figure 4. Representative histological images of kidney slices of male and female mice after 28-day repeated-dose toxicity study. Figures represent female (**A1**,**A2**) and male (**B1**,**B2**) control groups with normal histological aspects of renal tissue. Both female (**A3**,**A4**) and male mice (**B3**,**B4**) treated with 25 mg/kg of V 15 demonstrated vacuolization, cell peeling, cytoplasmic debris (**B3**,**A4**) (yellow arrows), focal tubular damage (**A4**,**B4**) (red arrows), and thickening of the basal membrane (A3,**B3**,**A4**,**B4**) (black arrows). The female (**A5**,**A6**) and male mice (**B5**,**B6**) treated with 50 mg/kg of V 15. (**A5**,**B5**) presents vacuolization, cytoplasmic debris (yellow arrows), and (**A6**,**B5**,**B6**) thickening of the basal membrane (black arrows) (H&E x40).

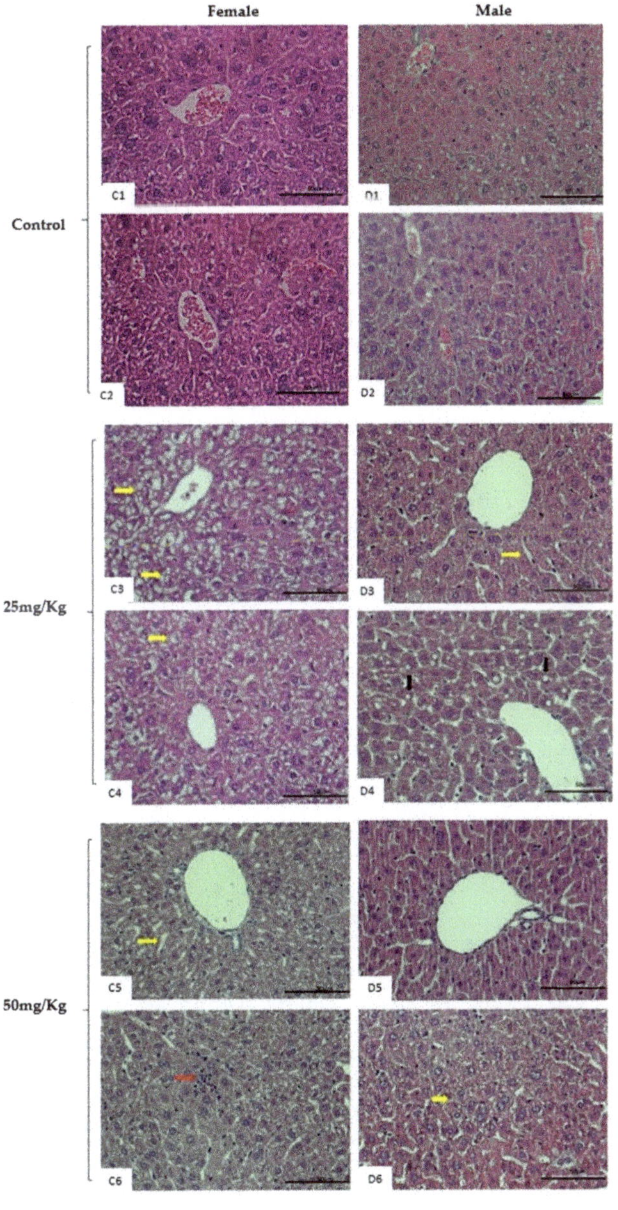

Figure 5. Representative histological images of liver slices of male and female mice after 28-day repeated-dose toxicity study. Figures represent (**C1**,**C2**) female and (**D1**,**D2**) male control groups with normal histological aspects of liver tissue. Both female (**C3**,**C4**) and (**D3**,**D4**) male mice treated with 25 mg/kg of V 15 showed (**C3**,**C4**,**D3**) focal degeneration, degeneration of hepatocytes (yellow arrows), and (**D4**) indications of local macrovesicular steatosis (black arrows). The female (**C5**,**C6**) and male animals (**D5**,**D6**) treated with 50 mg/kg of V 15, showing vacuolization (yellow arrow) (**C5**), (**D6**) hypertrophy of hepatocyte nuclei, and (**C6**) inflammation with mononuclear cell infiltrates (red arrow). (H&E x40) The scale bar used was 100 mm.

3. Discussion

Polyoxometalates, including POVs, have been studied for potential biomedical applications, including antibacterial, antidiabetic, antiprotozoal, antiviral, and anticancer applications [1–3]. Given the wide structural and electronic range of POVs, a wide range of biological activities is apparent [46–54], and it is important to evaluate the possible toxicological effects. In general, short-term toxicity of vanadium administration has been observed in several species, including mammals [35,55,56] and fish [57,58]. However, the oral treatment by a single dose of a POV, and particularly a multivalent POV, has not been reported previously. The present study is the first report of an oral dose of POV repeated for 28 days in mice. These studies are important because some of the POVs have been found to have undesirable side effects [33], whereas others have been found to have beneficial effects [1–3]. The detailed mechanism of action of most POVs, however, is not understood and likely will vary depending on the specific POV or POM. We recently reported that V_{15} was able to activate signal transduction [9,13] and suggested that the redox properties of V_{15} were important to that mode of action. V_{15} has also been found to be effective in inhibiting cell growth [31], and the mode of action involved bacterial excretion of a temperature-insensitive material. In this case, V_{15} behaves similar to V_{10}, which has several activities that are not attributed to its redox properties [1,22].

Several methods are useful to predict the toxicity of different types of substances [59,60], and the toxicity of the V_{15} cluster was screened against PMBC viability and a brine shrimp lethality bioassay [61,62]. Cell viability is one of the vital methods for toxicology analysis, which measures the cellular response to a toxic agent and provides information on survival and cell death. In this study, we found that V_{15} has a high cytotoxicity effect with low IC_{50} values in both PBMCs (17.5 μmol L^{-1}) and the brine shrimp assay (17.9 mg mL^{-1}). The studies demonstrated that molecules with IC_{50} values less than 100 μmol L^{-1} are considered very toxic [63]. Thus, comparable studies have shown the cytotoxicity of the vanadium compounds investigated [64–66]. Although the mechanism of vanadium toxicity remains to be elucidated and likely to vary depending on the compound examined, the cytotoxicity of vanadium is associated with the ability to accumulate vanadium ions and generate redox imbalance through an increase in intracellular ROS species [39].

The in vivo acute oral toxicity test represents a useful initial step in the toxicity assessment of compounds with potential biomedical applications [45]. Though the oral acute single administration of V_{15} did not cause any mortality, it showed a potential toxic effect through transitory behavior alterations, such as piloerection and lethargy after dosing, organ coefficients in the lung and liver, and biochemical parameters that indicated metabolic damage. Importantly, the toxicity evaluation is essential to identify a dose range and, hence, differences between efficacy and toxicity. This information would be useful to assess potential biomedical applications of a compound and identify any potential clinical symptoms that might be triggered by toxic levels of vanadium compounds under investigation.

In addition, there are no previous reports concerning repeated 28-day oral administration and toxicity of POVs. The Kaplan–Meier curve demonstrated that deaths was dose- and time-dependent. Survival analysis is important when the time between exposure and events is of clinical importance. Animal deaths occurred in both male and female Swiss mice sooner at higher dose treatments. Our data from the 28-day repeated-dose toxicity evaluation showed that male and female albino Swiss mice responded to the repeated exposure to V_{15} with observable side effects. All animals treated with 300 mg/kg of V_{15} died before the 12th day of the experiment (Figure 2). All of them demonstrated mortality/moribundity and clinical signs of toxicity, including lethargy, diarrhea, slimming, and abnormal gait and breathing.

Animals treated with a lower dose of 50 mg/kg of V_{15}, on the other hand, showed only a slight reduction in food and water intake. In addition, the weight of the kidney and testis was reduced in males treated with 50 mg/kg, and the weight of the skeletal muscles and reproductive organs was reduced in females treated with 25 and 50 mg/kg treatments. The observed reduction in organ weight could indicate that kidneys, skeletal muscles, and reproductive organs are potential targets following oral and repetitive treatments with soluble vanadium compounds. Kamboj and Kar showed that after two days of a single dose of vanadium exposure (0.08 nmol/kg), necrosis and a reduction in testes weight was detected in rats [67]. Furthermore, CD-1 male mice exposed to V_2O_5 inhalation for 12 weeks exhibited changes in the testes weight and the action of the testicular cytoskeleton, which might at least in part explain the reported reprotoxic effect of some vanadium compounds [68,69]. However, there are only a few reports concerning the effects of vanadium on the female reproductive system [70,71]. It has been shown that vanadium reduces glycogen content in the ovaries and uterus tissue, as well as protein content and acid phosphatase activity in the uterus. Furthermore, a microscopic evaluation of the ovaries revealed a decreasing mature follicle diameter, ovum disintegration, disorganized and hypertrophied growing follicles, and loose and fibrotic stroma [72]. Our results suggest that reprotoxic effects induced by POV are an important parameter that should be further investigated in future investigations of potential compounds for clinical use.

Hematological parameter evaluations were used to determine the extent of the harmful effect of V_{15} exposure. Repetitive dosing of V_{15} for 28 days at both doses caused similar toxic effects in male and female subjects. The impact of vanadium compounds on blood parameters has been demonstrated previously [73–75]. Sanchez-Gonzaléz et al. reported anemia in healthy rats treated with bis(maltolate)oxidovanadium(IV) dissolved in drinking water [73]. However, other studies have reported unchanged hematological parameters with treatment by different vanadium compounds. Specifically, Dai et al. (1994) showed that bis(maltolato)oxidovanadium(IV) (BMOV), ammonium metavanadate, and vanadyl sulfate in drinking water caused no hematological damage for 12 weeks [74]. Our results suggest a moderate normocytic normochromic anemia in both male and female mice treated with V_{15}. Some vanadium compounds could affect physiological iron concentration [73] and may explain the anemia observed in animals exposed to POV. Furthermore, the thrombocytosis detected in V_{15}-treated male and female mice may be due to inflammation, which is supported by the observed leukocyte count results. Our results agree with previous reports [75]. A higher predictive value for human toxicity for the hematological system is obtained upon the extrapolation of animal data. Thus, we suggest that V_{15} may be severely toxic to hematopoiesis.

An imbalance in oxidative stress is recognized as one potential cause of vanadium toxicity [76]. In addition, the great concern is the accumulation of vanadium in organs, such as the kidney or liver, which induces oxidative harm, lipid peroxidation, and toxicity [76–79]. Studies have demonstrated that some vanadium compounds may induce and exacerbate lipid peroxidation (LPO) and also affect human health. On the other hand, other vanadium compounds have protective effects on LPO [20,39]. Furthermore, vanadium compounds can sometimes generate free radicals, and thus form ROS/RNS directly, but they can also act indirectly through the effects on LPO [20,80–82]. Oster et al. (1993) showed that some vanadium compound treatments are correlated with TBAR production in both the liver and kidney in male rats [83]. Furthermore, Gândara et al. (2005) showed that decavanadate did not induce short-term LPO (12 h), but it later (24 h) increased LPO in the liver of *H. didactylus* [84]. Apart from this, rabbits exposed to vanadium pentoxide (V_2O_5) had increased LPO in the liver and kidney [85]. Our findings demonstrated an increase in LPO levels and a subsequent significant increase in liver lipid peroxidation in treated male animals at both doses and females at a higher dose of V_{15}.

The prooxidant and antioxidant response induced by V_{15} was different in male and female mice. SOD activity was depressed in male mice, whereas it was stimulated in female mice treated with POV. In addition, the activity of the GSH antioxidant system decreased, whereas CAT activity increased only in female mice. Our data suggest that V_{15} has a higher toxic effect in female mice compared with male mice. GSH is the largest endogenous cellular redox buffer, and thus the most significant intracellular non-enzymatic antioxidant source [86]. Therefore, it could be possible that the GSH depletion and changes in the GSH:GSSH ratio may represent the mechanism of oxidative intensity and a higher toxicity of V_{15} in female mice. Female mice appear to be less vulnerable to oxidative stress under physiological conditions, but during stressful conditions, it is unclear whether female mice are more protected against oxidative stress than males [87]. The literature is controversial regarding the effects of vanadium compounds on GSH levels. $NaVO_3$ in drinking water decreased GSH content in female rats [84]; on the other hand, ammonium metavanadate in drinking water increased GSH in the liver in the same animal model [88]. In 1993, Thompson and McNeill did not detect any noticeable differences in the amount of GSH in male rat livers treated with $VOSO_4$ in drinking water for 12 weeks [89]. Some of these differences have been traced to specific conditions of the interaction of the vanadium compound with the GSH [90]. In conclusion, our results demonstrate that oxidative stress is associated with the toxicity of POVs, and that multiple mechanisms are likely responsible for V_{15} oral toxicity. According to our results, decavanadate inhibits respiration and induces membrane depolarization of the mitochondria, acting as a strong inhibitor, most likely through interference with the respiratory chain, and specifically complex III of the mitochondrial inner membrane, in the heart and liver [19,21].

The assessment of kidney and liver function in response to a new drug has been supported by the measurements of serum marker enzymes and/or metabolites [91,92]. In this study, the exposure of Swiss mice (male and female) to repeated doses of V_{15} produced several treatment-related effects, particularly in the groups treated with high doses of V_{15}. Serum levels of ALT and AST have been observed as markers of liver injury, due to a variety of etiologies, such as medication toxicity and viral hepatitis [92]. In addition, a serum increase in ALT and AST levels is indicative of cell damage and a loss of the functional integrity of the hepatocyte membrane. Our data are comparable to previous reports on the liver toxicity of vanadium compounds in fish [57,84], broilers [93], and mammals [82,94,95]. These findings are related to an increase in liver histopathology with focal inflammation, diffused swelling of hepatocytes, and marked centrilobular hepatocellular injury. Pathological changes resulted from POV-induced free radical generation and the diminished antioxidant status of the compound. Similar histological alterations in the liver and kidney were seen in rabbits treated with vanadium pentoxide (V_2O_5) [85]. In addition, mice exposed to V_2O_5 inhalation for 6 weeks demonstrated inflammatory foci in liver parenchyma [95,96]. Cardiac myocytes also showed vacuolization and granular degeneration after 44 days of metavanadate (NH_4VO_3) exposure [97]. Taken together, previous results and observations in this paper suggest that oxidative stress is at the core of this histomorphology alteration and impairment of liver function and liver toxicity.

The kidney is a common target organ for toxicity and injury because of its exposure to drugs, including metallodrugs. Thus, renal function is usually assessed by serum levels of creatinine, uric acid, and BUN. These parameters may provide information on the effects of V_{15} on the tubular and/or glomerulus part of the kidney. V_{15} treatment increased serum creatinine levels in male mice at higher doses, which suggests the nephrotoxicity of this POV. In addition, hyperuricemia (HUA) may result from an imbalance in UA metabolism, which can lead to hyperuricemia and represent a very important signal of kidney toxicity. Both a reduction in glomerular filtration rate and an increase in net tubular absorption may result in HUA [91]. The kidney toxicity induced by vanadium compounds detected by an increasing UC has been demonstrated previously [94]. In this study, the increasing UA exerted by V_{15} plays an important role in inducing renal damage in both male and female mice. In addition, the microscopic appearance of kidney tissues of male and female mice dosed at 25 and 50 mg/kg of V_{15} showed that morphological alterations are compatible with nephrotoxicity induced by metals. Similar alterations were found in *H. didatylys* exposed to decavanadate oligomers [58] or ammonium metavanadate in broilers [93]. It has been demonstrated that vanadium may induce morphological damage to several organs, including kidneys. CD-1 mice exposed to vanadium oxide inhalation showed histological evidence of inflammatory foci [78]. Another study showed that V_{10} and a high-fat diet in Wistar rats caused harm to kidney cytoarchitecture [98]. Furthermore, vanadium compounds have the capacity to disturb the cell redox balance, which can result in damage to the antioxidant enzymatic and non-enzymatic systems. In general, these pathological changes in the liver and kidney resulted from POV-induced free radical generation and compromised the activity of the antioxidant status in these tissues.

4. Materials and Methods

4.1. Ethical Statement

The experimental strategy for these studies was approved by the human ethics committee of the Health Sciences Center/UFPE with the process number CAAE 04633218.6.0000.8807. Additionally, all animal care procedures were approved by the National Institutes of Health Guide for the Care and Use of Laboratory Animals (NIH Publication 8023, revised 1978) and followed the Ethics Committee on the Use of Animals of the Universidade Federal de Pernambuco (CEUA/UFPE, protocol 0021/2021).

4.2. Chemicals and General Methods

Ammonium metavanadate (NH_4VO_3, 99%) and D-mannitol ($C_6H_{14}O_6$, 98%), both from Sigma-Aldrich® (St. Louis, MO, USA), and tetramethylammonium chloride ($C_4H_{12}ClN$, 98%) from Fluka® were used without purification. Deuterium oxide (D_2O, 99.9 atom %D) was purchased from Oakwood Chemical and used as received. Ultrapure water (18.2 mΩ·cm) was used in the synthesis and characterization of V_{15}, excepted when otherwise stated. The ^{51}V NMR data were collected on a Bruker NEO 400 MHz spectrometer with a 105.2 MHz frequency for vanadium at 298 K, using 4096 scans in the f1 domain, 8 steady-state transitions, a spectra width of 800 ppm, a transmitter frequency offset of −548 ppm, 0.01 s relaxation delay, a 0.08192 s acquisition time, and a 16 µs pulse. Fourier-transform infrared (IR) spectra (400 to 4000 cm^{-1}) were recorded from KBr pellets on a Bruker VERTEX-70 spectrometer with a resolution of 4 cm^{-1}. Electron paramagnetic resonance in X-band (EPR) spectra (9.5 GHz) were recorded at 77 K on an X-band (9.5 GHz) Bruker EMX-Micro spectrometer using a pulverized sample. Electronic absorption spectra of 0.250 mmol L^{-1} aqueous solution of V_{15} were observed from 340 to 900 nm, using an AvaLight UV-Vis/NIR light source and an AvaSpex-UL S2048 Fiber-Optic Spectrometer.

4.2.1. Synthesis of V_{15}-$(Me_4N)_6[V_{15}O_{36}(Cl)]$

Mannitol (1.428 g, 7.839 mmol), tetramethylammonium chloride (0.7530 g, 6.870 mmol), and ammonium metavanadate (1.621 g, 13,86 mmol) were weighed and transferred to a round-bottom flask. Water (75 mL) was added, and the reaction mixture was stirred under reflux (which is 93 °C in Colorado) for 24 h. The solution changed from colorless to deep green in 2 h. A small amount of black powder started forming after 30 min. After 24 h, the reaction mixture was filtered to remove the black solid and kept at 4 °C to crystalize. Due to the small amount of black solid isolated, the insoluble black side product was confirmed by IR as $(NH_4)_2V_3O_8$ [29]. Pure V_{15} was obtained from the mother liquor after 4 days as dark green crystals, which were filtered off, washed with 10 mL of cold water, and dried under air. The average yield was 1.36 g, 81%, and the synthesis was reproduced 16 times to generate 10.864 g of V_{15}. The dark-green crystals of V_{15} were insoluble in common organic solvents and soluble in hot water.

4.2.2. Characterization of V_{15} in the Solid and Solution States

The solid V_{15}, with a formula $[(CH_3)_4N]_6[V^{IV}{}_8V^{V}{}_7O_{36}(Cl)]$, was confirmed by single-crystal X-ray diffraction analysis. One dark-green crystal of $\mathbf{V_{15}}$ was subjected to analysis at 300 K on a Bruker D8 Venture diffractometer equipped with a Photon 100 CMOS detector using Mo-Kα radiation (µ(Mo-K_α) = 0.711 mm^{-1}). *Crystal data:* crystal system hexagonal $P6_{3/mmc}$, a = b = 13.720 Å, c = 20.020 Å, α = β = 90°, and γ = 120°. The X-ray powder diffraction (PXRD) pattern of the V_{15} was registered at 40 kV and 30 mA on a Shimadzu XRD600 diffractometer equipped with a Cu-target tube (Cu-K_α, k = 1.5418 Å). The calculated diffractogram of V_{15} was generated from the single-crystal crystallographic information (CIF) file available at Cambridge Crystallographic Data Centre, CCDC code 794586, using Mercury 4.0 software [99].

To verify the purity of the compound in bulk solid, powder X-ray diffraction pattern of V_{15} was compared with the previously described single-crystal X-ray diffraction structure, showing good correspondence.

The IR spectrum of solid V_{15} contains characteristic bands of the polyoxoanion at 560, 658, 728, and 793 cm^{-1} assigned to $\nu_{as}(V-O-V)$ and 980 cm^{-1} to $\nu(V{=}O)$. Me_4N^+ bands appeared at 1288 cm^{-1} as $\rho_r(CH_3)$, strong δ_{as}, and partial reduction in the vanadium (V) and is compatible with the intra- and intermolecular exchange interaction among vanadium (IV) centers reported in the $[V^{IV}{}_8V^{V}{}_7O_{36}(Cl)]^{6-}$ (V_{15}) polyoxoanion [29,43]. The EPR spectrum of the pulverized V_{15} sample was measured at 77 K and found to be like those reported previously [29].

The electronic absorption spectra were recorded for a 0.025 mmol L^{-1} aqueous solution of V_{15} and showed a ligand-to-metal charge transfer (LMCT, $p(O){\rightarrow}d(V)$) band below 400 nm and a broad band that extended from 600 nm to the near-infrared assigned to an intervalence charge transfer transition (IVCT, $V^{IV} \rightarrow V^{V}$) in the polyoxoanion. The spectra changed little over a period of 24 h, like studies by ^{51}V NMR and EPR spectroscopy that monitored up to 34 h of bioassays with pH 7.4 [9,30]. These studies show that V_{15} slowly decomposes in aqueous solution to give V_{15} in the presence of species of oxovanadates ($H_2VO_4^-$, $H_2V_2O_7^{2-}$, $V_4O_{12}^{4-}$, $V_5O_{15}^{5-}$) and the oxidovanadium(IV) complex $[VO_2(OH_2)_4]^+$. Importantly, no precipitation of black solid $(NH_4)_2V_3O_8$ was observed, further consistent with the lack of formation of the fresnoite type oxide $(NH_4)_2VO_3$ and a solution containing hydrolytically stable V_{15}.

4.3. Peripheral Blood Mononuclear Cells (PBMC) and Cytotoxicity Assay by 3-(4,5-Dimethyl-2-thiazolyl)-2,5-diphenyl-2H-tetrazolium bromide (MTT) of V_{15}

PBMCs were obtained from the peripheral blood of healthy volunteers and collected in heparinized tubes. For PBMC separation, the peripheral blood was centrifuged using Ficoll Paque Plus (GE Healthcare Biosciences®, Pittsburgh, PA, USA). After separation, the cells were counted and placed in 96-well plates (5×10^5 cells/100 μL/well) in RPMI-1640 medium (Gibco, ThermoFischer scientific®, Waltham, MA, USA), supplemented with L-Glutamine, 10% fetal bovine serum (FBS) (Gibco, ThermoFischer scientific®), 10 mmol L^{-1} HEPES (4-(2-hydroxyethyl)-1-piperazineethanesulfonic acid) (Gibco, Thermo Fischer scientific®), and 200 U/mL penicillin/streptomycin (Gibco). After being isolated, cells were exposed to V_{15} at serial concentrations of 10, 25, 50, and 100 μmol L^{-1} and then kept for 48 h in a humid atmosphere containing 5% CO_2 at 37 °C. Cells treated with the solvent dimethyl sulfoxide (DMSO 0.1% *v*/*v*) were used as a control. The cytotoxicity was then quantified by the reduction of 3-(4,5-dimethyl-2-thiazolyl)-2,5-diphenyl-2H-tetrazolium bromide (MTT) to the purple reduction product formed in living cells. The absorbance was measured at 570 nm by the Elx808 (Biotek, Shoreline, WA, USA, USA apparatus®). Thus, the cytotoxic activity of V_{15} was quantified as the observed percentage of the control absorbance. The results are reported as the average value calculated from three separate experiments.

4.4. Toxicity against Larvae of Artemia Salina Leach

Artemia salina (Leach, 1819) (Tropical®, Rio de Janeiro, Brazil) dry cysts were used to evaluate V_{15}-$(Me_4N)_6[V_{15}O_{36}Cl]$ toxicity. Firstly, cysts were hydrated in artificial seawater (40 g sea salt/L, pH 8,0) and exposed to constant aeration at ambient temperature (25 ± 3 °C) for 48 h. After hatching, the larvae (nauplii) were transferred into vials (n = 10 per vial) containing seawater with 5, 10, 25, 50, 100, 125, 250, 500, and 1000 mg mL^{-1} final concentrations of V_{15} taken from the stock solutions of 1 mg mL^{-1} in DMSO 5% for 24 h at 25 ± 3 °C [61]. The control wells contained 10 nauplii in artificial seawater and DMSO 5%. Two experiments were performed in triplicate, with a total of 110 specimens per treatment, and assessments of mortality and survival of nauplii were carried out by observation of mobility for live nauplii identified after 24 h under a microscope.

4.5. Oral Acute Toxicity

The evaluation of acute oral toxicity was measured in accordance with Guideline 423, which describes the acute toxic class method reported by the Organization of Economic Co-operation and Development (OECD) for measuring acute oral toxicity of chemicals [45]. Nulliparous and non-pregnant female mice (three for each group) were acclimatized in a propylene cage for five days before dosage, fasted for 3 h, and weighed before the administration of a POV dose. The V_{15} was dissolved in NaCl 0.9% (*m*/*v*) and administered by gavage in mice in a single dose using animal feeding needles (100 µL/100 g b.w.). Animals were randomly divided into three groups with three animals each: (a) a control group treated with 0.9% (*m*/*v*) NaCl; (b) a low-dose group treated with 300 mg/kg of V_{15} (V_{15-300}); and (c) a high-dose group treated with 2000 mg/kg of V_{15} ($V_{15-2000}$).

All animals were closely monitored for the first four hours to identify piloerection as well as changes in eyes, skin, and fur, toxic effects on the mucous membranes, behavior pattern disorientation, asthenia, hypoactivity, hyperventilation, lethargy, lack of sleep, tremors, salivation, diarrhea, convulsion, coma, motor activity, or death. During the study, the animals were monitored daily for 14 days. Daily monitoring included measuring body weight, food intake, and water consumption. On the 14th day, the female mice were euthanized by intravenous injection of a solution containing both xylazine (80 mg/kg, i.p.) and ketamine (10 mg/kg, i.p.). The blood was drawn, and the organs collected for biochemical and macroscopic analysis. The LD_{50} was estimated based on mortality in each group as previously reported [100].

4.6. Toxicity Administered by the Repeated 28-Day Oral Toxicity Dose

The repeated oral dose 28-day toxicity test was carried out in accordance with Guideline 407 for rodents reported by the Organization of Economic Co-operation and Development (OECD) to evaluate chemicals for long-term oral toxicity [101]. Kaplan–Meier plots were used to assess the data on mice survival. The animals were acclimated in propylene cages for five days, at which point male, nulliparous, and non-pregnant female mice were fasted for three hours and weighed before they were given the compound dose. The mice were divided into groups of 10 (5 male and 5 female each) to be treated daily over 28 consecutive days by gavage using animal feeding needles (100 µL/100 g b.w.) The groups were divided as follows: (a) the control group was treated with 0.9% *m/v* NaCl; (b) the low-dose group was treated with 25 mg/kg of V_{15} (V_{15-25}); (c) the high-dose group was treated with 50 mg/kg of V_{15} (V_{15-50}); and (d) then treated with 300 mg/kg of V_{15} (V_{15-300}). The number of deaths was recorded, and a Kaplan–Maier survival probability curve was plotted [102].

Animals were weighed once a week, and their basic morphological parameters, food and water intake, and any behavioral alterations were tracked. At the end of the 28 days of treatment, all animals were anesthetized by intraperitoneal administration of a xylazine and ketamine mixture (10 mg and 80 mg/kg, respectively) and euthanized by cervical dislocation. Blood samples were collected for biochemical and hematological examination, and the extracted liver and kidney organs were weighed and kept for histopathology.

4.7. Biochemical and Hematological Analysis

Mice fasted overnight were anesthetized as described above in the 14 days and 28 days following treatment. Blood samples were collected by the retro-orbital technique with or without heparin for hematological and serum biochemical analysis, respectively.

Hematological parameters included red blood cells (RBC), red cell volume distribution (RDW), hemoglobin concentration (HBG), hematocrit (HCT), platelet count (PLT), mean platelet volume (MPV), mean corpuscular volume (MCV), mean corpuscular hemoglobin (MCH), mean corpuscular hemoglobin concentration (MCHC), white blood cells (WBC), and lymphocytes (LYM). The test was performed using a multiparameter automatic hematology analyzer SDH-20, Labtest®, Curitiba, Brazil) designed for hematology testing samples.

The serum levels of several parameters were measured, including enzyme levels of alanine (ALT) and aspartate aminotransferase (AST), as well as metabolite levels including blood urea nitrogen (BUN), creatinine (CRE), total cholesterol (TC), total protein (TP), triacylglycerol (TG), and uric acid (UA) with a colorimetric assay using commercial kits (Lab Test Diagnostic SA®, Santa Lagoa, Brazil). Optical densities were measured by spectrophotometry using a Varioskan TM Lux multimode microplate reader on a Thermo Scientific®, Waltham, MA, USA instrument at the wavelengths designated on the datasheet for each biochemical parameter. Baseline measurements of controls were obtained by comparing the optical densities of the samples with the measurements using the appropriate standards provided in the kits. Data are showed in terms of U/mL (ALT and AST) and mg/dL for the others [16].

4.8. Oxidative Stress Evaluation in the Liver

Oxidative stress markers and lipid peroxidation in the liver were measured using thiobarbituric acid (TBAR), which measures the malondialdehyde (MDA) formed when lipids are oxidized according to the method reported previously [103]. Glutathione peroxidase (GPx) activity was measured in an assay using glutathione reductase as reported by Paglia and Valentine (1967) [104]; glutathione-reduced (GSH) levels were measured using a fluorometrical method as described by Hissin and Hilf (1976) [105]; the antioxidant activity of superoxide dismutase (SOD) was measured by monitoring superoxide anions generated by xanthine oxidase as described by Misra and Fridovich (1972) [106]; and the catalase (CAT) activity was monitored by measuring the decomposition of H_2O_2 by absorbance at 240 nm as reported by Aebi (1984) [107].

4.9. Organ Mass and Histopathological Analysis

After the 14th and 28th days of treatment, all mice were necropsied after blood withdrawal for anatomical localization and visible examination of detectable macroscopic organ changes (color, aspect, and size). Selected organs examined included the heart, liver, kidney, lungs, testicles, uterus, ovaries, spleen, adipose tissue, stomach, and soleus and extensor digitorum longus (EDL) muscles. These organs were carefully excised and trimmed of fat and connective tissue before weighing. The organ weight values were converted to g/100 g of body weight (percentage of body weight).

Following the 28-day treatment plan, the liver and kidney were fixed in 10% formalin buffer solution for 24 h at ambient temperature before being rinsed with water for four hours. They were then dehydrated stepwise using 70% (*v*/*v*) of EtOH. Dehydrated tissues were rendered transparent by addition of xylene. Tissues were added to pre-heated paraffin for sufficient infiltration, and after cooling, paraffin blocks were formed before being cut into 5 μm sections using a microtome (Leica® RM 2025, Heidelberger, Germany). The paraffin was removed, and specimens were dehydrated and stained with hematoxylin-eosin for observation with an inverted optical microscope (Leica®, Heidelberger, Germany) connected to a video camera (Leica® DFC 280, Wetzlar, Germany) and a computer monitor. An expert pathologist took pictures of the liver and kidney samples and examined them for any signs of cellular harm or morphological changes.

4.10. Statistical Analysis

Unless otherwise specified, the data were reported in triplicates as mean ± standard error of the mean (S.E.M). The data of V_{15}-treated groups and their respective control groups was analyzed by one-way analysis of variance (ANOVA) followed by the Bonferroni test. The long-rank Kaplan–Meier survival test was used to compare the survival distribution of the different doses and treatment groups. A p value less than 0.05 was considered statistically significant. Graph Pad Prism® (GraphPad Software, San Diego, CA, USA) version 6.0 software was used for all statistical analysis.

5. Conclusions

Administration of the mixed-valence polyoxovanadate (POV) V_{15}, a nanosized anionic cluster, induces in vitro and in vivo toxicity in mice. The acute toxicity demonstrated that the LD_{50} value is more than 2000 mg/kg; thus, this POV has a low order of acute oral toxicity in mice and is a category 5 compound. However, the repeated 28-day assessment with systematic administration demonstrated toxic effects of this POV, especially in the liver and kidney tissues, and this is more toxic than in acute studies. The multivalent V_{15} was found to be stable under the condition resembling those used for the in vitro and in vivo studies. This suggests that the observed toxicity is mainly caused by the intact V_{15}, considering the multivalent nature of the cluster and the ease of its oxidation. The redox properties of the cluster are likely related to its mode of action. Accordingly, the toxicity is attributed to redox damage produced by V_{15}, which is consistent with induced hematological, hepatic, and renal toxicity observed in both male and female mice. Due to the oxidation of V_{15} and the resulting oxidative damage, tissues and metabolites are also likely involved in the process. Clinically, these results suggest that future research with any type of redox-active compounds may result in toxicity. Thus, there is a need to design non-toxic vanadium compounds, particularly POVs, for biomedical applications, which will be most beneficial if the delivery vehicle involved could control the reactivity of these compounds. These results should encourage researchers to develop new strategies to reduce the toxic effects of POVs, which could involve encapsulation in liposomes, other protective delivery systems, and complex-containing bioconjugate molecules.

Supplementary Materials: The following supporting information can be downloaded from: https://www.mdpi.com/article/10.3390/ph16091232/s1, Figure S1: Comparison between simulated and experimental powder X-ray diffraction patterns of V_{15}, in A. Crystal packing of $[(CH_3)_4N]_6[V_{15}O_{36}(Cl)]$ along with the c axis, in B. The weak C H···O interaction between the cation and the POV are shown in light blue; Figure S2: Infrared spectra recorded for products V_{15} (in black) and $(NH_4)_2V_3O_8$ (red) in KBr pellets; Figure S3: X-band EPR spectrum recorded for pulverized sample of V_{15} at 77 K; Figure S4: Electronic absorption spectra observed for a 0.025 mmol L^{-1} aqueous solution of V_{15} in t = 0 h, 20 min, 40 min, 1 h, 2 h, and 24 h.

Author Contributions: K.P., M.R., K.K., G.G.N. and D.C.C. designed and performed the chemical studies. M.d.M.B., L.M.A.d.L., E.K.B.d.L., W.A.d.S.A., L.A.d.S., T.D.d.S., M.C.P. and E.C.L. designed and performed the biological studies. E.C.L. and D.C.C. supervised the project. M.F.B., E.C.L., L.M.A.d.L. and D.C.C. wrote the first draft. G.G.N., M.F.B., W.E.d.S., E.C.L., D.A.G., K.K. and D.C.C. reviewed and edited the manuscript. All authors have read and agreed to the published version of the manuscript.

Funding: This research was funded by CNPq (Conselho Nacional de Desenvolvimento Científico e Tecnológico), grants 406946/2021-3 and 309691/2021-4 (to E.C.L.), CAPES (Coordenação de Aperfeiçoamento de Pessoal de Nível Superior, Finance Code 001) (to E.C.L.), PrInt/CAPES-UFPR (Internationalization Program for scholarships and financial support) (to K.P.), FACEPE (Fundação de Amparo à Ciência e Tecnologia do Estado de Pernambuco), grant AMD 0060-1.06/2019 (to E.C.L. and L.M.A.d.L), and Colorado State University (to D.C.C.).

Institutional Review Board Statement: The study was conducted according to the guidelines on the Ethical Principles in Animal Research, and the protocols for these experiments were approved by the Animal Ethics Committee of the Universidade Federal de Pernambuco with the process number CAAE #04633218.6.0000.8807.

Informed Consent Statement: Not applicable.

Data Availability Statement: All other data are available from the corresponding author upon reasonable request. Data is contained within the article or supplementary material.

Acknowledgments: The authors would like to thank UFRPE, UFPE, UFPR, and CSU for funding. L.M.L. thanks FACEPE for the scholarship with D.C.C in 2020 (AMD 0060-1.06/2019), and K.P. thanks CAPES for the scholarship with D.C.C in 2020.

Conflicts of Interest: The authors declare no conflict of interest.

Abbreviations

ALB	Albumin
ALT	Alanine aminotransferase
ANOVA	One-way analysis of variance
AST	Aspartate aminotransferase
BMOV	Bis(maltolato)oxidovanadium(IV)
BUN	Blood urea nitrogen
b.w.	Body weight
CAT	Catalase
CEUA	Ethics Committee on the Use of Animals
CRE	Creatinine
CHO	Chinese Hamster Ovary Cells
DMSO	Dimethyl sulfoxide
DNA	Deoxyribonucleic acid
EDL	Extensor digitorum longus
EPR	Eletronic paragmagnetic ressonance
FBS	Fetal bovine serum
GSH	Glutathione reduced
GSSG	Glutathione oxidized
GLB	Globulin
GPx	Glutathione peroxidase
Hb	Hemoglobin
HCT	Hematocrit
HEPES	4-(2-hydroxyethyl)-1-piperazineethanesulfonic acid
i.p	Intraperitoneal administration
LC50	Lethal concentration median dose
LD50	Median lethal dose
LPO	Lipid peroxidation
LYM	Lymphocytes
MCH	Mean corpuscular hemoglobin
MCHC	Mean corpuscular hemoglobin concentration
MDA	Malondialdehyde
MVP	Mixed-valence polyoxovanadates
MTT	3-(4,5- dimethyl-2-thiazolyl)-2,5-diphenyl-2H-Tetrazolium bromide
NMR	Nuclear magnetic ressonance
o.g.	Oral gavage
OECD	Organization for Economic Cooperation and Development
PBMC	Peripheral blood mononuclear cells
POM	Polyoxometalate
POV	Polyoxovanadate
RBC	Red blood cells
RNS	Reactive nitrogen species
ROS	Reactive oxygen species
Rpm	Rotation per minute
SD	Standard deviation
SEM	Standard error of the mean
SOD	Superoxide dismutase
STZ	Streptozotocin
TC	Total cholesterol
TP	Total protein
TBAR	Thiobarbituric acid
UA	Uric acid
WBC	White blood cells

References

1. Aureliano, M.; Gumerova, N.I.; Sciortino, G.; Garribba, E.; Rompel, A.; Crans, D.C. Polyoxovanadates with emerging biomedical activities. *Coord. Chem. Rev.* **2021**, *447*, 214143. [CrossRef]
2. Pope, M.T.; Müller, A. Polyoxometalate chemistry: An old field with new dimensions in several disciplines. *Angew. Chem. Int.* **1991**, *30*, 34–48. [CrossRef]
3. Gumerova, N.I.; Rompel, A. Speciation atlas of polyoxometalates in aqueous solutions. *Sci. Adv.* **2023**, *9*, eadi0814. [CrossRef]
4. Wang, X.; Wei, S.; Zhao, C.; Li, X.; Jin, J.; Shi, X.; Su, Z.; Li, J.; Wang, J. Promising application of polyoxometalates in the treatment of cancer, infectious diseases and Alzheimer's disease. *J. Biol. Inorg. Chem.* **2022**, *27*, 405–419. [CrossRef] [PubMed]
5. Pereira, M.J.; Carvalho, E.; Eriksson, J.W.; Crans, D.C.; Aureliano, M. Effects of decavanadate and insulin enhancing vanadium compounds on glucose uptake in isolated rat adipocytes. *J. Inorg. Biochem.* **2009**, *103*, 1687–1692. [CrossRef] [PubMed]
6. Treviño, S.; González-Vergara, E. Metformin-decavanadate treatment ameliorates hyperglycemia and redox balance of the liver and muscle in a rat model of alloxan-induced diabetes. *New J. Chem.* **2019**, *43*, 17850–17862. [CrossRef]
7. Colović, M.B.; Lacković, M.; Lalatović, J.; Mougharbel, A.S.; Kortz, U.; Krstić, D.Z. Polyoxometalates in biomedicine: Update and overview. *Curr. Med. Chem.* **2020**, *27*, 362–379. [CrossRef] [PubMed]
8. Samart, N.; Saeger, J.; Haller, K.J.; Aureliano, M.; Crans, D.C. Interaction of decavanadate with interfaces and biological model membrane systems: Characterization of soft oxometalate systems. *J. Mol. Eng. Mater.* **2014**, *2*, 1440007. [CrossRef]
9. Althumairy, D.; Pozeal, K.; Barisas, B.G.; Nunes, G.G.; Roess, D.A.; Crans, D.C. Polyoxometalates function as indirect activators of a G protein-coupled receptor. *Metallomics* **2020**, *12*, 1044–1061. [CrossRef]
10. Zhai, F.; Wang, X.; Li, D.; Zhang, H.; Li, R.; Song, L. Synthesis and biological evaluation of decavanadate $Na_4Co(H_2O)_6V_{10}O_{28}\cdot 18H_2O$. *Biomed. Pharmacother.* **2009**, *63*, 51–55. [CrossRef]
11. Ferretti, V.A.; León, I.E. An Overview of Vanadium and Cell Signaling in Potential Cancer Treatments. *Inorganics* **2022**, *10*, 47. [CrossRef]
12. Kostenkova, K.; Althumairy, D.; Rajan, A.; Kortz, U.; Barisas, B.G.; Roess, D.A.; Crans, D.C. Polyoxidovanadates $[Mo^{VI}V^{V}{}_9O_{28}]^{5-}$ and $[H_2Pt^{IV}V^{V}{}_9O_{28}]^{5-}$ interact with CHO cell plasma membrane lipids causing aggregation and activation of a G protein-coupled receptor. *Front. Chem. Biol.* **2023**, *2*, 1126975. [CrossRef]
13. Samart, N.; Althumairy, D.; Zhang, D.; Roess, D.A.; Crans, D.C. Initiation of a novel mode of membrane signaling: Vanadium facilitated signal transduction. *Coord. Chem. Rev.* **2020**, *416*, 213–286. [CrossRef]
14. Bulgheroni, A.; Kinsner-Ovaskainen, A.; Hoffmann, S.; Hartung, T.; Prieto, P. Estimation of acute oral toxicity using the No Observed Adverse Effect Level (NOAEL) from the 28 days repeated dose toxicity studies in rats. *Regul. Toxicol. Pharmacol.* **2009**, *53*, 16–19. [CrossRef] [PubMed]
15. Qu, X.; Xu, K.; Zhao, C.; Song, X.; Li, J.; Li, L.; Nie, W.; Bao, H.; Wang, J.; Niu, F.; et al. Genotoxicity and acute and subchronic toxicity studies of a bioactive polyoxometalate in Wistar rats. *BMC Pharmacol. Toxicol.* **2017**, *18*, 26. [CrossRef] [PubMed]
16. Lima, L.M.A.; Murakami, H.; Gaebler, D.J.; Silva, W.E.; Belian, M.F.; Lira, E.C.; Crans, D.C. Acute Toxicity Evaluation of Non-Innocent Oxidovanadium(V) Schiff Base Complex. *Inorganics* **2021**, *9*, 42. [CrossRef]
17. Cohen, M.D.; Sisco, M.; Prophete, C.; Yoshida, K.; Chen, L.C.; Zelikoff, J.T.; Smee, J.; Holder, A.A.; Stonehuerner, J.; Crans, D.C.; et al. Effects of metal compounds with distinct physicochemical properties on iron homeostasis and antibacterial activity in the lungs: Chromium and vanadium. *Inhal. Toxicol.* **2010**, *22*, 169–178. [CrossRef]
18. Cohen, M.D.; Sisco, M.; Prophete, C.; Chen, L.C.; Zelikoff, J.T.; Ghio, A.J.; Stonehuerner, J.D.; Smee, J.J.; Holder, A.A.; Crans, D.C. Pulmonary immunotoxic potentials of metals are governed by select physicochemical properties: Vanadium agents. *J. Immunotoxicol.* **2007**, *4*, 49–60. [CrossRef]
19. Soares, S.S.; Gutiérrez-Merino, C.; Aureliano, M. Decavanadate induces mitochondrial membrane depolarization and inhibits oxygen consumption. *J. Inorg. Biochem.* **2007**, *101*, 789–796. [CrossRef]
20. Soares, S.S.; Martins, H.; Duarte, R.O.; Moura, J.J.; Coucelo, J.; Gutiérrez-Merino, C.; Aureliano, M. Vanadium distribution, lipid peroxidation and oxidative stress markers upon decavanadate in vivo administration. *J. Inorg. Biochem.* **2007**, *101*, 80–88. [CrossRef]
21. Soares, S.S.; Gutiérrez-Merino, C.; Aureliano, M. Mitochondria as a target for decavanadate toxicity in *Sparus aurata* heart. *Aquat. Toxicol.* **2007**, *83*, 1–9. [CrossRef]
22. Samart, N.; Arhouma, Z.; Kumar, S.; Murakami, H.A.; Crick, D.C.; Crans, D.C. Decavanadate inhibits microbacterial growth more potently than other oxovanadates. *Front. Chem.* **2018**, *6*, 16. [CrossRef] [PubMed]
23. Arhouma, Z.; Murakami, H.A.; Koehn, J.T.; Li, X.; Roess, D.A.; Crick, D.C.; Crans, D.C. Exploring growth of *Mycobacterium smegmatis* treated with anticarcinogenic vanadium compounds. *Inorganics* **2022**, *10*, 50. [CrossRef]
24. Kostenkova, K.; Arhouma, Z.; Postal, K.; Rajan, A.; Kortz, U.; Nunes, G.G.; Crick, D.C.; Crans, D.C. PtIV- or MoVI- substituted decavanadate inhibits the growth of *Mycobacterium smegmatis*. *J. Inorg. Biochem.* **2021**, *217*, 111356. [CrossRef] [PubMed]
25. Kioseoglou, E.; Petanidis, S.; Gabriel, C.; Salifoglou, A. The chemistry and biology of vanadium compounds in cancer therapeutics. *Coord. Chem. Rev.* **2015**, *301–302*, 87–105. [CrossRef]
26. Qi, W.; Zhang, B.; Qi, Y.; Guo, S.; Tian, R.; Sun, J.; Zhao, M. The Anti-Proliferation Activity and Mechanism of Action of $K_{12}[V_{18}O_{42}(H_2O)]\cdot 6H_2O$ on Breast Cancer Cell Lines. *Molecules* **2017**, *22*, 1535. [CrossRef] [PubMed]

27. Roubatis, L.; Anastasiadis, N.C.; Paratriantafyllopoulou, C.; Moushi, E.; Tasiopoulos, A.J.; Karkabounas, S.C.; Veltsistas, P.G.; Perlepes, S.P.; Evangelou, A.M. A missing oxidation-state level in the family of polyoxo(azide)octadecavanadate(IV/V) clusters: Synthesis, structure and antitumoural properties of $[V^{IV}{}_{11}V^{V}{}_{7}O_{44}(N_3)]^{10-}$ in a sodium containing-3D architecture. *Inorg. Chem. Commun.* **2016**, *69*, 85–88. [CrossRef]
28. Carvalho, F.; Aureliano, M. Polyoxometalates Impact as Anticancer Agents. *Int. J. Mol. Sci.* **2023**, *24*, 5043. [CrossRef]
29. Nunes, G.G.; Bonatto, A.C.; Albuquerque, C.G.; Barison, A.; Ribeiro, R.R.; Back, D.F.; Andrade, A.V.C.; de Sá, E.L.; Pedrosa, F.D.O.; Soares, J.F.; et al. Synthesis, characterization and chemoprotective activity of polyoxovanadates against DNA alkylation. *J. Inorg. Biochem.* **2012**, *108*, 36–46. [CrossRef]
30. Postal, K.; Maluf, D.L.; Valdameri, A.; Rudiger, A.L.; Hughes, D.L.; de Sá, E.L.; Ribeiro, R.R.; de Souza, E.M.; Soares, J.F.; Nunes, G.G. Chemoprotective activity of mixed valence polyoxovanadates against diethylsulphate in *E. coli* cultures: Insights from solution speciation studies. *RSC Adv.* **2016**, *6*, 114955. [CrossRef]
31. Arhouma, Z.A. Antibacterial Growth Effects and Speciations of Several Vanadium Salts and Complexes. Ph.D. Thesis, Colorado State University, Fort Collins, CO, USA, 2022.
32. Kita, D.H.; de Andrade, G.A.; Missina, J.M.; Postal, K.; Boell, V.K.; Santana, F.S.; Zattoni, I.F.; Zanzarini, I.D.S.; Moure, V.R.; Rego, F.G.M.; et al. Polyoxovanadates as new P-glycoprotein inhibitors: Insights into the mechanism of inhibition. *FEBS Lett.* **2022**, *596*, 381–399. [CrossRef] [PubMed]
33. Fraqueza, G.; Aureliano, M. Polyoxovanadates Contribution to Pharmacological, Antimicrobial and Toxicological Actions of Vanadium. *Med. Sci. Forum* **2022**, *11*, 8. [CrossRef]
34. Doucette, K.A.; Hassell, K.N.; Crans, D.C. Selective Speciation Improves Efficacy and Lowers Toxicity of Platinum Anticancer and Vanadium Antidiabetic Drugs. *J. Inorg. Biochem.* **2016**, *165*, 56–70. [CrossRef] [PubMed]
35. Llobet, J.M.; Domingo, J.L. Acute Toxicity of Vanadium Compounds in Rats and Mice. *Toxicol. Lett.* **1984**, *23*, 227–231. [CrossRef] [PubMed]
36. Domingo, J.L.; Gomez, M.; Sanchez, D.J.; Llobet, J.M.; Keen, C.L. Toxicology of Vanadium Compounds in Diabetic Rats: The Action of Chelating Agents on Vanadium Accumulation. *Mol. Cell. Biochem.* **1995**, *153*, 233–240. [CrossRef] [PubMed]
37. Xi, W.S.; Li, J.B.; Liu, Y.Y.; Wu, H.; Cao, A.; Wang, H. Cytotoxicity and genotoxicity of low-dose vanadium dioxide nanoparticles to lung cells following long-term exposure. *Toxicology* **2021**, *459*, 152859. [CrossRef] [PubMed]
38. Crans, D.C.; Trujillo, A.M.; Pharazyn, P.S.; Cohen, M.D. How environment affects drug activity: Localization, compartmentalization and reactions of a vanadium insulin-enhancing compound, dipicolinatooxovanadium(V). *Coord. Chem. Rev.* **2011**, *19–20*, 2178–2192. [CrossRef]
39. Aureliano, M.; de Sousa-Coelho, A.L.; Dolan, C.C.; Roess, D.A.; Crans, D.C. Biological Consequences of Vanadium Effects on Formation of Reactive Oxygen Species and Lipid Peroxidation. *Int. J. Mol. Sci.* **2023**, *24*, 5382. [CrossRef]
40. Scibior, A.; Pietrzyk, L.; Plewa, Z.; Skiba, A. Vanadium: Risks and possible benefits in the light of a comprehensive overview of its pharmacological mechanisms and multi-application with a summary of further research trends. *J. Trace Elem. Med. Bio.* **2020**, *61*, 126508. [CrossRef]
41. Muller, A.; Krickemeyer, E.; Penk, M.; Walberg, H.J.; Bogge, H. Spherical Mixed-Valence $[V_{15}O_{36}]^{5\ominus}$, an Example from an Unusual Cluster Family. *Chem. Int. Ed.* **1987**, *26*, 1045–1046. [CrossRef]
42. Rhaiem, A.B.; Hlel, F.; Guidara, K.; Gargouri, M. Vibrational study of $[(CH_3)_4N]_2Cu_{0.5}Zn_{0.5}Cl_4$. *Spectrochim. Acta Part. A Mol. Biomol. Spectrosc.* **2007**, *66*, 1107–1109. [CrossRef] [PubMed]
43. Mabbs, F.E. Some aspects of the electron paramagnetic resonance spectroscopy of d-transition metal compounds. *Chem. Soc. Rev.* **1993**, *22*, 313–324. [CrossRef]
44. Postal, K.; Santana, F.S.; Hughes, D.L.; Rüdiger, A.L.; Ribeiro, R.R.; de Sá, E.L.; de Souza, E.M.; Soares, J.F.; Nunes, G.G. Stability in solution and chemoprotection by octadecavanadates(IV/V) in *E. coli* cultures. *J. Inorg. Biochem.* **2021**, *219*, 1114–1138. [CrossRef]
45. OECD. Test No. 423: Acute Oral Toxicity—Acute Toxic Class Method. In *OECD Guidelines for the Testing of Chemicals, Section 4*; OECD Publishing: Paris, France, 2002.
46. Rhule, J.T.; Hill, C.L.; Judd, D.A.; Schinazi, R.F. Polyoxometalates in medicine. *Chem. Rev.* **1998**, *98*, 327–358. [CrossRef] [PubMed]
47. Bijelic, A.; Aureliano, M.; Rompel, A. Polyoxometalates as potential next-generation metallodrugs in the combat against cancer. *Angew. Chem. Int. Ed.* **2019**, *58*, 2980–2999. [CrossRef] [PubMed]
48. Bijelic, A.; Aureliano, M.; Rompel, A. The antibacterial activity of polyoxometalates: Structures, antibiotic effects and future perspectives. *Chem. Commun.* **2018**, *54*, 1153–1169. [CrossRef] [PubMed]
49. Fang, Y.; Xing, C.; Zhan, S.; Zhao, M.; Li, M.; Liu, H. A polyoxometalate-modified magnetic nanocomposite: A promising antibacterial material for water treatment. *J. Mater. Chem. B* **2019**, *7*, 1933–1944. [CrossRef]
50. Kiss, T.; Jakusch, T.; Hollender, D.; Dörnyei, A.; Enyedy, E.A.; Pessoa, J.C.; Sakurai, H.; Sanz-Medel, A. Biospeciation of antidiabetic VO(IV) complexes. *Coord. Chem. Rev.* **2008**, *252*, 1153–1162. [CrossRef]
51. Pessoa, J.C.; Santos, M.F.A.; Correia, I.; Sanna, D.; Sciortino, G.; Garribba, E. Binding of vanadium ions and complexes to proteins and enzymes in aqueous solution. *Coord. Chem. Rev.* **2021**, *449*, 214192. [CrossRef]
52. Rehder, D. The potentiality of vanadium in medicinal applications. *Future Med. Chem.* **2012**, *4*, 1823–1837. [CrossRef]
53. Corona-Motolinia, N.D.; Martínez-Valencia, B.; Noriega, L.; Sánchez-Gaytán, B.L.; Melendez, F.J.; García-García, A.; Choquesillo-Lazarte, D.; Rodríguez-Diéguez, A.; Castro, M.E.; González-Vergara, E. Tris(2-Pyridylmethylamine)V(O)$_2$ Complexes as Counter Ions of Diprotonated Decavanadate Anion: Potential Antineoplastic Activity. *Front. Chem.* **2022**, *10*, 830511. [CrossRef] [PubMed]

54. García-García, A.; Noriega, L.; Meléndez-Bustamante, F.J.; Castro, M.E.; Sánchez-Gaytán, B.L.; Choquesillo-Lazarte, D.; González-Vergara, E.; Rodríguez-Diéguez, A. 2-Aminopyrimidinium Decavanadate: Experimental and Theoretical Characterization, Molecular Docking, and Potential Antineoplastic Activity. *Inorganics* **2021**, *9*, 67. [CrossRef]
55. Olaolorun, F.A.; Obasa, A.A.; Balogun, H.A.; Alnda, O.O.; Olopade, J.O. Lactational vitamin E protects against the histotoxic effects of systematically administers vanadium in neonatal rats. *Niger. J. Physiol. Sci.* **2014**, *29*, 125–129. Available online: https://pubmed.ncbi.nlm.nih.gov/26196578/ (accessed on 14 June 2023). [PubMed]
56. Adebiyi, O.E.; Obisesan, A.D.; Olayemie, F.O.; Olopase, J.O. Protective effect of ethanolic extract of *Grewia carpinifolia* leaves on vanadium induced toxicity. *Alex. J. Vet. Sci.* **2015**, *47*, 1–6. [CrossRef]
57. Beusen, J.M.; Neven, B. Toxicity of vanadium to different freshwater organisms. *Bull. Environ. Contam. Toxicol.* **1987**, *39*, 194–201. [CrossRef] [PubMed]
58. Borges, G.; Medonça, P.; Joaquim, N.; Coucelo, J.; Aureliano, M. Acute effects of vanadate oligomers on heart, kidney, and liver histology in the Lusitanian toadsifh (*Halobatrachus didactylys*). *Arch. Environ. Contam. Toxicol.* **2003**, *45*, 415–422. [CrossRef] [PubMed]
59. Roberts, R.; Callander, R.; Duffy, P.; Jacobesen, M.; Knight, R.; Boobis, A. Target organ profiles in toxicity studies supporting human dosing: Does severity progress with longer duration of exposure? *Regul. Toxicol. Pharmacol.* **2015**, *73*, 737–746. [CrossRef]
60. Lipnick, R.L.; Cotruvo, J.A.; Hill, R.N.; Bruce, R.D.; Stitzel, K.A.; Walker, A.P.; Chu, I.; Goddard, M.; Segal, L.; Springer, J.A.; et al. Comparison of the Up-and-down, Conventional LD50, and Fixed-Dose Acute Toxicity Procedures. *Food Chem. Toxicol.* **1995**, *33*, 223–231. [CrossRef]
61. Meyer, B.N.; Ferrigni, N.R.; Putnam, J.E.; Jacobsen, L.B.; Nichols, D.E.; McLaughilin, J.L. Brine shrimp: A convenient general bioassay for active plant constituents. *Planta Medica* **1982**, *45*, 31–34. [CrossRef]
62. Sarah, Q.S.; Anny, F.; Misbahuddin, M. Brine shrimp lethality assay. *Bangladesh J. Pharmacol.* **2017**, *12*, 5. [CrossRef]
63. David, J.P.; Silva, E.F.; Moura, D.L.; Guedes, M.L.; Assunção, R.J.; David, J.M. Lignanas e triterpenos do extrato citotóxico de *Eriope blanchetii*. *Quim. Nova* **2001**, *24*, 730–733. [CrossRef]
64. Yang, J.; Zhang, Z.; Jiang, S.; Zhang, M.; Lu, J.; Huang, L.; Zhang, T.; Gong, K.; Yan, S.; Yang, Z.; et al. Vanadate-induced antiproliferative and apoptotic response in esophageal squamous carcinoma cell line EC109. *J. Toxicol. Environ. Health* **2016**, *79*, 864–868. [CrossRef] [PubMed]
65. Montiel-Dávalos, A.; Gonzalez-Villava, A.; Rodriguéz-Lara, V.; Montaño, L.F.; Fortoul, T.I.; López-Marure, R. Vanadium pentoxide induces activation and death of endothelial cells. *J. Appl. Toxicol.* **2012**, *32*, 26–33. [CrossRef] [PubMed]
66. Rodrígues-Mercado, J.J.; Roldán-Reyes, E.; Altamirano-Lozano, M.A. Genotoxic effects of vanadium(IV) in human peripheral blood cells. *Toxicol. Lett.* **2003**, *144*, 359–369. [CrossRef] [PubMed]
67. Kamboj, V.P.; Kar, A.B. Antitesticular effect of metallic and rate earth salts. *J. Reprod. Fertil.* **1964**, *7*, 21–28. [CrossRef] [PubMed]
68. Rodríguez-Lara, V.; Morales-Rivero, A.; Rivera-Cambas, A.M.; Fortoul, T.I. Vanadium inhalation induces actin changes in mice testicular cells. *Toxicol. Ind. Health* **2016**, *32*, 367–374. [CrossRef] [PubMed]
69. Srivastava, A.K. Anti-Diabetic and Toxic Effects of Vanadium Compounds. *Mol. Cell. Biochem.* **2000**, *206*, 177–182. [CrossRef]
70. Wilk, A.; Szypulska-Koziarska, D.; Wiszniewska, B. The toxicity of vanadium on gastrointestinal, urinary and reproductive system, and its influence on fertility and fetuses malformations. *Adv. Hyg. Exp. Med.* **2017**, *71*, 850–859. [CrossRef]
71. Domingo, J.L. Vanadium: A review of the reproductive and developmental toxicity. *Reprod. Toxicol.* **1996**, *10*, 175–182. [CrossRef]
72. Shrivastava, S.; Joshi, D.; Bhadauria, M.; Shukla, S.; Mathur, R. Cotherapy of tiron and selenium Against vanadium induced toxic effects in lactating rats. *Iran. J. Reprod. Med.* **2011**, *9*, 229–238. Available online: https://pubmed.ncbi.nlm.nih.gov/26396569/ (accessed on 14 June 2023).
73. Sánchez-González, C.; Rivas-García, L.; López-Chaves, C.; Rodríguez-Nogales, A.; ALgieri, F.; Gálvez, J.; Gómez-Aracena, J.; Vera-Ramírez, L.; Montes-Bayon, M.; Sanz-Medel, L.; et al. Exposure to bis(maltolato)oxovanadium(IV) increases levels of hepcidin mRNA and impairs the homeostasis of iron but not that of manganese. *J. Food Chem. Toxicol.* **2014**, *73*, 113–118. [CrossRef] [PubMed]
74. Dai, S.; McNeill, J.H. One-year treatment of non-diabetic and streptozotocin-diabetic rats with vanadyl sulphate did not alter blood pressure or haematological indices. *Pharmacol. Toxicol.* **1994**, *74*, 110–115. [CrossRef] [PubMed]
75. Gonzalez-Villalva, A.; Fortoul, T.I.; Avila-Costa, M.R.; Piñon-Zarate, G.; Rodriguez-Lara, V.; Martinéz-Levi, G.; Rojas-Lemus, M.; Bizarro-Nevarez, P.; Díaz-Bech, P.; Mussali-Galante, P.; et al. Thrombocytosis induced in mice after subacute and subchronic V_2O_5 inhalation. *Toxicol. Ind. Health* **2006**, *22*, 113–116. [CrossRef] [PubMed]
76. Sanna, D.; Garribba, E. Pharmacologically active vanadium species: Distribution in biological media and interaction with molecular targets. *Curr. Med. Chem.* **2021**, *28*, 7339–7384. [CrossRef]
77. Soares, S.S.; Martins, H.; Aureliano, M. Vanadium distribution following decavanadate administration. *Arch. Environ. Contam. Toxicol.* **2006**, *50*, 60–64. [CrossRef] [PubMed]
78. Espionsa-Zurutuza, M.; González-Villalva, A.; Albarrán-Alonso, J.C.; Colín-Barenque, L.; Bizarro-Nevares, P.; Rojas-Lemus, M.; López-Valdez, N.; Fortoul, T.I. Oxidative stress as a mechanism involved in kidney damage after subchronic exposure to vanadium inhalation and oral sweetened beverages in a mouse model. *Int. J. Toxicol.* **2018**, *37*, 45–52. [CrossRef]
79. Ghosh, S.K.; Saha, R.; Saha, B. Toxicity of inorganic vanadium compounds. *Res. Chem. Intermed.* **2015**, *41*, 4873–4897. [CrossRef]
80. Donaldson, J.; Hemming, R.; Labella, F. Vanadium exposure enhances lipid peroxidation in the kidney of rats and mice. *Can. J. Physiol. Pharmacol.* **1984**, *63*, 1961–1999. [CrossRef]

81. Sanchez-Gonzalez, C.; Bermudez-Peña, C.; Trenzado, C.E.; Goenaga-Infante, H.; Montes-Bayon, M.; Sanz-Medel, A.; Llopis, J. Change in the antioxidant defense in selenium concentration in tissues of vanadium exposed rats. *Metallomics* **2012**, *4*, 814–819. [CrossRef]
82. Koubaa, F.G.; Chaâbane, M.; Turki, M.; Ayadi, F.M.; El Feki, A. Anti-oxidant and hepatoprotective effects of *Salvia officinalis* essential oil against vanadium-induced oxidative stress and histological changes in the rat liver. *Environ. Sci. Pollut. Res. Int.* **2021**, *28*, 11001–11015. [CrossRef]
83. Oster, M.H.; Llobet, J.M.; Domingo, J.L.; Bruce German, J.; Keen, C.L. Vanadium treatment of diabetic Sprague-Dawley rats results in tissue vanadium accumulation and pro-oxidant effects. *Toxicology* **1993**, *83*, 115–130. [CrossRef]
84. Gândara, R.M.C.; Soares, S.S.; Martins, H.; Gutiérrez-Merino, C.; Aureliano, M. Vanadate oligomers: In vivo effects in hepatic vanadium accumulation and stress markers. *J. Inorg. Biochem.* **2005**, *99*, 238–1244. [CrossRef] [PubMed]
85. Nnama, A.U.; Ekeh, F.N.; Aguzie, I.F.; Udegbuman, S.L.; Nwani, C.D. Vanadium pentoxide induces hematological, oxidative stress and histological chages in *Oryctolagus cuniculus*. *J. Hazard. Mater. Adv.* **2022**, *5*, 100048. [CrossRef]
86. Chen, Y.; Ji, L.L.; Liu, T.Y. Evaluation of gender-related differences in various oxidative stress in enzymes in mice. *Chin. J. Physiol.* **2011**, *54*, 385–390. [CrossRef] [PubMed]
87. Kander, M.C.; Cui, Y.; Liu, Z. Gender difference in oxidative stress: A new look at the mechanisms for cardiovascular diseases. *J. Cell. Mol. Med.* **2017**, *21*, 1024–1032. [CrossRef] [PubMed]
88. Bishayee, A.; Oinam, S.; Basu, M.; Chatterjee, M. Vanadium chemoprevention of 7,12-dimethylbenz(a)anthracene-induced rat mammary carcinogenesis: Probable involvement of representative hepatic phase I and II xenobiotic metabolizing enzymes. *Breast Cancer Res. Treat.* **2000**, *63*, 133–145. [CrossRef] [PubMed]
89. Thompson, K.H.; McNeill, J.H. Effect of vanadyl sulfate feeding on susceptibility to peroxidative change in diabetic rats. *Res. Commun. Chem. Pathol. Pharmacol.* **1993**, *80*, 187–200. Available online: https://pubmed.ncbi.nlm.nih.gov/8100638 (accessed on 14 June 2023).
90. Crans, D.C.; Zhang, B.; Gaidamauskas, E.; Keramidas, A.D.; Willsky, G.R.; Roberts, C.R. Is vanadate reduced by thiols under biological conditions?: Changing the redox potential of V(V)/V(IV) by complexation in aqueous solution. *Inorg. Chem.* **2010**, *49*, 4245–4256. [CrossRef]
91. Kang, D.H.; Nakagawa, T.; Feng, L.; Watanabe, S.; Han, L.; Mazzali, M.; Truong, L.; Harris, R.; Johnson, R.J. A role for uric acid in the progression of renal disease. *J. Am. Soc. Nephrol.* **2002**, *13*, 2888–2897. [CrossRef]
92. Ozer, J.; Ratner, M.; Shaw, M.; Bailey, W.; Schomaker, S. The current state of serum biomarkers of hepatotoxicity. *Toxicology* **2008**, *245*, 194–205. [CrossRef]
93. Liu, J.; Cui, H.; Liu, X.; Peng, X.; Deng, H.; Zuo, Z.; Cui, W.; Deng, Y.; Wang, K. Dietary high vanadium causes oxidative damage-induced renal and hepatic toxicity in broilers. *Biol. Trace Elem. Res.* **2012**, *145*, 189–200. [CrossRef] [PubMed]
94. Ma, Y.; Shi, Y.; Wu, Q.; Ma, W. Epigallocatechin-3-gallate alleviates vanadium-induced reduction of antioxidant capacity via keap1-Nrf2-sMaf pathway in the liver, kidney, and ovary of laying hens. *Biol. Trace Elem. Res.* **2021**, *199*, 2707–2716. [CrossRef] [PubMed]
95. Cano-Gutiérrez, G.; Acevedo-Nava, S.; Santamaría, A.; Altamirano-Lozano, M.; Cano-Rodríguez, M.C.; Fortoul, T.I. Hepatic megalocytosis due to vanadium inhalation: Participation of oxidative stress. *Toxicol. Ind. Health* **2012**, *28*, 353–360. [CrossRef] [PubMed]
96. Fortoul, T.; Rodriguez-Lara, V.; González-Villalva, A.; Rojas-Lemus, M.; Cano-Gutiérrez, G.; Ustarroz-Cano, M.; Colín-Barenque, L.; Bizarro-Nevares, P.; García-Pealez, I.; Montaño, L.; et al. Inhalation of vanadium pentoxide and its effects in a mouse model. *Inorg. Chim. Acta* **2014**, *420*, 9–15. [CrossRef]
97. Xiong, Z.; Xing, C.; Xu, T.; Yang, Y.; Liu, G.; Hu, G.; Cao, H.; Zhang, C.; Guo, X.; Yang, F. Vanadium Induces Oxidative Stress and Mitochondrial Quality Control Disorder in the Heart of Ducks. *Front. Vet. Sci.* **2021**, *26*, 756534. [CrossRef]
98. Wang, J.P.; Cui, R.Y.; Zhang, K.Y.; Ding, X.M.; Luo, Y.H.; Bai, S.P.; Zeng, Q.F.; Xuan, Y.; Su, Z.W. High-fat diet increased renal and hepatic oxidative stress induced by vanadium of Wistar rat. *Biol. Trace Elem. Res.* **2016**, *170*, 415–423. [CrossRef] [PubMed]
99. Macrae, C.F.; Sovago, I.; Cottrell, S.J.; Galek, P.T.; McCabe, P.; Pidcock, E.; Platings, M.; Shields, G.P.; Stevens, J.S.; Towler, M.; et al. Mercury 4.0: From visualization to analysis, design and prediction. *J. Appl. Crystallogr.* **2020**, *53*, 226–235. [CrossRef]
100. Barbosa, H.M.; Do Nascimento, J.N.; Araújo, T.A.S.; Duarte, F.S.; Albuquerque, U.P.; Vieira, J.R.C.; De Santana, E.R.B.; Yara, R.; Lima, C.S.A.; Gomes, D.A.; et al. Acute Toxicity and Cytotoxicity Effect of Ethanolic Extract of Spondias Tuberose Arruda Bark: Hematological, Biochemical and Histopathological Evaluation. *An. Acad. Bras. Cienc.* **2016**, *88*, 1993–2004. [CrossRef]
101. OECD. Test No 407: Repeated dose 28-day oral toxicity study in rodent. In *OECD Guidelines for the Testing of Chemicals, Section 4, Health Effects*; OECD Publishing: Paris, France, 2008.
102. Jiang, P.; Ni, Z.; Wang, B.; Ma, B.; Duan, H.; Li, X.; Ma, X.; Wei, Q.; Ji, X.; Liu, Q.; et al. Acute toxicity, twenty-eight days repeated dose toxicity and genotoxicity of vanadyl trehalose in Kunming mice. *Regul. Toxicol. Pharmacol.* **2017**, *85*, 86–97. [CrossRef]
103. Uchiyama, M.; Mihara, M. Determination of malonaldehyde precursor in tissues by thiobarbituric acid test. *Anal. Biochem.* **1978**, *86*, 271–278. [CrossRef]
104. Paglia, D.E.; Valentine, W.N. Studies on the quantitative and qualitative characterization of erythrocyte glutathione peroxidase. *J. Lab. Clin. Med.* **1967**, *70*, 158–169.
105. Hissin, P.J.; Hilf, R. A Fluorometric method for determination of oxidized and reduced glutathione in tissues. *Anal. Biochem.* **1976**, *74*, 214–226. [CrossRef]

106. Misra, H.P.; Fridovich, I. The role of superoxide anion in the autoxidation of epinephrine and a simple assay for superoxide dismutase. *J. Biol. Chem.* **1972**, *10*, 3170–3175. [CrossRef]
107. Aebi, H. Catalase in vitro. *Methods Enzymol.* **1984**, *105*, 121–126. [CrossRef]

pharmaceuticals

MDPI

Review

Potential Applications of Rare Earth Metal Nanoparticles in Biomedicine

Svetlana A. Titova [1], Maria P. Kruglova [1,*], Victor A. Stupin [2], Natalia E. Manturova [2] and Ekaterina V. Silina [1]

[1] I.M. Sechenov First Moscow State Medical University (Sechenov University), 119991 Moscow, Russia; honey.liebe@mail.ru (S.A.T.); silinaekaterina@mail.ru (E.V.S.)

[2] Pirogov Russian National Research Medical University, 117997 Moscow, Russia; stvictor@bk.ru (V.A.S.); manturovanatali@yandex.ru (N.E.M.)

* Correspondence: marykruglova@live.ru; Tel.: +7-906-071-97-27

Abstract: In recent years, the world scientific community has shown increasing interest in rare earth metals in general and their nanoparticles in particular. Medicine and pharmaceuticals are no exception in this matter. In this review, we have considered the main opportunities and potential applications of rare earth metal (gadolinium, europium, ytterbium, holmium, lutetium, dysprosium, erbium, terbium, thulium, scandium, yttrium, lanthanum, europium, neodymium, promethium, samarium, praseodymium, cerium) nanoparticles in biomedicine, with data ranging from single reports of effects found in vitro to numerous independent in vivo studies, as well as a number of challenges to their potential for wider application. The main areas of application of rare earth metals, including in the future, are diagnosis and treatment of malignant neoplasms, therapy of infections, as well as the use of antioxidant and regenerative properties of a number of nanoparticles. These applications are determined both by the properties of rare earth metal nanoparticles themselves and the need to search for new approaches to solve a number of urgent biomedical and public health problems. Oxide forms of lanthanides are most often used in biomedicine due to their greatest biocompatibility and nanoscale size, providing penetration through biological membranes. However, the existing contradictory or insufficient data on acute and chronic toxicity of lanthanides still make their widespread use difficult. There are various modification methods (addition of excipients, creation of nanocomposites, and changing the morphology of particles) that can reduce these effects. At the same time, despite the use of some representatives of lanthanides in clinical practice, further studies to establish the full range of pharmacological and toxic effects, as well as the search for approaches to modify nanoparticles remain relevant.

Keywords: nanoparticles; rare earth metals; lanthanides; rare earth oxides; contrast agents; antitumor effect; antibacterial effect; antioxidant effect; regenerative effect

1. Introduction

At present, there are a number of primary and urgent problems in modern healthcare that require new solutions and approaches. Such problems include the issues of diagnostics of malignant neoplasms and their therapy [1,2], the growth of antibiotic resistance to modern antibacterial drugs [3], as well as the issues of regeneration in skin and soft tissue damage, especially in cases of decreased immune resistance as in diabetes mellitus, tissue ischemia, and many others [4,5].

One of the promising directions for solving a number of these problems is the use of nanoparticles of rare earth metals. The widely developing field of nanotechnology, which is already widely used in various spheres [6–10], can be considered as an alternative platform for the development of new therapeutic agents for solving a number of urgent biomedical tasks [11,12].

Rare earth metals include metals with atomic masses from 57 to 71 atomic mass units (a.m.u.), as well as 21 and 39 a.m.u. Representatives of rare earth metals are characterized by variable oxidation degrees, which determines their redox properties [13]. The study of the possibility of their application in medicine is a relatively new direction, and it is limited by a number of factors such as the complexity of the synthesis process with the preservation of nanoscale, requiring special conditions and specific equipment [14,15], and most importantly, contradictory data on acute and chronic toxicity of metal nanoparticles [16]. But despite this, some rare earth metals have already gone beyond research interest and have been introduced into clinical practice [17]. For example, gadolinium is currently used as a contrast agent in diagnostic manipulations [18], europium is a promising candidate for the production of drugs for photodynamic therapy of glioma, breast cancer, and cervical cancer [19,20].

The most studied to date are cerium dioxide nanoparticles (nanoceria). According to the international database ScienceDirect, the number of publications on the query "cerium oxide nanoparticles", devoted to their biological effect, is 3069. More than 1800 of them are represented by original research. Other representatives of this chemical group are much less discussed in the medical and pharmaceutical literature. Thus, the number of publications on biomedical applications of lanthanides in the ScienceDirect database is represented by the following number of articles on similar requests: lanthanum is devoted to 106 publications, praseodymium to 25, neodymium to 48, gadolinium to 200, europium to 53, ytterbium to 18, holmium, lutetium, erbium, terbium, prometium, dysprosium, samarium to 11, 6, 18, 23, 0, 22, and 48 articles, respectively. The biological effects of yttrium, scandium, and thulium oxide nanoparticles are even more poorly discussed in the literature.

However, the question of safety and at the same time preserving or even enhancing the effectiveness of lanthanide nanoparticles, and therefore the possibility of their wider application, is still acute, and it depends not only on their own properties, but also on the possibility of using a number of modifications to ensure the realization of the necessary effects. First of all, it is the chemical composition. A general group feature of rare earth metal nanoparticles is the possibility of their use in the form of oxides characterized by a lower risk of unwanted reactions on the part of living organisms. Compared to other forms of nanoparticles, oxides have on average an order of magnitude smaller size, which promotes better permeability through cellular barriers [21,22]. Other modification methods include changing the degree of oxidation [23], using excipients (particularly polymers) [24], modifying morphology [25], preparing nanocomposites [26] and bioconjugation [27]. All of these may allow their use for biomedical purposes more widely in the future with a relevant level of safety.

Thus, the purpose of this review is to summarize the existing topical applications of rare earth metal oxide nanoparticles in biomedicine, characterize each of them in terms of biological activity, focusing on little-known representatives, as well as to outline possible prospects and limitations for this group of elements and existing methods of their modification. When writing this review, we analyzed publications of medical and pharmaceutical topics in the international databases ScienceDirect and PubMed. A brief characterization of rare earth metal nanoparticles in accordance with the objectives of our work is shown in Figure 1.

Figure 1. Rare earth metal nanoparticles as agents for biomedical applications.

2. Nanoparticles of the Yttrium Subgroup

The yttrium subgroup includes such rare earth metals as gadolinium, ytterbium, holmium, lutetium, dysprosium, erbium, terbium, yttrium, scandium, and thulium.

2.1. Gadolinium Nanoparticles

At present, gadolinium nanoparticles of different compositions have been developed: bisphosphonate [28], phosphate [29], gadolinium ascorbate [30], multicomponent nanocomposites [31], and gadolinium oxide nanoparticles, which are the most studied [32,33]. In current clinical practice, gadolinium drugs are represented by contrast agents such as Magnevist®, Omniscan®, and Gadovist® [34–36]. Despite high imaging efficacy, the use of registered gadolinium preparations is associated with a significant risk of severe adverse reactions such as nephrogenic systemic fibrosis [37,38]. The mechanism of development of this phenomenon is associated with the ability of gadolinium compounds to dissociate and further accumulate in body tissues [39]. The use of a non-ionized form of gadolinium, particularly an oxide form, has been proposed as a possible solution to this problem.

Gadolinium oxide is a stable gadolinium (Gd) derivative that exhibits structure polymorphism depending on the method of synthesis and the use of doping materials [40]. In an in vivo comparative study with mice by Dai Y. et al. [41], PEGylated gadolinium oxide nanoparticles had less hepato- and nephrotoxic effects compared to Magnevist®. At the same time, the efficiency of tumor tissue visualization exceeded that of the reference drug. However, it should be noted that it cannot be clearly established whether the observed changes in the efficacy and safety profile are related to the nanoscale of gadolinium or to its modification by the excipient [41,42]. A similar result was achieved when compared with Omniscan®, which was also confirmed in in vitro and in vivo experiments [43]. Other possible pharmaceutical compositions aimed at reducing the risks of gadolinium-containing agents also include biocompatible polymers in the composition—such as polycyclodextrin [44], dextran [45,46], polyacrylic acid [47], or polymethylvinyl ether of maleic acid [48], for example. In addition to numerous in vitro studies, the safety of formulations containing biocompatible polymers and gadolinium oxide has been confirmed in an experiment using nephrectomized rats. Hanieh Ashouri et al. concluded that due to such modifications of nanoparticles, minimal accumulation of gadolinium can be achieved even in patients with kidney disease [44].

However, current research is not limited to the development of safer forms of gadolinium for medical use. For example, the already described combination of gadolinium oxide and polyacrylic acid nanoparticles in complex with arginylglycylaspartic acid increase targeting of tumor cells by increasing permeability through blood vessels and increased accumulation in cells, which was confirmed by comparing accumulation in normal NCTC1469 liver cells and the U87M glioblastoma cell line in a mouse model [49]. Another illustration of the promise of gadolinium nanoparticles as an antitumor agent is the study by Shailja Kumar et al., in which in addition to surface modification by PEGylation, the nanoparticles were loaded with doxorubicin. Efficacy was demonstrated in in vitro experiments using tumor cell cultures of the A-549 alveolar carcinoma, PANC-1 pancreatic carcinoma, and U-87 glioblastoma cell lines [50]. Comparable results were presented with the improved combination of gadolinium oxide and doxorubicin. This composition was supplemented with β-cyclodextrin and folic acid, giving the properties of a system with pH-sensitive release of the active ingredient, extending the spectrum of potential applications to solid tumors [51].

Another area of research into the possibility of using gadolinium oxide in oncology was the study of the therapeutic properties of this nanoparticle [52]. A number of publications have described its radiosensitizing properties [53,54]. For example, the 2019 work demonstrated activity against B16F10 melanoma cells [55]. A possible mechanism explaining this property is the established induction of oxidative stress causing DNA damage. According to the researchers' conclusion, these damages exceeded the fragmentation caused by isolated irradiation by 52% [56]. An important feature is the retention of this effect when attempting to target glioblastoma cells not only in the GL261 cell line [57], but also in vivo, although with the need for additional functionalization [58]. Its ability to affect the reduction in PTX3 protein expression was also established, which indirectly indicates the possibility of slowing down the metastatic process [59]. However, it should be noted that the described properties are dose- and time-dependent, which, together with the ability to pass through the blood–brain barrier, may complicate the prediction of the intensity of apoptosis and necrotic processes when trying to apply this information in clinical practice [58,60,61]. This observation brings us back to the question of the need to balance risk and benefit in the use of metallic nanoparticles.

It should be noted that the prospects for the use of gadolinium oxide nanoparticles as a drug go beyond the diagnosis and therapy of malignant neoplasms [54,62–66]. In modern literature it is reported that gadolinium has antimicrobial properties. According to the results of studies, it exhibits antibacterial activity against *Escherichia coli*, *Staphylococcus aureus*, *S. Typhimurium*, and antifungal activity against *Candida albicans*, *C. glabrata*, and *Fusarium oxysporum* [67–69]. Also, according to 2023 studies, gadolinium oxide can be used as an immunosensor to detect Vibrio cholerae toxin [70,71].

One of the promising and relevant areas of application of gadolinium oxide nanoparticles is its possible application in the field of regenerative medicine. In the studies of Vijayan V. et al., it was found that gadolinium oxide in the hydrogel of collagen and dextran shows anti-inflammatory and regenerative properties in corneal explant and inhibits neovascularization in endothelial cells of umbilical vein EA.hy926 [72]. In addition, bisphosphonate-functionalized gadolinium nanoparticles were found to be suitable as bone tissue engineering agents. At concentrations up to 10×10^{-6} m, the nanoparticles induced cell proliferation of primary human osteoblast cells, promoting bone regeneration. The efficacy was confirmed by in vivo MRI imaging. [73]. Large-scale studies conducted on human cell cultures have established the antioxidant activity of gadolinium oxide nanoparticles, acceleration of the metabolism of fibroblasts BJ TERT (by 27%), and proliferation of keratinocytes HaCaT (by 35%) in a wound surface model [74]. It was also found that when gadolinium was incorporated into a hydrogel containing the biopolymers dextran, collagen, and gelatin, a synergistic regenerative effect ex vivo was achieved [72]. However, the application of gadolinium in the above directions requires further research. A significant feature of gadolinium oxide nanoparticles is the fact that no cytotoxic effects were found in studies using cultures of SK-MEL-28 melanoma cells, adipose-derived mesenchymal stromal cells (ASC) [45], A-549 lung carcinoma cells, U-87 glioblastoma cells, HEK-293 normal cells [46], and the DU145 human prostate cancer cell line [48].

Thus, at the moment, the most studied and widespread application of gadolinium nanoparticles is MRI diagnostics and therapy of malignant neoplasms.

2.2. Ytterbium Nanoparticles

Ytterbium is most often used in biomedicine in the form of oxide, which is a compound that has a variable morphology depending on the method of synthesis, as well as the ability to luminescence [75]. A significant number of references to ytterbium oxide nanoparticles in the literature are devoted to their use as a potential antitumor agent [76–78]. It should be noted that ytterbium oxide has not been studied in isolation and was usually an element of complex stimulus-sensitive compositions. Thus, in the experiment of Peng H. et al., performed on MCF-7 breast adenocarcinoma cell cultures, ytterbium was part of a multicomponent shell that also included β-cyclodextrin, zinc oxide, and erbium deposited on an iron oxide core. According to researchers, this shell provided microwave-sensitive release of the active ingredient accompanied by controlled toxicity [77]. In a study with a different model drug, camptothecin, loaded on ytterbium ferrite nanoparticles coated with a polylactide-co-glycoside functionalized with a β-cyclodextrin-folate conjugate, a pH-dependent release of the active component was detected, which increased significantly in the slightly acidic environment characteristic of the tumor microenvironment. It was found that this composition has the ability for sustained prolonged release targeting folate receptors, but the results were also limited to in vitro experiments [78].

Special attention should be paid to the existing publications that have revealed antimicrobial properties of ytterbium oxide nanoparticles. The greatest contribution in this direction was made by the team of V. Muthulakshmi et al. who carried out a series of large-scale experiments in vitro. According to their results, ytterbium oxide exhibited an antibacterial effect against *E. coli* and *S. aureus*. The level of activity depended on the presence of a nanoparticle coating. Thus, when 1-butyl-3-methylimidazolium hexafluorophosphate was added, the result was superior to the activity of amikacin [79,80]. In other works, developing this direction, the possibility of using ytterbium oxide with chitosan as a dressing material was investigated. According to the results, the presence of antibacterial activity against Gram-positive and Gram-negative bacteria was confirmed, as well as the biocompatibility of this composition [81]. In addition to the study of antimicrobial action, V. Muthulakshmi et al. identified the potential use of ytterbium nanoparticles as anti-inflammatory, antioxidant, and diabetes drugs. Despite the fact that the effect of coated nanoparticles was superior to that of reference drugs (in particular, diclofenac and ascorbic acid), at the moment it is difficult to assess the real potential since the experiments were carried out under in vitro conditions and require further development of these directions [79,80].

In modern scientific publications, ytterbium oxide nanoparticles are considered not only as a potential therapeutic agent, but also as a diagnostic agent for a number of pathological conditions. To date, electrochemical sensors have been developed to determine blood glucose levels [82], dopamine [83], including dopamine and uric acid in the presence of ascorbic acid [84], and methadone [85].

A number of publications provide data on the cytotoxicity and antiproliferative effect of ytterbium. Thus, in the study of Wysokińska E. et al. for mouse embryonic fibroblast cells NIH3T3 and mouse macrophages RAW264.7 it was revealed that a significant loss of viability was observed at nanoparticle concentrations of 26 μg/mL and 1.6 μg/mL, respectively, although it should be noted that coating nanoparticles with polyethylene glycol reduced this effect, although it did not get rid of the inhibitory effect on proliferation [86]. Similar results were obtained in studies conducted on human hepatocellular carcinoma HepG2 cells. Wang C. et al. identified the mitochondrial-mediated pathway as a putative mechanism of apoptosis. The scientists also noticed that the toxicity of ytterbium oxide was more dependent on its surface properties rather than particle size [87]. Data from published in vivo studies report additional unwanted effects of ytterbium oxide. In particular, when administered through the respiratory tract, it induced oxidative stress in the lungs of mice with a histological picture of chronic inflammation [88].

In the course of analyzing the above-mentioned works, we can conclude that the application spectrum of ytterbium in biomedicine (mainly antitumor and antimicrobial) is rather narrow, which is determined by the data on its cytotoxic properties.

2.3. Holmium Nanoparticles

Holmium nanoparticles are represented by phosphate [89], acetylacetonate [90], and holmium oxide [91]. Their biological effects are poorly reported in the literature. The attention of the modern scientific community is more focused on the selection and standardization of optimal synthesis methods [92].

At present, it is known that chemically unmodified holmium oxide particles are capable of cumulation in the liver and spleen, as well as induction of immunologic reactions in vivo [93]. Also, like other lanthanides, it can be a potential candidate for the development of diagnostic imaging agents [94]. There are also references to its pharmacological action. In an in vivo study by Luo Y. et al. it was found that holmium oxide nanoparticles promoted angiogenesis by regulating the EphrinB2/VAV2/CDC42 signaling pathway, which also led to improved regeneration on the model of bone defects in rats [95]. On the other hand, Ahmed Mohamed H. E. et. al. performed experiments on cultures of MCF-7 ductal adenocarcinoma breast adenocarcinoma cells, 3T3 mouse embryonic fibroblasts, and in ovo experiments to determine the antioxidant and antidiabetic potential and revealed antiangiogenic properties [96]. Thus, the role of holmium oxide as a regulator of angiogenesis remains unclear.

One of the current areas of development is the use of the holmium-166 isotope in topical dosage forms, patches and medical devices (nanofiber dressings), for the therapy of squamous cell skin cancer [97–99]. Under in vivo conditions, a statistically significant increase in survival rate and reduction in tumor size were achieved [99].

Information on modified holmium oxide nanoparticles is also quite contradictory. In particular, the toxicity of their compounds with various polymers has been investigated. For example, in a study using DU145 prostate cancer cell cultures, the cytotoxicity of nanoparticles coated with polyethylenimine 1200 and polyethylenimine 60,000 was evaluated. After 48 hours, the signs of cytotoxicity were minimal. At the same time, free polyethylenimine decreased the percentage of viable cells by 22% [100]. Experiments evaluating PEGylated holmium oxide also demonstrated a favorable toxicity profile for L-929 fibroblasts at nanoparticle concentrations within 16 μg/mL [101].

Thus, the available data are insufficient to draw an unequivocal conclusion, for example, because of the presence of opposing effects regarding angiogenesis. However, potentially possible chemical modifications of holmium may be promising and provide the possibility of its application in biomedicine.

2.4. Lutetium Nanoparticles

Lutetium nanoparticles are usually represented either by multicomponent nanocomposites [102–104] or lutetium oxide. An interesting feature of publications devoted to the medical use of lutetium is the presence of data on the properties of its isotopes, in particular, lutetium 177 [105,106]. A study by Luna-Gutiérrez M. et al. evaluated its use as a radiopharmaceutical for intra-arterial administration, which provided a 7.5-fold reduction in prostate tumor size relative to the control group [105]. A similar study was carried out with hepatocellular carcinoma HepG2 cells. Additionally, it revealed a preferential accumulation of the isotope in the liver and spleen, which is in full agreement with the data obtained for other rare earth metals [106].

Regarding the cytotoxicity of lutetium, there are data on its effects on endothelial cells. The cytotoxicity of lutetium oxide to human HECV endothelial cells was investigated at concentrations ranging from 0.1 μg/mL to 10 μg/mL. The absence of toxic effects was found only with the lowest tested concentrations. However, it should be noted that the nanoparticles were doped with samarium; therefore, its influence on the result cannot be excluded because of the lack of control samples without samarium in the cytotoxicity study [107]. Despite this, during the study of lutetium nanoparticles in vivo, encouraging results were obtained. The data obtained by Liu Z. et. al. demonstrated not only the suitability of lutetium for use as a CT imaging agent but also the relatively rapid and complete elimination of lanthanide from the mouse body within one month [108].

Thus, information on biomedical applications of lutetium oxide nanoparticles is limited to isolated reports on its toxicity, toxicokinetics, and prospects for the development of its isotope.

2.5. Dysprosium Nanoparticles

Dysprosium nanoparticles include phosphate [109] and vanadate [110] in addition to oxide, but biological effects have been studied specifically for the oxide form. Dysprosium nanoparticles have been described as a potential candidate for MRI imaging drugs [109,111,112]. Through carbon coating, Yue H. et al. were able to achieve ultra-small nanoparticles with an average diameter of 3.0 nm and a hydrodynamic radius of 22.4 nm. The coated nanoparticles showed significantly less cytotoxicity than uncoated nanoparticles in an experiment on DU145 human prostate cancer cells and NCTC1469 mouse hepatocyte cells. In the next stage of the experiment, which was carried out in vivo, it was found that dysprosium oxide is an effective tool for visualization, and due to the specifics of its excretion, it can be used for diagnostic manipulation of various organs and tissues [112]. A similar result, including cytotoxicity, was achieved in the synthesis of dysprosium nanoparticles and nanorods conjugated with glucuronic acid. An interesting observation was the fact that nanorods, due to their larger size than nanoparticles, showed a longer period of accumulation in the liver of the experimental mouse, which led to a lethal outcome [113]. Thus, we can make an assumption that nanoparticles, rather than other morphological variants of nano-objects, should be considered as the priority area of research on rare earth metal nanoparticles. When the compositions were further complicated, namely, mixed gadolinium-dysprosium nanoparticles coated with hyaluronic acid, minimal cytotoxicity was maintained and the mixed nanoparticle showed better contrast than isolated gadolinium and dysprosium nanoparticles [114]. Another example of successful development of mixed nanoparticles could be iron oxide-dysprosium oxide particles. The obtained nanoparticles had a diameter of 4 nm and showed satisfactory safety in vitro. In an in vivo study, it was found that the combination of iron oxide and dysprosium oxide, as a diagnostic agent, can reliably differentiate early and moderate stages of liver fibrosis [115].

The literature also contains some reports on the possibility of using dysprosium oxide in other areas of medicine. For example, according to modern studies, it is possible to use dysprosium as an electrochemical detector for standardization and detection of substandard drugs in pharmaceutical activities [116,117].

Thus, dysprosium oxide nanoparticles can be considered as a tool for MRI diagnostics. However, when working with animals and in clinical practice, the need for the addition of excipients that could reduce cytotoxicity cannot be excluded.

2.6. Erbium Nanoparticles

As in the case of other lanthanides, data on possible applications of erbium are presented by the application of its oxides. Erbium oxide nanoparticles are a compound averaging 5 to 30 nm and are capable of luminescence [118]. Like other rare earth metals, erbium has been studied to evaluate the possibility of use as an imaging agent in MRI diagnostics. In an in vivo experiment by Tse J. J. et al. it was found that erbium oxide is a suitable agent for visualization of vessels less than 10 μm in diameter, making it one of the most promising candidates for further research in this direction [119].

It should be noted that this nanoparticle also has a fairly wide range of pharmacological activity. For example, several studies have demonstrated its antibacterial effect against *Staphylococcus aureus*, *Escherichia coli*, *Desmodesmus subspicatus*, and *Pseudomonas aeruginosa* with no effect against *Enterococcus faecalis* [120,121]. An interesting observation was the fact that antibacterial activity can be enhanced by additional ultraviolet irradiation [121]. There are also references to the hypoglycemic properties of erbium oxide. They were studied in vitro using α-glucosidase and α-amylase, as well as in vivo, with no signs of acute toxicity in animals [122].

Data about the presence of antitumor effects were also published, but the experiments were conducted only under in vitro conditions [120,123]. In the study by Safwat G. et al. the mechanism of action of erbium oxide nanoparticles on Hep-G2 hepatocellular carcinoma cells was described. They were found to inhibit cell proliferation and induce oxidative stress, resulting in DNA damage leading to apoptosis [123]. Interestingly, when working with the same cell culture, but with additional carbon coating of the nanoparticle, cytotoxicity was minimal [124]. However, it should be noted that according to Mohamed H. R. H. et al., using data obtained by working with HSF fibroblast cell cultures, the cytotoxicity of nanoparticles containing erbium can be characterized as dose-dependent and occurring with prolonged exposure, which requires additional chronic toxicity studies [125].

Antibacterial, antitumor, and hypoglycemic properties have been reported for erbium oxide nanoparticles, but evaluation of the possibility of using erbium in the therapy and diagnosis of diseases requires additional research.

2.7. Terbium Nanoparticles

Terbium nanoparticles are nano-objects with the property of fluorescence [126–129]. The largest-scale experiments devoted to the evaluation of pharmacological effects were conducted by P. Senthil Pandi et al. in 2024. The researchers evaluated anti-inflammatory, antidiabetic, antitumor, and antioxidant activity. According to their findings, terbium nanoparticles demonstrated in vitro efficacy comparable to diclofenac sodium, acarbose, and ascorbic acid [130]. The literature also provides information on the antibacterial activity of terbium oxide against *Escherichia coli* and *Staphylococcus aureus*, as well as regenerative properties in vivo [131].

Most studies on the biological effects of terbium oxide nanoparticles address the issues of its potential antitumor activity. For example, in the experiments of Sana Iram et al. using osteosarcoma cell cultures MG-63 and Saos-2, a significant cytotoxicity was established, which was assumed to be caused by the generation of reactive oxygen species, but no such effect was observed for normal osteoblasts obtained ex vivo [132]. The concentration dependence of toxicity was investigated by Eman I. Khalaf et al. The scientists found that for MCF-7 breast cancer cells and HepG2 hepatocellular carcinoma cells, a significant decrease in viability was observed only at a concentration of 100 μg/mL, which is a sufficiently high dose for nanoparticles [133]. Terbium has also been described as an effective agent of photodynamic therapy, including the results of in vivo experiments [134,135]. In a comparative study of nanoparticles doped with gadolinium and terbium, the latter demonstrated a 35% more pronounced cytostatic effect on U-251 MG glioblastoma cells [136].

Thus, information on the pharmacological effects of terbium oxide is currently insufficient. Given that most of the experiments mentioned in this section have been conducted in the last few years, we can hope that the study of the properties of these nanoparticles by the international scientific community will continue.

2.8. Thulium Nanoparticles

Thulium nanoparticles are represented by oxide [137] and hexacyanoferrate II [138]. In modern clinical practice, thulium is used mainly as a fiber laser in urology for lithotripsy and enucleation of the prostate gland [139–142]. Actual developments of thulium nanoparticles are devoted to their study as a visualization agent for malignant neoplasms, including microscopic ones, due to their up-conversion properties [143,144]. High fluorescent activity was demonstrated by Yang Yiwei et al. in in vivo experiments [145]. Also, thulium oxide nanoparticles, due to their specific uptake by tumor tissue, confirmed in work with models of breast tumors 4175 and MCF-7 and ovarian cancer SKOV3 [146], can be used as a radiosensitizer [137]. It should be mentioned that thulium-containing nanoparticles have selective toxicity against malignant tumor cells, such as those overexpressing HER2 [147], but have no therapeutic effect against GL261 and U-251 MG glioma cells [148]. A possible mechanism for the development of therapeutic action is the ability of thulium oxide to generate reactive oxygen species induced by near-infrared radiation [149,150]. A similar effect underlies the antibacterial properties of thulium nanoparticles demonstrated by Zhou Kun et al. against MRSA [151].

In addition to oncological applications, thulium nanoparticles have been investigated as a sensor. In particular, the possibility of their application as a sensor of glucose [138], diethylsilbestrol, and bisphenol A [152] was described.

It should be noted that in most cases researchers use thulium as a doping material for other nanoparticles rather than as an independent therapeutic or diagnostic agent [143,153–156].

Thus, despite the lack of confirmed cytotoxic effect [155] and immunogenicity [157], an objective assessment of the possibility of using thulium nanoparticles for medical applications requires additional studies of its activity and kinetics.

2.9. Scandium and Yttrium Nanoparticles

In addition to the 15 lanthanides, the group of rare earth metals includes scandium and yttrium. Scandium oxide nanoparticles are currently poorly understood as a means for medical applications. As potential applications of scandium, researchers have proposed nanocomposites with antibacterial properties (against *Escherichia coli*) [158], as well as suitable for noninvasive determination of glucose levels in exhaled air for monitoring diabetes control [159].

The current literature reports that scandium oxide has satisfactory biocompatibility without adverse effects on cell proliferation, which was established when working with the HOS TE 85 osteoblast-like culture line [160]. Assessment of the distribution of scandium in the body, carried out in in vivo experiments, showed that the greatest accumulation of this metal occurs in the liver and lungs, while the cumulation of nanoparticles in the lungs is dose-dependent and exceeds that of soluble forms of scandium [161].

Yttrium has already found application in real clinical practice in the form of the isotope yttrium 90 and is used for selective radiation therapy of malignant liver neoplasms [162], as well as brachytherapy of nonmelanoma skin cancer in combination with strontium 90 [163]. Further development of the direction led to the development of patches loaded with yttrium 90 microspheres, but the assessment of the possibility of application requires additional in vivo studies [164]. It is worth noting that the stable isotope of yttrium can also be applied in this direction, demonstrating therapeutic potential as a means of photodynamic therapy [165].

A significant number of studies conducted on laboratory animals report pronounced antioxidant properties of yttrium oxide nanoparticles [166–169]. According to the results of in vivo experiments, it can be concluded that the antioxidant activity of nano yttrium can have a therapeutic effect on a number of pancreatic pathologies. For example, Khurana A. et al. found a reduction in oxidative-nitrosative stress in models of both acute and chronic pancreatitis [170,171]. The pursuit of anti-diabetic properties has also led to positive findings [172]. For example, an increase in viability, ATP, and insulin production was observed in the pancreas of rats pretreated with hydrogen peroxide [173,174]. It is noteworthy that the favorable effect was also observed upon administration of diazinon, an organophosphorus insecticide, the main target of which in warm-blooded organisms is the pancreas. Several independent in vivo experiments revealed that yttrium oxide nanoparticles normalize the ATP/ADP ratio and the activity of caspases 3 and 9 and reduce the manifestations of oxidative stress caused by diazinon [175,176].

The potential application of antioxidant properties of yttrium oxide nanoparticles is not limited by the therapy of pancreatic diseases. Almost a comparable number of studies are devoted to the neuroprotective effects of nano yttrium. Initially demonstrated on cultures of hippocampal HT22 and pheochromocytoma PC12 nerve cells, an antioxidant effect [177,178] was confirmed in vivo. In the work of Baghaee P. et al., a weakening of the manifestation of cognitive deficits in the model of Alzheimer's disease was noted, presumably due to the reduction in mitochondrial dysfunction and neuroinflammation [179]. Neuroprotective effects were also achieved in models of lead [180] and silver oxide [181,182] poisoning, as revealed by assessing caspase 3 expression, Bax/Bcl-2 proteins, ADP/ATP ratio [180], and levels of oxidative biomarkers (such as superoxide dismutase levels, glutathione, etc.), as well as neurotransmitters (acetylcholinesterase, dopamine, and serotonin) in the brains of mice [181] and people with behavioral disorders [182].

Other prospects for the use of yttrium oxide nanoparticles are currently less studied. The literature describes the possibility of using yttrium nanoparticles in retinal degeneration in vivo [183], and they also demonstrate a selective and intense cytotoxic effect against

MDA-MB-231 triple-negative breast cancer cells, which reports antitumor potential [184], as well as a means for visualization of malignant neoplasms [185].

It should be noted that yttrium oxide nanoparticles are characterized by high biocompatibility, even at a dosage of 200 μg/ml/day [186,187]. Some studies describe manifestations of cytotoxicity against HEK293 embryonic kidney cells [188] and neurotoxicity by astrocyte cuproptosis in vivo [189]. At the same time, most researchers are inclined to the fact that the use of yttrium nanoparticles is associated with a low risk of unwanted effects.

Thus, the most studied effects for scandium nanoparticles are antibacterial, and the most studied effects for nanoyttrium nanoparticles are antioxidant.

3. Nanoparticles of Metal Oxides of the Cerium Subgroup

The cerium subgroup includes lanthanum, europium, neodymium, promethium, samarium, praseodymium, and cerium.

3.1. *Lanthanum Nanoparticles*

Lanthanum nanoparticles are a heterogeneous group of compounds, including hydroxides [190,191] and borides [192,193], possessing variability of morphology [194] and exhibiting scintillation [195]. This property prompted scientists to investigate the possibility of using lanthanum nanoparticles as antitumor agents. The work of Lu V. M. et al. presented data on the ability of lanthanum oxide nanoparticles to cross the blood–brain barrier during intravenous administration, as well as the manifestation of cytotoxicity against glioblastoma cells by generating reactive oxygen species, which was demonstrated in an in vivo experiment [196]. Similar results were demonstrated for PEGylated lanthanum nanoplatforms. In this study, also performed on an in vivo model of glioblastoma, the sensitivity of the composition to the pH of the tumor microenvironment and a higher efficiency of the damaging effect due to an increase in the specific surface area of the nanoparticle were revealed [197]. On the other hand, lanthanum nanoparticles can be considered not only as an independent antitumor agent, but also as a delivery system. In particular, the use of systems containing lanthanum and various radium isotopes for targeting micro metastases has been proposed [198].

Given the risk of systemic reactions, an alternative application of lanthanum nanoparticles in medicine may be its use as a local agent. Published studies report its antibacterial (*S. aureus* and *E. Coli*) and antioxidant properties that have been demonstrated in vitro, as well as a relatively small antifungal effect [199,200].

In addition to the therapeutic potential, lanthanum hydroxide nanoparticles coated with cysteine have been investigated as an electrochemical sensor for the diagnosis of oral carcinoma by detecting the marker Cyfra-21-1 [201]. Perhaps priority should be given to developing the potential of lanthanum nanoparticles as a diagnostic agent, as the mechanism of the identified antitumor action may be an issue in future attempts to apply lanthanum nanoparticles in clinical practice. The reason for this may be the fact that similar negative effects were detected both in HuH-7 liver tumor cells and in normal CHANG cell lines, however, with more severity in the former case [202,203]. Toxic effects of lanthanum have also been demonstrated in in vivo studies. When lanthanum oxide was injected into mice, a significant increase in the levels of superoxide dismutase, glutathione peroxidase, and caspase-3—markers of oxidative stress—was observed. The damaging effect was also proved by histological examination of the liver [204]. When administered orally, cumulation of lanthanum in bone tissue at a depth of up to 5 μm was noted [205]. However, in the study by Vijayan V. et al., also conducted in vivo, favorable effects were presented, indicating the regenerative effect of lanthanum-containing nanocomposites on bone tissue [206].

It should be noted that there are published data on the toxic effects of lanthanum in studies involving people whose activities were related to work with rare earth metals, with longer contact (5 years or more) leading to their cumulation in the body [207,208]. A special cause for concern is potential neurotoxicity. In vivo experiments have shown that lanthanum oxide nanoparticles can cause behavioral disorders, damage to hippocampal neurons, and a decrease in plasma levels of norepinephrine and acetylcholine [209,210]. As a possible way to reduce this toxic effect, Lu Yuan et al. proposed the use of an antioxidant from the group of carotenoids, which, according to the authors, had a neuroprotective effect due to the induction of the PI3K/AKT signaling pathway [210].

Despite the available data on cytotoxicity, including neurotoxicity, lanthanum can be considered as a promising antitumor agent, such as in the case of glioblastoma, which remains the most common malignant neoplasm of the brain with an extremely poor prognosis and limited possibilities of existing therapy [211]. On this point, we believe that the ratio of potential benefits to risks is a sufficient incentive to continue development in this area.

3.2. Europium Nanoparticles

Europium nanoparticles are a compound with luminescent properties and regulated photocatalytic activity [212]. Nanoeuropium has been investigated in many areas of biomedicine. As for other lanthanides, studies of europium as a tool for MRI imaging of tumors have been described in the literature [213].

Further development of the field led to experiments devoted to evaluating the possibility of using these nanoparticles as a radiopharmaceutical to visualize tumor tissue during its surgical removal. In a study by Zhenhua Hu et al. using a tumor model in laboratory mice, it was described that the use of europium allows the detection of neoplasms smaller than 1 mm, as well as residual carcinoma tissue without histological signs of toxicity [214]. Less encouraging data are presented in the work of Min Li et al. It confirms the information about the preferential accumulation of lanthanide nanoparticles in the liver and spleen during injection, but at the same time, there was an observed change in the population of immune cells in vivo, which, accordingly, increased the risk of unwanted reactions [215]. One of the possible directions for solving the problem of potential systemic toxicity is precisely the local use of europium nanocomposites as a means for photodynamic therapy of skin cancer when combined with photosensitizers, such as methylene blue [216–218]. Interestingly, the europium content in the nanocomposite and the radiation intensity were not directly related, and in the concentration range from 1 to 50%, the highest value was observed for 10% [219].

A safer variant may be the use of europium oxide nanoparticles for topical application. For example, as an antibacterial agent. The antimicrobial activity of europium oxide was found against Gram-positive and Gram-negative bacteria *E. coli*, *S. aureus*, and *S. typhimurium* [220]. At the same time, europium in nanocomposites was also found to exhibit antioxidant and regenerative properties, making it a potential candidate for the development of topical wound healing agents [221–223].

Another possible application of europium nanoparticles in medicine is as an electrochemical sensor. In a study by Tuğçe Teker et al. a carbon platform with the addition of dysprosium and europium oxides showed high sensitivity and selectivity in the determination of papaverine not only in pure form but also in biological fluids, which possibly makes it suitable for use in pharmaceutical chemistry and toxicology [224].

Based on the above, the application of europium oxide nanoparticles as an agent for visualization of malignant neoplasms is the most studied one.

3.3. Neodymium Nanoparticles

Based on the available data, neodymium is also used predominantly in the form of oxides. Neodymium oxide is a nano compound with photocatalytic properties and, depending on the method of synthesis, can exist as nanoparticles prone to aggregation or nanosheet formation [225]. The structure, optical properties, and morphology are influenced by the type of precursor [226] as well as the use of polyelectrolytes for stabilization [227]. Due to the sufficiently deep study of this topic, it became possible to use neodymium nanoparticles to produce hydrogel contact lenses, which in combination with methacrylic acid demonstrated high strength and optimal transparency [228]. Another promising example of the application of neodymium oxide as a medical device is an electrochemical sensor for the determination of adrenaline and tyrosine in drugs and biological samples. According to Biswas S. et al., this sensor is characterized by high specificity, selectivity, and repeatability of measurement results, and as a consequence, it can be used in clinical practice [229].

The therapeutic effect of neodymium, according to the current literature, although poorly studied, is characterized by a fairly wide spectrum. Muthulakshmi V. et al. in an in vitro study from 2022 report the presence of neodymium oxide antibacterial (against *E. coli* and *S. aureus*), antidiabetic, anti-inflammatory (determined in a test with bovine serum albumin), antioxidant, and antitumor properties (when working with MCF-7 breast adenocarcinoma cells) [230]. Comparable results were reported from experiments by M. Sundrarajan V. et al. conducted under similar conditions, indicating a satisfactory comparability of data [231]. Despite the achievement of all previously described effects, in vivo studies are required to confirm the acceptability of neodymium as an active pharmaceutical substance. A possible barrier to further development of drugs based on neodymium oxide is information about its toxicity. It was found that inhalation administration of neodymium oxide nanoparticles in bronchial epithelium produces an inflammatory reaction accompanied by an increase in the expression of interleukins 6 and 8, as well as an increase in the level of lactate dehydrogenase [232]. Also, the results of several studies have reported the ability of neodymium oxide to have a damaging effect on the DNA of the bronchial epithelium. It is noteworthy that the conclusion about the adverse effects of neodymium was based not only on the findings obtained when working with cultures of human bronchial epithelial cells 16HBE or laboratory animals, but also on the basis of examinations of people (a sample of 60–69 people) working with these nanoparticles [233,234].

Thus, neodymium oxide nanoparticles can be considered as a potentially effective antibacterial, antioxidant, antitumor, and anti-inflammatory agent, but more research is required to make a definitive conclusion on their acceptability for medical use.

3.4. Promethium Oxide

Promethium is characterized by high radioactivity, which limits its application [235]. Data on the biological properties of its oxide nanoparticles are limited by information about its ability to cause alveolar proteinosis when administered intratracheally to laboratory animals, which, however, is not its specificity, since a similar reaction is also observed for europium, samarium, dysprosium, and terbium oxides [236]. Accordingly, the use of promethium nanoparticles in medical practice is currently not possible.

3.5. Samarium Nanoparticles

Samarium nanoparticles are represented by various nanocomposites [237–239] as well as samarium oxides. The most studied is the effectiveness of samarium oxide nanoparticles as an antibacterial agent. This effect was found against such microorganisms as *Escherichia coli*, *S. aureus*, *Bacillus subtilis* [240,241], as well as *Pseudomonas aeruginosa* [242]. A fungicidal effect was also revealed for a wide range of microscopic fungi such as *Alternaria brassicae*,

Fusarium acuminatum, *Aspergillus flavus*, *Aspergillus penicillioides*, *Candida albicans*, and many others. It should be noted that the independent antimicrobial activity of samarium oxide is less than the effect of silver nanoparticles, but in the composition of complex nanoparticles, silver and samarium show greater efficacy [241]. It is possible that the role of samarium is not limited to the realization of mechanisms characteristic of rare earth metals [241] but is also associated with its ability to inactivate efflux by affecting the MFS genes encoding transfer proteins [243].

The literature also describes antidiabetic, antitumor, anti-inflammatory, and antioxidant properties of samarium oxide nanoparticles, determined in vitro, and, in general, demonstrating the maintenance of the general trend to determine these properties of rare earth metal nanoparticles in the scientific environment in recent years [240,244]. For example, samarium is currently used in the form of the radiopharmaceutical samarium 153 for the treatment of metastatic lesions [245].

Of greater interest is the study by Navid Mohammadian et al. in which samarium oxide was proposed as a specific and highly sensitive electrochemical sensor for the detection of the gene responsible for 20% of cases of amyotrophic lateral sclerosis; this was also tested in blood plasma, which indicates its potential in diagnosis [246]. At the same time, we cannot fail to mention the cytotoxic properties of samarium oxide nanoparticles, revealed when working with MC3T3-E1 preosteoblast cells. It is currently unknown whether this problem can be completely solved by the creation of complex nanoparticles, as was performed in the experiment of Milli Suchita Kujur et al., or by the use of samarium in ultralow doses [247].

The most studied effect for samarium nanoparticles is the antimicrobial effect, but based on research in recent years, we can assume that the potential range of applications will be expanded in the near future.

3.6. Praseodymium Nanoparticles

Praseodymium nanoparticles can be presented not only as oxide but also as hydroxide [248], and they can be modified with other metals or polymers. The antitumor properties of praseodymium nanoparticles are most frequently mentioned in the literature. In a study by Bakht M. K., praseodymium is characterized as a potential dual therapeutic agent in non-small cell lung cancer. It was found that it can be used as a radiopharmaceutical drug that also induces massive tumor cell death [249]. Additionally, loading of nanoparticles with doxorubicin with its controlled release and intensification of the effect due to photothermal transformation of the particle is possible [250]. In the experiment of Varnitha Manikantan et al. conducted on cultures of MCF-7 breast adenocarcinoma cells and Vero kidney epithelium, it was revealed that praseodymium oxide coated with poly-β-cyclodextrin is able to provide selective cytotoxic effects of 5-fluorouracil for tumor cells [251]. In addition, praseodymium can also be used as a diagnostic tool for malignant neoplasms. For example, Zhang Q. C. et al. investigated the possibility of its use as a gas sensor to detect traces of breast cancer marker (acetophenone) in exhaled air [252].

Praseodymium oxide nanoparticles have also been studied in other areas of medicine. For example, activity has been found against a wide range of microorganisms such as *S. aureus*, *E. coli*, *P. aeruginosa*, and *S. typhi* [253]. In the current literature, there are publications reporting evidence of regenerative [254] and antioxidant [255] properties of praseodymium nanoparticles, but they were part of complex compositions that also included collagen [254] and cerium [255], so it is impossible to make a definitive conclusion regarding the effect of a particular component.

Thus, praseodymium oxide nanoparticles can be characterized primarily as a potential antitumor agent as well as a drug for diagnosis. In addition, its antibacterial effects should also be considered as promising.

3.7. Cerium Nanoparticles

As a potential agent for medical applications, the modern scientific community primarily considers cerium oxides [256–263]. Due to the antioxidant properties of nanoceria, it has therapeutic potential in a wide range of pathologies. Thus, positive effects have been demonstrated in a number of studies: hepatoprotective effect [264] on models of carbon tetrachloride poisoning, non-alcoholic fatty liver disease of rats [265–267] and doxorubicin-induced lipid peroxidation [268], reduction in functional and biochemical manifestations of pulmonary hypertension and cardiomyocyte apoptosis in a model of pulmonary hypertension [269], long-term protection against acute damage to retinal neurons [270] and ischemic hippocampus [271], and radiation-induced lung injury [272].

Antioxidant action is also the basis of regenerative properties of cerium dioxide. Thus, during experiments on human fibroblast cultures, a dose-dependent induction of cell proliferation was established, as well as intensification of metabolism, manifested by an increase in elastin and collagen synthesis without signs of cytotoxicity [273]. Similar results were achieved when working with cultures of mesenchymal stem cells and keratinocytes [274]. Moreover, efficacy has been demonstrated ex vivo on spinal cord tissue subjected to mechanical traumatization [275], for non-sterile wounds in vivo [276], and in numerous experiments reporting intensification of osteogenesis, cell differentiation, and mineralization [277], including in periodontal regeneration [278,279]. Thus, we can consider cerium dioxide nanoparticles as a potential agent for regeneration of sterile and contaminated wound surfaces both in cosmetic medicine and in traumatology and dentistry.

Nanoceria also has an antitumor effect [280]. This property is most studied in relation to MCF-7 breast cancer cells, the dose-dependent cytotoxic effect for which it was selective [281–283]. A similar result was obtained when studying the effect on HT-29 colorectal cancer cells. It is noteworthy that an increase in cytotoxicity was observed when a hollow nanoparticle coated with SiO_2 was formed [284]. Another option for realizing the properties is the use of cerium dioxide derivatives as a means of photodynamic therapy, which was confirmed in the experiment with irradiation of triple negative breast cancer MDA-MB-231 and melanoma A375 cells by radiation with a wavelength of 465 nm [285]. Also, nanoceria can increase the sensitivity of L3.6pl pancreatic cancer cells to radiation exposure, which may be due to differences in local pH value [286]. In addition, cerium oxide nanoparticles can be loaded with an antitumor drug and thus play the role of a drug delivery system [287].

The spectrum of antibacterial effect of cerium dioxide nanoparticles extends to both Gram-positive and Gram-negative bacteria. In particular, the literature describes the effectiveness against such microorganisms as *Escherichia coli* [273,288–290], *Bacillus subtilis* [290], *Staphylococcus aureus* [288,289], *Pseudomonas aeruginosa* [289,291,292], *Salmonella typhimurium*, *Listeria monocytogenes*, *Bacillus cereus* [288], *Porphyromonas gingivalis*, and *Prevotella intermedia* [293].

Based on this, we can conclude that cerium dioxide nanoparticles have a wide range of actions and, although not yet used in biomedicine, are promising for further development.

4. Possible Directions of Modification of Rare Earth Metal Nanoparticles

Rare earth metal nanoparticles can be modified in several ways: conversion to a different oxidation degree, addition of excipients, creation of nanocomposites (discussed in the sections on the corresponding metals), formation of bioconjugates, and modification of particle morphology.

The first method is applicable for metals having variable oxidation degrees. At the same time, their nanoparticles, with the exception of cerium, can be stable in an uncharacteristic degree of oxidation mainly under specific conditions, which makes it impossible to use them in biomedicine [294].

Several ways of modifying lanthanides with excipients have been described in the literature. For example, the addition of excipients such as glycols allows the morphology, solubility, density, and optical properties of nanoparticles to be tuned [295].

For gadolinium, which is insoluble in oxide and toxic in ionic form [18], PEGylation using a microwave method has been proposed, leading to its solubilization [296]. In the course of further experiments using gadolinium, nanoparticles incorporated into liposomes with a targeting effect on pathologically altered vascular tissue were produced [297]. It is worth noting that dendrimers are considered to be an equally promising option for gadolinium modification. It is likely that their size (5–10 nm) contributes to better accumulation of substances in the tumor microenvironment. These technological methods can be used both in targeted drug delivery systems and for diagnostic purposes [298,299].

More complex pharmaceutical compositions have been investigated using ytterbium, gadolinium, and europium as examples. Hybrid micelles chemically linked to a complex of lanthanides and cyclenes were prepared. In evaluating the efficiency of accumulation of these compounds in tumor cells, experiments were performed with cultures of HeLa-type cervical cancer cells and mice with HeLa cell xenografts. Positive results were achieved for gadolinium and ytterbium [300]. For all the promising effects of gadolinium, a risk of systemic inflammatory reaction and fibrosis has been identified. The presumed cause of this phenomenon is the degradation of complexes under in vivo conditions. As a consequence, there is a need not only to use carriers that provide targeted action but also form strong chelates with rare earth metals [301]. An example of such a high-strength compound is an organometallic ytterbium nanoparticle bound to ferrocin and mannose. Ytterbium retained antitumor activity, while mannose provided release only during binding to the corresponding specific receptors of hepatocellular carcinoma cell line HepG2. The authors of the study emphasize that the therapeutic efficacy of this combination has been established only in vitro and requires additional study [302]. It is also inappropriate to project the results obtained on any combination of carbohydrates with lanthanides. In a study of the degree of binding of chitosan-modified D-mannose, it was observed that the strength of the complex varies significantly for metals with the same degree of oxidation. In particular, ionized europium is inferior to other trivalent metals, which requires further comparison of stability with micro- and ultramicroelements of the human body before in vivo application [303]. A similar situation is observed in attempts to combine lanthanides with alginate [304]. Anionic derivatives of cyclodextrin, on the contrary, have high biocompatibility in combination with the ability to reliably chelate lanthanides. However, this modification has another limitation: it has only been studied for MRI imaging applications, so we can neither confirm nor deny the retention of any pharmacological effects [305]. The main results of studies of interaction between rare earth metal nanoparticles and excipients are summarized in Table 1.

As a possible way to increase the biocompatibility of rare earth metal nanoparticles, it should be proposed to create their pharmaceutical compositions with biopolymers. These include such polymers as dextran [273], chitosan [306], alginate [307], hyaluronic acid [308], carboxymethylcellulose [309], collagen [310], and gelatin [311]. According to the results of studies of compositions of nanoceria and biopolymers with human fibroblast cell cultures, no signs of cytotoxicity were revealed [273,306]. At the same time, not only the preservation of antioxidant activity, but also a pronounced regenerative effect was revealed. Notably, metabolic activation prevailed over proliferative processes [273]. At the same time, a study by Verdiana Marchianò et al. described the need to increase the nanoparticle content when developing complex formulations containing PLGA-encapsulated nanoceria in gelatin due to the presumed decrease in particle availability [311].

Table 1. The result of the interaction of excipients and rare earth metal nanoparticles.

Metal	Excipient	Class of Compound	Effect	Type of Study	Sources
Gadolinium	Polyethyleneglycol	Biopolymer	Solubilization	In vitro	[296]
Gadolinium	Liposomes	Fatty substance	Targeted action on pathologically altered tissue	bEnd.5, RAW	[297]
Gadolinium	Polyamidoamine dendrimer	Biopolymer	Increased accumulation in the tumor microenvironment	In vivo	[298,299]
Gadolinium	Cyclen	Hybrid micelles	Increased accumulation in tumor cells	HeLa, in vivo	[300]
Ytterbium	Cyclen	Hybrid micelles	Increased accumulation in tumor cells	HeLa, in vivo	[300]
Ytterbium	Ferrocene and mannose	Metallocene and monosaccharide	Targeted antitumor	In vitro HepG2	[302]
Ytterbium	Anionic cyclodextrins	Oligomer	High strength of the complex	In vitro	[305]
Europium	Cyclen	Hybrid micelles	No effect	HeLa, in vivo	[300]
Europium	Chitosan and mannose	Biopolymer and monosaccharide	Low strength of the complex	In vitro	[303]

Thus, we can conclude that the strategy for selecting excipients in the research of rare earth metal nanoparticles as agents for biomedical applications includes the following main points: (1) Ensuring nanoscale. For this purpose, it is advisable to use substances with stabilizing properties (e.g., dextran stabilizing nanoceria) [273]. Depending on the particle size, both high biocompatibility (75 nm) and non-selective cytotoxic effect (100–160 nm) can be observed [312]. (2) If the risk of adverse reactions is high, preference should be given to excipients that form strong complexes with nanoparticles and thus slow down the release of the active substance and, consequently, the risk of acute toxicity [307]. Alternatively, we can consider the development of targeting systems that ensure the accumulation of nanoparticles in pathologically altered tissues [297,302]. (3) Preference should be given to biocompatible and biodegradable excipients that give the surface a positive charge, and excessive excipients should be avoided so that the pharmacological effect is not reduced [274,313].

Rare earth metal nanoparticles can be modified by bioconjugation. Currently, bioconjugation using interleukins, antibodies, and RNA has been described. Bioconjugation with interleukins 17, 13, and 8 is mentioned in the literature. A gelatin hydrogel coated with interleukin 17 demonstrated anti-inflammatory activity in an animal model of encephalitis [314]. Coating of gadolinium nanoparticles with interleukin 13 peptide increased the capture of the particle by tumor cells in a mouse model of glioblastoma [315]. It is possible to develop more complex compositions including nanoparticles containing ytterbium, erbium, gadolinium, and yttrium electrostatically bound to particles loaded with docetaxel, interleukin 8, and small interfering RNA. This multimodal platform allows simultaneous tumor imaging, gene therapy, and chemotherapy [316]. Conjugation with RNA has also been described for nanoceria. Conjugated with miR-146a, nanoparticles enhanced diabetic wound regeneration according to Carlos Zgheib et al. [317].

Bioconjugation with antibodies of rare earth metal nanoparticles has been described for europium, lanthanum, and cerium. All of them were conjugated with antibodies to interleukin 6. This immunoassay method has been characterized by researchers as fast, reproducible, highly sensitive, specific, and suitable for serum assays [318–320].

An alternative direction of modification is to change the morphology of nanoparticles. In addition to nanoparticles, nano-objects also include nanorods [321], nanocubes [322], and nanotubes [323]. It is noteworthy that the production of nanorods and nanocubes follows an identical technique and can be controlled by microwave radiation time. However, the morphology of nanorods should be prioritized due to their higher redox activity [322].

Cerium-aspartic acid nanotubes have also been described in the literature. It was observed that it was the combination of asparagic acid addition and tube morphology that allowed the catalytic activity of the colorimetric reaction to be achieved in the determination of cysteine levels, which was not found for the Ce $(NO_3)_3$ precursor. Thus, this nano-object can be used as a highly sensitive nano sensor for the determination of cysteine content [323], which has diagnostic value in chronic kidney disease [324]. In addition to new applications, changes in morphology may enhance the already known activity of the rare earth metal. Michelle de P. et el. described an enhanced antibacterial effect of cerium-gallium titanate nanotubes against *S. aureus* and *E. coli* microorganisms. According to the authors of the study, this was due to the increase in specific surface area relative to the nanoparticles [325]. Cerium-silver nanorods are also currently known. According to the authors' conclusions, this approach contributes to the long-term preservation of photocatalytic activity of the object [326].

Hybrid gadolinium oxide nanorods were also developed, which have a pronounced antibacterial effect against *Escherichia coli, Klebsiella pneumonia, Staphylococcus aureus*, and a small effect against *Micrococcus luteus*. It should be noted that no comparison has been made with the nanoparticle form, so a comparison of efficacy is not possible [327]. Europium nanorods demonstrated an improved safety profile in in vivo experiments. There was no significant accumulation of europium in liver, spleen, kidney, and lung tissues [328], which we have previously mentioned for europium nanoparticles.

It should be noted that other morphological variants of nano-objects have also been developed for other representatives of lanthanides, for example, nanoneedles and nanopolyhedrons of lanthanum oxide, but they have not been considered in the context of biomedical applications [329].

Thus, we can conclude that currently the most promising direction of modification we can consider is the use of excipients, as they allow us to reduce the risk of toxicity, increase the stability and biocompatibility of particles, and ensure targeting. At the same time, the main disadvantage of this direction is the difficulty in selecting the excipient, as there are no unambiguous methodological recommendations on this issue at the moment. Modification by bioconjugation is likely to enhance the intrinsic pharmacological effects of nanoparticles, increase targeting, and in some cases, expand the scope of application. At the same time, the published information in this area is still very limited in terms of the range of possible applications. Another significant disadvantage is undoubtedly the great complexity of working with immunobiological substances in relation to excipients. The main advantage of morphology modification is the large specific surface area, which theoretically favors the increase in particle activity. However, this approach does not exclude the use of excipients and also requires additional comparative studies. A similar problem is observed in the development of nanocomposites, where an additional difficulty is the difficulty of assessing the contribution of each component to the therapeutic and unwanted effects.

Thus, we can assume that possible future directions for the development of drugs containing rare earth metal nanoparticles could be hybrid micelles, as well as modified formulations of oligomers and polymers.

5. Discussion

This work does not comparatively analyze the efficacy of rare earth metal nanoparticles and drugs, nor does it consider nanoparticles in the context of their application limited solely to the delivery of other drugs. After analyzing all the above results of published studies, we can come to several conclusions:

(1) The main applications of rare earth metal nanoparticles are MRI diagnostics [111,112] and therapy of malignant [59,230,286] and infectious diseases [68,69,273,288].

(2) Safety of rare earth metals applications in medicine. At present, the information about possible risks of application is contradictory [178,186,215]. There are references to both protective [177] and regenerative properties [72,95,206,292] and cytotoxicity [72,189]. Less frequently, we can find information about selective cytotoxicity, particularly against tumor cells [77,78]. At the same time, nanoparticles of rare earth metal oxides are safer than their ionic forms [18].

(3) Possible solutions to the problem by assessing risks and benefits. Based on the first point, we can identify the preferred directions for the development of metal nanoparticles for medical applications. Thus, rare earth metal nanoparticles can be used for external application as antibacterial agents [68,155,241,242], which significantly reduces the risks of systemic unwanted reactions. Another possible direction is the use of some lanthanides for MRI imaging, which is usually not performed on a regular basis, thus reducing the risk of at least manifestations of chronic toxicity [43,66]. Finally, it is reasonable to investigate the possibility of use as an antitumor agent, provided that clinical efficacy is consistent with results obtained on cell cultures and in vivo, or if targeting is possible [147,196].

(4) Improving the safety profile through changes in technology. The rational choice of excipients or their combinations, as well as the combination of nanoparticles with other modern dosage forms and targeted delivery systems, allows for an increase in the targeting of the target (e.g., tumor) [298,299,302]. This reduces the risk of dissociation and thus cumulation of metals in tissues [305] and reduces the dosage due to the increased accumulation of nanoparticles at the point of application of the effect [300]. The size of the nanoparticles themselves also plays an important role since their cytotoxicity depends on it.

Certainly, for many rare earth metal nanoparticles, additional independent in vitro and in vivo studies are required before it is possible to draw unequivocal conclusions about the prospects for their application in medicine and pharmacy. At the same time, for metals already introduced into clinical practice (such as gadolinium and yttrium), as well as the most studied by modern science (for example, cerium), the format of oxide nanoparticles may allow for an expansion in the range of therapeutic action, as well as to increase the safety and efficacy of therapy.

6. Conclusions

In accordance with all of the above, we can conclude that the most studied rare earth metals as agents for medical use are nanoparticles of gadolinium, yttrium, cerium, ytterbium, and lanthanum oxides. However, the question of the balance between the benefits of their use and the risks accompanying it is still acute. Despite this, the application of rare earth metal nanoparticles is quite explored in such areas as visualization and treatment of malignant neoplasms, as well as therapy of bacterial infectious processes. In this regard, further work to confirm these effects in clinical trials and the selection of safe doses is reasonable.

Author Contributions: Conceptualization, S.A.T. and E.V.S.; methodology, S.A.T., M.P.K. and V.A.S.; validation, M.P.K., V.A.S. and N.E.M.; formal analysis, S.A.T., V.A.S. and N.E.M.; investigation, S.A.T. and M.P.K.; data curation, N.E.M. and E.V.S.; visualization, S.A.T. and M.P.K.; supervision, V.A.S. and N.E.M.; project administration, E.V.S.; writing—original draft preparation, S.A.T., M.P.K. and E.V.S.; writing—review and editing, N.E.M. and V.A.S. All authors have read and agreed to the published version of the manuscript.

Funding: This work was carried out with the support of the I.M. Sechenov First Moscow State Medical University (Sechenov University) as part of the development of the federal academic leadership program "Priority 2030", project "Possibilities of biomedical application of rare earth metal compounds", project code 04.000.y.128.

Institutional Review Board Statement: Not applicable.

Informed Consent Statement: Not applicable.

Data Availability Statement: Data sharing is not applicable.

Conflicts of Interest: The authors declare no conflicts of interest.

References

1. Uthamacumaran, A. Cancer: A Turbulence Problem. *Neoplasia* **2020**, *22*, 759–769. [CrossRef]
2. Lee, R.J.; Madan, R.A.; Kim, J.; Posadas, E.M.; Yu, E.Y. Disparities in Cancer Care and the Asian American Population. *Oncologist* **2021**, *26*, 453–460. [CrossRef]
3. Baquero, F. Threats of Antibiotic Resistance: An Obliged Reappraisal. *Int. Microbiol.* **2021**, *24*, 499–506. [CrossRef] [PubMed]
4. Sen, C.K. Human Wound and Its Burden: Updated 2020 Compendium of Estimates. *Adv. Wound Care* **2021**, *10*, 281–292. [CrossRef] [PubMed]
5. Kobayashi, N.; Hirano, K.; Nakano, M.; Muramatsu, T.; Tsukahara, R.; Ito, Y.; Ishimori, H. Wound Healing and Wound Location in Critical Limb Ischemia Following Endovascular Treatment. *Circ. J.* **2014**, *78*, 1746–1753. [CrossRef] [PubMed]
6. Yang, B.; Cao, Y.; Zhang, R.; Yu, X. Experimental Investigation on the Stability and Heat Transfer Enhancement of Phase Change Materials Composited with Nanoparticles and Metal Foams. *J. Energy Storage* **2024**, *89*, 111826. [CrossRef]
7. Kutlu, A.; Aykut, Y.; Eren, R. Thermal Hysteresis Enhancement and Dispersion Thermal Stability in Paraffin Actuators: Comparison of Pan-Nanofibers vs Metal Oxide Nanoparticles Use. *Sens. Actuators A Phys.* **2024**, *379*, 115965. [CrossRef]
8. Tang, X.; Jiang, Y.; Song, L.; Van Der Eycken, E.V. Metal Nanoparticle-Catalyzed Alkyne Cyclization for the Synthesis of Heterocycles. *Adv. Synth. Catal.* **2024**, *366*, 3085–3104. [CrossRef]
9. Prabakaran, E.; Pillay, K. Nanomaterials for Latent Fingerprint Detection: A Review. *J. Mater. Res. Technol.* **2021**, *12*, 1856–1885. [CrossRef]
10. Sharma, V.; Choudhary, S.; Mankotia, P.; Kumari, A.; Sharma, K.; Sehgal, R.; Kumar, V. Nanoparticles as Fingermark Sensors. *TrAC Trends Anal. Chem.* **2021**, *143*, 116378. [CrossRef]
11. Yadav, N. Cerium Oxide Nanostructures: Properties, Biomedical Applications and Surface Coatings. *3 Biotech* **2022**, *12*, 121. [CrossRef]
12. Jain, A.; Fournier, P.G.J.; Mendoza-Lavaniegos, V.; Sengar, P.; Guerra-Olvera, F.M.; Iñiguez, E.; Kretzschmar, T.G.; Hirata, G.A.; Juárez, P. Functionalized Rare Earth-Doped Nanoparticles for Breast Cancer Nanodiagnostic Using Fluorescence and CT Imaging. *J. Nanobiotechnology* **2018**, *16*, 26. [CrossRef] [PubMed]
13. Celardo, I.; De Nicola, M.; Mandoli, C.; Pedersen, J.Z.; Traversa, E.; Ghibelli, L. Ce^3+ Ions Determine Redox-Dependent Anti-Apoptotic Effect of Cerium Oxide Nanoparticles. *ACS Nano* **2011**, *5*, 4537–4549. [CrossRef] [PubMed]
14. Placke-Yan, C.; Bendt, G.; Salamon, S.; Landers, J.; Wende, H.; Hagemann, U.; Schulz, S. Versatile Synthesis of Sub-10 Nm Sized Metal-Doped $M_xCo_{3-x}O_4$ Nanoparticles and Their Electrocatalytic OER Activity. *Mater. Adv.* **2024**, *5*, 3482–3489. [CrossRef]
15. Alameen, A.S.; Undre, S.B.; Undre, P.B. Synthesis, Dispersion, Functionalization, Biological and Antioxidant Activity of Metal Oxide Nanoparticles: Review. *Nano-Struct. Nano-Objects* **2024**, *39*, 101298. [CrossRef]
16. Wang, Y.-L.; Lee, Y.-H.; Chou, C.-L.; Chang, Y.-S.; Liu, W.-C.; Chiu, H.-W. Oxidative Stress and Potential Effects of Metal Nanoparticles: A Review of Biocompatibility and Toxicity Concerns. *Environ. Pollut.* **2024**, *346*, 123617. [CrossRef] [PubMed]

17. Khan, S.; Cho, W.C.; Jaragh-Alhadad, L.A.; Tarharoudi, R.; Bloukh, S.H.; Edis, Z.; Sari, S.; Falahati, M.; Ten Hagen, T.L.M.; Khan, R.H.; et al. Nano-Bio Interaction: An Overview on the Biochemical Binding of DNA to Inorganic Nanoparticles for the Development of Anticancer and Antibacterial Nano-Platforms. *Int. J. Biol. Macromol.* **2023**, *225*, 544–556. [CrossRef]
18. Oluwasola, I.E.; Ahmad, A.L.; Shoparwe, N.F.; Ismail, S. Gadolinium Based Contrast Agents (GBCAs): Uniqueness, Aquatic Toxicity Concerns, and Prospective Remediation. *J. Contam. Hydrol.* **2022**, *250*, 104057. [CrossRef] [PubMed]
19. Master, A.; Livingston, M.; Sen Gupta, A. Photodynamic Nanomedicine in the Treatment of Solid Tumors: Perspectives and Challenges. *J. Control. Release* **2013**, *168*, 88–102. [CrossRef]
20. Silva, J.M.; Silva, E.; Reis, R.L. Light-Triggered Release of Photocaged Therapeutics—Where Are We Now? *J. Control. Release* **2019**, *298*, 154–176. [CrossRef] [PubMed]
21. Kumar, S.; Tripathy, S.; Singh, O.K.; Singh, S.G. Cerium Oxide Nanofiber Based Electroanalytical Sensor for TNF-α Detection: Improved Interfacial Stability with Nafion. *Bioelectrochemistry* **2021**, *138*, 107725. [CrossRef] [PubMed]
22. Kang, D.-W.; Cha, B.G.; Lee, J.H.; Yang, W.; Ki, S.K.; Han, J.H.; Cho, H.Y.; Park, E.; Jeon, S.; Lee, S.-H. Ultrasmall Polymer-Coated Cerium Oxide Nanoparticles as a Traumatic Brain Injury Therapy. *Nanomed. Nanotechnol. Biol. Med.* **2022**, *45*, 102586. [CrossRef] [PubMed]
23. Kozlova, T.O.; Vasilyeva, D.N.; Kozlov, D.A.; Kolesnik, I.V.; Teplonogova, M.A.; Tronev, I.V.; Sheichenko, E.D.; Protsenko, M.R.; Kolmanovich, D.D.; Ivanova, O.S.; et al. A Comparative Study of Cerium(III) and Cerium(IV) Phosphates for Sunscreens. *Molecules* **2024**, *29*, 2157. [CrossRef]
24. Yang, J.; Chu, Z.; Jiang, Y.; Zheng, W.; Sun, J.; Xu, L.; Ma, Y.; Wang, W.; Shao, M.; Qian, H. Multifunctional Hyaluronic Acid Microneedle Patch Embedded by Cerium/Zinc-Based Composites for Accelerating Diabetes Wound Healing. *Adv. Healthc. Mater.* **2023**, *12*, e2300725. [CrossRef]
25. De Santis, S.; Sotgiu, G.; Porcelli, F.; Marsotto, M.; Iucci, G.; Orsini, M. A Simple Cerium Coating Strategy for Titanium Oxide Nano-Tubes' Bioactivity Enhancement. *Nanomaterials* **2021**, *11*, 445. [CrossRef] [PubMed]
26. Priyadarshini, B.; Stango, A.X.; Balasubramanian, M.; Vijayalakshmi, U. In Situ Fabrication of Cerium-Incorporated Hydroxyapatite/Magnetite Nanocomposite Coatings with Bone Regeneration and Osteosarcoma Potential. *Nanoscale Adv.* **2023**, *5*, 5054–5076. [CrossRef] [PubMed]
27. Cimini, A.; D'Angelo, B.; Das, S.; Gentile, R.; Benedetti, E.; Singh, V.; Monaco, A.M.; Santucci, S.; Seal, S. Antibody-Conjugated PEGylated Cerium Oxide Nanoparticles for Specific Targeting of Aβ Aggregates Modulate Neuronal Survival Pathways. *Acta Biomater.* **2012**, *8*, 2056–2067. [CrossRef] [PubMed]
28. Zhang, S.; Wu, Y.; Yu, J.; Ma, C.; Wang, Y.; Wang, Y.; Li, L.; Zhang, L.W. Gadolinium-Bisphosphonate Nanoparticle-Based Low-Dose Radioimmunotherapy for Osteosarcoma. *ACS Biomater. Sci. Eng.* **2022**, *8*, 5329–5337. [CrossRef]
29. Yoon, Y.; Lee, B.-I.; Lee, K.S.; Heo, H.; Lee, J.H.; Byeon, S.-H.; Lee, I.S. Fabrication of a Silica Sphere with Fluorescent and MR Contrasting $GdPO_4$ Nanoparticles from Layered Gadolinium Hydroxide. *Chem. Commun.* **2010**, *46*, 3654. [CrossRef] [PubMed]
30. Dey, U.; Chattopadhyay, A. The Potential of Gadolinium Ascorbate Nanoparticles as a Safer Contrast Agent. *J. Phys. Chem. B* **2023**, *127*, 346–358. [CrossRef]
31. Wang, S.; Zhou, Z.; Yu, G.; Lu, N.; Liu, Y.; Dai, Y.; Fu, X.; Wang, J.; Chen, X. Gadolinium Metallofullerene-Polypyrrole Nanoparticles for Activatable Dual-Modal Imaging-Guided Photothermal Therapy. *ACS Appl. Mater. Interfaces* **2018**, *10*, 28382–28389. [CrossRef]
32. Nosrati, H.; Salehiabar, M.; Charmi, J.; Yaray, K.; Ghaffarlou, M.; Balcioglu, E.; Ertas, Y.N. Enhanced In Vivo Radiotherapy of Breast Cancer Using Gadolinium Oxide and Gold Hybrid Nanoparticles. *ACS Appl. Bio Mater.* **2023**, *6*, 784–792. [CrossRef]
33. Park, J.Y.; Baek, M.J.; Choi, E.S.; Woo, S.; Kim, J.H.; Kim, T.J.; Jung, J.C.; Chae, K.S.; Chang, Y.; Lee, G.H. Paramagnetic Ultrasmall Gadolinium Oxide Nanoparticles as Advanced T_1 MRI Contrast Agent: Account for Large Longitudinal Relaxivity, Optimal Particle Diameter, and In Vivo T_1 MR Images. *ACS Nano* **2009**, *3*, 3663–3669. [CrossRef]
34. Barkhausen, J.; Ebert, W.; Debatin, J.F.; Weinmann, H.-J. Imaging of Myocardial Infarction: Comparison of Magnevist and Gadophrin-3 in Rabbits. *J. Am. Coll. Cardiol.* **2002**, *39*, 1392–1398. [CrossRef] [PubMed]
35. Pintaske, J.; Martirosian, P.; Graf, H.; Erb, G.; Lodemann, K.-P.; Claussen, C.D.; Schick, F. Relaxivity of Gadopentetate Dimeglumine (Magnevist), Gadobutrol (Gadovist), and Gadobenate Dimeglumine (MultiHance) in Human Blood Plasma at 0. *2, 1.5, and 3 Tesla: Investig. Radiol.* **2006**, *41*, 213–221. [CrossRef]
36. Takanezawa, Y.; Nakamura, R.; Ohshiro, Y.; Uraguchi, S.; Kiyono, M. Gadolinium-Based Contrast Agents Suppress Adipocyte Differentiation in 3T3-L1 Cells. *Toxicol. Lett.* **2023**, *383*, 196–203. [CrossRef] [PubMed]
37. Xue, X.; Bo, R.; Qu, H.; Jia, B.; Xiao, W.; Yuan, Y.; Vapniarsky, N.; Lindstrom, A.; Wu, H.; Zhang, D.; et al. A Nephrotoxicity-Free, Iron-Based Contrast Agent for Magnetic Resonance Imaging of Tumors. *Biomaterials* **2020**, *257*, 120234. [CrossRef] [PubMed]
38. Lange, S.; Mędrzycka-Dąbrowska, W.; Zorena, K.; Dąbrowski, S.; Ślęzak, D.; Malecka-Dubiela, A.; Rutkowski, P. Nephrogenic Systemic Fibrosis as a Complication after Gadolinium-Containing Contrast Agents: A Rapid Review. *Int. J. Environ. Res. Public Health* **2021**, *18*, 3000. [CrossRef]

39. Bernstein, E.J.; Schmidt-Lauber, C.; Kay, J. Nephrogenic Systemic Fibrosis: A Systemic Fibrosing Disease Resulting from Gadolinium Exposure. *Best Pract. Res. Clin. Rheumatol.* **2012**, *26*, 489–503. [CrossRef] [PubMed]
40. Thanikaivelan, E.; Gokul Raj, S.; Murugakoothan, P. Synthesis and Characterisation of Erbium Doped Gadolinium Oxide (Er:Gd_2O_3) Nanorods. *Mater. Today Proc.* **2023**, S221478532305054X. [CrossRef]
41. Dai, Y.; Wu, C.; Wang, S.; Li, Q.; Zhang, M.; Li, J.; Xu, K. Comparative Study on in Vivo Behavior of PEGylated Gadolinium Oxide Nanoparticles and Magnevist as MRI Contrast Agent. *Nanomed. Nanotechnol. Biol. Med.* **2018**, *14*, 547–555. [CrossRef]
42. Abrikossova, N.; Skoglund, C.; Ahrén, M.; Bengtsson, T.; Uvdal, K. Effects of Gadolinium Oxide Nanoparticles on the Oxidative Burst from Human Neutrophil Granulocytes. *Nanotechnology* **2012**, *23*, 275101. [CrossRef]
43. Shalviri, A.; Foltz, W.D.; Cai, P.; Rauth, A.M.; Wu, X.Y. Multifunctional Terpolymeric MRI Contrast Agent with Superior Signal Enhancement in Blood and Tumor. *J. Control. Release* **2013**, *167*, 11–20. [CrossRef]
44. Ashouri, H.; Riyahi Alam, N.; Khoobi, M.; Haghgoo, S.; Rasouli, Z.; Gholami, M. NSF Evaluation of Gadolinium Biodistribution in Renally Impaired Rats: Using Novel Metabolic Gd_2O_3 Nanoparticles Coated with β-Cyclodextrin (Gd_2O_3@PCD) in MR Molecular Imaging. *Magn. Reson. Imaging* **2024**, *107*, 120–129. [CrossRef]
45. Brune, N.; Mues, B.; Buhl, E.M.; Hintzen, K.-W.; Jockenhoevel, S.; Cornelissen, C.G.; Slabu, I.; Thiebes, A.L. Dual Labeling of Primary Cells with Fluorescent Gadolinium Oxide Nanoparticles. *Nanomaterials* **2023**, *13*, 1869. [CrossRef] [PubMed]
46. Kumar, S.; Meena, V.K.; Hazari, P.P.; Sharma, S.K.; Sharma, R.K. Rose Bengal Attached and Dextran Coated Gadolinium Oxide Nanoparticles for Potential Diagnostic Imaging Applications. *Eur. J. Pharm. Sci.* **2018**, *117*, 362–370. [CrossRef] [PubMed]
47. Shen, Z.; Fan, W.; Yang, Z.; Liu, Y.; Bregadze, V.I.; Mandal, S.K.; Yung, B.C.; Lin, L.; Liu, T.; Tang, W.; et al. Exceedingly Small Gadolinium Oxide Nanoparticles with Remarkable Relaxivities for Magnetic Resonance Imaging of Tumors. *Small* **2019**, *15*, 1903422. [CrossRef] [PubMed]
48. Ahmad, M.Y.; Ahmad, M.W.; Yue, H.; Ho, S.L.; Park, J.A.; Jung, K.-H.; Cha, H.; Marasini, S.; Ghazanfari, A.; Liu, S.; et al. In Vivo Positive Magnetic Resonance Imaging Applications of Poly(Methyl Vinyl Ether-Alt-Maleic Acid)-Coated Ultra-Small Paramagnetic Gadolinium Oxide Nanoparticles. *Molecules* **2020**, *25*, 1159. [CrossRef] [PubMed]
49. Ho, S.L.; Choi, G.; Yue, H.; Kim, H.-K.; Jung, K.-H.; Park, J.A.; Kim, M.H.; Lee, Y.J.; Kim, J.Y.; Miao, X.; et al. In Vivo Neutron Capture Therapy of Cancer Using Ultrasmall Gadolinium Oxide Nanoparticles with Cancer-Targeting Ability. *RSC Adv.* **2020**, *10*, 865–874. [CrossRef] [PubMed]
50. Kumar, S.; Meena, V.K.; Hazari, P.P.; Sharma, R.K. PEG Coated and Doxorubicin Loaded Multimodal Gadolinium Oxide Nanoparticles for Simultaneous Drug Delivery and Imaging Applications. *Int. J. Pharm.* **2017**, *527*, 142–150. [CrossRef] [PubMed]
51. Mortezazadeh, T.; Gholibegloo, E.; Khoobi, M.; Alam, N.R.; Haghgoo, S.; Mesbahi, A. In Vitro and in Vivo Characteristics of Doxorubicin-Loaded Cyclodextrine-Based Polyester Modified Gadolinium Oxide Nanoparticles: A Versatile Targeted Theranostic System for Tumour Chemotherapy and Molecular Resonance Imaging. *J. Drug Target.* **2020**, *28*, 533–546. [CrossRef]
52. Sun, X.; Kou, B. Biocompatibility and Potential Anticancer Activity of Gadolinium Oxide (Gd_2O_3) Nanoparticles against Nasal Squamous Cell Carcinoma. *BMC Biotechnol.* **2024**, *24*, 53. [CrossRef] [PubMed]
53. Wu, C.; Cai, R.; Zhao, T.; Wu, L.; Zhang, L.; Jin, J.; Xu, L.; Li, P.; Li, T.; Zhang, M.; et al. Hyaluronic Acid-Functionalized Gadolinium Oxide Nanoparticles for Magnetic Resonance Imaging-Guided Radiotherapy of Tumors. *Nanoscale Res. Lett.* **2020**, *15*, 94. [CrossRef]
54. Rajaee, A.; Wang, S.; Zhao, L.; Wang, D.; Liu, Y.; Wang, J.; Ying, K. Multifunction Bismuth Gadolinium Oxide Nanoparticles as Radiosensitizer in Radiation Therapy and Imaging. *Phys. Med. Biol.* **2019**, *64*, 195007. [CrossRef] [PubMed]
55. Mahdavi, B.; Shokrani, P.; Hejazi, S.H.; Talebi, A.; Taheri, A. Doxorubicin-Loaded PVP Coated Gd_2O_3 NPs for Effective Chemoradiotherapy in Melanoma. *J. Drug Deliv. Sci. Technol.* **2019**, *53*, 101189. [CrossRef]
56. Yu, B.; Lu, X.; Feng, X.; Zhao, T.; Li, J.; Lu, Y.; Ye, F.; Liu, X.; Zheng, X.; Shen, Z.; et al. Gadolinium Oxide Nanoparticles Reinforce the Fractionated Radiotherapy-Induced Immune Response in Tri-Negative Breast Cancer via cGAS-STING Pathway. *Int. J. Nanomed.* **2023**, *18*, 7713–7728. [CrossRef]
57. Da Silva, B.A.; Nazarkovsky, M.; Padilla-Chavarría, H.I.; Mendivelso, E.A.C.; Mello, H.L.D.; Nogueira, C.D.S.C.; Carvalho, R.D.S.; Cremona, M.; Zaitsev, V.; Xing, Y.; et al. Novel Scintillating Nanoparticles for Potential Application in Photodynamic Cancer Therapy. *Pharmaceutics* **2022**, *14*, 2258. [CrossRef]
58. Shen, Z.; Liu, T.; Yang, Z.; Zhou, Z.; Tang, W.; Fan, W.; Liu, Y.; Mu, J.; Li, L.; Bregadze, V.I.; et al. Small-Sized Gadolinium Oxide Based Nanoparticles for High-Efficiency Theranostics of Orthotopic Glioblastoma. *Biomaterials* **2020**, *235*, 119783. [CrossRef] [PubMed]
59. Aloy, M.-T.; Sidi Boumedine, J.; Deville, A.; Kryza, D.; Gauthier, A.; Brichart-Vernos, D.; Ollier, G.; La Padula, V.; Lux, F.; Tillement, O.; et al. Proof of Concept of the Radiosensitizing Effect of Gadolinium Oxide Nanoparticles in Cell Spheroids and a Tumor-Implanted Murine Model of Chondrosarcoma. *Int. J. Nanomed.* **2022**, *17*, 6655–6673. [CrossRef]
60. Akhtar, M.J.; Ahamed, M.; Alhadlaq, H. Gadolinium Oxide Nanoparticles Induce Toxicity in Human Endothelial HUVECs via Lipid Peroxidation, Mitochondrial Dysfunction and Autophagy Modulation. *Nanomaterials* **2020**, *10*, 1675. [CrossRef]

61. Akhtar, M.J.; Ahamed, M.; Alhadlaq, H.; Alrokayan, S. Toxicity Mechanism of Gadolinium Oxide Nanoparticles and Gadolinium Ions in Human Breast Cancer Cells. *Curr. Drug Metab.* **2019**, *20*, 907–917. [CrossRef]
62. Liu, S.; Yue, H.; Ho, S.L.; Kim, S.; Park, J.A.; Tegafaw, T.; Ahmad, M.Y.; Kim, S.; Saidi, A.K.A.A.; Zhao, D.; et al. Enhanced Tumor Imaging Using Glucosamine-Conjugated Polyacrylic Acid-Coated Ultrasmall Gadolinium Oxide Nanoparticles in Magnetic Resonance Imaging. *Int. J. Mol. Sci.* **2022**, *23*, 1792. [CrossRef]
63. Shi, X.; Cao, C.; Zhang, Z.; Tian, J.; Hu, Z. Radiopharmaceutical and Eu^{3+} Doped Gadolinium Oxide Nanoparticles Mediated Triple-Excited Fluorescence Imaging and Image-Guided Surgery. *J. Nanobiotechnol.* **2021**, *19*, 212. [CrossRef] [PubMed]
64. Mortezazadeh, T.; Gholibegloo, E.; Alam, N.R.; Dehghani, S.; Haghgoo, S.; Ghanaati, H.; Khoobi, M. Gadolinium (III) Oxide Nanoparticles Coated with Folic Acid-Functionalized Poly(β-Cyclodextrin-Co-Pentetic Acid) as a Biocompatible Targeted Nano-Contrast Agent for Cancer Diagnostic: In Vitro and in Vivo Studies. *Magn. Reson. Mater. Phys.* **2019**, *32*, 487–500. [CrossRef]
65. Cheng, Y.; Tan, X.; Wang, J.; Wang, Y.; Song, Y.; You, Q.; Sun, Q.; Liu, L.; Wang, S.; Tan, F.; et al. Polymer-Based Gadolinium Oxide Nanocomposites for FL/MR/PA Imaging Guided and Photothermal/Photodynamic Combined Anti-Tumor Therapy. *J. Control. Release* **2018**, *277*, 77–88. [CrossRef] [PubMed]
66. Kumar, S.; Meena, V.K.; Hazari, P.P.; Sharma, R.K. FITC-Dextran Entrapped and Silica Coated Gadolinium Oxide Nanoparticles for Synchronous Optical and Magnetic Resonance Imaging Applications. *Int. J. Pharm.* **2016**, *506*, 242–252. [CrossRef] [PubMed]
67. Aashima; Pandey, S.K.; Singh, S.; Mehta, S.K. Ultrasonication Assisted Fabrication of L-Lysine Functionalized Gadolinium Oxide Nanoparticles and Its Biological Acceptability. *Ultrason. Sonochemistry* **2018**, *49*, 53–62. [CrossRef]
68. Aashima; Pandey, S.K.; Singh, S.; Mehta, S.K. Biocompatible Gadolinium Oxide Nanoparticles as Efficient Agent against Pathogenic Bacteria. *J. Colloid Interface Sci.* **2018**, *529*, 496–504. [CrossRef]
69. Kumar, S.; Chaudhary, S.; Chaudhary, G.R. Effect of Fabrication Strategies on the In-Vitro Antimicrobial and Antifungal Activities of Pr^{3+} Doped Gb_2O_3 Nanoparticles. *Environ. Nanotechnol. Monit. Manag.* **2021**, *16*, 100518. [CrossRef]
70. Kumar, A.; Sarkar, T.; Solanki, P.R. Amine Functionalized Gadolinium Oxide Nanoparticles-Based Electrochemical Immunosensor for Cholera. *Biosensors* **2023**, *13*, 177. [CrossRef]
71. Kumar, A.; Sarkar, T.; Kumar, R.; Panda, A.K.; Solanki, P.R. Electrochemical Detection of Vibrio Cholerae by Amine Functionalized Biocompatible Gadolinium Oxide Nanoparticles. *Micromachines* **2023**, *14*, 995. [CrossRef] [PubMed]
72. Vijayan, V.; Kiran, M.S. Hybrid Nanostructured Gadolinium Oxide-Collagen-Dextran Polymeric Hydrogel for Corneal Repair and Regeneration. *Int. J. Biol. Macromol.* **2023**, *224*, 1423–1438. [CrossRef]
73. Mastrogiacomo, S.; Kownacka, A.E.; Dou, W.; Burke, B.P.; De Rosales, R.T.M.; Heerschap, A.; Jansen, J.A.; Archibald, S.J.; Walboomers, X.F. Bisphosphonate Functionalized Gadolinium Oxide Nanoparticles Allow Long-Term MRI/CT Multimodal Imaging of Calcium Phosphate Bone Cement. *Adv. Healthc. Mater.* **2018**, *7*, 1800202. [CrossRef] [PubMed]
74. Silina, E.V.; Manturova, N.E.; Chuvilina, E.L.; Gasanov, A.A.; Andreeva, O.I.; Pugachevskii, M.A.; Kochura, A.V.; Kryukov, A.A.; Suzdaltseva, Y.G.; Stupin, V.A. Biomedical Application Prospects of Gadolinium Oxide Nanoparticles for Regenerative Medicine. *Pharmaceutics* **2024**, *16*, 1627. [CrossRef]
75. Tcibulnikova, A.V.; Slezhkin, V.A.; Bruykhanov, V.V.; Samusev, I.G.; Kozhevnikov, A.S.; Savin, V.V.; Medvedskaya, P.N.; Lyatun, I.I. Cooperative Luminescence of Yb 3 + Ions of the Ytterbium Oxide Porous Surface. *Opt. Commun.* **2020**, *459*, 125006. [CrossRef]
76. Hannachi, E.; Khan, F.; Slimani, Y.; Rehman, S.; Trabelsi, Z.; Akhtar, S.; Al-Suhaimi, E. In Vitro Antimicrobial and Anticancer Peculiarities of Ytterbium and Cerium Co-Doped Zinc Oxide Nanoparticles. *Biology* **2022**, *11*, 1836. [CrossRef]
77. Peng, H.; Cui, B.; Li, G.; Wang, Y.; Li, N.; Chang, Z.; Wang, Y. A Multifunctional β-CD-Modified Fe_3O_4@ZnO:Er^{3+},Yb^{3+} Nanocarrier for Antitumor Drug Delivery and Microwave-Triggered Drug Release. *Mater. Sci. Eng. C* **2015**, *46*, 253–263. [CrossRef] [PubMed]
78. Kaliyamoorthy, K.; Pillai, A.S.; Alexander, A.; Arivarasu, A.; Enoch, I.V.M.V.; Ramasamy, S. β-Cyclodextrin-Folate Functionalized Poly(Lactic-Co-Glycolide)–Superparamagnetic Ytterbium Ferrite Hybrid Nanocarrier for Targeted Delivery of Camptothecin. *Mater. Sci. Eng. C* **2021**, *122*, 111796. [CrossRef]
79. Muthulakshmi, V.; Kumar, P.; Sundrarajan, M. Green Synthesis of Ionic Liquid Mediated Ytterbium Oxide Nanoparticles by Andrographis Paniculata Leaves Extract for Structural, Morphological and Biomedical Applications. *J. Environ. Chem. Eng.* **2021**, *9*, 105270. [CrossRef]
80. Muthulakshmi, V.; Sundrarajan, M. Green Synthesis of Ionic Liquid Assisted Ytterbium Oxide Nanoparticles by Couroupita Guianensis Abul Leaves Extract for Biological Applications. *J. Environ. Chem. Eng.* **2020**, *8*, 103992. [CrossRef]
81. Almaieli, L.M.A.; Khalaf, M.M.; Gouda, M.; Elmushyakhi, A.; Abou Taleb, M.F.; Abd El-Lateef, H.M. Fabrication of Bio-Based Film Comprising Metal Oxide Nanoparticles Loaded Chitosan for Wound Dressing Applications. *Polymers* **2022**, *15*, 211. [CrossRef]
82. Mazloum-Ardakani, M.; Shaker-Ardakani, N.; Farbod, F. Highly Efficient Enzyme Less Glucose Sensor Based on Cross-Linked Nitrogen-Doped Graphene Aerogel Incorporated with Ytterbium Oxide and Decorated with Nickel Nitride Nanoparticles. *Sens. Int.* **2024**, *5*, 100290. [CrossRef]
83. Kumar, S.; Ram Chaudhary, G.; Chaudhary, S. Designing of Surface Engineered Ytterbium Oxide Nanoparticles as Effective Electrochemical Sensing Platform for Dopamine. *J. Mol. Liq.* **2022**, *355*, 118929. [CrossRef]

84. Jafari, H.; Ganjali, M.R.; Dezfuli, A.S.; Faridbod, F. Long Term Determination of Dopamine and Uric Acid in the Presence of Ascorbic Acid Using Ytterbia/Reduced Graphene Oxide Nanocomposite Prepared through a Sonochemical Route. *Appl. Surf. Sci.* **2018**, *427*, 496–506. [CrossRef]
85. Hosseini, M.; Pur, M.R.K.; Norouzi, P.; Moghaddam, M.R.; Ganjali, M.R. An Enhanced Electrochemiluminescence Sensor Modified with a Ru(Bpy)32+/Yb_2O_3 Nanoparticle/Nafion Composite for the Analysis of Methadone Samples. *Mater. Sci. Eng. C* **2017**, *76*, 483–489. [CrossRef] [PubMed]
86. Wysokińska, E.; Cichos, J.; Zioło, E.; Bednarkiewicz, A.; Strządała, L.; Karbowiak, M.; Hreniak, D.; Kałas, W. Cytotoxic Interactions of Bare and Coated NaGdF4:Yb^{3+}:Er^{3+} Nanoparticles with Macrophage and Fibroblast Cells. *Toxicol. Vitr.* **2016**, *32*, 16–25. [CrossRef] [PubMed]
87. Wang, C.; He, M.; Chen, B.; Hu, B. Study on Cytotoxicity, Cellular Uptake and Elimination of Rare-Earth-Doped Upconversion Nanoparticles in Human Hepatocellular Carcinoma Cells. *Ecotoxicol. Environ. Saf.* **2020**, *203*, 110951. [CrossRef] [PubMed]
88. Adeel, M.; Tingting, J.; Hussain, T.; He, X.; Ahmad, M.A.; Irshad, M.K.; Shakoor, N.; Zhang, P.; Changjian, X.; Hao, Y.; et al. Bioaccumulation of Ytterbium Oxide Nanoparticles Insinuate Oxidative Stress, Inflammatory, and Pathological Lesions in ICR Mice. *Environ. Sci. Pollut. Res.* **2020**, *27*, 32944–32953. [CrossRef]
89. Gómez-González, E.; Caro, C.; Martínez-Gutiérrez, D.; García-Martín, M.L.; Ocaña, M.; Becerro, A.I. Holmium Phosphate Nanoparticles as Negative Contrast Agents for High-Field Magnetic Resonance Imaging: Synthesis, Magnetic Relaxivity Study and in Vivo Evaluation. *J. Colloid Interface Sci.* **2021**, *587*, 131–140. [CrossRef] [PubMed]
90. Bult, W.; Varkevisser, R.; Soulimani, F.; Seevinck, P.R.; De Leeuw, H.; Bakker, C.J.G.; Luijten, P.R.; Van Het Schip, A.D.; Hennink, W.E.; Nijsen, J.F.W. Holmium Nanoparticles: Preparation and In Vitro Characterization of a New Device for Radioablation of Solid Malignancies. *Pharm. Res.* **2010**, *27*, 2205–2212. [CrossRef]
91. Marasini, S.; Yue, H.; Ho, S.L.; Park, J.A.; Kim, S.; Jung, K.-H.; Cha, H.; Liu, S.; Tegafaw, T.; Ahmad, M.Y.; et al. Synthesis, Characterizations, and 9.4 Tesla T2 MR Images of Polyacrylic Acid-Coated Terbium(III) and Holmium(III) Oxide Nanoparticles. *Nanomaterials* **2021**, *11*, 1355. [CrossRef] [PubMed]
92. Shirzadi-Ahodashti, M.; Mortazavi-Derazkola, S.; Ebrahimzadeh, M.A. Nanostructures of Rare Earth Oxides (Ho_2O_3 and Nd_2O_3): Synthesis Methods, Properties, and Comparative Analysis. *J. Mater. Res. Technol.* **2023**, *27*, 1843–1856. [CrossRef]
93. Cerqueira-Coutinho, C.; Vidal, L.P.; Pinto, S.R.; Santos-Oliveira, R. Drug Metabolism: Comparison of Biodistribution Profile of Holmium in Three Different Compositions in Healthy Wistar Rats. *Appl. Radiat. Isot.* **2016**, *112*, 27–30. [CrossRef] [PubMed]
94. Zhu, J.; Shao, X.-J.; Li, Z.; Lin, C.-H.; Wang, C.-W.-Q.; Jiao, K.; Xu, J.; Pan, H.-X.; Wu, Y. Synthesis of Holmium-Oxide Nanoparticles for Near-Infrared Imaging and Dye-Photodegradation. *Molecules* **2022**, *27*, 3522. [CrossRef]
95. Luo, Y.; Zheng, Y.; Chen, Z.; Mo, M.; Xie, J.; Zhou, X.; Wu, Y.; Yang, Q.; Zheng, M.; Hu, X.; et al. Proangiogenic Effect and Underlying Mechanism of Holmium Oxide Nanoparticles: A New Biomaterial for Tissue Engineering. *J. Nanobiotechnol.* **2024**, *22*, 357. [CrossRef] [PubMed]
96. Ahmed Mohamed, H.E.; Khalil, A.T.; Hkiri, K.; Ayaz, M.; Usman, A.; Sadiq, A.; Ullah, F.; Hussain, I.; Maaza, M. Phyto-Fabrication of Ultrafine Nanoscale Holmium Oxide HT-Ho_2 O_3 NPs and Their Biomedical Potential. *RSC Adv.* **2023**, *13*, 27912–27922. [CrossRef] [PubMed]
97. Munaweera, I.; Levesque-Bishop, D.; Shi, Y.; Di Pasqua, A.J.; Balkus, K.J. Radiotherapeutic Bandage Based on Electrospun Polyacrylonitrile Containing Holmium-166 Iron Garnet Nanoparticles for the Treatment of Skin Cancer. *ACS Appl. Mater. Interfaces* **2014**, *6*, 22250–22256. [CrossRef]
98. Lee, J.D.; Park, K.K.; Lee, M.G.; Kim, E.H.; Rhim, K.J.; Lee, J.T.; Yoo, H.S.; Kim, Y.M.; Park, K.B.; Kim, J.R. Radionuclide Therapy of Skin Cancers and Bowen's Disease Using a Specially Designed Skin Patch. *J. Nucl. Med.* **1997**, *38*, 697–702. [PubMed]
99. Osipitan, O.O.; Sun, M.; Gordish-Dressman, H.; Wendt, R.; Wight-Carter, M.; Balkus, K.J.; Di Pasqua, A.J. Laminated Holmium-166-Containing Electrospun Bandages for Use against Skin Cancer. *Nucl. Med. Biol.* **2022**, *114–115*, 78–85. [CrossRef] [PubMed]
100. Liu, S.; Yue, H.; Ho, S.L.; Kim, S.; Park, J.A.; Tegafaw, T.; Ahmad, M.Y.; Kim, S.; Saidi, A.K.A.A.; Zhao, D.; et al. Polyethylenimine-Coated Ultrasmall Holmium Oxide Nanoparticles: Synthesis, Characterization, Cytotoxicities, and Water Proton Spin Relaxivities. *Nanomaterials* **2022**, *12*, 1588. [CrossRef]
101. Atabaev, T.; Shin, Y.; Song, S.-J.; Han, D.-W.; Hong, N. Toxicity and T2-Weighted Magnetic Resonance Imaging Potentials of Holmium Oxide Nanoparticles. *Nanomaterials* **2017**, *7*, 216. [CrossRef] [PubMed]
102. Kim, H.; Yoon, J.; Kim, H.K.; Lee, W.T.; Nguyen, N.T.; Le, X.T.; Lee, E.-H.; Lee, E.S.; Oh, K.T.; Choi, H.-G.; et al. Upconverting Nanoparticle-Containing Erythrocyte-Sized Hemoglobin Microgels That Generate Heat, Oxygen and Reactive Oxygen Species for Suppressing Hypoxic Tumors. *Bioact. Mater.* **2023**, *22*, 112–126. [CrossRef] [PubMed]
103. Hernández-Jiménez, T.; Cruz-Nova, P.; Ancira-Cortez, A.; Gibbens-Bandala, B.; Lara-Almazán, N.; Ocampo-García, B.; Santos-Cuevas, C.; Morales-Avila, E.; Ferro-Flores, G. Toxicity Assessment of [177Lu]Lu−iFAP/iPSMA Nanoparticles Prepared under GMP-Compliant Radiopharmaceutical Processes. *Nanomaterials* **2022**, *12*, 4181. [CrossRef]
104. Tran, T.A.; Kappelhoff, J.; Jüstel, T.; Anderson, R.R.; Purschke, M. UV Emitting Nanoparticles Enhance the Effect of Ionizing Radiation in 3D Lung Cancer Spheroids. *Int. J. Radiat. Biol.* **2022**, *98*, 1484–1494. [CrossRef] [PubMed]

105. Luna-Gutiérrez, M.; Ocampo-García, B.; Jiménez-Mancilla, N.; Ancira-Cortez, A.; Trujillo-Benítez, D.; Hernández-Jiménez, T.; Ramírez-Nava, G.; Hernández-Ramírez, R.; Santos-Cuevas, C.; Ferro-Flores, G. Targeted Endoradiotherapy with Lu_2O_3-iPSMA/-iFAP Nanoparticles Activated by Neutron Irradiation: Preclinical Evaluation and First Patient Image. *Pharmaceutics* **2022**, *14*, 720. [CrossRef] [PubMed]
106. Ancira-Cortez, A.; Ferro-Flores, G.; Jiménez-Mancilla, N.; Morales-Avila, E.; Trujillo-Benítez, D.; Ocampo-García, B.; Santos-Cuevas, C.; Escudero-Castellanos, A.; Luna-Gutiérrez, M. Synthesis, Chemical and Biochemical Characterization of Lu_2O_3-iPSMA Nanoparticles Activated by Neutron Irradiation. *Mater. Sci. Eng. C* **2020**, *117*, 111335. [CrossRef]
107. Locardi, F.; Gianotti, E.; Nelli, I.; Caratto, V.; Martinelli, A.; Ferretti, M.; Costa, G.A.; Canesi, L.; Balbi, T.; Fasoli, M.; et al. Facile Synthesis of NIR and Visible Luminescent Sm^{3+} Doped Lutetium Oxide Nanoparticles. *Mater. Res. Bull.* **2017**, *86*, 220–227. [CrossRef]
108. Liu, Z.; Dong, K.; Liu, J.; Han, X.; Ren, J.; Qu, X. Anti-Biofouling Polymer-Decorated Lutetium-Based Nanoparticulate Contrast Agents for In Vivo High-Resolution Trimodal Imaging. *Small* **2014**, *10*, 2429–2438. [CrossRef]
109. Gómez-González, E.; Caro, C.; García-Martín, M.L.; Becerro, A.I.; Ocaña, M. Outstanding MRI Contrast with Dysprosium Phosphate Nanoparticles of Tuneable Size. *Nanoscale* **2022**, *14*, 11461–11470. [CrossRef] [PubMed]
110. Gómez-Gónzalez, E.; Núñez, N.O.; Caro, C.; García-Martín, M.L.; Fernández-Afonso, Y.; De La Fuente, J.M.; Balcerzyk, M.; Ocaña, M. Dysprosium and Holmium Vanadate Nanoprobes as High-Performance Contrast Agents for High-Field Magnetic Resonance and Computed Tomography Imaging. *Inorg. Chem.* **2021**, *60*, 152–160. [CrossRef]
111. Yue, H.; Park, J.Y.; Chang, Y.; Lee, G.H. Ultrasmall Europium, Gadolinium, and Dysprosium Oxide Nanoparticles: Polyol Synthesis, Properties, and Biomedical Imaging Applications. *Mini-Rev. Med. Chem.* **2020**, *20*, 1767–1780. [CrossRef]
112. Yue, H.; Park, J.A.; Ho, S.L.; Ahmad, M.Y.; Cha, H.; Liu, S.; Tegafaw, T.; Marasini, S.; Ghazanfari, A.; Kim, S.; et al. New Class of Efficient T2 Magnetic Resonance Imaging Contrast Agent: Carbon-Coated Paramagnetic Dysprosium Oxide Nanoparticles. *Pharmaceuticals* **2020**, *13*, 312. [CrossRef] [PubMed]
113. Kattel, K.; Park, J.Y.; Xu, W.; Kim, H.G.; Lee, E.J.; Bony, B.A.; Heo, W.C.; Jin, S.; Baeck, J.S.; Chang, Y.; et al. Paramagnetic Dysprosium Oxide Nanoparticles and Dysprosium Hydroxide Nanorods as T2 MRI Contrast Agents. *Biomaterials* **2012**, *33*, 3254–3261. [CrossRef] [PubMed]
114. Tegafaw, T.; Xu, W.; Ahmad, M.W.; Baeck, J.S.; Chang, Y.; Bae, J.E.; Chae, K.S.; Kim, T.J.; Lee, G.H. Dual-Mode T_1 and T_2 Magnetic Resonance Imaging Contrast Agent Based on Ultrasmall Mixed Gadolinium-Dysprosium Oxide Nanoparticles: Synthesis, Characterization, and in Vivo Application. *Nanotechnology* **2015**, *26*, 365102. [CrossRef]
115. Balachandran, Y.L.; Wang, W.; Yang, H.; Tong, H.; Wang, L.; Liu, F.; Chen, H.; Zhong, K.; Liu, Y.; Jiang, X. Heterogeneous Iron Oxide/Dysprosium Oxide Nanoparticles Target Liver for Precise Magnetic Resonance Imaging of Liver Fibrosis. *ACS Nano* **2022**, *16*, 5647–5659. [CrossRef]
116. Vignesh, K.; Senthil Kumar, A.; Arumugam Napoleon, A.; Govindasamy, M.; Yusuf, K. Ag-La(OH)3@Dy_2O_3 Hybrid Composite Modified Laser-Induced Graphene Surface for Simultaneous Electrochemical Detection of Bisphenol A and Tartrazine. *Appl. Surf. Sci.* **2024**, *676*, 160901. [CrossRef]
117. Baytak, A.K.; Akbaş, E.; Aslanoglu, M. A Novel Voltammetric Platform Based on Dysprosium Oxide for the Sensitive Determination of Sunset Yellow in the Presence of Tartrazine. *Anal. Chim. Acta* **2019**, *1087*, 93–103. [CrossRef]
118. Que, W.; Buddhudu, S.; Zhou, Y.; Lam, Y.L.; Zhou, J.; Chan, Y.C.; Kam, C.H.; Gan, L.H.; Roshan Deen, G. Preparation and Characterization of Erbium Oxalate and Erbium Oxide Nanoparticles by Microemulsion Technique. *Mater. Sci. Eng. C* **2001**, *16*, 51–54. [CrossRef]
119. Tse, J.J.; Dunmore-Buyze, P.J.; Drangova, M.; Holdsworth, D.W. Erbium-Based Perfusion Contrast Agent for Small-Animal Microvessel Imaging. *Contrast Media Mol. Imaging* **2017**, *2017*, 7368384. [CrossRef]
120. Azhagar, S.; Kumar Chinnalagu, D.; Mayakrishnan, A.; Sundrarajan, M. Study on the Therapeutic Activity of [BMIM]-PF6-IL Assisted Er_2O_3 NPs Synthesized by Polyol Method against Pathogenic Bacterium as Well as MCF-7 Breast Tumor Cells. *Inorg. Chem. Commun.* **2023**, *155*, 110988. [CrossRef]
121. Dědková, K.; Kuzníková, L'.; Pavelek, L.; Matějová, K.; Kupková, J.; Čech Barabaszová, K.; Váňa, R.; Burda, J.; Vlček, J.; Cvejn, D.; et al. Daylight Induced Antibacterial Activity of Gadolinium Oxide, Samarium Oxide and Erbium Oxide Nanoparticles and Their Aquatic Toxicity. *Mater. Chem. Phys.* **2017**, *197*, 226–235. [CrossRef]
122. Mohamed, H.E.A.; Khalil, A.T.; Hkiri, K.; Ayaz, M.; Abbasi, J.A.; Sadiq, A.; Ullah, F.; Nawaz, A.; Ullah, I.; Maaza, M. Physicochemical and Nanomedicine Applications of Phyto-Reduced Erbium Oxide (Er_2O_3) Nanoparticles. *AMB Expr.* **2023**, *13*, 24. [CrossRef] [PubMed]
123. Safwat, G.; Soliman, E.S.M.; Mohamed, H.R.H. Induction of ROS Mediated Genomic Instability, Apoptosis and G0/G1 Cell Cycle Arrest by Erbium Oxide Nanoparticles in Human Hepatic Hep-G2 Cancer Cells. *Sci. Rep.* **2022**, *12*, 16333. [CrossRef] [PubMed]
124. Saifullah, A.; Ul Hassan, S.M.; Khurshid, A. Carbon-Coated (Eu/Dy-Enabled) Erbium Hydroxide Nanostructures for Biomedical Imaging. *Materialia* **2024**, *35*, 102108. [CrossRef]

125. Mohamed, H.R.H.; Ibrahim, M.M.H.; Soliman, E.S.M.; Safwat, G.; Diab, A. Estimation of Calcium Titanate or Erbium Oxide Nanoparticles Induced Cytotoxicity and Genotoxicity in Normal HSF Cells. *Biol. Trace Elem. Res.* **2023**, *201*, 2311–2318. [CrossRef] [PubMed]
126. Cardoso Dos Santos, M.; Goetz, J.; Bartenlian, H.; Wong, K.-L.; Charbonnière, L.J.; Hildebrandt, N. Autofluorescence-Free Live-Cell Imaging Using Terbium Nanoparticles. *Bioconjugate Chem.* **2018**, *29*, 1327–1334. [CrossRef]
127. Charpentier, C.; Cifliku, V.; Goetz, J.; Nonat, A.; Cheignon, C.; Cardoso Dos Santos, M.; Francés-Soriano, L.; Wong, K.; Charbonnière, L.J.; Hildebrandt, N. Ultrabright Terbium Nanoparticles for FRET Biosensing and in Situ Imaging of Epidermal Growth Factor Receptors**. *Chem. A Eur. J.* **2020**, *26*, 14602–14611. [CrossRef] [PubMed]
128. Chen, Y.; Chi, Y.; Wen, H.; Lu, Z. Sensitized Luminescent Terbium Nanoparticles: Preparation and Time-Resolved Fluorescence Assay for DNA. *Anal. Chem.* **2007**, *79*, 960–965. [CrossRef] [PubMed]
129. Mohanto, S.; Biswas, A.; Gholap, A.D.; Wahab, S.; Bhunia, A.; Nag, S.; Ahmed, M.G. Potential Biomedical Applications of Terbium-Based Nanoparticles (TbNPs): A Review on Recent Advancement. *ACS Biomater. Sci. Eng.* **2024**, *10*, 2703–2724. [CrossRef]
130. Pandi, P.S.; Krishnan, P.; Sathish, S.; Sudhahar, S. Biomedical Applications of Terbium Oxide Nanoparticles by Couroupita Guianensis Aubl Leaves Extract: A Greener Approach. *Nano-Struct. Nano-Objects* **2024**, *40*, 101341. [CrossRef]
131. Li, C.; Sun, Y.; Li, X.; Fan, S.; Liu, Y.; Jiang, X.; Boudreau, M.D.; Pan, Y.; Tian, X.; Yin, J.-J. Bactericidal Effects and Accelerated Wound Healing Using Tb4O7 Nanoparticles with Intrinsic Oxidase-like Activity. *J. Nanobiotechnol.* **2019**, *17*, 54. [CrossRef] [PubMed]
132. Iram, S.; Khan, S.; Ansary, A.A.; Arshad, M.; Siddiqui, S.; Ahmad, E.; Khan, R.H.; Khan, M.S. Biogenic Terbium Oxide Nanoparticles as the Vanguard against Osteosarcoma. *Spectrochim. Acta Part A Mol. Biomol. Spectrosc.* **2016**, *168*, 123–131. [CrossRef]
133. Khalaf, E.I.; El-Shafai, N.M.; Nassar, A.M.; Assem, E.E.; Yahia, I.S.; El-Mehasseb, I.M. Enhancing the Photoinduced via a Novel Nano-Combination of Terbium Oxide and Nickel Oxide on Graphene Oxide Surface: Cytotoxicity and Water Treatment. *Spectrochim. Acta Part A Mol. Biomol. Spectrosc.* **2023**, *286*, 121949. [CrossRef] [PubMed]
134. Fetouh, H.A.; El-Mossalamy, E.H.; El Desouky, J.M.; Batouti, M.E. Synthesis and Characterization of New Organometallic Lanthanides Metal Complexes for Photodynamic Therapy. *Sci. Rep.* **2024**, *14*, 26184. [CrossRef]
135. Yu, X.; Liu, X.; Wu, W.; Yang, K.; Mao, R.; Ahmad, F.; Chen, X.; Li, W. CT/MRI-Guided Synergistic Radiotherapy and X-ray Inducible Photodynamic Therapy Using Tb-Doped Gd-W-Nanoscintillators. *Angew. Chem. Int. Ed.* **2019**, *58*, 2017–2022. [CrossRef] [PubMed]
136. Daouk, J.; Iltis, M.; Dhaini, B.; Béchet, D.; Arnoux, P.; Rocchi, P.; Delconte, A.; Habermeyer, B.; Lux, F.; Frochot, C.; et al. Terbium-Based AGuIX-Design Nanoparticle to Mediate X-Ray-Induced Photodynamic Therapy. *Pharmaceuticals* **2021**, *14*, 396. [CrossRef] [PubMed]
137. Perry, J.; Minaei, E.; Engels, E.; Ashford, B.G.; McAlary, L.; Clark, J.R.; Gupta, R.; Tehei, M.; Corde, S.; Carolan, M.; et al. Thulium Oxide Nanoparticles as Radioenhancers for the Treatment of Metastatic Cutaneous Squamous Cell Carcinoma. *Phys. Med. Biol.* **2020**, *65*, 215018. [CrossRef] [PubMed]
138. Meng, Z.; Zheng, J.; Sheng, Q.; Zheng, X. In Situ Synthesis of Thulium(III) Hexacyanoferrate(II) Nanoparticles and Its Application for Glucose Detection. *Anal. Chim. Acta* **2011**, *689*, 47–51. [CrossRef]
139. Juliebø-Jones, P.; Somani, B.K.; Gjengstø, P.; Æsøy, M.S.; Beisland, C.; Ulvik, Ø. Holmium and Thulium Fiber Laser Safety in Endourological Practice: What Does the Clinician Need to Know? *Curr. Urol. Rep.* **2023**, *24*, 409–415. [CrossRef]
140. Gismondi, J.P.M.; Bento, A.D.S.A.; Mazzucchi, E.; Nahas, W.C. Thulium Fiber Laser in Cystine Calculi. *Int. Braz. J. Urol.* **2023**, *49*, 519–520. [CrossRef] [PubMed]
141. Spinos, T.; Tatanis, V.; Peteinaris, A.; Somani, B.; Kartalas Goumas, I.; Liatsikos, E.; Kallidonis, P. Thulium Fiber Laser Enucleation of the Prostate: A Systematic Review of the Current Outcomes. *Minerva Urol. Nephrol.* **2024**, *76*, 157–165. [CrossRef]
142. Forlini, V.; Pellegrino, C.; Lena, F.; Capitanucci, M.L.; Van Uitert, A.; Mosiello, G. Thulium Laser for the Treatment of Posterior Urethral Valves in Infants. *J. Endourol.* **2023**, *37*, 1276–1281. [CrossRef] [PubMed]
143. Peng, T.; Pu, R.; Wang, B.; Zhu, Z.; Liu, K.; Wang, F.; Wei, W.; Liu, H.; Zhan, Q. The Spectroscopic Properties and Microscopic Imaging of Thulium-Doped Upconversion Nanoparticles Excited at Different NIR-II Light. *Biosensors* **2021**, *11*, 148. [CrossRef] [PubMed]
144. Guryev, E.L.; Smyshlyaeva, A.S.; Shilyagina, N.Y.; Shanwar, S.; Kostyuk, A.B.; Shulga, A.A.; Konovalova, E.V.; Zvyagin, A.V.; Deyev, S.M.; Petrov, R.V. Multifunctional Complexes Based on Photoluminescent Upconversion Nanoparticles for Theranostics of the HER2-Positive Tumors. *Dokl. Biochem. Biophys.* **2020**, *491*, 73–76. [CrossRef] [PubMed]
145. Yang, Y.; Chen, Y.; Pei, P.; Fan, Y.; Wang, S.; Zhang, H.; Zhao, D.; Qian, B.-Z.; Zhang, F. Fluorescence-Amplified Nanocrystals in the Second near-Infrared Window for in Vivo Real-Time Dynamic Multiplexed Imaging. *Nat. Nanotechnol.* **2023**, *18*, 1195–1204. [CrossRef] [PubMed]

146. Kantamneni, H.; Barkund, S.; Donzanti, M.; Martin, D.; Zhao, X.; He, S.; Riman, R.E.; Tan, M.C.; Pierce, M.C.; Roth, C.M.; et al. Shortwave Infrared Emitting Multicolored Nanoprobes for Biomarker-Specific Cancer Imaging in Vivo. *BMC Cancer* **2020**, *20*, 1082. [CrossRef] [PubMed]
147. Guryev, E.L.; Smyshlyaeva, A.S.; Shilyagina, N.Y.; Sokolova, E.A.; Shanwar, S.; Kostyuk, A.B.; Lyubeshkin, A.V.; Schulga, A.A.; Konovalova, E.V.; Lin, Q.; et al. UCNP-Based Photoluminescent Nanomedicines for Targeted Imaging and Theranostics of Cancer. *Molecules* **2020**, *25*, 4302. [CrossRef] [PubMed]
148. Mishchenko, T.A.; Mitroshina, E.V.; Smyshlyaeva, A.S.; Guryev, E.L.; Vedunova, M.V. Comparative Analysis of the Effects of Upconversion Nanoparticles on Normal and Tumor Brain Cells. *Acta Naturae* **2020**, *12*, 86–94. [CrossRef]
149. Duosiken, D.; Yang, R.; Dai, Y.; Marfavi, Z.; Lv, Q.; Li, H.; Sun, K.; Tao, K. Near-Infrared Light-Excited Reactive Oxygen Species Generation by Thulium Oxide Nanoparticles. *J. Am. Chem. Soc.* **2022**, *144*, 2455–2459. [CrossRef] [PubMed]
150. Zuo, Y.; Huo, C.; Chen, Y.; Ding, P.; Tong, S.; Xue, W.; Zhu, J. Cancer-Thylakoid Hybrid Membrane Camouflaged Thulium Oxide Nanoparticles with Oxygen Self-Supply Capability for Tumor-Homing Phototherapy. *Adv. Healthc. Mater.* **2024**, *13*, 2303779. [CrossRef]
151. Zhou, K.; Qiu, X.; Xu, L.; Li, G.; Rao, B.; Guo, B.; Pei, D.; Li, A.; He, G. Poly(Selenoviologen)-Assembled Upconversion Nanoparticles for Low-Power Single-NIR Light-Triggered Synergistic Photodynamic and Photothermal Antibacterial Therapy. *ACS Appl. Mater. Interfaces* **2020**, *12*, 26432–26443. [CrossRef]
152. Wu, J.; Wei, W.; Ahmad, W.; Li, S.; Ouyang, Q.; Chen, Q. Enhanced Detection of Endocrine Disrupting Chemicals in On-Chip Microfluidic Biosensors Using Aptamer-Mediated Bridging Flocculation and Upconversion Luminescence. *J. Hazard. Mater.* **2023**, *458*, 132025. [CrossRef]
153. Soares, A.C.C.; Sales, T.O.; Ximendes, E.C.; Jaque, D.; Jacinto, C. Lanthanide Doped Nanoparticles for Reliable and Precise Luminescence Nanothermometry in the Third Biological Window. *Nanoscale Adv.* **2023**, *5*, 3664–3670. [CrossRef] [PubMed]
154. Przybylska, D.; Ekner-Grzyb, A.; Grześkowiak, B.F.; Grzyb, T. Upconverting SrF_2 Nanoparticles Doped with Yb^{3+}/Ho^{3+}, Yb^{3+}/Er^{3+} and Yb^{3+}/Tm^{3+} Ions—Optimisation of Synthesis Method, Structural, Spectroscopic and Cytotoxicity Studies. *Sci. Rep.* **2019**, *9*, 8669. [CrossRef]
155. Shwetabh, K.; Banerjee, A.; Poddar, R.; Kumar, K. Assessment of NIR-Triggered PEG-Coated $NaGdF_4$:Tm^{3+}/Yb^{3+} Bio-Compatible Upconversion Nanoparticles for Contrast Enhancement in OCT Imaging and Optical Thermometry. *Biomed. Mater.* **2024**, *19*, 055001. [CrossRef]
156. Zhao, J.; Ellis-Davies, G.C.R. Intracellular Photoswitchable Neuropharmacology Driven by Luminescence from Upconverting Nanoparticles. *Chem. Commun.* **2020**, *56*, 9445–9448. [CrossRef]
157. Li, Q.; Wang, Z.; Chen, Y.; Zhang, G. Elemental Bio-Imaging of PEGylated $NaYF_4$:Yb/Tm/Gd Upconversion Nanoparticles in Mice by Laser Ablation Inductively Coupled Plasma Mass Spectrometry to Study Toxic Side Effects on the Spleen, Liver and Kidneys. *Metallomics* **2017**, *9*, 1150–1156. [CrossRef]
158. Liu, Y.; Li, X.; Liu, Y.; Huang, Y.; Wang, F.; Qian, Y.; Wang, Y. Preparation and Properties of (Sc_2O_3-MgO)/Pcl/Pvp Electrospun Nanofiber Membranes for the Inhibition of Escherichia Coli Infections. *Int. J. Mol. Sci.* **2023**, *24*, 7649. [CrossRef]
159. Chu, B.H.; Kang, B.S.; Hung, S.C.; Chen, K.H.; Ren, F.; Sciullo, A.; Gila, B.P.; Pearton, S.J. Aluminum Gallium Nitride (GaN)/GaN High Electron Mobility Transistor-Based Sensors for Glucose Detection in Exhaled Breath Condensate. *J. Diabetes Sci. Technol.* **2010**, *4*, 171–179. [CrossRef]
160. Herath, H.M.T.U.; Silvio, L.D.; Evans, J.R.G. Scandia—A Potential Biomaterial? *J. Mater. Sci Mater. Med.* **2005**, *16*, 1061–1065. [CrossRef]
161. Nnomo Assene, A.; Dieme, D.; Jomaa, M.; Côté, J.; Bouchard, M. Toxicokinetic Study of Scandium Oxide in Rats. *Toxicol. Lett.* **2024**, *392*, 56–63. [CrossRef]
162. Shaker, T.M.; Chung, C.; Varma, M.K.; Doherty, M.G.; Wolf, A.M.; Chung, M.H.; Assifi, M.M. Is There a Role for Ytrrium-90 in the Treatment of Unresectable and Metastatic Intrahepatic Cholangiocarcinoma? *Am. J. Surg.* **2018**, *215*, 467–470. [CrossRef]
163. Ferreira, C.; Johnson, D.; Rasmussen, K.; Leinweber, C.; Ahmad, S.; Jung, J.W. A Novel Conformal Superficial High-Dose-Rate Brachytherapy Device for the Treatment of Nonmelanoma Skin Cancer and Keloids. *Brachytherapy* **2017**, *16*, 215–222. [CrossRef] [PubMed]
164. Pashazadeh, A.; Landes, R.; Boese, A.; Kreissl, M.C.; Klopfleisch, M.; Friebe, M. Superficial Skin Cancer Therapy with Y-90 Microspheres: A Feasibility Study on Patch Preparation. *Ski. Res. Technol.* **2020**, *26*, 25–29. [CrossRef]
165. Zeki, K.; Yabaş, E.; Erden, F.; Salih, B.; Canlıca, M. Lu, Sm, and Y-Based Double-Decker Phthalocyanines with Enhanced Photodynamic Therapy Performance. *Polyhedron* **2024**, *261*, 117138. [CrossRef]
166. Kassem, S.; Mohamed, M.; Sayour, H.; Canfarotta, F.; Piletsky, S.; Soliman, M.A.M. Functionalized Core-Shell Yttrium Oxide Nanoparticles as Antioxidants Agents in Heat Stressed Rats. *Biol. Trace Elem. Res.* **2020**, *198*, 189–197. [CrossRef]
167. Kassem, S.; Arafa, M.M.; Yehya, M.M.; Soliman, M.A.M. In Vivo Study of Dose-Dependent Antioxidant Efficacy of Functionalized Core–Shell Yttrium Oxide Nanoparticles. *Naunyn-Schmiedeberg's Arch. Pharmacol.* **2022**, *395*, 593–606. [CrossRef]

168. Mohamed, H.R.H.; Farouk, A.H.; Elbasiouni, S.H.; Nasif, K.A.; Safwat, G.; Diab, A. Genotoxicity and Oxidative Stress Induction by Calcium Hydroxide, Calcium Titanate or/and Yttrium Oxide Nanoparticles in Mice. *Sci. Rep.* **2023**, *13*, 19633. [CrossRef] [PubMed]
169. Song, X.; Shang, P.; Sun, Z.; Lu, M.; You, G.; Yan, S.; Chen, G.; Zhou, H. Therapeutic Effect of Yttrium Oxide Nanoparticles for the Treatment of Fulminant Hepatic Failure. *Nanomedicine* **2019**, *14*, 2519–2533. [CrossRef]
170. Khurana, A.; Saifi, M.A.; Godugu, C. Yttrium Oxide Nanoparticles Attenuate L-Arginine Induced Chronic Pancreatitis. *Biol. Trace Elem. Res.* **2023**, *201*, 3404–3417. [CrossRef]
171. Khurana, A.; Anchi, P.; Allawadhi, P.; Kumar, V.; Sayed, N.; Packirisamy, G.; Godugu, C. Yttrium Oxide Nanoparticles Reduce the Severity of Acute Pancreatitis Caused by Cerulein Hyperstimulation. *Nanomed. Nanotechnol. Biol. Med.* **2019**, *18*, 54–65. [CrossRef]
172. Hosseini, A.; Abdollahi, M. Through a Mechanism-Based Approach, Nanoparticles of Cerium and Yttrium May Improve the Outcome of Pancreatic Islet Isolation. *J. Med. Hypotheses Ideas* **2012**, *6*, 4–6. [CrossRef]
173. Hosseini, A.; Baeeri, M.; Rahimifard, M.; Navaei-Nigjeh, M.; Mohammadirad, A.; Pourkhalili, N.; Hassani, S.; Kamali, M.; Abdollahi, M. Antiapoptotic Effects of Cerium Oxide and Yttrium Oxide Nanoparticles in Isolated Rat Pancreatic Islets. *Hum. Exp. Toxicol.* **2013**, *32*, 544–553. [CrossRef] [PubMed]
174. Tavoosi, S.; Baghsheikhi, A.H.; Shetab-Boushehri, S.V.; Navaei-Nigjeh, M.; Sarvestani, N.N.; Karimi, M.Y.; Ranjbar, A.; Ebadollahi-Natanzi, A.; Hosseini, A. Cerium and Yttrium Oxide Nanoparticles and Nano-Selenium Produce Protective Effects Against H_2O_2-Induced Oxidative Stress in Pancreatic Beta Cells by Modulating Mitochondrial Dysfunction. *Pharm. Nanotechnol.* **2020**, *8*, 63–75. [CrossRef]
175. Khaksar, M.R.; Rahimifard, M.; Baeeri, M.; Maqbool, F.; Navaei-Nigjeh, M.; Hassani, S.; Moeini-Nodeh, S.; Kebriaeezadeh, A.; Abdollahi, M. Protective Effects of Cerium Oxide and Yttrium Oxide Nanoparticles on Reduction of Oxidative Stress Induced by Sub-Acute Exposure to Diazinon in the Rat Pancreas. *J. Trace Elem. Med. Biol.* **2017**, *41*, 79–90. [CrossRef]
176. Navaei-Nigjeh, M.; Daniali, M.; Rahimifard, M.; Khaksar, M.R. Multi-Organ Toxicity Attenuation by Cerium Oxide and Yttrium Oxide Nanoparticles: Comparing the Beneficial Effects on Tissues Oxidative Damage Induced by Sub-Acute Exposure to Diazinon. *Pharm. Nanotechnol.* **2020**, *8*, 225–238. [CrossRef] [PubMed]
177. Schubert, D.; Dargusch, R.; Raitano, J.; Chan, S.-W. Cerium and Yttrium Oxide Nanoparticles Are Neuroprotective. *Biochem. Biophys. Res. Commun.* **2006**, *342*, 86–91. [CrossRef]
178. Ghaznavi, H.; Najafi, R.; Mehrzadi, S.; Hosseini, A.; Tekyemaroof, N.; Shakeri-zadeh, A.; Rezayat, M.; Sharifi, A.M. Neuro-Protective Effects of Cerium and Yttrium Oxide Nanoparticles on High Glucose-Induced Oxidative Stress and Apoptosis in Undifferentiated PC12 Cells. *Neurol. Res.* **2015**, *37*, 624–632. [CrossRef]
179. Baghaee, P.; Yoonesi, M.; Esfahani, D.E.; Beirami, E.; Dargahi, L.; Rashidi, F.S.; Valian, N. Yttrium Oxide Nanoparticles Alleviate Cognitive Deficits, Neuroinflammation, and Mitochondrial Biogenesis Impairment Induced by Streptozotocin. *Neurosci. Lett.* **2024**, *837*, 137895. [CrossRef] [PubMed]
180. Hosseini, A.; Sharifi, A.M.; Abdollahi, M.; Najafi, R.; Baeeri, M.; Rayegan, S.; Cheshmehnour, J.; Hassani, S.; Bayrami, Z.; Safa, M. Cerium and Yttrium Oxide Nanoparticles Against Lead-Induced Oxidative Stress and Apoptosis in Rat Hippocampus. *Biol. Trace Elem. Res.* **2015**, *164*, 80–89. [CrossRef] [PubMed]
181. Abu-Taweel, G.M.; Al-Mutary, M.G.; Albetran, H.M. Yttrium Oxide Nanoparticles Moderate the Abnormal Cognitive Behaviors in Male Mice Induced by Silver Nanoparticles. *Oxidative Med. Cell. Longev.* **2022**, *2022*, 9059371. [CrossRef]
182. Abu-Taweel, G.M.; Albetran, H.M.; Al-Mutary, M.G.; Ahmad, M.; Low, I.M. Alleviation of Silver Nanoparticle-Induced Sexual Behavior and Testicular Parameters Dysfunction in Male Mice by Yttrium Oxide Nanoparticles. *Toxicol. Rep.* **2021**, *8*, 1121–1130. [CrossRef] [PubMed]
183. Mitra, R.N.; Merwin, M.J.; Han, Z.; Conley, S.M.; Al-Ubaidi, M.R.; Naash, M.I. Yttrium Oxide Nanoparticles Prevent Photoreceptor Death in a Light-Damage Model of Retinal Degeneration. *Free Radic. Biol. Med.* **2014**, *75*, 140–148. [CrossRef] [PubMed]
184. Emad, B.; WalyEldeen, A.A.; Hassan, H.; Sharaky, M.; Abdelhamid, I.A.; Ibrahim, S.A.; Mohamed, H.R. Yttrium Oxide Nanoparticles Induce Cytotoxicity, Genotoxicity, Apoptosis, and Ferroptosis in the Human Triple-Negative Breast Cancer MDA-MB-231 Cells. *BMC Cancer* **2023**, *23*, 1151. [CrossRef] [PubMed]
185. Zako, T.; Yoshimoto, M.; Hyodo, H.; Kishimoto, H.; Ito, M.; Kaneko, K.; Soga, K.; Maeda, M. Cancer-Targeted near Infrared Imaging Using Rare Earth Ion-Doped Ceramic Nanoparticles. *Biomater. Sci.* **2015**, *3*, 59–64. [CrossRef]
186. Akhtar, M.J.; Ahamed, M.; Alrokayan, S.A.; Ramamoorthy, M.M.; Alaizeri, Z.M. High Surface Reactivity and Biocompatibility of Y_2O_3 NPs in Human MCF-7 Epithelial and HT-1080 Fibro-Blast Cells. *Molecules* **2020**, *25*, 1137. [CrossRef]
187. Sayour, H.; Kassem, S.; Canfarotta, F.; Czulak, J.; Mohamed, M.; Piletsky, S. Biocompatibility and Biodistribution of Surface-Modified Yttrium Oxide Nanoparticles for Potential Theranostic Applications. *Environ. Sci. Pollut. Res.* **2020**, *27*, 19095–19107. [CrossRef] [PubMed]
188. Selvaraj, V.; Bodapati, S.; Murray, E.; Blough, E.; Winston, N.; Rice, K.; Shokufar, T.; Zhao, Y. Cytotoxicity and Genotoxicity Caused by Yttrium Oxide Nanoparticles in HEK293 Cells. *Int. J. Nanomed.* **2014**, *9*, 1379–1391. [CrossRef]

189. Chen, Z.; Liu, J.; Zheng, M.; Mo, M.; Hu, X.; Liu, C.; Pathak, J.L.; Wang, L.; Chen, L. TRIM24-DTNBP1-ATP7A Mediated Astrocyte Cuproptosis in Cognition and Memory Dysfunction Caused by Y_2O_3 NPs. *Sci. Total Environ.* **2024**, *954*, 176353. [CrossRef] [PubMed]
190. Jiang, D.; Wang, X.; Feng, L.; Yu, Y.; Hu, J.; Liu, X.; Wu, H. Structural Insight into the Alginate Derived Nano-La$(OH)_3$/Porous Carbon Composites for Highly Selective Adsorption of Phosphate. *Int. J. Biol. Macromol.* **2022**, *200*, 172–181. [CrossRef]
191. Wei, Y.; Yuan, P.; Zhou, J.; Liu, J.; Losic, D.; Wu, H.; Bu, H.; Tan, X.; Li, Z. Direct Atomic-Scale Insight into the Precipitation Formation at the Lanthanum Hydroxide Nanoparticle/Solution Interface. *J. Phys. Chem. Lett.* **2023**, *14*, 3995–4003. [CrossRef] [PubMed]
192. Lai, B.-H.; Chen, D.-H. LaB6 Nanoparticles with Carbon-Doped Silica Coating for Fluorescence Imaging and near-IR Photothermal Therapy of Cancer Cells. *Acta Biomater.* **2013**, *9*, 7556–7563. [CrossRef] [PubMed]
193. Jiang, F.; Leong, Y.-K.; Saunders, M.; Martyniuk, M.; Faraone, L.; Keating, A.; Dell, J.M. Uniform Dispersion of Lanthanum Hexaboride Nanoparticles in a Silica Thin Film: Synthesis and Optical Properties. *ACS Appl. Mater. Interfaces* **2012**, *4*, 5833–5838. [CrossRef] [PubMed]
194. Sobhani-Nasab, A.; Zahraei, Z.; Akbari, M.; Maddahfar, M.; Hosseinpour-Mashkani, S.M. Synthesis, Characterization, and Antibacterial Activities of ZnLaFe 2 O 4 /NiTiO 3 Nanocomposite. *J. Mol. Struct.* **2017**, *1139*, 430–435. [CrossRef]
195. Dinakaran, D.; Sengupta, J.; Pink, D.; Raturi, A.; Chen, H.; Usmani, N.; Kumar, P.; Lewis, J.D.; Narain, R.; Moore, R.B. PEG-PLGA Nanospheres Loaded with Nanoscintillators and Photosensitizers for Radiation-Activated Photodynamic Therapy. *Acta Biomater.* **2020**, *117*, 335–348. [CrossRef]
196. Lu, V.M.; Jue, T.R.; McDonald, K.L. Cytotoxic Lanthanum Oxide Nanoparticles Sensitize Glioblastoma Cells to Radiation Therapy and Temozolomide: An in Vitro Rationale for Translational Studies. *Sci. Rep.* **2020**, *10*, 18156. [CrossRef]
197. Zheng, G.; Wu, S.; Deng, X.; Wang, A.; Ying, Y.; Li, S.; Wang, F.; Liu, X.; Wang, P.; Wei, D. Lanthanum-Based Dendritic Mesoporous Nanoplatform for Tumor Microenvironment Activating Synergistic Anti-Glioma Efficacy. *Mater. Today Bio* **2024**, *28*, 101223. [CrossRef] [PubMed]
198. Rojas, J.V.; Woodward, J.D.; Chen, N.; Rondinone, A.J.; Castano, C.H.; Mirzadeh, S. Synthesis and Characterization of Lanthanum Phosphate Nanoparticles as Carriers for 223Ra and 225Ra for Targeted Alpha Therapy. *Nucl. Med. Biol.* **2015**, *42*, 614–620. [CrossRef]
199. Muthulakshmi, V.; Dhilip Kumar, C.; Sundrarajan, M. Biological Applications of Green Synthesized Lanthanum Oxide Nanoparticles via Couroupita Guianensis Abul Leaves Extract. *Anal. Biochem.* **2022**, *638*, 114482. [CrossRef]
200. Zhang, L.; Zhou, L.; Li, Q.X.; Liang, H.; Qin, H.; Masutani, S.; Yoza, B. Toxicity of Lanthanum Oxide Nanoparticles to the Fungus Moniliella Wahieum Y12T Isolated from Biodiesel. *Chemosphere* **2018**, *199*, 495–501. [CrossRef]
201. Tiwari, S.; Gupta, P.K.; Bagbi, Y.; Sarkar, T.; Solanki, P.R. L-Cysteine Capped Lanthanum Hydroxide Nanostructures for Non-Invasive Detection of Oral Cancer Biomarker. *Biosens. Bioelectron.* **2017**, *89*, 1042–1052. [CrossRef]
202. Almukhlafi, H.; Ali, D.; Almutairi, B.; Yaseen, K.N.; Alyami, N.; Almeer, R.; Alkahtani, S.; Alarifi, S. Role of Oxidative Stress in La_2O_3 Nanoparticle-Induced Cytotoxicity and Apoptosis in CHANG and HuH-7 Cells. *Int. J. Nanomed.* **2021**, *16*, 3487–3496. [CrossRef]
203. Alsubaie, S.M.; Ali, D.; Almutairi, B.O.; Almeer, R.; Alarifi, S. Evaluation of Cyto—And Genotoxic Influence of Lanthanum Dioxide Nanoparticles on Human Liver Cells. *Dose-Response* **2022**, *20*, 15593258221128428. [CrossRef]
204. Alyami, N.M.; Alobadi, H.; Maodaa, S.; Alothman, N.S.; Almukhlafi, H.; Yaseen, K.N.; Alnakhli, Z.A.; Alshiban, N.M.; Elnagar, D.M.; Rady, A.; et al. Determination of Dose- and Time-Dependent Hepatotoxicity and Apoptosis of Lanthanum Oxide Nanoparticles in Female Swiss Albino Mice. *Environ. Sci. Pollut. Res.* **2024**, *31*, 17124–17139. [CrossRef]
205. Dressler, V.L.; Ogunmodede, O.T.; Heidrich, G.M.; Neves, V.M.; Schetinger, M.R.C.; Morsch, V.M. Investigative Analysis of Lanthanum Oxide Nanoparticles on Elements in Bone of Wistar Rats After 30 Days of Repeated Oral Administration. *Biol. Trace Elem. Res.* **2020**, *196*, 153–167. [CrossRef] [PubMed]
206. Vijayan, V.; Sreekumar, S.; Ahina, K.M.; Lakra, R.; Kiran, M.S. Lanthanum Oxide Nanoparticles Reinforced Collagen ƙ-Carrageenan Hydroxyapatite Biocomposite as Angio-Osteogenic Biomaterial for In Vivo Osseointegration and Bone Repair. *Adv. Biol.* **2023**, *7*, 2300039. [CrossRef]
207. Li, Y.; Yu, H.; Li, P.; Bian, Y. Assessment the Exposure Level of Rare Earth Elements in Workers Producing Cerium, Lanthanum Oxide Ultrafine and Nanoparticles. *Biol. Trace Elem. Res.* **2017**, *175*, 298–305. [CrossRef]
208. Li, Y.; Yu, H.; Zheng, S.; Miao, Y.; Yin, S.; Li, P.; Bian, Y. Direct Quantification of Rare Earth Elements Concentrations in Urine of Workers Manufacturing Cerium, Lanthanum Oxide Ultrafine and Nanoparticles by a Developed and Validated ICP-MS. *Int. J. Environ. Res. Public Health* **2016**, *13*, 350. [CrossRef]
209. Xiao, X.; Yong, L.; Jiao, B.; Yang, H.; Liang, C.; Jia, X.; Liu, Z.; Sang, Y.; Song, Y. Postweaning Exposure to Lanthanum Alters Neurological Behavior during Early Adulthood in Rats. *NeuroToxicology* **2021**, *83*, 40–50. [CrossRef]
210. Yuan, L.; Qu, Y.; Li, Q.; An, T.; Chen, Z.; Chen, Y.; Deng, X.; Bai, D. Protective Effect of Astaxanthin against La_2O_3 Nanoparticles Induced Neurotoxicity by Activating PI3K/AKT/Nrf-2 Signaling in Mice. *Food Chem. Toxicol.* **2020**, *144*, 111582. [CrossRef]

211. Dundar, B.; Alsawas, M.; Masaadeh, A.; Conway, K.; Snow, A.N.; Sompallae, R.R.; Bossler, A.D.; Ma, D.; Lopes Abath Neto, O. Molecular Characterization and Survival Analysis of a Cohort of Glioblastoma, IDH-Wildtype. *Pathol.—Res. Pract.* **2024**, *257*, 155272. [CrossRef]
212. Jing, Y.; Kang, T.; Hu, P.; Fan, H.; Teng, F.; Zhao, X.; Song, J. Surfactant-Induced Photocatalytic Performance Enhancement of Europium Oxide Nanoparticles. *Process Saf. Environ. Prot.* **2023**, *171*, 888–894. [CrossRef]
213. Teng, M.; Liang, X.; Liu, H.; Li, Z.; Gao, X.; Zhang, C.; Cheng, H.; Chen, H.; Liu, G. Cerenkov Radiation Shining a Light for Cancer Theranostics. *Nano Today* **2024**, *55*, 102174. [CrossRef]
214. Hu, Z.; Chi, C.; Liu, M.; Guo, H.; Zhang, Z.; Zeng, C.; Ye, J.; Wang, J.; Tian, J.; Yang, W.; et al. Nanoparticle-Mediated Radiopharmaceutical-Excited Fluorescence Molecular Imaging Allows Precise Image-Guided Tumor-Removal Surgery. *Nanomed. Nanotechnol. Biol. Med.* **2017**, *13*, 1323–1331. [CrossRef] [PubMed]
215. Li, M.; Gao, J.; Yao, L.; Zhang, L.; Li, D.; Li, Z.; Wu, Q.; Wang, S.; Ding, J.; Liu, Y.; et al. Determining Toxicity of Europium Oxide Nanoparticles in Immune Cell Components and Hematopoiesis in Dominant Organs in Mice: Role of Lysosomal Fluid Interaction. *Sci. Total Environ.* **2024**, *937*, 173482. [CrossRef] [PubMed]
216. Chávez-García, D.; Juarez-Moreno, K.; Calderón-Osuna, I.; Navarro, P.; Hirata, G.A. Nanotoxicological Study of Downconversion Y_2O_3:Eu^{3+} Luminescent Nanoparticles Functionalized with Folic Acid for Cancer Cells Bioimaging. *J. Biomed. Mater. Res.* **2020**, *108*, 2396–2406. [CrossRef]
217. Chen, Y.; Huang, S.; Wang, Y.; Liu, Y.; Wu, T.; Chen, T. Ligand-Functionalization of BPEI-Coated YVO_4:Bi^{3+},Eu^{3+} Nanophosphors for Tumor-Cell-Targeted Imaging Applications. *Chem. Asian J.* **2013**, *8*, 2652–2659. [CrossRef] [PubMed]
218. Jia, J.; Zhang, Y.; Zheng, M.; Shan, C.; Yan, H.; Wu, W.; Gao, X.; Cheng, B.; Liu, W.; Tang, Y. Functionalized Eu(III)-Based Nanoscale Metal–Organic Framework To Achieve Near-IR-Triggered and -Targeted Two-Photon Absorption Photodynamic Therapy. *Inorg. Chem.* **2018**, *57*, 300–310. [CrossRef]
219. Gadzhimagomedova, Z.; Polyakov, V.; Pankin, I.; Butova, V.; Kirsanova, D.; Soldatov, M.; Khodakova, D.; Goncharova, A.; Mukhanova, E.; Belanova, A.; et al. BaGdF5 Nanophosphors Doped with Different Concentrations of Eu^{3+} for Application in X-Ray Photodynamic Therapy. *Int. J. Mol. Sci.* **2021**, *22*, 13040. [CrossRef]
220. Abd Elkader, M.F.; Almogbel, M.S.; Elabbasy, M.T. Survival and Reduction in Foodborne Bacteria Using Methyl Cellulose Film Doped with Europium Oxide Nanoparticles. *Food Sci. Nutr.* **2020**, *8*, 291–298. [CrossRef]
221. Fu, M.; Zhao, Y.; Wang, Y.; Li, Y.; Wu, M.; Liu, Q.; Hou, Z.; Lu, Z.; Wu, K.; Guo, J. On-Demand Removable Self-Healing and pH-Responsive Europium-Releasing Adhesive Dressing Enables Inflammatory Microenvironment Modulation and Angiogenesis for Diabetic Wound Healing. *Small* **2023**, *19*, e2205489. [CrossRef] [PubMed]
222. Fernández-Villa, D.; Aguilar, M.R.; Rojo, L. Europium-Tannic Acid Nanocomplexes Devised for Bone Regeneration under Oxidative or Inflammatory Environments. *J. Mater. Chem. B* **2024**, *12*, 7153–7170. [CrossRef]
223. Wiatrak, B.; Sobierajska, P.; Szandruk-Bender, M.; Jawien, P.; Janeczek, M.; Dobrzynski, M.; Pistor, P.; Szelag, A.; Wiglusz, R.J. Nanohydroxyapatite as a Biomaterial for Peripheral Nerve Regeneration after Mechanical Damage-In Vitro Study. *Int. J. Mol. Sci.* **2021**, *22*, 4454. [CrossRef] [PubMed]
224. Teker, T.; Aslanoglu, M. Development of a Dy_2O_3@Eu_2O_3-Carbon Nanofiber Based Electrode for Highly Sensitive Detection of Papaverine. *Anal. Chim. Acta* **2021**, *1183*, 338972. [CrossRef] [PubMed]
225. Narasimharao, K.; Ali, T.T. Influence of Synthesis Conditions on Physico-Chemical and Photocatalytic Properties of Rare Earth (Ho, Nd and Sm) Oxides. *J. Mater. Res. Technol.* **2020**, *9*, 1819–1830. [CrossRef]
226. Zawadzki, M.; Kępiński, L. Synthesis and Characterization of Neodymium Oxide Nanoparticles. *J. Alloys Compd.* **2004**, *380*, 255–259. [CrossRef]
227. Dorris, A.; Sicard, C.; Chen, M.C.; McDonald, A.B.; Barrett, C.J. Stabilization of Neodymium Oxide Nanoparticles via Soft Adsorption of Charged Polymers. *ACS Appl. Mater. Interfaces* **2011**, *3*, 3357–3365. [CrossRef] [PubMed]
228. Lee, M.-J.; Park, S.-Y.; Sung, A.-Y. Characterization of Biocompatible Hydrogel Lenses Using Methacrylic Acid with Neodymium Oxide Nanoparticles. *Polymers* **2021**, *13*, 1575. [CrossRef] [PubMed]
229. Biswas, S.; Naskar, H.; Pradhan, S.; Wang, Y.; Bandyopadhyay, R.; Pramanik, P. Simultaneous Voltammetric Determination of Adrenaline and Tyrosine in Real Samples by Neodymium Oxide Nanoparticles Grafted Graphene. *Talanta* **2020**, *206*, 120176. [CrossRef]
230. Muthulakshmi, V.; Dhilip Kumar, C.; Sundrarajan, M. Green Synthesis of Ionic Liquid Mediated Neodymium Oxide Nanoparticles via *Couroupita Guianensis* Abul Leaves Extract with Its Biological Applications. *J. Biomater. Sci. Polym. Ed.* **2022**, *33*, 1063–1082. [CrossRef] [PubMed]
231. Sundrarajan, M.; Muthulakshmi, V. Green Synthesis of Ionic Liquid Mediated Neodymium Oxide Nanoparticles by Andrographis Paniculata Leaves Extract for Effective Bio-Medical Applications. *J. Environ. Chem. Eng.* **2021**, *9*, 104716. [CrossRef]
232. Yu, F.; Zhang, X.; Gao, L.; Xue, H.; Liu, L.; Wang, S.; Chen, S.; Huang, L. LncRNA Loc105377478 Promotes NPs-Nd_2O_3-Induced Inflammation in Human Bronchial Epithelial Cells through the ADIPOR1/NF-κB Axis. *Ecotoxicol. Environ. Saf.* **2021**, *208*, 111609. [CrossRef]

233. Liu, L.; Cui, J.; Chen, S.; Zhang, X.; Wang, S.; Huang, L. Circ_002363 Is Regulated by the RNA Binding Protein BCAS2 and Inhibits Neodymium Oxide Nanoparticle-Induced DNA Damage by Non-Homologous End-Joining Repair. *Sci. Total Environ.* **2023**, *863*, 160819. [CrossRef]
234. Gao, L.; Zhang, X.; Cui, J.; Liu, L.; Tai, D.; Wang, S.; Huang, L. Transcription Factor TP63 Mediates LncRNA CNTFR-AS1 to Promote DNA Damage Induced by Neodymium Oxide Nanoparticles via Homologous Recombination Repair. *Environ. Pollut.* **2023**, *334*, 122191. [CrossRef]
235. May, E.; Thoennessen, M. Discovery of Cesium, Lanthanum, Praseodymium and Promethium Isotopes. *At. Data Nucl. Data Tables* **2012**, *98*, 960–982. [CrossRef]
236. Jeon, S.; Kim, S.-H.; Jeong, J.; Lee, D.-K.; Lee, S.; Kim, S.; Kim, G.; Maruthupandy, M.; Cho, W.-S. ABCG1 and ABCG4 as Key Transporters in the Development of Pulmonary Alveolar Proteinosis by Nanoparticles. *J. Hazard. Mater.* **2021**, *420*, 126595. [CrossRef]
237. Bartenbach, D.; Popescu, R.; Gerthsen, D.; Feldmann, C. $[Sm_6O_4(Cbz)_{10}(Thf)_6]{\cdot}2C_7H_8$: A Polynuclear Samarium Oxo Cluster Obtained from Carbazole-Driven Oxidation of Samarium Nanoparticles. *Inorg. Chem.* **2022**, *61*, 3072–3077. [CrossRef] [PubMed]
238. Damirchi, Z.; Firoozbakhtian, A.; Hosseini, M.; Ganjali, M.R. Ti_3C_2/Ni/Sm-Based Electrochemical Glucose Sensor for Sweat Analysis Using Bipolar Electrochemistry. *Microchim. Acta* **2024**, *191*, 137. [CrossRef] [PubMed]
239. Gilnezhad, J.; Firoozbakhtian, A.; Hosseini, M.; Adel, S.; Xu, G.; Ganjali, M.R. An Enzyme-Free Ti_3C_2/Ni/Sm-LDH-Based Screen-Printed-Electrode for Real-Time Sweat Detection of Glucose. *Anal. Chim. Acta* **2023**, *1250*, 340981. [CrossRef]
240. Muthulakshmi, V.; Balaji, M.; Sundrarajan, M. Biomedical Applications of Ionic Liquid Mediated Samarium Oxide Nanoparticles by Andrographis Paniculata Leaves Extract. *Mater. Chem. Phys.* **2020**, *242*, 122483. [CrossRef]
241. Alshahrani, A.A.; Alqarni, L.S.; Alghamdi, M.D.; Alotaibi, N.F.; Moustafa, S.M.N.; Nassar, A.M. Phytosynthesis via Wasted Onion Peel Extract of Samarium Oxide/Silver Core/Shell Nanoparticles for Excellent Inhibition of Microbes. *Heliyon* **2024**, *10*, e24815. [CrossRef]
242. Zahmatkesh, H.; Mirpour, M.; Zamani, H.; Rasti, B.; Rahmani, F.A.; Padasht, N. Effect of Samarium Oxide Nanoparticles on Virulence Factors and Motility of Multi-Drug Resistant Pseudomonas Aeruginosa. *World J. Microbiol. Biotechnol.* **2022**, *38*, 209. [CrossRef] [PubMed]
243. Chanket, W.; Pipatthana, M.; Sangphukieo, A.; Harnvoravongchai, P.; Chankhamhaengdecha, S.; Janvilisri, T.; Phanchana, M. The Complete Catalog of Antimicrobial Resistance Secondary Active Transporters in Clostridioides Difficile: Evolution and Drug Resistance Perspective. *Comput. Struct. Biotechnol. J.* **2024**, *23*, 2358–2374. [CrossRef] [PubMed]
244. Mohamed, H.E.A.; Khalil, A.T.; Hkiri, K.; Ayaz, M.; Usman, A.; Sadiq, A.; Ullah, F.; Khan, M.A.; Ullah, I.; Maaza, M. Potential Nanomedicinal Applications and Physicochemical Nature of *Hyphaene thebaica*-reduced Nano-samaria. *Microsc. Res. Tech.* **2024**, *87*, 2829–2841. [CrossRef]
245. Smith, A.W.; Greenberger, B.A.; Den, R.B.; Stock, R.G. Radiopharmaceuticals for Bone Metastases. *Semin. Radiat. Oncol.* **2021**, *31*, 45–59. [CrossRef] [PubMed]
246. Mohammadian, N.; Faridbod, F. ALS Genosensing Using DNA-Hybridization Electrochemical Biosensor Based on Label-Free Immobilization of ssDNA on Sm_2O_3 NPs-rGO/PANI Composite. *Sens. Actuators B Chem.* **2018**, *275*, 432–438. [CrossRef]
247. Kujur, M.S.; Manakari, V.; Parande, G.; Prasadh, S.; Wong, R.; Mallick, A.; Gupta, M. Effect of Samarium Oxide Nanoparticles on Degradation and Invitro Biocompatibility of Magnesium. *Mater. Today Commun.* **2021**, *26*, 102171. [CrossRef]
248. Shrestha, S.; Yeung, C.M.Y.; Nunnerley, C.; Tsang, S.C. Comparison of Morphology and Electrical Conductivity of Various Thin Films Containing Nano-Crystalline Praseodymium Oxide Particles. *Sens. Actuators A Phys.* **2007**, *136*, 191–198. [CrossRef]
249. Bakht, M.K.; Sadeghi, M.; Ahmadi, S.J.; Sadjadi, S.S.; Tenreiro, C. Preparation of Radioactive Praseodymium Oxide as a Multifunctional Agent in Nuclear Medicine: Expanding the Horizons of Cancer Therapy Using Nanosized Neodymium Oxide. *Nucl. Med. Commun.* **2013**, *34*, 5–12. [CrossRef]
250. Jang, H.; Choi, M.; Yim, Y.; Kim, Y.; Min, D. Dual-Wavelength Irradiation and Dox Delivery for Cancer Cell Ablation with Photocatalytic Pr Doped TiO_2 /NGO Hybrid Nanocomposite. *Adv. Healthc. Mater.* **2015**, *4*, 1833–1840. [CrossRef] [PubMed]
251. Manikantan, V.; Varalakshmi, G.S.; Pillai, A.S.; Alexander, A.; Lucas, A.; Kathiravan, E.; Akash, B.A.; Enoch, I.V.M.V. 5-Fluorouracil-Loaded Designed Praseodymium Oxide—Poly-β-Cyclodextrin Nanorods for Effectively Inhibiting Breast Cancer Cells. *Inorg. Chem. Commun.* **2023**, *153*, 110830. [CrossRef]
252. Zhang, Q.-C.; Yan, W.-L.; Jiang, L.; Zheng, Y.-G.; Wang, J.-X.; Zhang, R.-K. Synthesis of Nano-Praseodymium Oxide for Cataluminescence Sensing of Acetophenone in Exhaled Breath. *Molecules* **2019**, *24*, 4275. [CrossRef] [PubMed]
253. Sharma, P.; Kaur, S.; Chaudhary, S.; Umar, A.; Kumar, R. Bare and Nonionic Surfactant-Functionalized Praseodymium Oxide Nanoparticles: Toxicological Studies. *Chemosphere* **2018**, *209*, 1007–1020. [CrossRef] [PubMed]
254. Vijayan, V.; Sreekumar, S.; Singh, F.; Srivatsan, K.V.; Lakra, R.; Sai, K.P.; Kiran, M.S. Nanotized Praseodymium Oxide Collagen 3-D pro-Vasculogenic Biomatrix for Soft Tissue Engineering. *Nanomed. Nanotechnol. Biol. Med.* **2021**, *33*, 102364. [CrossRef]

255. Zou, S.; Guo, F.; Wu, L.; Ju, H.; Sun, M.; Cai, R.; Xu, L.; Gong, Y.; Gong, A.; Zhang, M.; et al. One-Pot Synthesis of Cerium and Praseodymium Co-Doped Carbon Quantum Dots as Enhanced Antioxidant for Hydroxyl Radical Scavenging. *Nanotechnology* **2020**, *31*, 165101. [CrossRef] [PubMed]
256. Zheng, H.; Tang, J.; Wei, Y.; Deng, X.; Zhang, Y.; Ma, X.; Jiang, X.; Xu, Z.P.; Liao, H. Antibacterial Properties of Cerium Oxide Nanoparticles: Recent Progresses and Future Challenges. *Particuology* **2024**, *93*, 264–283. [CrossRef]
257. Yadav, S.; Chamoli, S.; Kumar, P.; Maurya, P.K. Structural and Functional Insights in Polysaccharides Coated Cerium Oxide Nanoparticles and Their Potential Biomedical Applications: A Review. *Int. J. Biol. Macromol.* **2023**, *246*, 125673. [CrossRef]
258. Nosrati, H.; Heydari, M.; Khodaei, M. Cerium Oxide Nanoparticles: Synthesis Methods and Applications in Wound Healing. *Mater. Today Bio* **2023**, *23*, 100823. [CrossRef] [PubMed]
259. S Jairam, L.; Chandrashekar, A.; Prabhu, T.N.; Kotha, S.B.; Girish, M.S.; Devraj, I.M.; Dhanya Shri, M.; Prashantha, K. A Review on Biomedical and Dental Applications of Cerium Oxide Nanoparticles — Unearthing the Potential of This Rare Earth Metal. *J. Rare Earths* **2023**, *41*, 1645–1661. [CrossRef]
260. Chai, W.F.; Tang, K.S. Protective Potential of Cerium Oxide Nanoparticles in Diabetes Mellitus. *J. Trace Elem. Med. Biol.* **2021**, *66*, 126742. [CrossRef] [PubMed]
261. Rajeshkumar, S.; Naik, P. Synthesis and Biomedical Applications of Cerium Oxide Nanoparticles—A Review. *Biotechnol. Rep.* **2018**, *17*, 1–5. [CrossRef]
262. Meng, X.; Wang, W.-D.; Li, S.-R.; Sun, Z.-J.; Zhang, L. Harnessing Cerium-Based Biomaterials for the Treatment of Bone Diseases. *Acta Biomater.* **2024**, *183*, 30–49. [CrossRef] [PubMed]
263. Zandi, M.; Hosseini, F.; Adli, A.H.; Salmanzadeh, S.; Behboudi, E.; Halvaei, P.; Khosravi, A.; Abbasi, S. State-of-the-Art Cerium Nanoparticles as Promising Agents against Human Viral Infections. *Biomed. Pharmacother.* **2022**, *156*, 113868. [CrossRef]
264. Hashem, R.M.; Rashd, L.A.; Hashem, K.S.; Soliman, H.M. Cerium Oxide Nanoparticles Alleviate Oxidative Stress and Decreases Nrf-2/HO-1 in D-GALN/LPS Induced Hepatotoxicity. *Biomed. Pharmacother.* **2015**, *73*, 80–86. [CrossRef] [PubMed]
265. Abbasi, E.; Vafaei, S.A.; Naseri, N.; Darini, A.; Azandaryani, M.T.; Ara, F.K.; Mirzaei, F. Protective Effects of Cerium Oxide Nanoparticles in Non-Alcoholic Fatty Liver Disease (NAFLD) and Carbon Tetrachloride-Induced Liver Damage in Rats: Study on Intestine and Liver. *Metab. Open* **2021**, *12*, 100151. [CrossRef] [PubMed]
266. Oró, D.; Yudina, T.; Fernández-Varo, G.; Casals, E.; Reichenbach, V.; Casals, G.; González De La Presa, B.; Sandalinas, S.; Carvajal, S.; Puntes, V.; et al. Cerium Oxide Nanoparticles Reduce Steatosis, Portal Hypertension and Display Anti-Inflammatory Properties in Rats with Liver Fibrosis. *J. Hepatol.* **2016**, *64*, 691–698. [CrossRef] [PubMed]
267. Kobyliak, N.; Virchenko, O.; Falalyeyeva, T.; Kondro, M.; Beregova, T.; Bodnar, P.; Shcherbakov, O.; Bubnov, R.; Caprnda, M.; Delev, D.; et al. Cerium Dioxide Nanoparticles Possess Anti-Inflammatory Properties in the Conditions of the Obesity-Associated NAFLD in Rats. *Biomed. Pharmacother.* **2017**, *90*, 608–614. [CrossRef]
268. Ibrahim, H.G.; Attia, N.; Hashem, F.E.Z.A.; El Heneidy, M.A.R. Cerium Oxide Nanoparticles: In Pursuit of Liver Protection against Doxorubicin-Induced Injury in Rats. *Biomed. Pharmacother.* **2018**, *103*, 773–781. [CrossRef] [PubMed]
269. Nassar, S.Z.; Hassaan, P.S.; Abdelmonsif, D.A.; ElAchy, S.N. Cardioprotective Effect of Cerium Oxide Nanoparticles in Monocrotaline Rat Model of Pulmonary Hypertension: A Possible Implication of Endothelin-1. *Life Sci.* **2018**, *201*, 89–101. [CrossRef] [PubMed]
270. Tisi, A.; Passacantando, M.; Lozzi, L.; Riccitelli, S.; Bisti, S.; Maccarone, R. Retinal Long Term Neuroprotection by Cerium Oxide Nanoparticles after an Acute Damage Induced by High Intensity Light Exposure. *Exp. Eye Res.* **2019**, *182*, 30–38. [CrossRef] [PubMed]
271. Estevez, A.Y.; Pritchard, S.; Harper, K.; Aston, J.W.; Lynch, A.; Lucky, J.J.; Ludington, J.S.; Chatani, P.; Mosenthal, W.P.; Leiter, J.C.; et al. Neuroprotective Mechanisms of Cerium Oxide Nanoparticles in a Mouse Hippocampal Brain Slice Model of Ischemia. *Free Radic. Biol. Med.* **2011**, *51*, 1155–1163. [CrossRef]
272. Kadivar, F.; Haddadi, G.; Mosleh-Shirazi, M.A.; Khajeh, F.; Tavasoli, A. Protection Effect of Cerium Oxide Nanoparticles against Radiation-Induced Acute Lung Injuries in Rats. *Rep. Pract. Oncol. Radiother.* **2020**, *25*, 206–211. [CrossRef] [PubMed]
273. Silina, E.V.; Manturova, N.E.; Ivanova, O.S.; Baranchikov, A.E.; Artyushkova, E.B.; Medvedeva, O.A.; Kryukov, A.A.; Dodonova, S.A.; Gladchenko, M.P.; Vorsina, E.S.; et al. Cerium Dioxide–Dextran Nanocomposites in the Development of a Medical Product for Wound Healing: Physical, Chemical and Biomedical Characteristics. *Molecules* **2024**, *29*, 2853. [CrossRef] [PubMed]
274. Silina, E.V.; Stupin, V.A.; Manturova, N.E.; Ivanova, O.S.; Popov, A.L.; Mysina, E.A.; Artyushkova, E.B.; Kryukov, A.A.; Dodonova, S.A.; Kruglova, M.P.; et al. Influence of the Synthesis Scheme of Nanocrystalline Cerium Oxide and Its Concentration on the Biological Activity of Cells Providing Wound Regeneration. *Int. J. Mol. Sci.* **2023**, *24*, 14501. [CrossRef]
275. Fang, X.; Song, H. Synthesis of Cerium Oxide Nanoparticles Loaded on Chitosan for Enhanced Auto-Catalytic Regenerative Ability and Biocompatibility for the Spinal Cord Injury Repair. *J. Photochem. Photobiol. B Biol.* **2019**, *191*, 83–87. [CrossRef]
276. Silina, E.V.; Manturova, N.E.; Vasin, V.I.; Artyushkova, E.B.; Khokhlov, N.V.; Ivanov, A.V.; Stupin, V.A. Efficacy of A Novel Smart Polymeric Nanodrug in the Treatment of Experimental Wounds in Rats. *Polymers* **2020**, *12*, 1126. [CrossRef]

277. Wei, F.; Neal, C.J.; Sakthivel, T.S.; Kean, T.; Seal, S.; Coathup, M.J. Multi-Functional Cerium Oxide Nanoparticles Regulate Inflammation and Enhance Osteogenesis. *Mater. Sci. Eng. C* **2021**, *124*, 112041. [CrossRef] [PubMed]
278. Ren, S.; Zhou, Y.; Zheng, K.; Xu, X.; Yang, J.; Wang, X.; Miao, L.; Wei, H.; Xu, Y. Cerium Oxide Nanoparticles Loaded Nanofibrous Membranes Promote Bone Regeneration for Periodontal Tissue Engineering. *Bioact. Mater.* **2022**, *7*, 242–253. [CrossRef] [PubMed]
279. Nilawar, S.; Yadav, P.; Jain, N.; Saini, D.K.; Chatterjee, K. Protective Role of Nanoceria-Infused Nanofibrous Scaffold toward Bone Tissue Regeneration with Senescent Cells. *Biomacromolecules* **2024**, *25*, 4074–4086. [CrossRef]
280. Foroutan, Z.; Afshari, A.R.; Sabouri, Z.; Mostafapour, A.; Far, B.F.; Jalili-Nik, M.; Darroudi, M. Plant-Based Synthesis of Cerium Oxide Nanoparticles as a Drug Delivery System in Improving the Anticancer Effects of Free Temozolomide in Glioblastoma (U87) Cells. *Ceram. Int.* **2022**, *48*, 30441–30450. [CrossRef]
281. Parvathy, S.; Manjula, G.; Balachandar, R.; Subbaiya, R. Green Synthesis and Characterization of Cerium Oxide Nanoparticles from Artabotrys Hexapetalus Leaf Extract and Its Antibacterial and Anticancer Properties. *Mater. Lett.* **2022**, *314*, 131811. [CrossRef]
282. Sridharan, M.; Kamaraj, P.; Vennilaraj; Arockiaselvi, J.; Pushpamalini, T.; Vivekanand, P.A.; Hari Kumar, S. Synthesis, Characterization and Evaluation of Biosynthesized Cerium Oxide Nanoparticle for Its Anticancer Activity on Breast Cancer Cell (MCF 7). *Mater. Today Proc.* **2021**, *36*, 914–919. [CrossRef]
283. Mousaiyan, S.; Baharara, J.; Es-haghi, A. Biopreparation of Cerium Oxide Nanoparticles Using Alginate: Characterization and Estimation of Antioxidant and Its Activity against Breast Cancer Cell Lines (MCF7). *Results Chem.* **2024**, *7*, 101468. [CrossRef]
284. Rasouli, Z.; Yousefi, M.; Torbati, M.B.; Samadi, S.; Kalateh, K. Synthesis and Characterization of Nanoceria-Based Composites and in Vitro Evaluation of Their Cytotoxicity against Colon Cancer. *Polyhedron* **2020**, *176*, 114297. [CrossRef]
285. Hosseinzadeh, R.; Khorsandi, K.; Esfahani, H.S.; Habibi, M.; Hosseinzadeh, G. Preparation of Cerium-Curcumin and Cerium-Quercetin Complexes and Their LEDs Irradiation Assisted Anticancer Effects on MDA-MB-231 and A375 Cancer Cell Lines. *Photodiagnosis Photodyn. Ther.* **2021**, *34*, 102326. [CrossRef]
286. Wason, M.S.; Colon, J.; Das, S.; Seal, S.; Turkson, J.; Zhao, J.; Baker, C.H. Sensitization of Pancreatic Cancer Cells to Radiation by Cerium Oxide Nanoparticle-Induced ROS Production. *Nanomed. Nanotechnol. Biol. Med.* **2013**, *9*, 558–569. [CrossRef]
287. Cheng, F.; Wang, S.; Zheng, H.; Yang, S.; Zhou, L.; Liu, K.; Zhang, Q.; Zhang, H. Cu-Doped Cerium Oxide-Based Nanomedicine for Tumor Microenvironment-Stimulative Chemo-Chemodynamic Therapy with Minimal Side Effects. *Colloids Surf. B Biointerfaces* **2021**, *205*, 111878. [CrossRef] [PubMed]
288. Pop, O.L.; Mesaros, A.; Vodnar, D.C.; Suharoschi, R.; Tăbăran, F.; Mageruşan, L.; Tódor, I.S.; Diaconeasa, Z.; Balint, A.; Ciontea, L.; et al. Cerium Oxide Nanoparticles and Their Efficient Antibacterial Application In Vitro against Gram-Positive and Gram-Negative Pathogens. *Nanomaterials* **2020**, *10*, 1614. [CrossRef] [PubMed]
289. Abid, S.A.; Taha, A.A.; Ismail, R.A.; Mohsin, M.H. Antibacterial and Cytotoxic Activities of Cerium Oxide Nanoparticles Prepared by Laser Ablation in Liquid. *Environ. Sci. Pollut. Res.* **2020**, *27*, 30479–30489. [CrossRef] [PubMed]
290. Senthilkumar, R.P.; Bhuvaneshwari, V.; Ranjithkumar, R.; Sathiyavimal, S.; Malayaman, V.; Chandarshekar, B. Synthesis, Characterization and Antibacterial Activity of Hybrid Chitosan-Cerium Oxide Nanoparticles: As a Bionanomaterials. *Int. J. Biol. Macromol.* **2017**, *104*, 1746–1752. [CrossRef]
291. Zamani, K.; Allah-Bakhshi, N.; Akhavan, F.; Yousefi, M.; Golmoradi, R.; Ramezani, M.; Bach, H.; Razavi, S.; Irajian, G.-R.; Gerami, M.; et al. Antibacterial Effect of Cerium Oxide Nanoparticle against Pseudomonas Aeruginosa. *BMC Biotechnol.* **2021**, *21*, 68. [CrossRef]
292. Silina, E.V.; Stupin, V.A.; Manturova, N.E.; Chuvilina, E.L.; Gasanov, A.A.; Ostrovskaya, A.A.; Andreeva, O.I.; Tabachkova, N.Y.; Abakumov, M.A.; Nikitin, A.A.; et al. Development of Technology for the Synthesis of Nanocrystalline Cerium Oxide Under Production Conditions with the Best Regenerative Activity and Biocompatibility for Further Creation of Wound-Healing Agents. *Pharmaceutics* **2024**, *16*, 1365. [CrossRef] [PubMed]
293. Chatzimentor, I.; Tsamesidis, I.; Ioannou, M.-E.; Pouroutzidou, G.K.; Beketova, A.; Giourieva, V.; Papi, R.; Kontonasaki, E. Study of Biological Behavior and Antimicrobial Properties of Cerium Oxide Nanoparticles. *Pharmaceutics* **2023**, *15*, 2509. [CrossRef] [PubMed]
294. Lu, J.; Luo, J.; Li, J.; Fu, S.; Ran, Y.; Li, J.; Zhao, Y.; Hao, Y. Fluorescent Pirfenidone-Cerium(III) Nanocomplexes Protect against Radiation-Induced Pulmonary Fibrosis and Inhibit Tumor Cell Growth. *J. Drug Deliv. Sci. Technol.* **2023**, *86*, 104651. [CrossRef]
295. Chaudhary, S.; Umar, A. Glycols Functionalized Fluorescent Eu_2O_3 Nanoparticles: Functionalization Effect on the Structural and Optical Properties. *J. Alloys Compd.* **2016**, *682*, 160–169. [CrossRef]
296. Wang, X.; Wu, J.; Jia, H.; Liu, Z.; Liu, C.; Qi, Y.; Wang, Z. PEG Coated Nano-Phosphor $GdPO_4{\bullet}H{\cdot}H_2O$:12%$Eu^{3+}$ and Its Performance Analysis. *J. Alloys Compd.* **2022**, *905*, 164223. [CrossRef]
297. Paulis, L.E.; Jacobs, I.; Van Den Akker, N.M.; Geelen, T.; Molin, D.G.; Starmans, L.W.; Nicolay, K.; Strijkers, G.J. Targeting of ICAM-1 on Vascular Endothelium under Static and Shear Stress Conditions Using a Liposomal Gd-Based MRI Contrast Agent. *J. Nanobiotechnol.* **2012**, *10*, 25. [CrossRef] [PubMed]

298. Gao, X.; Guo, L.; Li, J.; Thu, H.E.; Hussain, Z. Nanomedicines Guided Nanoimaging Probes and Nanotherapeutics for Early Detection of Lung Cancer and Abolishing Pulmonary Metastasis: Critical Appraisal of Newer Developments and Challenges to Clinical Transition. *J. Control. Release* **2018**, *292*, 29–57. [CrossRef] [PubMed]
299. Kukowska-Latallo, J. Targeted Gadolinium-Loaded Dendrimer Nanoparticles for Tumor-Specific Magnetic Resonance Contrast Enhancement. *Int. J. Nanomed.* **2008**, *3*, 201–210. [CrossRef]
300. Sikder, A.; Vambhurkar, G.; Amulya, E.; Bagasariya, D.; Famta, P.; Shah, S.; Khatri, D.K.; Singh, S.B.; Sinha, V.R.; Srivastava, S. Advancements in Redox-Sensitive Micelles as Nanotheranostics: A New Horizon in Cancer Management. *J. Control. Release* **2022**, *349*, 1009–1030. [CrossRef]
301. Idée, J.-M.; Port, M.; Medina, C.; Lancelot, E.; Fayoux, F.; Ballet, S.; Corot, C. Possible Involvement of Gadolinium Chelates in the Pathophysiology of Nephrogenic Systemic Fibrosis: A Critical Review. *Toxicology* **2008**, *248*, 77–88. [CrossRef] [PubMed]
302. Meng, X.; Sun, Z.; Chu, H.; Wang, Y. Self-Assembled Nanoplatforms for Chemodynamic Therapy. *Chem. Eng. J.* **2024**, *479*, 147702. [CrossRef]
303. Jayakumar, R.; Prabaharan, M.; Reis, R.L.; Mano, J.F. Graft Copolymerized Chitosan—Present Status and Applications. *Carbohydr. Polym.* **2005**, *62*, 142–158. [CrossRef]
304. Guo, X.; Wang, Y.; Qin, Y.; Shen, P.; Peng, Q. Structures, Properties and Application of Alginic Acid: A Review. *Int. J. Biol. Macromol.* **2020**, *162*, 618–628. [CrossRef]
305. Sehgal, V.; Pandey, S.P.; Singh, P.K. Prospects of Charged Cyclodextrins in Biomedical Applications. *Carbohydr. Polym.* **2024**, *323*, 121348. [CrossRef]
306. Petrova, V.A.; Dubashynskaya, N.V.; Gofman, I.V.; Golovkin, A.S.; Mishanin, A.I.; Aquino, A.D.; Mukhametdinova, D.V.; Nikolaeva, A.L.; Ivan'kova, E.M.; Baranchikov, A.E.; et al. Biocomposite Films Based on Chitosan and Cerium Oxide Nanoparticles with Promising Regenerative Potential. *Int. J. Biol. Macromol.* **2023**, *229*, 329–343. [CrossRef] [PubMed]
307. Shurfa, M.K.; Girigoswami, A.; Sakthi Devi, R.; Harini, K.; Agraharam, G.; Deepika, B.; Pallavi, P.; Girigoswami, K. Combinatorial Effect of Doxorubicin Entrapped in Alginate-Chitosan Hybrid Polymer and Cerium Oxide Nanocomposites on Skin Cancer Management in Mice. *J. Pharm. Sci.* **2023**, *112*, 2891–2900. [CrossRef] [PubMed]
308. Lord, M.S.; Farrugia, B.L.; Yan, C.M.Y.; Vassie, J.A.; Whitelock, J.M. Hyaluronan Coated Cerium Oxide Nanoparticles Modulate CD44 and Reactive Oxygen Species Expression in Human Fibroblasts. *J. Biomed. Mater. Res.* **2016**, *104*, 1736–1746. [CrossRef] [PubMed]
309. Orhan, B.; Karadeniz, D.; Kalaycıoğlu, Z.; Kaygusuz, H.; Torlak, E.; Erim, F.B. Foam-Based Antibacterial Hydrogel Composed of Carboxymethyl Cellulose/Polyvinyl Alcohol/Cerium Oxide Nanoparticles for Potential Wound Dressing. *Int. J. Biol. Macromol.* **2025**, *291*, 138924. [CrossRef]
310. Zhao, Y.; Song, L.; Li, M.; Peng, H.; Qiu, X.; Li, Y.; Zhu, B.; Liu, C.; Ren, S.; Miao, L. Injectable CNPs/DMP1-Loaded Self-Assembly Hydrogel Regulating Inflammation of Dental Pulp Stem Cells for Dentin Regeneration. *Mater. Today Bio* **2024**, *24*, 100907. [CrossRef]
311. Marchianò, V.; Matos, M.; Marcet, I.; Carmen Blanco-López, M.; Gutiérrez, G.; Cioffi, N.; Ditaranto, N. Preparation and Characterization of Biodegradable Gelatine and Starch Films Embedding Cerium Oxide Nanoparticles Stabilized by PLGA Micelles for Antibiofilm Applications. *J. Mol. Liq.* **2024**, *398*, 124215. [CrossRef]
312. Varlamova, E.G.; Gudkov, S.V.; Turovsky, E.A. Differential Effect of Cerium Nanoparticles on the Viability, Redox-Status and Ca2+-Signaling System of Cancer Cells of Various Origins. *Arch. Biochem. Biophys.* **2025**, *764*, 110261. [CrossRef] [PubMed]
313. Chaudhary, S.; Sharma, P.; Kumar, S.; Alex, S.A.; Kumar, R.; Mehta, S.K.; Mukherjee, A.; Umar, A. A Comparative Multi-Assay Approach to Study the Toxicity Behaviour of Eu_2O_3 Nanoparticles. *J. Mol. Liq.* **2018**, *269*, 783–795. [CrossRef]
314. Hekmatimoghaddam, S.; Iman, M.; Shahdadi Sardo, H.; Jebali, A. Gelatin Hydrogel Containing Cerium Oxide Nanoparticles Covered by Interleukin-17 Aptamar as an Anti- Inflammatory Agent for Brain Inflammation. *J. Neuroimmunol.* **2019**, *326*, 79–83. [CrossRef] [PubMed]
315. Li, T.; Murphy, S.; Kiselev, B.; Bakshi, K.S.; Zhang, J.; Eltahir, A.; Zhang, Y.; Chen, Y.; Zhu, J.; Davis, R.M.; et al. A New Interleukin-13 Amino-Coated Gadolinium Metallofullerene Nanoparticle for Targeted MRI Detection of Glioblastoma Tumor Cells. *J. Am. Chem. Soc.* **2015**, *137*, 7881–7888. [CrossRef]
316. Bulmahn, J.C.; Kutscher, H.L.; Cwiklinski, K.; Schwartz, S.A.; Prasad, P.N.; Aalinkeel, R. A Multimodal Theranostic Nanoformulation That Dramatically Enhances Docetaxel Efficacy Against Castration Resistant Prostate Cancer. *J. Pharm. Sci.* **2020**, *109*, 2874–2883. [CrossRef]
317. Zgheib, C.; Hilton, S.A.; Dewberry, L.C.; Hodges, M.M.; Ghatak, S.; Xu, J.; Singh, S.; Roy, S.; Sen, C.K.; Seal, S.; et al. Use of Cerium Oxide Nanoparticles Conjugated with MicroRNA-146a to Correct the Diabetic Wound Healing Impairment. *J. Am. Coll. Surg.* **2019**, *228*, 107–115. [CrossRef] [PubMed]
318. Liu, X.-P.; Xie, X.-L.; Wei, Y.-P.; Mao, C.-J.; Chen, J.-S.; Niu, H.-L.; Song, J.-M.; Jin, B.-K. Photoelectrochemical Immunoassay for Human Interleukin 6 Based on the Use of Perovskite-Type $LaFeO_3$ Nanoparticles on Fluorine-Doped Tin Oxide Glass. *Mikrochim. Acta* **2017**, *185*, 52. [CrossRef] [PubMed]

319. Huang, D.; Ying, H.; Jiang, D.; Liu, F.; Tian, Y.; Du, C.; Zhang, L.; Pu, X. Rapid and Sensitive Detection of Interleukin-6 in Serum via Time-Resolved Lateral Flow Immunoassay. *Anal. Biochem.* **2020**, *588*, 113468. [CrossRef]
320. Peng, J.; Guan, J.; Yao, H.; Jin, X. Magnetic Colorimetric Immunoassay for Human Interleukin-6 Based on the Oxidase Activity of Ceria Spheres. *Anal. Biochem.* **2016**, *492*, 63–68. [CrossRef] [PubMed]
321. Zhang, D.-E.; Ni, X.-M.; Zheng, H.-G.; Zhang, X.-J.; Song, J.-M. Fabrication of Rod-like CeO_2: Characterization, Optical and Electrochemical Properties. *Solid. State Sci.* **2006**, *8*, 1290–1293. [CrossRef]
322. Clavijo-Chaparro, S.L.; Hernández-Gordillo, A.; Camposeco-Solis, R.; Rodríguez-González, V. Water Splitting Behavior of Copper-Cerium Oxide Nanorods and Nanocubes Using Hydrazine as a Scavenging Agent. *J. Mol. Catal. A Chem.* **2016**, *423*, 143–150. [CrossRef]
323. Li, Y.; Zhang, Z.; Tao, Z.; Gao, X.; Wang, S.; Liu, Y. A Asp/Ce Nanotube-Based Colorimetric Nanosensor for H_2O_2-Free and Enzyme-Free Detection of Cysteine. *Talanta* **2019**, *196*, 556–562. [CrossRef] [PubMed]
324. Kruglova, M.P.; Ivanov, A.V.; Fedoseev, A.N.; Virus, E.D.; Stupin, V.A.; Parfenov, V.A.; Titova, S.A.; Lazareva, P.I.; Kubatiev, A.A.; Silina, E.V. The Diagnostic and Prognostic Roles Played by Homocysteine and Other Aminothiols in Patients with Chronic Kidney Disease. *J. Clin. Med.* **2023**, *12*, 5653. [CrossRef] [PubMed]
325. Madeira, M.d.P.; Costa, K.R.B.d.S.; de Lima, I.S.; Barreto, H.M.; Garcia, R.R.P.; Triboni, E.R.; Silva-Filho, E.C.; Viana, B.C.; Cecilia, J.A.; Osajima, J.A. Unveiling the Bacterial Photoinactivation through Cerium-Gallium Titanate Nanotubes. *J. Photochem. Photobiol. A Chem.* **2024**, *451*, 115483. [CrossRef]
326. Vanasundari, K.; Ponnarasi, P.; Mahalakshmi, G. A Green Approach to Synthesis of Ag-Doped CeO_2 Nanorods Embedded Reduced Graphene Oxide Nanocomposite for Excellent Photocatalytic and Antimicrobial Activity. *Inorg. Chem. Commun.* **2024**, *165*, 112523. [CrossRef]
327. Chaturvedi, K.; Singhwane, A.; Jaiswal, A.; Mili, M.; Tilwari, A.; Mohapatra, R.K.; Srivastava, A.K.; Verma, S. Robust Synthesis and Characteristics of Novel One-Dimensional Gadolinium Oxide Nanorods Decorated Multiwalled Carbon Nanotubes Based Antibacterial Nanocomposites for Health Care Applications. *Inorg. Chem. Commun.* **2023**, *157*, 111324. [CrossRef]
328. Patra, C.R.; Abdel Moneim, S.S.; Wang, E.; Dutta, S.; Patra, S.; Eshed, M.; Mukherjee, P.; Gedanken, A.; Shah, V.H.; Mukhopadhyay, D. In Vivo Toxicity Studies of Europium Hydroxide Nanorods in Mice. *Toxicol. Appl. Pharmacol.* **2009**, *240*, 88–98. [CrossRef]
329. He, Q.; Zhao, H.; Teng, Z.; Wang, Y.; Li, M.; Hoffmann, M.R. Phosphate Removal and Recovery by Lanthanum-Based Adsorbents: A Review for Current Advances. *Chemosphere* **2022**, *303*, 134987. [CrossRef]

MDPI AG
Grosspeteranlage 5
4052 Basel
Switzerland
Tel.: +41 61 683 77 34

Pharmaceuticals Editorial Office
E-mail: pharmaceuticals@mdpi.com
www.mdpi.com/journal/pharmaceuticals

www.ingramcontent.com/pod-product-compliance
Lightning Source LLC
Chambersburg PA
CBHW061809210926
51850CB00006B/652